AF786783

Modern Approaches to Wettability

Theory and Applications

Modern Approaches to Wettability

Theory and Applications

Edited by

Malcolm E. Schrader

Institute of Chemistry
The Hebrew University of Jerusalem
Jerusalem, Israel

and

George I. Loeb

Geomar Associates
Bethesda, Maryland

formerly of

David Taylor Research Center
Ship Materials Engineering Department
Annapolis, Maryland

Springer Science+Business Media, LLC

Library of Congress Cataloging-in-Publication Data

Modern approaches to wettability : theory and applications / edited by
Malcolm E. Schrader and George I. Loeb.
p. cm.
Includes bibliographical references and index.
1. Wetting. I. Schrader, Malcolm E. II. Loeb, George I.
QD506.M63 1992
541.3'3--dc20 92-15930
CIP

ISBN 978-1-4899-1178-0 ISBN 978-1-4899-1176-6 (eBook)
DOI 10.1007/978-1-4899-1176-6

© Springer Science+Business Media New York 1992
Originally published by Plenum Press, New York in 1992
Softcover reprint of the hardcover 1st edition 1992

All rights reserved

No part of this book may be reproduced, stored in a retrieval system, or transmitted in any form or by any means, electronic, mechanical, photocopying, microfilming, recording, or otherwise, without written permission from the Publisher

Contributors

Willard D. Bascom • Department of Materials Science and Engineering, 304 EMRO, University of Utah, Salt Lake City, Utah 84112

H. J. Busscher • Laboratory for Materia Technica, University of Groningen, 9713 AV Groningen, The Netherlands

H. K. Christenson • Department of Applied Mathematics, Institute of Advanced Studies, Research School of Physical Sciences, Australian National University, Canberra ACT 2601, Australia

Gregory S. Ferguson • Department of Chemistry, Lehigh University, Bethlehem, Pennsylvania 18015

Robert J. Good • Department of Chemical Engineering, State University of New York at Buffalo, Buffalo, New York 14260

Frank J. Holly • Dry Eye Institute, Lubbock, Texas 79499

Joseph L. Katz • Department of Chemical Engineering, The Johns Hopkins University, Baltimore, Maryland 21218

D. Li • Department of Mechanical Engineering, University of Toronto, Toronto, Ontario, Canada M5S 1A4

George I. Loeb • David Taylor Research Center, Ship Materials Engineering Department, Annapolis, Maryland 21402

Jer Ru Maa • Department of Chemical Engineering, National Cheng Kung University, Tainan, Taiwan, 70101, R.O.C.

Abraham Marmur • Department of Chemical Engineering, Technion–Israel Institute of Technology, 32000 Haifa, Israel

James W. Mihm • David Taylor Research Center, Ship Materials Engineering Department, Annapolis, Maryland 21402

Michel Nardin • Centre de Recherches sur la Physico-Chimie des Surfaces Solides, CNRS, F-68200 Mulhouse, France *and* Laboratoire de Recherches sur la Physico-Chimie des Interfaces de l'Ecole National Supérieure de Chimie de Mulhouse, 68093 Mulhouse Cedex, France

A. W. Neumann • Department of Mechanical Engineering, University of Toronto, Toronto, Ontario, Canada M5S 1A4

Eli Ruckenstein • Department of Chemical Engineering, State University of New York at Buffalo, Buffalo, New York 14260

Clifford K. Schoff • Coatings Research and Development, PPG Industries, Inc., Allison Park, Pennsylvania 15101

Malcolm E. Schrader • Institute of Chemistry, The Hebrew University of Jerusalem, Jerusalem, Israel

Jacques Schultz • Centre de Recherches sur la Physico-Chimie des Surfaces Solides, CNRS, F-68200 Mulhouse, France *and* Laboratoire de Recherches sur la Physico-Chimie des Interfaces de l'Ecole Nationale Supérieure de Chimie de Mulhouse, 68093 Mulhouse Cedex, France

Jin Sheng Sheu • Department of Chemical Engineering, National Cheng Kung University, Tainan, Taiwan, 70101, R.O.C.

O. Smigelschi • Lummus Crest, GmbH, Weisbaden, Germany

J. K. Spelt • Department of Mechanical Engineering, University of Toronto, Toronto, Ontario, Canada M5S 1A4

D. G. Suciu • Lummus Crest, Inc., Bloomfield, New Jersey

Atsushi Takahara • Department of Chemical Science and Technology, Faculty of Engineering Kyushu University, Higashi-ku, Fukuoka 812, Japan

Carel J. van Oss • Department of Microbiology, State University of New York at Buffalo, Buffalo, New York 14214

George M. Whitesides • Department of Chemistry, Harvard University, Cambridge, Massachusetts 02138

Shmuel Yariv • Department of Inorganic and Analytical Chemistry, The Hebrew University of Jerusalem, Jerusalem 91904, Israel

Foreword: In Appreciation of William A. Zisman

On July 21, 1986, Dr. William A. Zisman succumbed to pneumonia at age 80. His obituary in the *Washington Post* two days later read in part: "... resident of Silver Spring [Maryland] ... worked at the research laboratory from 1939 until he retired in 1975." This formal notice of Dr. Zisman's death recalled that he was the former head of the Surface Chemistry Branch of the Naval Research Laboratory in Washington, D.C., an authority in the field of lubrication, and that he received his government's Distinguished Civilian Service Award in 1964 as well as his peers' Hillebrand Award, the top honor of the Chemical Society of Washington, in 1955.

Most surface scientists also know of his Kendall Award from the Division of Surface and Colloid Chemistry of the ACS, the volume of the *Advances in Chemistry Series* dedicated to him in 1964, and the general admiration of colleagues around the world expressed in numerous awards and accolades. However, these facts do not present a complete picture of the man. As his postdoctoral apprentice from 1966 to 1968, and as late as 1970 (when Dr. Zisman was still calling about overdue manuscripts), I have some more pithy remembrances of my mentor.

Dr. Zisman's most noteworthy attributes, as viewed by this fledgling physicist given 105 technical manuscripts to read before a first formal appointment, were his intensity and sternness. Over the years, these impressions were reinforced, while I also came to recognize his fierce loyalty and enormous competence. Others can tell of his love of family, musical interests, and out-of-the-lab activities, so I will not dwell on them here. Rather, I reflect on the lessons he taught in our standing Friday afternoon appointments, including the personal lessons that usually began, "Young man, I am afraid these data are just not good enough..." as I was engaged in the measurement of over 10,000 contact angles before he (and I) actually came to believe in the merit of any one of them.

This was the setting: The Surface Chemistry Branch of the Chemistry Division of the U.S. Naval Research Laboratory (NRL), of which William Zisman served as Division Superintendent after many years as Branch Head. The Branch had 26 full-time employees—most of whom had been with Dr. Zisman throughout the challenging years of World War II. They had, obviously, not published much during that time, but in the decade afterward, the Zisman group put the fruits of 100 person-years of dedicated experimental effort into the literature. Dr. Zisman and his associates ultimately published 150 scientific articles and obtained 36 patents. In the mid-1960s, they were still working very hard but also basking (a bit) in the glory of belated international recognition of their achievements and concepts, not least of which was the "critical surface tension" as a new and useful correlating parameter for solid surface energies. (The discovery of the interesting surface properties of Teflon had been a helpful step along the way.)

This environment was as intense, productive, and protective as that in any major university department at the crest of its discipline's acceptance. I came to think of it then, and I still do, as Zisman University. It is sad that not many were graduated from this "University," and most of them stayed in government service. I write as one of the few who emerged into the outside world of commerce and academia. Valued "visiting professors" to the NRL campus and colleagues met at ACS symposia, Gordon Conferences, and the like, including the surface scientists Robert Good, Fred Fowkes, A. Zettlemoyer, and many others whom Dr. Zisman considered cherished peers. He was well aware of their efforts to understand the nature of surface forces and had given serious consideration to their dissection of surface energy into "polar," "dispersive," H-bonding, and other components. However, as he showed me in his notebooks, he had concluded that these formulations did not sufficiently match the measurements then made with numerous solid–liquid combinations. I would like to think that he still follows the efforts of his colleagues to refine and rationalize these approaches, and that this empirical "critical surface tension" concept remains useful albeit theoretically undignified.

Dr. Zisman particularly valued the work of Rulon E. (Ted) Johnson of Du Pont, who brought new understanding to the role of surface texture as an important influence on contact angle values, and who was a younger ally in discounting receding contact angle measurements for material surface-compositional analysis. He relished his interactions with Tomlinson Fort, then also of Du Pont, and predicted the rewarding career that "young Tom" would have in surface science. He also continually gave credit to his longtime associates in the experimental arena, including Harry Fox, Al Ellison, Elaine Shafrin, and Mia Bernett. His admiration and gratitude for their skills would come out in unlikely circumstances, such as while he was intensely focused on one of his "young man" lectures to me or another disciple who had not properly heeded the warning from his secretary: "Don't get trapped at the blackboard! Stand with your back to the door!"

Dr. Zisman's respect for experimental "ground truth" showed itself frequently in pronouncements made on occasions such as these: At the conclusion of a theory-rich, but data-poor, lecture from an MIT postdoctoral visitor to NRL, "Young man, in the world of surface science, a pound of experiment is worth a ton of theory!" On hanging up the telephone in a fit of pique, after debating for the previous hour the problems with surface energy dissection into various "terms" based on spreading rates and inconstant contact angle values, "Kinetics! ... Thermodynamics! Fie! A pox on both their camps!"

He could feel personally hurt and saddened when he felt that the quality of "his" chosen field could have been diminished by what he perceived to be bad data, and this feeling often expressed itself in outspoken criticism.

His righteous defense of the "scientific faith" was carried over into his management tasks at his home institution, where he deliberately accepted the role of "common enemy" of those in the branches and sections that reported to him.

The exceptional and beneficial result of this polarization was the elimination of almost all hostillity and jealousy among workers at the bench, where cooperation and mutual assistance became the general rule. His opportunity to mete out (constructive, always constructive) criticism and the occasional (some would say, rare) grudging note of approval came frequently: Branch review was held every Thursday morning, and each laboratory scientist could expect to come up in the rotation three or four times per year. This in-house critical assessment was coupled with a public display of intense loyalty that would see Dr. Zisman "defend to the death" the Division staff at national conferences and other forums. And the staff reluctantly agreed among themselves that they really could do no better than continue to "back a winning horse," testimony that his scientific calls were right more than 90% of the time. He himself made almost all the hard decisions to drop projects, reassign personnel, and shift priorities, again welding his team into a coherent whole antagonistic only to himself. Since he brought in almost all the required funding, and saw to it that the funding supported the production of solid results for his Navy sponsors, the economic, political, and technical health of the organization he led was always excellent. Dr. Zisman's personal advice to me upon departing government employment was: "Deliver to your sponsors as your first responsibility and priority." His personal model was splendid in this regard, and I can cite the positive results of two projects he supervised that each saved the U.S. Navy enough money to pay the bills of NRL's Chemistry Division at least until the year 2000 (to say nothing about their critical value to the defense posture of the United States).

Dr. Zisman also advised on my last day as his apprentice, "When you have built a 'shelf of capability' as an experimentalist, don't rush off into new fields too hastily. Give yourself time to roam around on that shelf."

Our relationship was a special one because I had come to his tutelage fresh from medical school training for a Ph.D. in biophysics, and had completed two years of surgery as a hospital technician during the period I had dropped out as a chemical engineering student. He was in the later years of a successful career as a "hard" physical-chemical scientist, and truly longed to make a contribution to the "soft" but compelling biological sciences which were just emerging into real disciplines. The field of molecular biology was just becoming fashionable, and the phrase "genetic engineering" had mainly negative and racial implications while "biotechnology" meant little more than beermaking and sewage treatment. Dr. Zisman saw, in me, his opportunity to send a missionary into the biological and medical sciences. Although the possibility of using surface chemistry to cure cancer did not escape his vision, his ever-practical nature suggested that I first address problems dealing with prosthetic medical and dental implants. Indeed, my most vivid memory of this great man and teacher is his mood upon returning from service as an ad hoc reviewer for the National Heart Institute and the National Institute of Dental Research in Bethesda. That mood was angry disappointment that the surface properties of materials, those provided both by nature and by manufacture, were so completely neglected as to make the interfaces of prosthetic reconstructions with body tissues almost destined to fail. He urged me to carry his concepts and standards to these critical fields, and to refuse to accept the lesser standards then common in biosurface research.

It is with Dr. Zisman's missionary zeal that I still approach such topics, and it is with a reluctant but real love for the man that I—as do all his colleagues and students—still revere his memory.

Robert E. Baier

Buffalo, New York

Preface

This volume is dedicated to the memory of a pioneer in the study of wettability and contact angle, as well as surface properties of materials in general: Dr. William A. Zisman. His Naval Research Laboratory (NRL) group's work in this field and the closely related field of tribology was of great value to the U.S. Navy and the Allied cause during the troubled days of the 1940s and 1950s, as well as to surface science in a more general sense. As is true in most pioneering laboratories, many young investigators passed through his group, and the years after the war saw the spread of his influence to other laboratories and groups as recognition of the importance of this discipline grew. We are fortunate that several of his former colleagues were in a position to contribute to this volume, and we appreciate their contributions. Dr. Baier entered the group during a time when Dr. Zisman's interest was attracted to biological problems, and he describes the unique experience of his association with the group at NRL during that period, in his foreword.

The work of the NRL surface chemistry group spanned a broad range of surface energies. A wealth of data was produced in confirmation of the concept that high-energy surfaces, e.g., glasses, ceramics, and metals, spread all room temperature liquids except mercury. Of particular importance was their finding that many apparent exceptions to this rule were in reality the result of the formation of an "autophobic" layer by the spreading liquid. For example, certain esters would hydrolyze on the surfaces of glass, leaving behind a chemisorbed organic fragment that would create a low-energy hydrophobic surface.

For low-energy surfaces, such as organic materials and plastics, they discovered an empirical relationship in which $\cos\theta$ was found to increase linearly with the decrease in γ_L of members of a set of liquids. In addition, they found that although different sets gave straight lines with different slopes on a given solid surface, they all extrapolated to yield an intercept on the $\cos\theta = 1$ (i.e., $\theta = 0°$) line at more or less the same value of γ_L. This value of γ_L was therefore a characteristic of the solid surface whose wettability was being measured, and was named the "critical surface tension," or γ_C, of a surface of that constitution. This discovery enabled the quantitative comparison of the wettability characteristics of various low-energy solid surfaces as a function of constitution.

In 1957, the ability to predict wettability from solid surface characterization took another important step forward when researchers treated contact angles on a more theoretical level by assuming interfacial interaction to be equal to the geometric mean of the surface tensions of two interacting condensed phases. Since these values did not always agree with measured interfacial tensions, a multiplier Φ was introduced because specific surface interactions could be significant. In 1962, another advance took place when the surface tension of a given liquid or solid was written as the sum of components due to different forces. This enabled interfacial tensions resulting from dispersion force interactions to be predicted from knowledge of the dispersion components of surface tension of solids and liquids.

More recently, acid–base theory has been adapted to surface interactions to enable liquids and solids to be characterized by parameters describing their acid–base reactivity as well as the dispersion reactivity. This last development is the subject of Chapter 1.

Chapter 2 reviews a phenomenon called the "hydrophobic interaction," which occurs in aqueous media between low-energy substrates expected to attract each other through dispersion forces only. Emphasis is placed on the results of direct measurements of this force. In Chapter 3, the determination of the properties of high- and medium-energy surfaces is reviewed. For high-energy surfaces, such as glasses and metals, ultrahigh vacuum is the only applicable technique, since contamination commences rapidly with molecular collisions from the gas phase. For medium-energy surfaces, such as the basal plane of graphite, the relevant contamination may be due to deposition of large organic molecules suspended in air, perhaps in clusters, on the smooth surface. A monolayer of that sort can take more than a minute to form and, in principle, should be preventable by far less stringent means than vacuum. In practice, however, the ultraclean conditions of ultrahigh vacuum provide the method of choice for eliminating this type of contamination.

In the fourth chapter, the method of two-liquid phase contact angles is reviewed. The capabilities of the contact angle method on low-energy surfaces are expanded by measuring the contact angle of a given liquid on a solid surface in an environment consisting of another liquid instead of air. On "high-energy" surfaces, the method, in principle, can yield finite contact angles rather than the zero angles observed with ordinary liquids. It must be kept in mind, however, that these are real surfaces derived from high-energy materials; i.e., the atomic constitution of the surface reflects the results of adsorption of components of the liquids as well as ambient contamination. The results are useful, however, because these are the surfaces as they exist in technological applications.

All of the chapters mentioned thus far use the "separation of forces" approach for interpretation of the experimental results with the methods employed. Subsequent chapters involving application of contact angle techniques to obtain fundamental information for technological application also rely heavily on this approach. In line with the policy of this book to present alternative points of view as well as a variety of approaches, however, we are pleased to include the "equation of state" approach in Chapter 5, authored by its originator and his colleagues. The approach, which eschews "separation of forces," proposes that an equation of state must exist that gives the solid–liquid interfacial tension as a function of the solid–vapor and liquid–vapor surface tensions, develops the resulting relationship, and illustrates its application to specific systems.

Chapter 6 reviews the subject of interface reconstruction of polymeric materials. The phenomenon was first reported in the middle of the 1970s. Although the nature of a solid surface interacting with liquid had usually been considered to be constant, there are situations in which the surface of a polymer can undergo a rearrangement due to the presence of the liquid probe. Whereas the dry surface had reached equilibrium by minimizing the polymer–air surface tension, the presence of liquid can result in a new equilibrium that minimizes the polymer–liquid interfacial tension. The process is generally slow and can be investigated by pretreating a polymer surface with various liquids under various conditions. The result can then be determined by other surface analytic techniques or by contact angles with appropriate liquid probes. The seventh chapter reviews methods of deliberately creating hydrophilicity on a polymer surface. This is fundamental to various applications, including biocompatibility.

The eighth chapter gives an extensive review, ranging from fundamental considerations to the highly applied, of wettability and bioadhesion in ophthalmology. Basic considerations involving static and dynamic aspects of tear film formation are presented. The theory and practice of biomaterial selection for contact lenses is discussed. The important role of wettability in adhesion and erosion of the corneal epithelium, in retinal adhesion and its failure, and in intraocular surgery is elucidated. Chapter 9 reviews the important influence of the wettability concept in various aspects of dental care. Examples include its effect on adhesion of conventional and modern fillings and on formation and prevention of deleterious bacteria adhesion, plaque formation, and so on. In Chapter 10 the authors review the correlation between adhesion of aquatic organisms to surfaces and the critical surface tension of the surface in question. The discussion in this chapter considers both bacteria and larger organisms, and strength of adhesion as well as its frequency.

Chapter 11 reviews water adsorption to clay, a class of minerals of geological and biological significance. The review describes various types of physical interaction of H_2O with the silicate framework, including hydration of the oxygen plane, hydrogen bond formation with exposed hydroxyl planes and with exposed O or OH groups on broken bounds, and hydration of exchangeable cations. Chapter 12 reviews capillarity, with emphasis on the special characteristics of the capillary penetration of small drops, a recent development in the theory and experimental aspect of capillarity. In Chapter 13 the special features of wettability determinations on single fibers and their application to fabrication of composite materials are reviewed. The fourteenth chapter discusses the application of wettability concepts to coatings problems. Included are properties of the outside surface of the coating, as well as those of the interface between the coating and substrate. Chapter 15 discusses work on the Marangoni effect, involving spreading phenomena resulting from the dynamics of surface tension changes during the mixing of liquids. Chapter 16 concerns recent developments in the theory of drop formation, during heterogeneous nucleation of H_2O on solid surfaces.

Dr. Zisman was always conscious of the benefit of maintaining a healthy balance between basic and applied research, realizing that their close association is synergistic. In agreement with that concept, as well as in tribute to his scientific leadership, we have tried to follow that tradition in this book.

G.I.L. is grateful to the David Taylor Research Center for the facilities and opportunities extended to him to participate in this work.

M. E. Schrader
G. I. Loeb

Contents

Chapter 1

The Modern Theory of Contact Angles and the Hydrogen Bond Components of Surface Energies

Robert J. Good and Carel J. van Oss

Chapter 2

The Long-Range Attraction between Macroscopic Hydrophobic Surfaces

H. K. Christenson

Chapter 7
Block Copolymers and Hydrophilicity
Atsushi Takahara

Chapter 8
Wettability and Bioadhesion in Ophthalmology
Frank J. Holly

Chapter 9
Wettability of Surfaces in the Oral Cavity
H. J. Busscher

Chapter 10
Wettability as a Surface Signal for Sessile Aquatic Organisms
James W. Mihm and George Loeb

Chapter 11
Wettability of Clay Minerals
Shmuel Yariv

Modern Approaches to Wettability

Theory and Applications

1

The Modern Theory of Contact Angles and the Hydrogen Bond Components of Surface Energies

Robert J. Good and Carel J. van Oss

1. INTRODUCTION

We owe a great debt to W. A. Zisman and his colleagues at the Naval Research Laboratory for their extensive, pioneering work that opened up the field of contact angles and made possible the development of the modern theory of wetting and adhesion. Their data on the wetting of solids by apolar liquids and by hydrogen bonding liquids pointed the way to the recent introduction of a theory of hydrogen bond interactions across interfaces. We will devote this chapter to a review of this new theory.

Hydrogen bonds have been recognized as chemical entities ever since 1920.[1] Their contribution to phenomena such as the mechanical strength of proteins, the high boiling point and electric conductivity of water, the information-carrying structure of DNA, and many other phenomena, have been known for several decades. The fact that hydrogen bonds influence adsorbed films on water was mentioned by Harkins and Meyers[2] and Harkins,[3] and the influence of orientation due to hydrogen bonds on surface entropy at liquid–vapor interfaces was noted by Good[4] over 30 years ago. There is no question that hydrogen bonding is involved in hydrophilic behavior of solid surfaces and that its absence is correlated strongly with hydrophobic behavior. But, until recently, no quantitative theory has been available by which these surface manifestations of hydrogen bonding could be evaluated.

Pimentel and McClellan[5] have given the most authoritative discussion of hydrogen bonds, saying (on p. 63) that "no really definitive paper connecting surface

---

Robert J. Good • Department of Chemical Engineering, State University of New York at Buffalo, Buffalo, New York 14260. *Carl J. van Oss* • Department of Microbiology, State University of New York at Buffalo, Buffalo, New York 14214.

Modern Approach to Wettability: Theory and Applications, edited by Malcolm E. Schrader and George Loeb. Plenum Press, New York, 1991.

tension to hydrogen bonding has appeared." They mentioned, however, the correlation found by Good and Girifalco[6] between the ratio Φ of the free energy of adhesion to the geometric mean of the free energies of cohesion of the separate phases and the presence or absence of hydrogen bonding in one or both phases. Jensen[7] has discussed the Lewis[8] and Brönsted–Lowry[9,10] theories of acids and bases, but says little on hydrogen bonds. Coulson[11] provided a theoretical basis for hydrogen bonds in systems in which π electrons constitute the electron donor system. Good and Elbing[12] gave the first quantitative treatment of liquid–liquid interfacial hydrogen bonds involving π electrons, showing that a bond energy of about 2.4 kcal/mole would account for the difference between the water–benzene interfacial tension and that between water and saturated hyrocarbons. See also Ref. 13 in regard to polystyrene.

Hydrogen bonding is the most common type of molecular interacting that leads to an acid–base component of the free energy of adhesion and of the interfacial tension between two phases. Hydrogen bonding can be well described in terms of the Brönsted–Lowry acid–base (proton donor–proton acceptor) theory. However, it is possible to describe hydrogen bonding via the (more general) Lewis acid–base theory, in which an acid is an electron acceptor and a base is an electron donor. Thus, we will use the terminology *Lewis base*, which is general, together with *Lewis acid*, which means either an electron acceptor or the subclass of electron acceptors that comprises proton donors. We may refer to the creation of an adduct by hydrogen bond formation as an example of *Lewis neutralization.*

A quantitative theory has recently been developed[14–19] for estimating the contribution due to the hydrogen bonding character of a *single* substance, to surface and interfacial tension, and to free energy of adhesion at interfaces in binary systems. The progenitor of this theory was Fowkes's[20,21] resolution of surface free energy γ into an apolar component and an acid–base component. A treatment of the apolar component had been proposed earlier by Good and Girifalco[4,12,13,22,23,24] and is by now well understood. The physical basis of that treatment lies in the molecular theory of intermolecular forces between *nonpolar* molecules in the *gaseous* state, due to Berthelot[25] and London[26] (the theory of the dispersion force), and in the Scatchard–Hildebrand theory of the solubility of nonelectrolytes.[27] (It was Berthelot who first proposed a geometric mean combining rule for intermolecular interactions.)

Hildebrand and Scott have pointed out[27] that, in contrast to the London dispersion force, hydrogen bonds cannot be fitted into the Berthelot–London–Scatchard–Hildebrand formalism, with regard to the behavior of solutions. So, it might be expected that the geometric mean combining rule would not be successful for describing the hydrogen bond components of interfacial energy. This is indeed the case.

The reason for this failure lies in the complementarity of functions in acid–base interactions, which is in strong contrast to the symmetry of functions in the dispersion force field that acts between any two small molecules or molecular segments. In the dispersion force interaction of two unlike molecules, i and j, the contributions of i and of j are, formally, of the same kind as the interaction of two i molecules with each other or of two j molecules with each other. The molecular polarizabilities α_i and α_j, and the ionization energies I_i and I_j, enter the equations for the London dispersion force in a perfectly symmetrical way. But in an acid–base interaction, that symmetry is totally lacking; instead, a complementarity of functions exists. A basic moiety, e.g., a carbonyl group, does not interact *as a base* with another basic moiety, and there is a similar lack of interaction of acidic moieties with each other *as acids*. It is only when both kinds of molecules or groups, having complementary functions, are present that an acid–base interaction can take place. In the case of $CHCl_3$ and acetone, a hydrogen bonded adduct, Cl_3CH---$OC(CH_3)_2$, is formed.[6,27] (Experimental evidence has also been reported[28] for a three-molecule complex in acetone–chloroform solutions in which two chloroform molecules are bonded to the carbonyl oxygen of acetone. There are, of course, two unshared pairs of electrons on the oxygen.) For an alcohol *acting as a proton donor* (acid) the coordination number for hydrogen bonding is 1; when acting as a proton *acceptor*, the coordination number of an oxygen in a hydroxyl is 1 or, at most, 2. This is far lower than the coordination number in dispersion force interactions, which is limited only by the geometry of packing and can be as large as 12.

We have elsewhere[14] discussed the dipole–dipole (Keesom) and dipole–induced dipole (Debye) components of the van der Waals force. These components turned out to be small *in liquids* (as distinct from the gaseous state) relative to the dispersion force. (This conclusion, which supersedes the theory of Good and Elbing,[12] was reached on the basis of the Lifshitz theory of forces across an interface between condensed phases. Fowkes[21] reached this conclusion earlier on the basis of a more limited argument.) It was shown that the London,[26] Keesom,[29] and Debye components[30] can be lumped together as the "apolar" (LW) component of attraction. (The designation LW stands for Lifshitz–van der Waals.)

The new acid–base principle in interfacial interactions, which we now review, has important consequences with regard to contact angles and surface energies. We will discuss some of those consequences and some recent results later in this chapter.

2. *ACID–BASE INTERACTIONS AND THE NEW SURFACE PARAMETERS*

Following the form suggested by Fowkes,[20] we write for phase i,

$$\gamma_i = \gamma_i^{LW} + \gamma_i^{AB} \tag{1}$$

Thus, surface free energy* is the sum of an apolar component and an acid–base (AB) component. The free energy of cohesion ΔG_i^c likewise has two components:

$$\Delta G_i^c = -2\gamma_i \tag{2}$$

$$= \Delta G_i^{cLW} + \Delta G_i^{cAB} \tag{3}$$

* Thermodynamic discussions of adhesion in solid–liquid systems should be carried out in terms of surface free energy rather than surface tension. Discussions that involve the *shape* of liquid–gas or liquid–liquid interfaces can be carried out either in terms of surface tension or surface free energy.

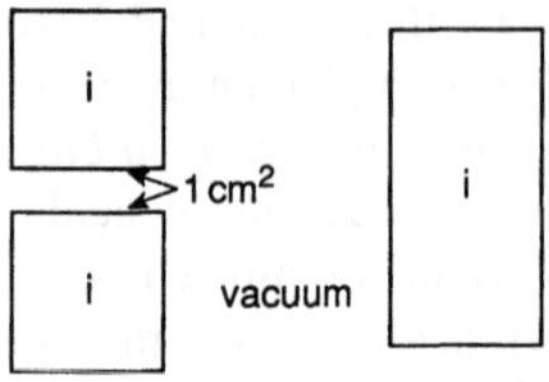

FIGURE 1. Diagram of the process corresponding to Eq. (2); the formation of a "cohesive" union of two bodies of the same material.

as does the free energy of adhesion $\Delta G_{ij}^{\mathrm{a}}$ for the interface between condensed phases i and j, given by

$$\Delta G^{\mathrm{a}} = \gamma_{ij} - \gamma_i - \gamma_j \tag{4}$$

$$\Delta G_{ij}^{\mathrm{a}} = \Delta G_{ij}^{\mathrm{aLW}} + \Delta G_{ij}^{\mathrm{aAB}} \tag{5}$$

See Figs. 1 and 2.

For the LW component of free energy per unit area, the Berthelot geometric mean rule holds:[13,25]

$$\gamma_{ij}^{\mathrm{LW}} - \gamma_i^{\mathrm{LW}} - \gamma_j^{\mathrm{LW}} = -2\sqrt{\gamma_i^{\mathrm{LW}}\gamma_j^{\mathrm{LW}}} \tag{6a}$$

$$\gamma_{ij}^{\mathrm{LW}} = \left(\sqrt{\gamma_i^{\mathrm{LW}}} - \sqrt{\gamma_j^{\mathrm{LW}}}\right)^2 \tag{6b}$$

$$\Delta G_{ij}^{\mathrm{aLW}} = -2\sqrt{\gamma_i^{\mathrm{LW}}\gamma_j^{\mathrm{LW}}} \tag{7}$$

γ^{AB} and ΔG^{aAB} require a notably different combining rule treatment from γ^{LW} and ΔG^{aLW} because of the complementarity in acid–base interactions. To obtain a form capable of yielding a quantitative, predictive treatment of acid–base interactions, we must postulate a resolution of the acid–base term γ^{AB} into an acidic surface parameter and a basic surface parameter. We therefore define two monopolar surface parameters:

$$\gamma^{\oplus} \equiv \text{(Lewis) acid component of surface interaction}$$

$$\gamma^{\ominus} \equiv \text{(Lewis) base component of surface interaction}$$

If both the acidic and basic components for a substance i are negligible, then i is referred to as *apolar*. If either the acidic or the basic component is negligible and the other component is appreciable, then i is referred to as *monopolar*.[16] For example, $CHCl_3$ is monopolar acidic, and an ether or carbonyl group is monopolar basic due to the unshared pair electrons on the oxygen. If the acidic and basic components are appreciable, the substance is *bipolar*. (Bipolar compounds are sometimes described as self-associated. Water is the best-known example.)

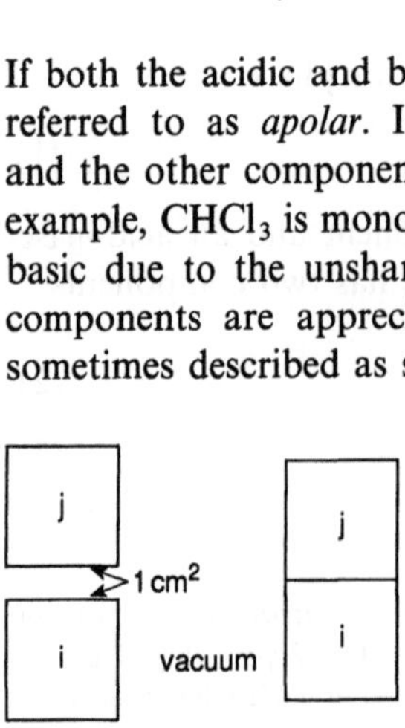

FIGURE 2. Diagram of the process corresponding to Eq. (3); the formation of an "adhesive" union of two unlike bodies in vacuum.

It would, in principle, be possible to introduce two or even four parameters besides $\gamma^{\oplus}$ and $\gamma^{\ominus}$. This increase in complexity would be required in order to take into account the theory of hard acids and bases versus soft acids and bases[7] and the employment of covalent as well as electrostatic terms in the Fowkes–Drago theory.[21] We will not comment on the unresolved theoretical questions involved in the hard–soft and covalent–electrostatic theories. However, it appears to be possible to represent surface phenomena with only two new parameters, so there may be no need to introduce two or four additional terms to characterize acid–base interfacial interactions.

The relation among the terms $\gamma^{\oplus}$, $\gamma^{\ominus}$, and γ^{AB} is given in the form of combining rules that connect the surface tension components γ_i^{AB} and γ_j^{AB}, the interfacial tension component γ_{ij}^{AB}, and the component of the free energy of adhesion ΔG_{ij}^{aAB}. Thus,

$$\Delta G_{ij}^{aAB} = -2\sqrt{\gamma_i^{\oplus}\gamma_j^{\ominus}} - 2\sqrt{\gamma_i^{\ominus}\gamma_j^{\oplus}} \tag{8a}$$

$$= -2\sqrt{\gamma_i^{\oplus}\gamma_j^{\ominus}} \quad \text{if } \gamma_j^{\oplus}=0 \text{ or } \gamma_i^{\ominus}=0 \quad (j \text{ a monopolar base or } i \text{ a monopolar acid}) \tag{8b}$$

$$= -2\sqrt{\gamma_i^{\ominus}\gamma_j^{\oplus}} \quad \text{if } \gamma_i^{\oplus}=0 \text{ or } \gamma_j^{\ominus}=0 \quad (i \text{ a monopolar base or } j \text{ a monopolar acid}) \tag{8c}$$

$$\gamma_i^{AB} = 2\sqrt{\gamma_i^{\oplus}\gamma_i^{\ominus}} \quad \text{if substance } i \text{ bipolar} \tag{9a}$$

$$= 0 \quad \text{if substance } i \text{ monopolar or apolar} \tag{9b}$$

$$\gamma_{ij}^{AB} = 2\{\sqrt{\gamma_i^{\oplus}\gamma_i^{\ominus}} + \sqrt{\gamma_j^{\oplus}\gamma_j^{\ominus}} - \sqrt{\gamma_i^{\oplus}\gamma_j^{\ominus}} - \sqrt{\gamma_i^{\ominus}\gamma_j^{\oplus}}\} \tag{10a}$$

$$= 2(\sqrt{\gamma_i^{\oplus}} - \sqrt{\gamma_j^{\oplus}})(\sqrt{\gamma_i^{\ominus}} - \sqrt{\gamma_j^{\ominus}}) \tag{10b}$$

The parameters $\gamma^{\oplus}$ and $\gamma^{\ominus}$ for a material have *operational* significance only when used in these combining rules.

The formal difference between Eqs. (6b) and (10b) is itself an interesting matter. For example, Eq. (6) can only lead to positive values of γ_{ij}^{LW}, while, according to Eq. (10), the value of γ_{ij}^{AB} will be negative if either $\gamma_j^{\oplus} > \gamma_i^{\oplus}$ and $\gamma_j^{\ominus} < \gamma_i^{\ominus}$ or if $\gamma_j^{\oplus} < \gamma_i^{\oplus}$ and $\gamma_j^{\ominus} > \gamma_i^{\ominus}$. We will discuss the physical significance of negative interfacial free energy below.

We may illustrate another of the characteristics of $\gamma^{\oplus}$ and $\gamma^{\ominus}$, as brought out in Eqs. (3)–(10). Eq. (9) correctly describes the acid–base surface tension component for chloroform: the surface tension of $CHCl_3$ is almost exactly the same as that of CCl_4, which is apolar. Yet the free energy of adhesion of (monopolar) $CHCl_3$ to water has a larger magnitude (-68.3 mJ/m^2) than that of CCl_4 with water (-54.7 mJ/m^2). The difference is due to the hydrogen bond structure, $Cl_3C{-}H{-}{-}{-}OH_2$; it cannot be accounted for by theories that ignore acid–base interactions.

Equations (6)–(10) also lead to a method of treating certain systems of 3 condensed phases. The free energy of spreading–displacement (Fig. 3a) is

$$\Delta G_{s21}^{s} = \gamma_{12} + \gamma_{s2} - \gamma_{s1} \tag{11}$$

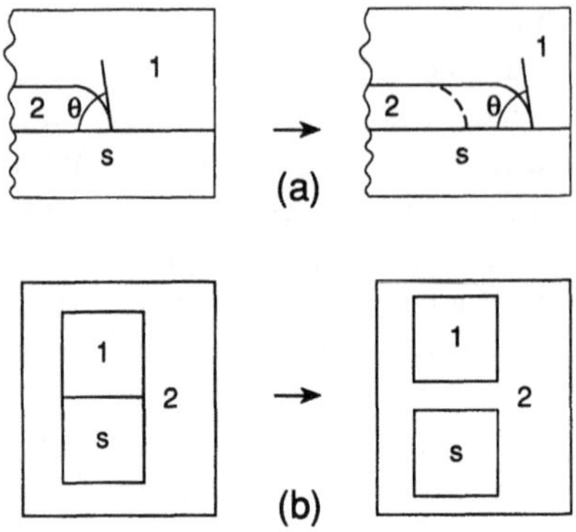

FIGURE 3. (a) Spreading displacement: Liquid 2 on solid s, under liquid 1, makes a nonzero contact angle θ. Liquid 2 is made to advance over s, displacing 1 from 1 cm^2 of solid surface. (b) Adhesional displacement: For a 1-cm^2 area of interface between solids and liquid 1 (at left) the two phases are separated (at right) and phase 2 is interposed, forming a stable layer.

This expression is the same as that for the process illustrated by Fig. 3b, which can be described as *adhesional* displacement. Curiously, this identity of form does not hold when phase 2 is replaced by vacuum. Then, the free energy of adhesion of liquid l to solid s (compare Fig. 2) is

$$\Delta G^{a}_{sl} = \gamma_{sl} - \gamma_{lv} - \gamma_{sv} \tag{12}$$

The free energy of spreading liquid l over solid s is

$$\Delta G^{s}_{sl} = \gamma_{sl} + \gamma_{lv} - \gamma_{sv} = -S_{l/s} \tag{13}$$

where $S_{l/s}$ is the Harkins spreading coefficient[3]. (See also Ref. 31, p. 104.)

The acid–base component of the free energy for the spreading–displacement process is

$$\begin{aligned}\Delta G^{sAB}_{s21} = -2[&\sqrt{\gamma_2^{\oplus}}(\sqrt{\gamma_s^{\ominus}} + \sqrt{\gamma_1^{\ominus}} - \sqrt{\gamma_2^{\ominus}}) \\ &+ \sqrt{\gamma_2^{\ominus}}(\sqrt{\gamma_s^{\oplus}} + \sqrt{\gamma_1^{\oplus}} - \sqrt{\gamma_2^{\oplus}}) - \sqrt{\gamma_s^{\oplus}\gamma_1^{\ominus}} - \sqrt{\gamma_s^{\ominus}\gamma_1^{\oplus}}]\end{aligned} \tag{14}$$

The acid–base component is, generally, of a much larger magnitude than the apolar component, which is given by

$$\Delta G^{sLW}_{s21} = 2(\gamma_2^{LW} - \sqrt{\gamma_s^{LW}\gamma_2^{LW}} - \sqrt{\gamma_1^{LW}\gamma_2^{LW}} + \sqrt{\gamma_s^{LW}\gamma_1^{LW}}) \tag{15a}$$

$$= 2(\sqrt{\gamma_1^{LW}} - \sqrt{\gamma_s^{LW}})(\sqrt{\gamma_2^{LW}} - \sqrt{\gamma_1^{LW}}) \tag{15b}$$

The entire free energy of spreading–displacement is

$$\Delta G^{s}_{s21} = \Delta G^{sLW}_{s21} + \Delta G^{AB}_{s} \tag{16}$$

The term ΔG^{sAB}_{s21}, like γ^{AB}_{ij}, may be positive or negative; the displacement process occurs spontaneously when ΔG_{s21} is negative.

It has been demonstrated[32] that the LW component of ΔG^{a}_{s21} will be negative when the LW component of γ_k lies between the LW components of γ_i and γ_j. This treatment has recently been reviewed and extended by van Oss.[33]

It may be noticed that we have *not* employed the interaction parameter Φ of Good and Girifalco.[6,12,13,23] That is because the modern treatment, Refs. 14–19, reviewed above, supersedes the theory that was given in terms of Φ. While Φ can still be used as an empirical function, it cannot in general be computed *a priori* from currently available molecular properties of the chemical components. The theory for the calculation of Φ, given by Good and Elbing,[12] ignores hydrogen bonding; hence, that theory is, at best, seriously incomplete.

3. CONTACT ANGLES

The contact angle θ_{ls} of a liquid l on a solid s is given by the Young equation

$$\cos\theta = \frac{\gamma_{sv} - \gamma_{sl}}{\gamma_{lv}} \tag{17a}$$

provided $\gamma_{sv} - \gamma_{sl} \leqq \gamma_{lv}$. Young's equation may also be written

$$\cos\theta = \frac{\gamma_{s} - \pi_{esl} - \gamma_{sl}}{\gamma_{l} - \pi_{elv}} \tag{17b}$$

where γ_{sv} is the surface free energy of the solid in equilibrium with the saturated vapor of the liquid, γ_{lv} is the surface tension of the liquid in equilibrium with the solid, and γ_{sl} is the solid–liquid interfacial free energy. The π terms are defined by Eq. (18):

$$\gamma_{s} \equiv \gamma_{sv} + \pi_{esv} \tag{18a}$$

$$\gamma_{l} \equiv \gamma_{lv} + \pi_{elv} \tag{18b}$$

Here, γ_s and γ_l are the surface free energies of the separate phases in contact only with their own vapor, and π_e is the equilibrium spreading pressure. With perhaps no more than one paper as an exception,[34] π_{elv} has been tacitly assumed to be zero.

The term π_{esv} was introduced in conjunction with the postulated geometric mean combining rule theory in Ref. 23. It has received a statistical–mechanical treatment in this laboratory in Ref. 35. On molecularly smooth low-energy solids, in the case of dispersion force liquids that form nonzero contact angles (i.e., liquids with $\gamma_l > \gamma_c$, where γ_c is the critical surface tension for wetting[36]), it was shown that π_{esv} should be negligible. This conclusion has been supported by Fowkes *et al.*[37] The derivation in Ref. 35 was based exclusively on dispersion force interactions. It is likely that, for some cases of hydrogen bonding liquids on solids that contain acidic or basic groups, π_{esv} may be appreciable, even when the contact angle is greater than zero. The "autophobic" systems described by Zisman *et al.* constitute an example of such behavior.

"Combining rule" equations such as (6) to (10) connect the free energy of cohesion [Eqs. (2) and (3)] of *separate* phases, *i* and *j*, and hence the surface free energies (for example, γ_s and γ_l) with γ_{sl}. Note that the geometric mean combining rule for apolar components, Eqs. (5) and (6), written for a solid–liquid system, does not employ γ_{sv} or γ_{lv} directly. These functions can be used for substitution into Eq. (17) only when the form (17b) is used.

The two spreading pressures in Eqs. (17) and (18) are properties of the particular liquid–solid combination. They are the result of the adsorption of the vapor of the liquid (at saturation) at the solid–vapor interface and of molecules from the solid at the liquid–vapor interface. These terms can, in principle, be determined by experiment.

However, the measurement of π_e is not, in general, a simple matter on a macroscopic solid surface; and its theoretical estimation is difficult, except in cases that could be idealized, such as those discussed in Ref. 35. Besides, it is not yet clear how to make a unique or general assignment of components of π_e as between apolar interactions and acid–base interactions. This question must be left open for future study. So it is appropriate, for the present, to explore the case in which the two π_e terms are negligible.

In the approximation that π_{esv} and π_{elv} can be neglected, Young's equation becomes

$$\cos\theta \cong \frac{\gamma_s - \gamma_{sl}}{\gamma_l} \qquad \gamma_s - \gamma_{sl} \leqq \gamma_l \tag{17c}$$

$$\gamma_s - \gamma_{sl} \geqq \gamma_l \qquad \text{if } \cos\theta = 1 \tag{17d}$$

We should also call attention to the neglect of roughness in equations such as (17c). This subject has been reviewed in Ref. 38.

For the case where both phases are apolar, which has been discussed extensively by Fowkes, the Young equation can be written

$$\cos\theta = \frac{\gamma_s^{LW} - \gamma_{sl}^{LW}}{\gamma_l^{LW}} = \frac{\gamma_s - \gamma_{sl}}{\gamma_l} \tag{19}$$

Combining Eqs. (19) and (6), we obtain

$$\gamma_l^{LW}(1+\cos\theta) = 2\sqrt{\gamma_s^{LW}\gamma_l^{LW}} \tag{20a}$$

$$\cos\theta = -1 + 2\sqrt{\gamma_s^{LW}/\gamma_l^{LW}} \tag{20b}$$

Useful, alternative forms of Eq. (20) are

$$\gamma_l^{LW} = \frac{4\gamma_s^{LW}}{(1+\cos\theta)^2} \tag{20c}$$

$$\gamma_s^{LW} = \frac{\gamma_l^{LW}(1+\cos\theta)^2}{4} \tag{20d}$$

$$= \frac{\gamma_{l(\text{apolar liquid})}(1+\cos\theta)^2}{4} \tag{20e}$$

Equation (20) is used in determining γ_s^{LW} for apolar solids such as Teflon*, and Eq. (20c) can be used in determining the LW component of γ_l, for a polar liquid, by measuring the contact angle on an apolar solid whose γ_s^{LW} is known.

It has been pointed out[23] that these equations provide an interpretation of the Fox–Zisman principle, which is that there exists a critical liquid surface tension γ_c (or γ_{lc}) for a solid s, such that for any liquid with $\gamma_l < \gamma_c$, $\theta = 0$. It is immediately evident from Eq. (20) that, for dispersion force liquids,

$$\gamma_c = \gamma_l^{LW}(\cos\theta \to 1) = \gamma_s^{LW} \tag{21}$$

This conclusion can be extended somewhat (as we will show), and the question of its application to the spreading of *polar* liquids on *polar* solids can be given a fresh examination.

Using the inequality (17d) and combining with Eq. (6), we obtain, when $\cos\theta = 1$, or

$$\gamma_s^{LW} \geqq \gamma_{lc}^{LW} \qquad \text{if liquid } l \text{ is apolar} \tag{22}$$

Equation (20b) has important implications when the determination of the critical surface tension requires an extrapolation. If the extrapolation is long, then the *linear* extrapolation, i.e., use of the Fox–Zisman equation

$$\cos\theta = 1 + b(\gamma_l - \gamma_c)$$

is unreliable. But a graph of $\cos\theta$ versus $1/\sqrt{\gamma_l}$ (which may be called the modified Fox–Zisman plot) is much more reliable because it has a theoretical foundation. This point has been discussed at some length in Ref. 13. Thus, there is a formal, theoretical foundation underneath the modified Fox–Zisman methodology *for apolar liquids on apolar and polar solids.*

As will be shown, there is not so firm a theoretical foundation when *polar* liquids are employed—particularly, when the solid is polar.

For the general case, in which $\gamma^{\oplus}$ and $\gamma^{\ominus}$ components may be present in either phase, we may employ (5), (7), and (8), written in the following forms:

$$\Delta G_{sl}^{a} = \Delta G_{sl}^{aLW} + \Delta G_{sl}^{aAB} \tag{5}$$

$$\Delta G_{sl}^{aLW} = -2\sqrt{\gamma_s^{LW}\gamma_l^{LW}} \tag{7}$$

$$\Delta G_{sl}^{aAB} = -2\sqrt{\gamma_s^{\oplus}\gamma_l^{\ominus}} - 2\sqrt{\gamma_s^{\ominus}\gamma_l^{\oplus}} \tag{8a}$$

* Teflon is effectively apolar despite the fact that $(-CF_2-)$ and $(-CF_3)$ groups have dipole moments that are in the neighborhood of 1.2 D. The locally opposed orientation of dipoles, causing mutual cancellation of fields *in a condensed phase*, has been discussed at some length in Ref. 14. The calculations reported there, which were made using the Lifshitz theory of electromagnetic interactions, showed that *dipole–dipole* interactions across a boundary between *condensed* phases are small. Solubility studies[27] have shown that fluorine atoms in fluorocarbons have negligible capacity to take part in hydrogen bonding. Hence, in regard to interfacial behavior, Teflon and other fluorocarbons are apolar.

These, together with the Young–Dupré equation

$$\Delta G^{a}_{sl} = -\gamma_l(1+\cos\theta) \tag{23}$$

yield the important, general result

$$\gamma_l(1+\cos\theta) = 2\sqrt{\gamma_l^{LW}\gamma_s^{LW}} + 2\sqrt{\gamma_l^{\oplus}\gamma_s^{\ominus}} + 2\sqrt{\gamma_l^{\ominus}\gamma_s^{\oplus}} \tag{24a}$$

If either the liquid or the solid is monopolar, then ΔG^{aAB}_{sl} is given by

$$\Delta G^{aAB}_{sl} = -2\sqrt{\gamma_i^{\oplus}\gamma_j^{\ominus}} \tag{25}$$

where $i = s$ and $j = l$, or vice versa; and one or the other of the two terms on the right in Eq. (8) is zero. [Compare Eq. (9).] If the solid is apolar, then we obtain the generalization of Eq. (20a) for a *polar* liquid on the solid:

$$\gamma_l(1+\cos\theta) = 2\sqrt{\gamma_s^{LW}\gamma_l^{LW}} \qquad \text{if} \quad \gamma_s^{\oplus} = \gamma_s^{\ominus} = 0 \tag{24b}$$

If in (24b) we let $\cos\theta \to 1$, we obtain

$$2\gamma_l(\cos\theta \to 1) = 2\gamma_c = 2\sqrt{\gamma_s^{LW}\gamma_c^{LW}}$$

$$\gamma_s^{LW} = \frac{\gamma_c^2}{\gamma_c^{LW}} \tag{26}$$

If a series of polar liquids is used to determine a critical surface tension for wetting, using the Fox–Zisman extrapolation to $\cos\theta = 1$, a unique γ_c will be obtained only if the fractional polarity, $(\gamma_l - \gamma_l^{LW})/\gamma_l$, is constant for all members of the series. This is rare. So it is not to be expected that the Fox–Zisman extrapolation will, in general, yield a reliable γ_c value for a series of polar liquids on an apolar solid.

Finally, if both the solid and the liquid are polar, we must use Eq. (24a) if $\theta > 0$. If the contact angle goes to zero, then (24a) reduces to

$$\gamma_l(\theta = 0) \leqq \sqrt{\gamma_s^{LW}\gamma_l^{LW}} + \sqrt{\gamma_s^{\oplus}\gamma_l^{\ominus}} + \sqrt{\gamma_s^{\ominus}\gamma_l^{\oplus}} \tag{24c}$$

Combining Eqs. (1) and (9a), we obtain

$$\gamma_s = \gamma_s^{LW} + \gamma_s^{AB} = \gamma_s^{LW} + 2\sqrt{\gamma_s^{\oplus}\gamma_s^{\ominus}} \tag{27}$$

Equations (24c) and (27) contain four more parameters than are present in Eqs. (20a)–(20d) and (21)–(22). Physically, Eqs. (24c) and (27) mean that the condition, $\theta = 0$, for polar liquid l on solid s, may be achieved in a number of different ways. For example, $\gamma_l^{\ominus}$ might be large enough that the second term on the right would dominate Eq. (24c), or $\gamma_s^{\oplus}$ might be large enough to do so, and so on. It appears that observations of zero contact angles of polar liquids on a solid s do not lead directly to any estimate of γ_s.

Therefore, the Fox–Zisman method of handling contact angle data, *as applied to polar liquids on polar solids*, is at best an empiricism, unlike the cases of apolar liquids on solids (whether polar or apolar). A "critical surface tension for wetting" will, in general, exist for any series of liquids, on any solid, but it is not, in general, related to the surface free energy of the solid in any simple way.

Zisman, himself, said many times that he did not interpret a γ_c value as being anything more than "a *measure* of the surface free energy of the solid." And in his later work, he used the Fox and Zisman extrapolation only with a homologous series of hydrocarbons.

4. APPLICATION OF THE LEWIS ACID–BASE PARAMETERS TO CONTACT ANGLES IN PRACTICAL SYSTEMS

It should be possible to apply Eq. (24) to practical systems, if a set of γ^{LW}, $\gamma^{\oplus}$, and $\gamma^{\ominus}$ parameters of liquids can be established. The γ^{LW} terms can be determined in a straightforward manner. By definition, apolar solids such as Teflon, and apolar liquids such as alkanes, have surface free energies that are purely of the Lifshitz–van der Waals type:

$$\gamma_i(\text{apolar}) \equiv \gamma_i^{LW}$$

Fowkes has pointed out[20] that polar forces, such as those due to hydrogen bonds, do not contribute to the free energy of adhesion, ΔG^a, of a polar liquid versus an *apolar* liquid or solid:

$$\Delta G_{ij}^{aLW} = \Delta G_{ij}^{a} \tag{28}$$

Equation (6b), for γ^{LW}, is valid regardless of whether phase i or phase j has the capability of forming hydrogen bonds. If both phases are liquid, and if j is apolar (so that $\gamma_j = \gamma_j^{LW}$ and $\gamma_{ij} = \gamma_{ij}^{LW}$), we may determine γ_i^{LW} by measuring the interfacial tension, γ_{ij}, and using Eq. (6), solved for γ_i^{LW}. Then

$$\gamma_i^{LW} = \gamma_{ij} - \gamma_j + 2\sqrt{\gamma_i^{LW}\gamma_j} \qquad j \text{ apolar} \tag{29}$$

The apolar component of the surface free energy of a polar solid can be determined by measuring the contact angle of an apolar liquid and using Eq. (20d). As already noted, we ignore (for the present) the equilibrium spreading pressure π_e.

An important, practical restriction, for Eqs. (20)–(24) to be usable, is that the liquid must have a surface tension larger than the critical surface tension of the solid. The γ_s^{LW} values for a large number of important polymers are in the neighborhood of 40 mJ/m^2. For use in determining γ_s^{LW} of such solids, two liquids are available, which have high surface tensions but very small propensities for AB interactions; these are diiodomethane ($\gamma = 50.8$ mJ/m^2) and α-bromonaphthalene ($\gamma = 44.4$ mJ/m^2). The γ^{LW} values of seven liquids are given in Table 1. In practical studies of the surface parameters of a solid, γ_s^{LW} is first determined with the aid of an apolar liquid and Eq. (20).

TABLE 1. Surface Tension Components (in mJ/m²) of Some Liquids[a]

	γ_l	γ_l^{LW}	γ_l^{AB}	$\gamma_l^{\oplus}$	$\gamma_l^{\ominus}$	Reference
Water (W)	72.8	21.8	51.0	25.5	25.5	
Glycerol (Gl)	64	34	30	3.92	57.4	39
Ethylene glycol (EG)	48.0	29	19.0	1.92	47.0	39[c]
Formamide (Fo)	58	39	19	2.28	39.6	39
Dimethyl sulfoxide (DMSO)	44	36	8	0.5[b]	32[b]	18
α-Bromonaphthalene (ABN)	44.4	43.5	≅0			
Diiodomethane (DIM)	50.8	50.8	≅0			

[a] $\gamma^{\oplus}$ and $\gamma^{\ominus}$ values based on $\gamma_W^{\oplus} = \gamma_W^{\ominus}$ reference convention.
[b] Values of $\gamma^{\oplus}$ for DMSO range from 0.07 to 0.7 mJ/m². The values reported here are the average of several determinations. The extreme hygroscopic character of DMSO makes it a difficult liquid to use in contact angle measurements.
[c] Revision in proof; unpublished results (1992) of C. J. van Oss.

To explain the determination of the $\gamma^{\oplus}$ and $\gamma^{\ominus}$ values of reference liquids, we must first develop the set of relations connecting these parameters to the contact angles that can be measured. For a particular solid s, the contact angles of an apolar liquid and of two polar liquids, l_1 and l_2, whose γ^{LW}'s are known, are determined.* We may rearrange Eq. (24a) into the form with the "unknowns" $\gamma_s^{\oplus}$ and $\gamma_s^{\ominus}$ on the right:

$$\gamma_{l_1}(1+\cos\theta_{l_1/s}) - 2\sqrt{\gamma_{l_1}^{LW}\gamma_s^{LW}} = 2\sqrt{\gamma_{l_1}^{\ominus}\gamma_s^{\oplus}} + 2\sqrt{\gamma_{l_1}^{\oplus}\gamma_s^{\ominus}} \tag{30a}$$

$$\gamma_{l_2}(1+\cos\theta_{l_2/s}) - 2\sqrt{\gamma_{l_2}^{LW}\gamma_s^{LW}} = 2\sqrt{\gamma_{l_2}^{\ominus}\gamma_s^{\oplus}} + 2\sqrt{\gamma_{l_2}^{\oplus}\gamma_s^{\ominus}} \tag{30b}$$

To solve for $\gamma_s^{\oplus}$ and $\gamma_s^{\ominus}$, we need four more relations. Two of these can be obtained with the aid of other cantact angle measurements. These are the ratio $(\gamma_{l_1}^{\oplus}/\gamma_{l_2}^{\oplus})^{1/2}$, which we may call δ, and the corresponding ratio of basic components. We may explain the determination of the $(\gamma_{l_1}^{\oplus}/\gamma_{l_2}^{\oplus})^{1/2}$ and $(\gamma_{l_1}^{\ominus}/\gamma_{l_2}^{\ominus})^{1/2}$ ratios by illustration with the liquid pair water and glycerol.

The AB component of the free energy of adhesion of a bipolar liquid to a *monopolar* solid, such as poly(methylmethacrylate) (PMMA), may be obtained by use of Eqs. (4), (7), (8), and (20), from which

$$\Delta G_{1s}^{aAB} = -\gamma_l(1+\cos\theta_{l/s}) + 2\sqrt{\gamma_l^{LW}\gamma_s^{LW}} \tag{31}$$

$$-2\sqrt{\gamma_l^{\oplus}\gamma_s^{\ominus}} = -\gamma_l(1+\cos\theta_{l/s}) + 2\sqrt{\gamma_l^{LW}\gamma_s^{LW}} \quad \text{(solid a monopolar base)} \tag{32a}$$

$$-2\sqrt{\gamma_l^{\ominus}\gamma_s^{\oplus}} = -\gamma_l(1+\cos\theta_{l/s}) + 2\sqrt{\gamma_l^{LW}\gamma_s^{LW}} \quad \text{(solid a monopolar acid)} \tag{32b}$$

Several monopolar solids, for which γ_s^{LW} has already been determined, are employed; and the contact angle of each of the two liquids is determined on each

* Mathematically, it is possible to use three *polar* liquids and a set of *three* equations in the form of Eq. (24a) rather than two equations in the form of Eq. (24a) together with Eq. (20d). Such tactics work if the values of the parameters (e.g., $\gamma^{\ominus}$) for the three liquids are not too close together. If they are too close, the calculated values of the three parameters for the solid will be unduly sensitive to small errors in the values of the parameters of the liquids, and in the measured contact angles.

solid. Then, for water (W) and glycerol (Gl), on each of the (basic) solids, by Eq. (32a),

$$\frac{\gamma_{\mathrm{W}}^{\oplus}}{\gamma_{\mathrm{Gl}}^{\oplus}}=\left[\frac{\gamma_{\mathrm{W}}(1+\cos\theta_{\mathrm{W/s}})-2\sqrt{\gamma_{\mathrm{W}}^{\mathrm{LW}}\gamma_{\mathrm{s}}^{\mathrm{LW}}}}{\gamma_{\mathrm{Gl}}(1+\cos\theta_{\mathrm{Gl/s}})-2\sqrt{\gamma_{\mathrm{Gl}}^{\mathrm{LW}}\gamma_{\mathrm{s}}^{\mathrm{LW}}}}\right]^2 \tag{33}$$

$$\equiv\delta_{\mathrm{WGl}}^{2} \tag{34}$$

Thus, for arbitrary liquids l_1 and l_2 we obtain the first two of the four needed equations:

$$\gamma_{l_1}^{\oplus}=\gamma_{l_2}^{\oplus}(\delta_{l_1 l_2}^{\oplus})^2 \tag{35a}$$

and the related expression

$$\gamma_{l_1}^{\ominus}=\gamma_{l_2}^{\ominus}(\delta_{l_1 l_2}^{\ominus})^2 \tag{35b}$$

or, combining Eqs. (35a) and (9a),

$$\gamma_{l_1}^{\ominus}=(\gamma_{l_1}^{\mathrm{AB}})^2\,\frac{\delta_{l_1 l_2}}{4\gamma_{l_2}^{\oplus}} \tag{36}$$

Some values of $\delta^{\oplus}$ and $\delta^{\ominus}$ for selected liquids are tabulated in Ref. 18.

There does not seem to be any method available, at this time, to obtain the last two independent equations. So we will resort to the selection of a reference bipolar liquid that has a high surface tension, for which an arbitrary ratio, α, of $\gamma^{\oplus}$ to $\gamma^{\ominus}$ is chosen. This methodology is analogous to the tactics used in thermodynamics, in which an arbitrary zero of energy is chosen. (The pH scale is an example.) The use of an arbitrary value of α that is reasonably close to unity does not affect the determination of operational quantities.

We will use the convention that, for water as the reference liquid,

$$\alpha=1, \qquad \gamma_{\mathrm{W}}^{\oplus}=\gamma_{\mathrm{W}}^{\ominus}$$

Since $\gamma_{\mathrm{W}}^{\mathrm{LW}}=21.8\ \mathrm{mJ/m^2}$ and $\gamma_{\mathrm{W}}^{\mathrm{AB}}=\gamma_{\mathrm{W}}-21.8=51.0\ \mathrm{mJ/m^2}$, this convention, together with Eq. (9a), yields

$$\gamma_{\mathrm{W}}^{\oplus}=25.5\ \mathrm{mJ/m^2} \tag{37a}$$

$$\gamma_{\mathrm{W}}^{\ominus}=25.5\ \mathrm{mJ/m^2} \tag{37b}$$

These are the fifth and sixth needed relations.

The direct employment of the dimensionless δ's throughout the theory (which would eliminate the need for $\gamma^{\oplus}$ and $\gamma^{\ominus}$ parameters) would be mathematically equivalent to using the convention $\gamma_{\mathrm{W}}^{\oplus}=\gamma_{\mathrm{W}}^{\ominus}$, but the convention has the advantage that dimensional, numerical values of $\gamma_i^{\oplus}$ and $\gamma_i^{\ominus}$ for liquids and solids give specific character to these acid–base parameters.

We may illustrate the use of Eqs. (33)–(37) to determine δ for water and glycerol, using PMMA as substrate. (For convenience, the additional parameter β may be introduced, here: $\beta=\delta^2$; $\beta=\gamma_W^{\oplus}/\gamma_l^{\oplus}$.) On one sample of PMMA, the contact angle of α-bromonaphthalene was found to be 26°. Using Eq. (20d) (i.e., assuming α-bromonaphthalene to be apolar) and $\gamma_{\alpha BN}=44.4$ mJ/m^2, the value $\gamma_{PMMA}^{LW}=40.0$ mJ/m^2 was obtained. The contact angle of glycerol was 69.0°, that of water 74.5°. Since γ^{LW} of water is 21.8 mJ/m^2 and γ^{LW} of glycerol is 34 mJ/m^2, Eq. (33) yields $\beta=6.45$. The average value of β for water–glycerol obtained from measurements of this type for 11 solids, was 6.505. Details of the measurements are given in Ref. 39. For formamide, β was found to be 11.2, and for dimethyl sulfoxide (DMSO), $\beta=36.4$. Then

$$\gamma_{Gl}^{\ominus}=\frac{6.505(\gamma_{Gl}^{AB})^2}{4\gamma_W^{\oplus}}=57.4$$

A similar treatment for formamide and DMSO yielded the values of $\gamma^{\ominus}$ for these two liquids, shown in Table 1, and use of Eq. (9a) yielded the values of $\gamma^{\oplus}$.

Using these values of β, e.g., 6.505 for glycerol versus water, together with the convention that $\alpha=1$ and $\gamma_W^{\oplus}=25.5$ mJ/m^2, we have obtained the data in Table 1.

5. *ACIDIC AND BASIC PARAMETERS OF SELECTED SOLID SURFACES*

As indicated, Eq. (32a) can be used for monopolar solids that are Lewis bases, and Eq. (32b) can be used for monopolar solids that are Lewis acids. An independent criterion for identifying solids that are monopolar bases (or acids) is that contact angle measurements *with a monopolar liquid that has the same polarity* yield the same apparent γ_s^{LW} as does an *apolar liquid.* Thus, the value of γ_s^* is calculated, using

$$\gamma_s^*=\frac{\gamma_{l(\text{monopolar})}(1+\cos\theta)^2}{4} \tag{38}$$

where the liquid l must be a monopolar base if the solid is being tested for basic character, and l must be a monopolar acid if the liquid is being tested for acidic character. This result is compared with the value obtained with an apolar liquid, using Eq. (20d). If $\gamma_s^*=\gamma_s^{LW}$, then the solid is either apolar or has the same polarity as the monopolar liquid. If $\gamma_s^*>\gamma_s^{LW}$, then the solid is either monopolar with the opposite polarity, or else it is bipolar. Dimethyl sulfoxide can be employed as a test liquid, if strict precautions can be taken to exclude water vapor (although its small $\gamma^{\oplus}$, shown in Table 1, may be a cause of uncertainty; on the basis of its structure, it should be a preponderantly Lewis base monopole). Using DMSO and the apolar liquids DIM or ABN as test liquids, we have found that the following polymers appear to be $\gamma^{\ominus}$ monopoles: PMMA, cellulose acetate, polyoxyethylene, sucrose, zein, gelatin, RNA, human serum albumin, and many other proteins. Bipolar solids were: glucose, cellulose, and DNA. Several solids that were $\gamma^{\ominus}$ monopoles when fully dried exhibited $\gamma^{\oplus}$ polarity as well when hydrated.[16,40,41]

We may make quantitative estimates of the $\gamma^{\oplus}$ and $\gamma^{\ominus}$ parameters of solid surfaces, from measured contact angles, using Eqs. (20d), (30a), and (30b). While these equations are mathematically elementary, it is desirable to show explicitly, here, how their solution can be carried out.

The γ_s^{LW} term is determined first, using the contact angle of either diiodomethane or α-bromonaphthalene, in Eq. (20d). Equations (30a) and (30b) may be put in the form

$$A = C\sqrt{\gamma_s^{\oplus}} + D\sqrt{\gamma_s^{\ominus}} \tag{39a}$$

$$B = E\sqrt{\gamma_s^{\oplus}} + F\sqrt{\gamma_s^{\ominus}} \tag{39b}$$

For water (W) as l_1 and glycerol (Gl) as l_2,

$$A = \gamma_W(1 + \cos\theta_{W/s}) - 2\sqrt{\gamma_s^{LW}\gamma_W^{LW}}$$
$$B = \gamma_{Gl}(1 + \cos\theta_{Gl/s}) - 2\sqrt{\gamma_s^{LW}\gamma_{Gl}^{LW}}$$
$$C = 2\sqrt{\gamma_W^{\ominus}} \qquad E = 2\sqrt{\gamma_{Gl}^{\ominus}}$$
$$D = 2\sqrt{\gamma_W^{\oplus}} \qquad F = 2\sqrt{\gamma_{Gl}^{\oplus}}$$

TABLE 2. Surface Parameters for Some Solids, from Contact Angles of DIM or ABN and of Glycerol and Water

Solid	γ^{LW} (mJ/m^2)	$\gamma^{\oplus}$ (mJ/m^2)	$\gamma^{\ominus}$ (mJ/m^2)	Reference
	(a) Based on advancing angles			
Poly(methylmethacrylate), PMMA cast film[a]	39 to 43	(0)	9.5 to 22.4[b]	40, 42
Poly(vinylchloride)	43	0.04	3.5	40, 42
Poly(oxyethylene): PEG 6000	45	(0)	66[b]	40, 42
Cellulose acetate	35	0.3	22.7	19, 41
Cellulose nitrate	45	(0)	16	19, 41
Agarose	41	0.1	24	19, 41
Gelatin	38	(0)	19	19, 41
Human serum albumin (dry)	41	0.15	18	40, 42
Polystyrene	42	(0)	1.1	19, 41
	(b) Based on advancing and retreating angles			
Polyethylene (corona-treated commercial film)[c]				
Based on advancing angles	33	(0)	0.1	[b]
Based on retreating angles	42	2.1	30	[b]

[a] Kindly furnished by Prof. J. A. Gardella, Jr.
[b] Recent measurements in this laboratory.
[c] Kindly furnished by the Kendall Corp., Lexington, MA. Contact angle results on corona-treated polyethylene are not representative of purified polyethylene. Measurements for polyethylene that has been given various treatments are in progress and are to be published.

Then

$$\sqrt{\gamma_s^{\ominus}} = \frac{AF - BD}{CF - DE} \tag{40a}$$

$$\sqrt{\gamma_s^{\ominus}} = \frac{BC - AE}{CF - DE} \tag{40b}$$

Some results of this treatment of contact angle data are given in Table 2.

6. *DISCUSSION OF THE GENERAL MAGNITUDES OF THE $\gamma^{\oplus}$ AND $\gamma^{\ominus}$ PARAMETERS*

The numbers in Table 2 were obtained with samples that were readily available. For example, the samples of PMMA furnished by Prof. Gardella were from an electron spectroscopy for chemical analyses (ESCA) and Fourier transform infrared (FTIR) study that was in progress, in his laboratory.[(43)] Details of our part of this study are to be published.

The contact angles measured on a single sample (e.g., PMMA) were self-consistent, but varied with the treatment (e.g., RF discharge bombardment) that various samples received. The same was true of polyethylene (PE); see Table 2b. It is well known that contact angles on polymers such as polyethylene and PMMA are sensitive to pretreatment, and that such measurements can be used for "control analysis" of pretreatment processes such as corona discharge, which are employed to enhance adhesion.

The polymers PMMA and PE are included in Table 2 in order to illustrate two conclusions. First, the determination of the $\gamma^{\oplus}$ and $\gamma^{\ominus}$ parameters introduces a new dimension, in regard to the surface properties of solids. Correlations with the results of other studies (e.g., FTIR) and with the presence of specific groups (carbonyls, etc.) can be undertaken, but will not be discussed here.

Second, the retreating angles on polyethylene yield appreciably different information than the advancing angles. According to the advancing angles, corona-treated polyethylene does not differ very markedly from the untreated polymer. But the retreating angles appear to describe a different set of chemical species or groups, which can take part in hydrogen bonding both as acids and bases. As studied with retreating angles, treated and untreated surfaces were notably different. (It has been suggested[(38,44)] that the different species may be present as patches.) It is obvious that these two conclusions have considerable significance for the investigation of adhesion of coatings and adhesives to polymers, and for many other applications.

The measurement of the contact angle of a *single* liquid on a solid does not reveal very much about the acid–base character of the solid. Why this is so is quite clear: the acid parameter $\gamma^{\oplus}$, the basic parameter $\gamma^{\ominus}$, and the apolar parameter γ^{LW}, are not closely coupled or correlated. It is likely that the three parameters are independent of each other when they are considered with respect to different solids, or to the same solid that has been given different treatments. So, in general, we recommend that three contact angles be measured to evaluate the three independent parameters. It is clear that, in general, no one of the three parameters can be taken as representative of all three for any solid or liquid.

We recommend that retreating angles (θ_r) as well as advancing angles (θ_a) be measured when possible. The fact that the scatter of data is poorer (e.g., the standard error being larger) for retreating angles than for advancing angles does not mean that the information available in the retreating angles can be disregarded. The comparison of the two sets of data in Table 2b demonstrates that to fail to measure retreating angles would be to fail to determine the most interesting properties of the treated surface. A similar conclusion can be reached on the basis of very recent measurements at Buffalo on coal and on graphite.[45] It has long been suspected that coal has a mosaic surface structure; the θ_r measurements appear to reveal a variation of properties of a higher-energy component, with coal rank, that is not accessible by means of θ_a measurements.

A notable characteristic of Table 2a is that none of these solids has a large $\gamma^{\oplus}$; five of the solids may be classed asmonopolar Lewis bases, and for three others $\gamma^{\oplus}$ is very small. Also see the list in Section 5 of polarities as determined using DMSO. In terms of whether the properties that were observed could reasonably have been expected on the basis of structure, PMMA and PEG 6000 (polyethylene glycol) are "well behaved," i.e., monopolar basic. It is to be expected that ester groups would be Lewis bases, and $\gamma^{\oplus}$ should be zero. This is, indeed, what was observed for these two polymers. For polyoxyethylene (PEG 6000) the polar moiety is the ether group. (The ratio of hydroxyls to ether groups in this polymer should be less than 0.01.) Hence, dominant Lewis base behavior, with $\gamma^{\ominus}$ large, is unsurprising. It is to be expected that the few hydroxyls in the polymer be buried below the surface, forming hydrogen bonds (by Lewis neutralization) to ether groups, which are present in a large excess.

It is surprising to find that certain carbohydrates such as sucrose and agarose, and also cellulose acetate and nitrate, are effectively $\gamma^{\ominus}$ monopoles. On the basis of structure, one might expect these to be bipolar; the protons of the hydroxyl groups should be electron acceptors. There are several possible explanations. One is that, in certain carbohydrate structures, there may be no hydroxyls pointed "outward" to the adjacent phase. In these structures, all the hydroxyls of the surface carbohydrate rings are Lewis-neutralized by hydrogen bonding to the Lewis base oxygen atoms of adjacent hydroxyls, etc. When a stronger Lewis base is present in the other phase, or in a solvent, it might be expected that hydroxyls so bonded to oxygen atoms could be taken away by the external Lewis base, with the hydroxyls turning from "inward-to-the bulk phase" to "outward" orientation.

A second possible explanation is surface hydration. Adsorbed water molecules may be bound sufficiently tightly to the surfaces so that they are not easily desorbed and the Lewis acid character of underlying groups may not be manifested.

A third "explanation" is that, quite conceivably, it may be a general law that $\gamma^{\oplus}$ for a solid is appreciably smaller than $\gamma^{\ominus}$ for the same substance, even in the absence of self-association.

Regarding poly(vinylchloride), we may note that the value $\gamma^{\oplus} = 0.04\ \mathrm{mJ/m^2}$ in Table 2 is considerably smaller than the $\gamma^{\ominus}$ value 3.5. A hydrogen on the same carbon atom as a chlorine of this polymer should be less acidic than the hydrogen of chloroform. The larger magnitude of $\gamma^{\ominus}$ is probably due to the chlorine atoms acting as electron donors. However, with chloroform, no Lewis basic character is exhibited in the surface tension of the pure liquid. The behavior of poly(vinylchloride) deserves further investigation.

Gelatin and serum albumin are also effectively $\gamma^{\ominus}$ monopoles. This may well be due to the orientation of NH groups toward Lewis base groups in the bulk, as suggested earlier in regard to OH groups in certain carbohydrates. It is unlikely that the absence of $\gamma^{\oplus}$ character has the same cause as the lack of Lewis acid behavior of NH_3, which has recently been reported[46] on the basis of microwave spectroscopic experiments. Amides have not yet been subjected to the same kind of microwave spectroscopic study as ammonia, so we cannot be sure whether behavior analogous to ammonia will be present. The existence of amide hydrogen bonds is commonly inferred from the mechanical strength of proteins and nylon polymers. Moreover, we can draw on the difference of boiling points of acetamide (222°) vs. propylamine (48.7°) and acetone (56.6°) to conclude that the NH group of an amide behaves differently in self-association hydrogen bonding from an NH in an amine.

Polystyrene has an appreciable $\gamma^{\ominus}$ character. This is, no doubt, due to the π electrons of the aromatic ring, which we have mentioned above.

7. *NEGATIVE INTERFACIAL FREE ENERGY*

7.1. *The Difference between Interfacial Tension and Interfacial Free Energy*

We alluded to this question in Section 2. The units force/length and energy/area are the same, but the physical functions are different. A liquid–vapor interface, or a liquid–liquid interface, can be mechanically stable only if the interfacial *tension* is positive. The free energy of cohesion, ΔG_i^c, for a stable body of liquid is negative; see Eq. (2). The free energy of adhesion for a liquid–liquid system must be negative for the two-phase system to be stable.

True liquids have zero shear strength, so a liquid–liquid interface can have only a transient existence if γ_{ij} is negative. Either the interface spontaneously increases in area so that emulsification occurs, or, if mutual solubility is appreciable, interdiffusion may occur, resulting in formation of a solution.

A solid, on the other hand, has a nonzero shear strength. The surface area of a solid body of arbitrary shape generally does not decrease spontaneously by a flow process. (The exception to this generality is in the unusual conditions where negative creep occurs. Positive creep is slow deformation in the direction of an applied external force, which is generally a tensile force, but may also be in shear. Negative creep is slow, spontaneous deformation in a direction opposite to the external applied force.) It is possible to devise experiments in which the reduction of surface and grain boundary free energy can cause certain solids to flow. But with the widest variety of solids, the shear strength is sufficient that nondiffusional transport of atoms or molecules, from the interior to surface or interfacial regions, does not occur at any detectable rate—in contrast to the behavior of liquids. Therefore, a solid cannot be said to have a surface *tension* in the same sense that a liquid has a surface tension. So the question of whether a solid–liquid interface has a negative interfacial *tension* has no meaning.

A solid surface always has a surface free energy. A state of surface stress, σ^s, will usually also exist. Elastic energy is stored in a strain field in the surface region. For a liquid, the state of surface stress is zero, because of the lack of shear strength. Hence, there is no connection between the surface tension for a liquid, γ, and the state of surface stress. For a solid, the state of surface stress does contribute one component to the total surface free energy, due to the stored elastic energy. The main component of surface free energy of a solid is due to the force field of the atoms or molecules, i.e., the unbalanced force directed normal to the surface. In addition, there is a density gradient across the phase boundary, and there are orientation (excess entropy) effects. It is only the component of surface free energy due the normal force field (and not the components due to surface stress, to density gradients, and to orientation) that is estimated by contact angles in equilibrium systems, using the theory discussed in this chapter.

7.2. Liquid–Solid Interfaces

We have already noted that for a liquid–vapor interface, the surface free energy is always positive. For a liquid–solid interface, the issue of whether the interfacial free energy is positive or negative is dominated by the attractive forces across the interface. The magnitude of the cohesive free energy of the liquid may be less than the magnitude of the free energy of adhesion of the liquid to the solid that is due to these forces. Equation (12) may be put in the form

$$\Delta G^{a}_{ij} < -\gamma_i - \gamma_j \tag{41}$$

If ΔG^{a}_{ij}, which is negative, has a large enough magnitude, γ_{ij} will be negative, as may be seen by rearranging Eq. (12):

$$\gamma_{ij} = \Delta G^{a}_{ij} + \gamma_i + \gamma_j \tag{42a}$$

$$\gamma^{LW}_{ij} = \Delta G^{aLW}_{ij} + \gamma^{LW}_{i} + \gamma^{LW}_{j} \tag{42b}$$

$$\gamma^{AB}_{ij} = \Delta G^{aAB}_{ij} + \gamma^{AB}_{i} + \gamma^{AB}_{j} \tag{42c}$$

This discussion allows us to expand a remark made in Section 2. The *apolar* component of the free energy of adhesion [see Eq. (7)] is always negative, and the apolar component of the interfacial free energy γ^{LW}_{ij} [see Eq. (6)] is always positive and smaller than the larger of the two surface free energies γ^{LW}_{i} and γ^{LW}_{j}. The acid–base, or hydrogen bonding, component of the free energy of adhesion is also always negative and often larger than the apolar component. The condition for γ^{AB}_{ij} to be zero is

$$\Delta G^{aAB}_{ij} = -\gamma^{AB}_{i} - \gamma^{AB}_{j} \tag{43}$$

If ΔG^{aAB}_{ij} is any more negative than the value given by Eq. (43), then γ^{AB}_{ij} will be negative. And if ΔG^{aAB}_{ij} is less negative, then γ^{AB}_{ij} will be positive.

It might be expected, then, that a solid with a very large $\gamma^{\ominus}$ or $\gamma^{\oplus}$ would interact so strongly with water that ΔG^{aAB}_{ij} would be large and negative. This would lead to a negative value of γ^{AB}_{ij}. Thus, for PEG 6000, $\gamma^{\ominus}$ is 66 mJ/m^2. So it would be expected that there would be a considerably greater interaction energy across the interface with water than the cohesive energy within either phase:

$\Delta G^{a}_{W/PEG} = -2(\gamma^{LW}_{W}\gamma^{LW}_{PEG})^{1/2} - \gamma^{AB}_{W} - \gamma^{AB}_{PEG} = -179.6\ \text{mJ/m}^2$. This is less negative than the sum of the free energies of the separate phases: $\Delta G^{c}_{W} = -144$ and $\Delta G^{c}_{PEG} = \Delta G^{cLW}_{PEG} = -90\ \text{mJ/m}^2$, so $\Delta G^{c}_{W} + \Delta G^{c}_{PEG} = -234\ \text{mJ/m}^2$. Using Eqs. (1), (6), and (14) and the data in Table 2, the calculated value of γ_{sl} is found to be $-26.5\ \text{mJ/m}^2$.

The contact angle of water on PEG 6000 was found[41] to be 21.5°. It is not zero, despite the negative value of γ_{sl}, because PEG 6000 has a zero $\gamma^{\oplus}$. See Table 2.

8. AN APPARENT ANOMALY THAT ARISES FROM CERTAIN OBSERVED CONTACT ANGLE DATA

An examination of Eqs. (40a) and (40b) shows that certain possible combinations of the observed data that make up the parameters A to F can lead to negative values of $(\gamma^{\ominus}_{s})^{1/2}$ or $(\gamma^{\oplus}_{s})^{1/2}$. For example, if $BD > AF$ and if $CF > DE$, $(\gamma^{\oplus}_{s})^{1/2}$ is negative, though the square of this quantity, $\gamma^{\oplus}_{s}$ is positive. It is, however, questionable as to what the physical meaning would be for this condition. It appears to be anomalous.

Experimentally, we have mainly observed this condition[47] in some cases when contact angles were measured on polyolefins that had been treated with corona or RF discharges, or flame, in order to improve their adhesion behavior. (Adhesive bonding and coating or printing on polyolefin surfaces are often ineffective if such modification is not employed.) Negative square roots were obtained almost exclusively among advancing contact angle results, and in $\gamma^{\oplus}_{s}$ computations.

There are at least three or four possible explanations for the finding of negative square roots. The simplest is experimental error in the contact angle measurement. However, the systematic occurrence of this observation in $\gamma^{\oplus}_{s(a)}$ values argues for the effect not being an artifact. If it were a result of experimental error, then one would expect it to appear with roughly equal frequency in all four acid–base parameters, $\gamma^{\oplus}_{a}$ $\gamma^{\ominus}_{a}$ $\gamma^{\oplus}_{r}$ and $\gamma^{\ominus}_{r}$. Another consideration is that the anomaly can be eliminated by "adjustment" of the observed contact angles. In our recent work at Buffalo, we have seen that, occasionally, when the negative magnitude of $(\gamma^{\oplus})^{1/2}$ was small (e.g., $0.02\ \text{mJ/m}^2$) a not implausible "correction" in one or more of the θ values would make $(\gamma^{\oplus})^{1/2}$ positive. But it turns out that this calculation is rather "robust," in the sense that a large change in the input data (the measured contact angles) is needed to make a small change in the output, i.e., in the values of $(\gamma^{\oplus}_{s})^{1/2}$. In a number of cases, the required fudging of the observations to bring $(\gamma^{\oplus}_{s})^{1/2}$ from, say, -0.5 into the positive range was larger than 20°, which is quite implausible.

However, a very appreciable error may be made in the water contact angle on highly hydrophilic solids such as polyoxyethylene, dextran, and some hydrophilic proteins in a dry condition. (Any liquid that swells the solid on which the contact angle is measured will give this trouble.[48]) We have observed that the contact angle of water may be time dependent in a different way from solids that are not highly hydrophilic. In ideal measurements of the water-advancing angle, water is added slowly to the drop, from a micrometer syringe, through a hypodermic needle. (See Section 10.) Addition of water is stopped, and the telemicroscope crosshairs are moved (by a horizontal traverse control) to the edge of the drop so that the observer can see when the forward motion of the edge stops. For low-viscosity liquids, this usually takes about 10 s, during which time one crosshair can be lined up as the tangent to the drop to make the actual measurement. With the hydrophilic solids mentioned above, if there is too long a delay before lining up the crosshair, the angle begins to decrease. (This occurs more rapidly than the decrease in angle due to evaporation of water.) Apparently, water is dissolving in the solid and swelling it. On close examination, it may be possible to detect a decrease in the drop height; on removing the drop from the solid, a slightly elevated region of the solid may be observable where the drop had been. We have found that, in some cases, the water contact angle reached values below those of glycerol or formamide. On solids that do not absorb water, the water contact angle is ordinarily above that of glycerol and formamide, and we can suggest that the observation that $\theta_W < \theta_{GL}$ may be a clue to the presence of this error.

Retreating water contact angles on highly hydrophilic solids do not exhibit this kind of time dependence, for two reasons. The first is that the retreating angle is usually zero and remains zero. The second is that, if $\theta_r > 0$, the retreat ordinarily occurs over an area of solid surface that has already been swollen by the liquid. So further swelling, while liquid is being removed, would be unlikely.

A second possible cause of the negative square roots is the neglect of the equilibrium spreading pressure π_e. For each contact angle of the three liquids (e.g., water, glycerol, and diiodomethane) on a given solid, a π_{esv} term may exist. See the first few paragraphs of Section 3. Also, in some cases, the surface tension of the liquid phase may be reduced by adsorption of material from the solid phase. Again, as in the previous paragraph, it is highly improbable that appreciable spreading pressures should appear only in the values of $(\gamma^{\oplus}_{s(a)})^{1/2}$ and, particularly, *not* in parameters deduced from *retreating* angles.

Next, we must consider the possibility that negative values of, e.g., $(\gamma^{\oplus}_s)^{1/2}$ are empirically valid, and that we need to examine the general consequences of such values. Now, the solutions to Eqs. (39a) and (b), given by Eqs. (40a) and (40b), have significance only when used in Eq. (9). When one of the square root terms is negative, these must be combined in the operational form

$$\gamma_i^{AB} = 2\sqrt{\gamma_i^{\oplus}}\sqrt{\gamma_i^{\ominus}} \tag{9c}$$

with the result that γ_i^{AB} is negative. By contrast, $((\gamma_i^{\oplus})^{1/2})^2$ has only formal significance. As noted in the discussion of Eqs. (8)–(10) in Section 2, there is no *operational* significance in the value (or sign) of a surface parameter such as $\gamma_i^{\oplus}$ if it is not in combination with a $\gamma^{\ominus}$ parameter, whether $\gamma_i^{\ominus}$ or $\gamma_j^{\ominus}$.

A negative value of γ_i^{AB} is physically allowable for a mechanically stable condensed phase, provided $\gamma_i^{AB} < \gamma_i^{LW}$, i.e., provided that the total surface free energy (surface tension) is positive.

Finally, we may recall that in studies of polymer surfaces, negative square roots have arisen (so far) only with surfaces that have been exposed to corona discharge or flame. The result of this treatment is degradation of surface chains or chain segments, including (within a short time, under exposure to gases such as oxygen or water vapor) the development of oxygen-containing groups such as carbonyls, hydroxyls, etc. The resulting physical state or structure may require a more complex mathematical model of the physical interactions of surfaces. See the following section, where we comment on the adsorption of components from liquid mixtures.

9. CRITIQUE OF SOME OTHER METHODOLOGIES

The theory that we have proposed carries with it the implicit criticism of certain existing methods of treating contact angles. In this section, we make those criticisms explicit.

9.1. Use of Liquid Mixtures

W. A. Zisman was adamant against the use of mixtures of liquids in the Fox–Zisman method of determining γ_c, and he was unhappy with the American Society for Testing and Materials (ASTM) recommendation[49] to do so. We strongly support the condemnation of the use of liquid mixtures, particularly when one of the liquids is capable of hydrogen bonding. Adsorption of at least one component from the liquid onto the solid is likely to occur, and this will lower γ_{sl} or γ_{sv} or both. Moreover, the surface tension of a liquid mixture is seldom independent of composition, and adsorption at the liquid–gas interface is almost always significant. The Fox–Zisman linear equation (see Section 3) was empirically validated only for *pure* liquids, for which adsorption of one component of the liquid from *solution* cannot take place.

The Good–Girifalco–Fowkes equations, i.e., Eqs. (6) and (20), make it an essential premise of the model that the composition of each phase is homogeneous right up to the physical surfaces that separate the phases. If the surface region consists of a layer of close-packed, oriented molecules, such that the outer surface is "autonomous," i.e., effectively uninfluenced by the substrate, then the conditions for validity of the Good–Girifalco theory are fulfilled. But if the outer surface is not close-packed; for example, if it is a *partial* monolayer or a liquidlike adsorbed film, then the conditions will not be fulfilled. In such cases, there may be no grounds for believing *a priori* that the geometric mean relations will hold for the LW component of surface energy; and this denial will also hold with respect to the acid–base combining rules, such as Eqs. (8)–(10).

The ASTM methodology is empirically usable; for example, as a "control" method for testing the level of treatment of a polymer surface, say by flame or corona discharge. The resulting contact angle data, or the existence of complete ($\theta = 0$) wetting, can in many cases be correlated with the adhesion of ink or coatings or adhesives. But for every chemical system, a careful validation must be carried out; for example, validity with polypropylene film from one source does not imply validity with (nominally) the same polymer from a different source.

9.2. The "Equation of State" Interfacial Theory

An approach to contact angles has been proposed[50,51] in which a universal expression is employed:

$$\cos\theta = \frac{[0.015\gamma_{sv} - 2.00(\gamma_{sv}\gamma_{lv})]^{1/2} + \gamma_{lv}}{\gamma_{lv}[0.015(\gamma_{sv}\gamma_{lv})^{1/2} - 1]} \tag{44}$$

The corresponding expression for interfacial tension is

$$\gamma_{ij} = \frac{[(\gamma_i)^{1/2} - (\gamma_j)^{1/2}]^2}{1 - 0.015(\gamma_i\gamma_j)^{1/2}} \tag{45}$$

We must caution against the use of these expressions. This warning has a basis both in experiment and theory.

Experimentally, Eq. (45) does not lead to correct predictions of oil–water interfacial tensions. See Refs. 52 and 53. For example, for water versus benzaldehyde, the observed value of γ_{12} is 15.5 mJ/m^2 versus 25.7 calculated, based on $\gamma_l = 40.4$ mJ/m^2.[6] This level of disagreement between prediction and observation is not an isolated case. Fowkes's paper[54] contains interfacial tension data for 26 liquids versus squalane. These results[52–54] give conclusive empirical refutation of the "equation of state" as a method of handling interfacial energy and contact angle data. Fowkes also included an extensive review of other studies that support the acid–base theory over the "equation of state" hypothesis.

In Ref. 52, we pointed out the reason why these equations cannot be used: they do not take any account of the complementarity of acidic and basic behavior. They were derived by Neumann from the Good–Girifalco Φ function approach,[6,12] which, as we remarked in Section 2 ignored hydrogen bonding. Equations (44) and (45) do not have enough material parameters (e.g., molecular properties of the interacting chemical species) to account for the different kinds of interaction that occur.

Moreover, Morrison[55] has given a clear, rigorous refutation of the derivation employed[50,51] to obtain Eqs. (44) and (45). Thus, there is ample reason for the warning that we have given against pursuit of this theory.

10. A NOTE ON CONTACT ANGLE MEASUREMENT

We have already alluded to experimental aspects of contact angles. We do not have space for a detailed review of measurement techniques. An extensive, 10-year-old discussion is available.[44] We will confine the treatment in this section to the experimental measurement of hysteresis (i.e., the question of advancing and retreating angles) and to the drop size effect, for which we describe some significant new results for practical measurements.

10.1. Hysteresis

Our first point is an enlargement on the recommendation (Section 6) that retreating angles, in addition to advancing angles, be measured wherever possible. *Experimenters need not waste time trying to measure equilibrium angles.*

If the experimenter insists on making only one angle measurement, the determination should be made in such a way as to ensure that the advancing case is studied. (The captive drop method[44] is the most convenient technique for this purpose. A micrometer syringe and hypodermic needle provide the most convenient technique of depositing a sessile drop and holding it "captive.") Evaporation of liquid from a drop causes the liquid front to retreat, and even if no visible retreat occurs, the drop will shrink and the angle will lie somewhere *between* the advancing angle and the lower angle at which actual retreat starts. This is, quite possibly, one of the two most common sources of error in contact angle determinations.

It is true that in the experimental technique developed in Zisman's laboratory a drop was deposited from a wire tip. The method followed a very detailed protocol, however, in order for the angles to be reproducible. (The contact angles of such drops have been found to approximate the advancing angles, when comparisons have been made.) A laboratory observer who uses the deposited drop technique, as opposed to the captive drop method, usually is unaware of the detailed Zisman protocol. It is easier to train operators in the advancing and retreating contact angle techniques than in the Zisman method, and the disadvantage of the Zisman method—that retreating contact angles cannot be measured—is avoided. The disturbance of the liquid surface by the needle does not affect the angle made by the liquid with the solid surface.

It often happens, when the hysteresis on a solid is large, that the retreating contact angle is zero. [See Section 3, especially Eqs. (20)–(26).] We urge experimenters to record this observation ($\theta_r = 0$) if it occurs: there is empirical information in it.

10.2. The Drop Size Effect

Good and Koo[56] observed a dependence of contact angles on drop size for polar liquids on apolar and polar solids, both in the advancing and retreating angles. (Apolar liquids on apolar solids, e.g., *n*-decane on Teflon FEP (a copolymer of perfluoroethylene and perfluoropropylene) showed no drop size dependence; and in the advancing angle, ethylene glycol on Teflon showed no drop size dependence.) Above a critical drop diameter *d*, around 0.5 cm, θ_a was independent of diameter.

Recently, we have extended the study of the drop size effect in the case of retreating angles.[57] Figure 4 shows some measurements with water on a PMMA sheet. A decrease in θ_r with decreasing drop size appears to start at about 0.5 cm versus about 0.9 cm in the previous study. (The difference is probably due to some degree of better surface quality in the newer samples.) The rise in θ_r with drop diameter above 1.0 cm, in the series in which the measurement was made 10 s after stopping withdrawing liquid from the drop, was not recognizable in the earlier work.[56] (In the previous study, the scatter of data was somewhat larger, in θ_r, and only a very few data points were taken with drops larger than 1.1 cm diameter. Much less of an attempt was made to measure θ at a specified time after stopping the withdrawal of liquid.) When the delay time was increased to 30 s, most of the rise in angle above the plateau value was eliminated in the present study; only a small increment, for diameters near 1.20 cm, remained.

Bulk swelling of PMMA by water is negligible or zero, so *ab*sorption of water by the solid was probably not involved in this phenomenon. See Section 8.

The difference between the 10- and 30 s angles (Fig. 4) points to the conclusion that adsorption or desorption kinetic factors are influencing the retreating contact angle. A supersaturated adsorbed film of molecules from the liquid may remain on

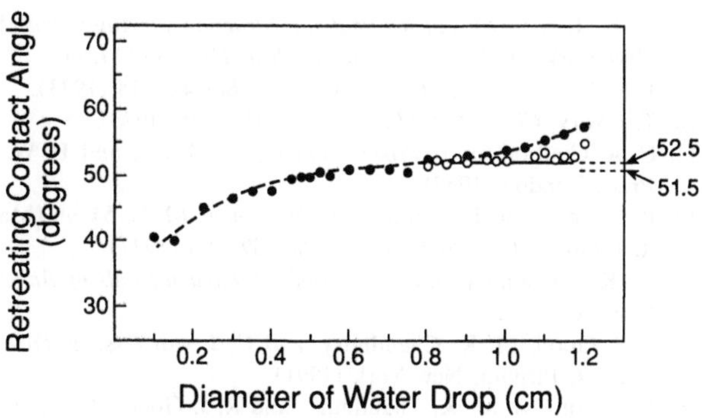

FIGURE 4. Retreating contact angle of water on PMMA in air as a function of drop size and delay time before making measurement: ● 10 s delay; ○ 30-s delay.

the solid, in a narrow region close to the three-phase line, when the liquid is made to retreat. This film would lower the surface free energy of the solid, and as it evaporated or as molecules moved to the liquid by surface diffusion, γ_{sv} would rise. This would lead to a decrease in the contact angle, such as we have observed. That this effect (or, for that matter, any drop size effect at all) was not found with apolar liquids on apolar solids may have been due to the low energy of adsorption and desorption in the absence of hydrogen bonding. We do not at present have an explanation for the specific drop size at which the rise in contact angle started.

Perhaps the answer to this puzzle may be found in the "gradient theory" that recently came out of Scriven's laboratory.[58] While that theory was developed for equilibrium systems near the critical point of the liquid, it could provide a starting point for developing a theory of films whose thickness is greater than the equilibrium value.

We may conclude from our experiments that the drop size effect is more complex in the retreating angle than in the advancing angle. The tactics of measurement, using the captive drop method, should take into account this difference between the effects of increasing drop size and of decreasing it.

ACKNOWLEDGMENT

One of us (RJG) wishes to thank the U.S. Department of Energy, Pittsburgh Energy Technology Center, for support in research on contact angle measurement, upon which some of this material is based.

REFERENCES

1. W. M. Latimer and W. H. Rhodebusch, *J. Am. Chem. Soc.* **42**, 1419–1433 (1920).
2. W. D. Harkins and R. J. Meyers, *J. Am. Chem. Soc.* **58**, 1817 (1936); *J. Phys. Chem.* **40**, 959 (1936).
3. W. D. Harkins, *The Physical Chemistry of Surface Films*, Reinhold, New York (1952).
4. R. J. Good, *J. Phys. Chem.* **61**, 810–812 (1957).
5. G. C. Pimentel and A. L. McClellan, *The Hydrogen Bond*, Freeman, San Francisco (1960).
6. L. A. Girifalco and R. J. Good, *J. Phys. Chem.* **61**, 904 (1957).
7. W. B. Jensen, *The Lewis Acid-Base Concept*, Wiley-Interscience, New York (1980).

8. G. N. Lewis, (a) *Valence and the Structure of Atoms and Molecules*, The Chemical Catalog Co., New York (1923); (b) *J. Franklin Inst.* **226**, 293 (1938).
9. J. N. Brönsted, *Rec. Trav. Chim. Pays-Bas* **42**, 718 (1923).
10. T. Lowry, *Chem. Ind. (London)* **42**, 43, 1048 (1923).
11. C. A. Coulson, in *Hydrogen Bonding* (D. Hadzi and H. W. Thompson, eds.), p. 349, Pergammon Press, London (1959).
12. R. J. Good and E. Elbing, *Ind. Eng. Chem.* **62**(3), 54–96 (March 1970).
13. R. J. Good, *J. Colloid Interface Sci.* **59**, 398 (1977).
14. M. K. Chaudhury and R. J. Good, in *Fundamentals of Adhesion* (L. H. Lee, ed.), Chap. 3, Plenum, New York (1991).
15. R. J. Good, M. K. Chaudhury, and C. J. van Oss, in *Fundamentals of Adhesion* (L. H. Lee, ed.), Chap. 4, Plenum, New York (1991).
16. C. J. van Oss, M. K. Chaudhury, and R. J. Good, *Adv. Colloid Interface Sci.* **28**, 35–64 (1987).
17. C. J. van Oss, M. K. Chaudhury, and R. J. Good, *Chem. Rev.* **88**, 927–991 (1988).
18. C. J. van Oss, L. Ju, M. K. Chaudhury, and R. J. Good, *J. Colloid Interface Sci.* **128**, 313 (1988).
19. C. J. van Oss, M. K. Chaudhury, and R. J. Good, *J. Chromatogr.* **391**, 53 (1987).
20. F. M. Fowkes, (a) *J. Phys. Chem.* **66**, 682 (1962); (b) *Adv. Chem. Ser. No.* 43, American Chemical Society, Washington, DC (1964).
21. F. M. Fowkes and M. A. Mostafa, *I.E.C. Prod. Res. Dev.* **17**, 3 (1978).
22. R. J. Good, L. A. Girifalco and G. Kraus, *J. Phys. Chem.* **62**, 1418 (1958).
23. R. J. Good and L. A. Girifalco, *J. Phys. Chem.* **64**, 561 (1960).
24. R. J. Good and C. J. Hope, (a) *J. Chem. Phys.* **53**, 540 (1970); (b) **55**, 111 (1971); (c) *J. Colloid Interface Sci.* **35**, 171 (1971).
25. D. Berthelot, *Compte Rend.* **126**, 1703, 1857 (1898).
26. F. London, *Trans. Faraday Soc.* **33**, 8 (1937).
27. J. H. Hildebrand and R. L. Scott, *The Solubility of Nonelectrolytes*, 3rd ed., Van Nostrand Reinhold, New York (1950).
28. A. Apelblat, A. Tamir, and M. Wagner, *Fluid Phase Equilibria* **4**, 229–255 (1980).
29. W. H. Keesom, *Phys. Z.* **22**, 126 (1921); **23**, 225 (1922).
30. P. Debye, *Phys. Z.* **21**, 178 (1920); **22**, 302 (1921).
31. A. W. Adamson, *Physical Chemistry of Surfaces*, 4th ed., Wiley, (1982).
32. (a) C. J. van Oss, A. W. Neumann, and S. N. Omenyi, *J. Colloid Polym. Sci.* **257**, 737 (1979). (b) C. J. van Oss, D. R. Absolom, and A. W. Neumann, *Colloids Surfaces* **1**, 45–56 (1980).
33. C. J. van Oss, *J. Dispersion Sci. Technol.* **11**, 491 (1990).
34. R. J. Good, *Nature* **212**, 276 (1966).
35. R. J. Good, (a) *J. Colloid Interface Sci.* **52**, 308 (1975); (b) ACS Symposium Series no. 8, American Chemical Society, Washington, DC (1975).
36. H. W. Fox and W. A. Zisman, *J. Collloid Sci.* **7**, 428 (1952).
37. F. M. Fowkes, D. C. McCarthy, and M. A. Mostafa, *J. Colloid Interface Sci.* **78**, 200 (1980).
38. R. J. Good, in *Colloid and Surface Science*, Vol. 11 (R. J. Good and R. R. Stromberg, eds.), Chap. 1, Plenum, New York (1979).
39. C. J. van Oss, R. J. Good, and H. J. Busscher, *J. Dispersion Sci. Technol.* **11**, 75–81 (1990).
40. C. J. van Oss, R. J. Good, and M. K. Chaudhury, *J. Protein Chem.* **5**, 385, (1986).
41. C. J. van Oss and R. J. Good, *J. Macromol. Sci.-Chem.* **A26**, 1183 (1989).
42. C. J. van Oss, R. J. Good, and M. K. Chaudhury, *J. Colloid Interface Sci.* **111**, 378 (1986).
43. T. G. Vargo, J. A. Gardella, Jr., and L. Salvati, *J. Polym. Sci. A: Polym. Chem.* **27**, 1267 (1989).
44. A. W. Neumann and R. J. Good, in *Colloid and Surface Science*, Vol. 11 (R. J. Good and R. R. Stromberg, eds.), Chap. 2, Plenum, New York (1979).
45. R. J. Good, N. R. Srivatsa, M. Islam, H. T. L. Huang, and C. J. van Oss, *J. Adhesion Sci. Technol.* **4**, 608–617 (1990).
46. D. D. Nelson, G. T. Fraser, and W. Klemperer, *Science* **238**, 1670 (1987).
47. R. J. Good and L. K. Shu, unpublished work (1990).
48. (a) R. J. Good and E. D. Kotsidas, *J. Colloid Interface Sci.* **66**, 360 (1978); (b) R. J. Good and E. D. Kotisidas, *J. Adhesion* **10**, 17 (1979).
49. ASTM Standard D2578-84, *Annual Book of ASTM Standards*, Vol. 08.02, American Society for Testing and Materials, Philadelphia, PA.

50. O. Dreidger, A. W. Neumann, and P.-J. Sell, *Kolloid-Z. Z. Polym.* 101 (1965).
51. A. W. Neumann, R. J. Good, C. J. Hope, and M. Sejpal, *J. Colloid Interface Sci.* **49**, 291 (1974).
52. C. J. van Oss, R. J. Good, and M. K. Chaudhury, *Langmuir* **4**, 884 (1988).
53. R. E. Johnson and R. H. Dettre, *Langmuir* **5**, 293 (1989).
54. F. M. Fowkes, F. L. Riddle, W. E. Pastore, and A. A. Weber, *Colloids Surfaces* **43**, 367–387 (1990).
55. I. D. Morrison, *Langmuir* **5**, 540 (1989).
56. R. J. Good and M. N. Koo, *J. Colloid Interface Sci.* **71**, 283 (1979).
57. R. J. Good and H. T.-L. Huang, unpublished results (1990).
58. G. F. Teletzke, L. E. Scriven, and H. T. Davis, *J. Colloid Interface Sci.*, **87** 550–571 (1982).

2

The Long-Range Attraction between Macroscopic Hydrophobic Surfaces

H. K. Christenson

1. INTRODUCTION

1.1. Background

The very strong and long-range attraction found between macroscopic hydrophobic surfaces occupies a unique place in surface science. Not only is its discovery very recent but its origin is also as yet unexplained. Repulsive electrical double-layer forces, attractive van der Waals forces, forces due to steric stabilization by adsorbed polymers, and oscillatory solvation forces were all the subject of considerable theoretical work long before accurate experimental data were available. Not so with the attractive forces between macroscopic hydrophobic surfaces. The first direct measurements were only carried out five years ago, and there is currently no theory that can explain the results in terms of a physically reasonable model. Recent years have seen a few attempts to provide a mathematical description of the forces, but the connection to the physical world is missing. Computer simulations are at present not a help. Apart from the usual problems with accurate modeling of the water molecule there is another, possibly even more formidable obstacle. Because of the very long range nature of the interaction any simulation would have to be carried out on systems with prohibitively large surface separations. These are equivalent to hundreds of diameters of the water molecule, and not even the fastest computers can currently handle the necessary number of water molecules.

H. K. Christenson • Department of Applied Mathematics, Institute of Advanced Studies, Research School of Physical Sciences, Australian National University, Canberra ACT 2601, Australia.

Modern Approaches to Wettability: Theory and Applications, edited by Malcolm E. Schrader and George Loeb. Plenum Press, New York, 1992.

1.2. The Hydrophobic Effect between Nonpolar Solute Molecules

The hydrophobic effect, or the interactions between small, nonpolar solute molecules in water, is a well-known concept.[1,2] The low solubility of hydrocarbons and inert gases in water is a consequence of the hydrogen bonding properties of water. In order to minimize the number of hydrogen bonds that have to be broken to accommodate a solute molecule, the water forms a cagelike structure around it. This leads to an increase in hydrogen bonding in the vicinity of the solute molecule and hence a decrease in entropy (increased order). The enthalpy of transfer of the nonpolar solute from bulk to water is negative because of the formation of additional hydrogen bonds. The result is a free energy of transfer ΔG from bulk to water that consists of two terms, a favorable enthalpy term ΔH (negative) and an unfavorable entropy term ΔS (negative). This can be written as

$$\Delta G = \Delta H - T\,\Delta S \tag{1}$$

At normal temperatures the entropy term dominates; the total free energy of transfer is positive and therefore unfavorable, which leads to the very low solubility of hydrocarbon in water.

The hydrophobic effect is responsible for the vast array of aggregation structures formed by amphiphilic compounds in water, as well as for the conformation of proteins and other biological molecules in aqueous solution. Such molecules attempt to minimize the contact of their hydrophobic regions with water either by aggregating or by adjusting their conformation.

One consequence of the hydrophobic effect is an enhanced attraction between two small, nonpolar solute molecules in water. It is more favorable to have the molecules close together so that one larger cage of water molecules can form instead of two small ones. This has been extensively studied by computer simulation[3] and various theoretical computations.[4,5] We are here concerned with macroscopic hydrophobic surfaces, which leads us to pose the question: What happens as the size of the solute molecule increases so that eventually, as the radius goes to infinity, we have the case of two surfaces?

It is clear that one cannot in any simple way extrapolate the cage formation. Above a certain size cage formation is no longer a viable proposition for the water molecules, and the result of the introduction of a hydrophobic entity into water must be the net loss of hydrogen bonds. In both cases the contact with water is unfavorable from a free-energy point of view, but the breakdown into entropy and enthalpy are completely opposite. In the case of a hydrophobic surface there is a positive enthalpy and a positive entropy contribution (which means that the entropy term gives a negative, or favorable, contribution to the total free energy) since the introduction of a surface must lead to a decrease in hydrogen bonding. The interfacial energy between hydrocarbon and water, which is a measure of the free energy needed to create a hydrocarbon–water interface, decreases with temperature because of the increasing importance of the favorable entropy term with temperature. The solubility of nonpolar species in water, however, decreases with temperature because the entropy term is here unfavorable [see Eq. (1)]. It is thus quite wrong to assume that the energetics of a hydrophobic surface and a hydrophobic solute molecule in water are similar, and that one can in any meaningful way make quantitative comparisons.[6,7]

1.3. The Interfacial Energy between Water and Hydrocarbon

The interfacial energy between water and hydrocarbon is much larger than expected from continuum Lifshitz theory. The disagreement is obviously related, in some way, to water structure. Bulk dielectric data, which are used in Lifshitz theory, do not allow for structural changes close to interfaces, such as the considerable reorientation of water molecules that must take place near a hydrophobic surface. In various theories dealing with surface and interfacial tensions, the matter is often taken care of by dividing the total surface tension of water γ_t into a dispersion force contribution γ^d and a contribution due to specific, polar effects γ^p, so that[8]

$$\gamma_t = \gamma^d + \gamma^p \tag{2}$$

For hydrocarbons $\gamma_t = \gamma^d$. There is a voluminous literature dealing with the application of equations such as Eq. (2) to surface energies and contact angles.

The adhesion between two hydrophobic surfaces of unit area in water is equal to twice the interfacial energy as the process of pulling the surfaces apart to "infinite" separation creates two hydrophobic-water interfaces, each of unit area. Since the interaction energy at contact, or the adhesion energy, is anomalously large, it is reasonable to assume that this extra attraction, above and beyond the predictions of continuum theory, has some finite range. Otherwise we would need to postulate some discontinuity in the force law between two hydrophobic surfaces in water, not a physically reasonable proposition.

The magnitude of the interfacial energy between water and a hydrophobic surface thus points at the existence of a strongly attractive force between macroscopic hydrophobic surfaces in water. Further evidence is provided by the stability of thin films of water on hydrophobic surfaces. Continuum theory predicts that a thin film of water on a hydrophobic surface should be stable due to repulsive van der Waals forces across the water film. Such a situation occurs in mineral flotation, where the efficiency of the process relies on the attachment of air bubbles to hydrophobed mineral surfaces. Attachment would not occur unless there were an *attraction* across the water film between the mineral particle and bubble. (Air may be considered as a very hydrophobic surface for $\gamma = 72\ \mathrm{mJ/m^2}$!)

1.4. Cavitation

If we define a hydrophobic surface as one on which water has an advancing contact angle Θ_a greater than 90°, it follows from the Young equation that a water film confined between two such surfaces will be metastable below a certain thickness. If $\gamma_{L/V}$, $\gamma_{S/V}$, $\gamma_{S/L}$ are the surface energy of water, the surface energy of the hydrophobic solid, and the interfacial energy, respectively, we have

$$\gamma_{S/V} = \gamma_{S/L} + \gamma_{L/V} \cos \theta_a < \gamma_{S/L} \tag{3}$$

It is thus energetically favorable to replace water between two hydrophobic surfaces by vapor. On purely thermodynamic grounds we should expect cavitation, or formation of vapor cavities, to occur between hydrophobic surfaces once the separation becomes small enough.[9] The requirements are that the increase in energy due to the formation of a vapor–water interface and the necessary work of expansion (*PV* work) do not exceed the energy gain on replacing water by vapor. The exact surface separation below which the water film is metastable depends on the geometry of the two surfaces.

This phenomenon of cavitation between two hydrophobic surfaces in water was indirectly inferred by Rabinovich *et al.* in a series of experiments on the interaction between hydrophobed glass filaments in aqueous solution.[10] The hysteresis observed on approach and separation indicated a completely different force law between the filaments in the two cases. It seemed as if cavitation only occurred, however, after the surfaces were first brought into contact, and not at any finite separation. The phenomenon was subsequently theoretically analyzed by Yushchenko *et al.*[9] They postulated the existence of an energy barrier, or activation energy for the process. The size of this activation energy was calculated by assuming that for any given separation the cavity would have to grow from a vanishingly small lateral extension, thereby passing through unfavorable stability conditions related to the contact angle, which determines the curvature of the vapor–water interface. Once formed, the cavities or vapor bridges are indefinitely stable over large separation ranges, and at very large separations there is an energy barrier toward their disappearance. We shall return to direct observations of cavitation between hydrophobic surfaces in Sections 4.3 and 5.3.

An analogous phenomenon occurs around the contact zone of two glass surfaces immersed in mercury, as reported by Yaminsky *et al.*[11] The same wetting conditions (mercury has a very high contact angle on glass) lead to a similar vaporization of mercury, and this can be directly observed if a test tube is pressed against the glass bottom of a beaker full of mercury.

2. *INDIRECT OBSERVATIONS OF THE HYDROPHOBIC ATTRACTION*

The earliest discussion of the possible existence of a long-range attraction between hydrophobic surfaces seems to have been published by Laskowski and Kitchener.[12] They made the observation that methylation of silica neither changes its zeta potential (surface charge) nor affects the stability of a dispersion of silica. In other words, the forces at *long range* between silica particles are not affected. On methylation the contact angle of water on such a surface, however, changes from wetting ($\theta \approx 0$) to 70 or 80°. This indicates that the interaction between vapor and substrate across the water film is attractive for certain separations, because the film is unstable for small thicknesses. Since both the double-layer and the van der Waals forces across the water film are repulsive and essentially unchanged by methylation, they surmized the existence of some additional, attractive force between the hydrophobic surface and the vapor–water interface. The origin of the force was supposed to lie in a perturbation of the water structure next to hydrophobic surfaces. They concluded by stating that "the question of the structure and extent of this perturbed layer is, of course, still unanswered." Now, more than 20 years later, we at least know something about the extent of the effect, even if its true nature remains obscure.

Blake and Kitchener subsequently gave an upper bound for the range of this hydrophobic attraction by observing the rupture of air bubbles against hydrophobed silica surfaces.[13] They obtained a value of 64 nm, below which rupture occurred. The authors cautioned that "the above result must not, however, be taken to show that hydrophobic forces extend to 64 nm." As we shall see, they might well have afforded to be less cautious.

3. DIRECT FORCE MEASUREMENTS

3.1. The Surface Force Apparatus

Almost all direct measurements of the interaction between hydrophobic surfaces have been carried out with the surface force apparatus developed at the Australian National University.[14] (To the author's knowledge the only exception is the work by Derjaguin and co-workers in Moscow.)

In a surface force apparatus the force between two molecularly smooth sheets of mica mounted in a crossed-cylinder configuration can be measured with a resolution of 10^{-7} N, while independently controlling and measuring the separation between the surfaces to within 0.1 nm (1 Å). The mica surfaces are silvered on the back sides, and if white light impinges normally on the surfaces only discrete wavelengths are passed by this interferometer. These appear as sharp lines when resolved in a spectrometer and the wavelengths of these fringes of equal chromatic order (FECO) allow the separation of the surfaces to be calculated.[15,16] If the wavelength of two such fringes is measured, the refractive index of the medium between the mica surfaces may also be determined. The force acting between the surfaces is computed by calibrating the movement of the surfaces at large separations where no force is present and then monitoring the difference between expected and real position of the surfaces, which is a result of the deflection of a (double-cantilever) spring on which one of the surfaces is mounted.

3.2. Preparation of Hydrophobic Surfaces

The mica substrate can be rendered hydrophobic in a variety of ways. The simplest is adsorption of surfactants from solution. Cationic surfactants adsorb strongly to mica by exchanging with the surface ions. The driving force for adsorption is thus largely electrostatic. Another method is the deposition of insoluble monolayers of surfactants or lipids (Langmuir–Blodgett deposition). The most stable monolayers are formed by depositing cationic surfactants, preferably double-chain quaternary ammonium compounds. Finally, chemical modification of the mica surface is also possible. Standard methylation techniques are not very successful; unlike the silica surface mica is inert and contains no reactive hydroxyl groups. Some degree of surface polymerization of chlorosilanes may occur, but surfaces thus prepared are not stable in water. By first subjecting the surface to water vapor plasma, however, it is possible to activate the surface by a combination of sputtering and chemical reaction.[17] This surface will then react with chlorosilane derivatives to form very hydrophobic surfaces with contact angles of the order of 100°.

3.3. Interpretation of Direct Force Measurements

It can be shown with a simple integration procedure that the force F between crossed cylinders of radius R ($\approx$ a sphere and a flat) is proportional to the free energy of interaction per unit area G of two parallel, flat surfaces.[18] This "Derjaguin approximation" is very useful for relating measured forces to theoretical results on the energy of interaction of flat surfaces.

$$\frac{F}{R} = 2\pi G \tag{4}$$

The gradient of the force law between crossed cylinders is hence proportional to the pressure P between flat surfaces, or

$$\frac{\partial F}{\partial D} = 2\pi R \times P \tag{5}$$

Because one of the surfaces is mounted on a spring, the system consisting of the two surfaces and the spring is susceptible to mechanical instabilities. If the gradient of the force law exceeds the value of the spring constant, the surfaces will jump to the next stable position. This means that a monotonically attractive force like a van der Waals force will pull the two surfaces into molecular contact from a finite distance, and in the intermediate regime it is not possible to measure the force, unless the spring constant is increased. A series of measurements of the separation at which such a jump occurs, as a function of the spring constant, allows one to determine the gradient of the force law, which is easily converted to the pressure between flat surfaces via Eq. (5).

Another feature of the experimental system is the occurrence of surface deformations. The mica surfaces, of thickness 1–3 μm, are mounted on silica disks with an epoxy resin that is quite soft and much thicker than the mica. Under the influence of large forces, especially adhesive interactions, this glue layer deforms and may flatten considerably. The effective radius of curvature of the surfaces will thus increase, and eventually one is measuring the force between flat surfaces rather than between curved surfaces. This is equivalent to going from measuring the energy between flat surfaces to measuring the pressure between flat surfaces. The extent of flattening depends not only on the magnitude of the force but also on the gradient of the force: the steeper the force the more substantial the deformations. In the case of hydrophobic surfaces surface deformations occur particularly when the surfaces are in adhesive contact, but, fortunately, the deformations appear to be largely elastic.[19] In other words, the surfaces return to their original, undeformed shape after separation from contact, as has been verified experimentally.

Using the Derjaguin approximation [Eq. (4)] it is easy to show that the adhesion force is related to the interfacial energy by

$$F = 4\pi R\gamma \tag{6}$$

provided no surface deformation occurs. There is an as yet unresolved controversy surrounding the corresponding relation in the presence of elastic surface deformations. Theoretically, it is expected that the relationship should change to[20,21]

$$F = 3\pi R\gamma \tag{7}$$

but experimental verification of this has proved difficult.[19,22,23]

4. THE HYDROPHOBIC ATTRACTION BETWEEN ADSORBED MONOLAYERS

4.1. Initial Experiments—CTAB Monolayers

The first direct indications that the interaction between macroscopic hydrophobic surfaces is stronger than the continuum van der Waals force were provided by Pashley and Israelachvili.[24] Using a surface force apparatus, they studied the adsorption of cetyltrimethylammonium bromide (CTAB) to mica surfaces from solution. By measuring the interaction (force versus distance curves) for various bulk concentrations of CTAB, from 10^{-6} M to 4×10^{-3} M [four times the critical micellization concentration (cmc)] they identified concentration regimes corresponding to submonolayer adsorption, with an effect mainly on the surface charge of the mica, monolayer adsorption, and bilayer adsorption. At concentrations where a packed monolayer of CTAB adsorbed to the surfaces they found a double-layer repulsion at large separations, but closer to contact the attraction was clearly greater than expected from (DLVO) Derjaguin–Landau–Verwey–Overbeek-theory. These results clearly suggested the presence of an "extra attractive force" between hydrophobic surfaces in aqueous solution.

Subsequently, a detailed study of this "hydrophobic interaction" was undertaken by the same authors.[25,26] An example of the results obtained is reproduced in Fig. 1, where the force as a function of separation between monolayers of CTAB adsorbed at 2.5×10^{-5} M is shown. Note the excellent fit to the nonlinear Poisson–Boltzmann equation at separations beyond about 6 nm from monolayer contact. At smaller separations the force turns over and becomes much more attractive than expected from DLVO theory. Their final results were extracted from a series of measurements at different electrolyte concentrations and using different force-measuring techniques, i.e., by measuring both the force and the gradient of the force (see Section 3.3). The conclusion was that the "extra" attraction is an exponentially decaying function of separation given by

$$\frac{F}{R} = C \exp\left(\frac{-D}{D_0}\right) \tag{8}$$

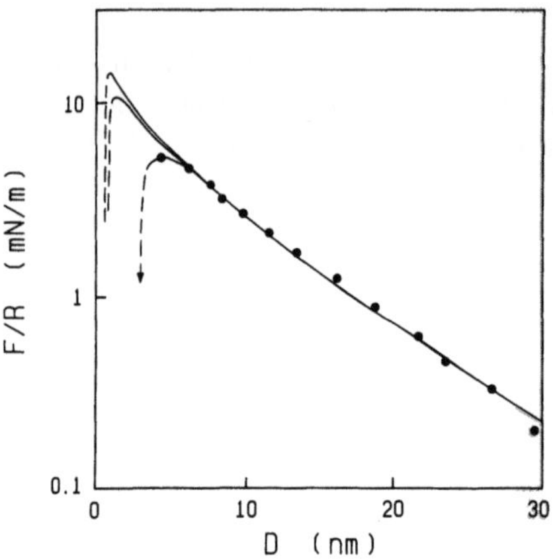

FIGURE 1. Measured force F (normalized by the radius of curvature R of the surfaces) as a function of surface separation D between mica surfaces in 2.5×10^{-5} M CTAB solution (with a background of 10^{-3} M NaCl) plotted on a logarithmic force scale. At this concentration the mica surfaces are covered with a hydrophobic monolayer of CTA^+ ions. Note the excellent agreement with the DLVO theory (solid line) at separations beyond 5 nm and the additional attraction—the hydrophobic attraction—observed at smaller separations. Parameters for the theoretical line are: surface potential at infinite separation $\Psi_0^\infty = 140$ mV; exponential decay length (Debye length) $\kappa^{-1} = 8.5$ nm; Hamaker constant $A_{121} = 1.5 \times 10^{-20}$ J. The upper part of the curve near the (theoretical) force maximum is the interaction under constant surface charge conditions, the lower line shows the corresponding constant potential interaction. [Data supplied by R. M. Pashley.]

where the preexponential factor $C = -0.14 \pm 0.02$ N m^{-1}, and the decay length $D_0 = 1.0 \pm 0.1$ nm. This hydrophobic attraction was moreover found to be relatively insensitive to ionic strength (concentrations of monovalent electrolyte from 10^{-5} M to 0.04 M), pH (5.4 to 10.4), the thickness of the CTAB monolayers, double-layer charge, and temperature (20 to 50°C).

The advancing contact angle of an aqueous solution of CTAB on the monolayers used in these experiments was 64°. The measured adhesion force between the surfaces in water was 140 ± 25 mN/m, which corresponds to an interfacial energy of only 11 ± 2 mJ/m^2 [using Eq. (6)]. This is obviously substantially less than expected for a hydrocarbon–water interface. Although it was founs that the measured adhesion forces were consistent with the Young–Dupré equation, later measurements by Pashley *et al.*[27] have shown that the contact angle on pure CTAB monolayers is over 90°, and the adhesion force in water is at least a factor of 2 greater. It seems likely that the CTAB used in the original experiments was contaminated. Subsequent experiments by Pashley *et al.*[27] and Kékicheff *et al.*[28] have indeed demonstrated that both the extent of adsorption and surface charge of the adsorbed layers are different for pure CTAB. Nevertheless, these first experiments did initiate a number of investigations of the hydrophobic attraction and certainly gave a qualitative indication of the interactions expected across thin films confined between hydrophobic surfaces.

4.2. Further Work on Adsorbed Monolayers

The results of a second experiment on measurements of the attraction between hydrophobic surfaces were published by Pashley *et al.* in 1985.[29] This time the surfaces were prepared by adsorption from solution of a double-chain quaternary ammonium surfactant (dihexadecyldimethylammonium acetate). Analysis of the results was facilitated by the fact that the surfaces thus obtained appeared to be neutral, within the accuracy of the measurements. No uncertainty was introduced by the subtraction of a double-layer repulsion. Although qualitatively similar, the measured attraction was somewhat stronger, with a slightly longer exponential decay (1.4 nm) and, in particular, a much larger magnitude of the attraction, about three times stronger (note that this refers to the *total* attraction, without subtraction of any assumed continuum van der Waals force, as was done in the case of the CTAB measurements). The advancing contact angle of water on these surfaces was 95°, and the measured force of adhesion (pull-off force) more reasonable for a hydrocarbon–water interface [$\approx 40\ \text{mJ/m}^2$, using Eq. (6)].

There are numerous, more recent observations of long-range attractive forces between surfactant monolayers adsorbed from solution. This includes interactions between dodecylamine oxide monolayers,[30] dodecylamine and octylamine monolayers,[31] as well as the repeat experiments with purified CTAB samples mentioned above. Attractive forces of extreme range have been measured between monolayers of a fluorocarbon surfactant adsorbed from solution.[32] In view of the very much stronger attractive forces found between deposited monolayers, recent attention has centered on this and on explaining the difference between adsorbed and deposited monolayers rather than obtaining accurate force curves for hydrophobic surfaces consisting of adsorbed surfactant monolayers.

4.3. Cavitation between Adsorbed Surfactant Monolayers

Pashley *et al.* observed that when the dihexadecyldimethylammonium acetate-coated surfaces were separated from molecular contact in water a vapor cavity formed as a neck joining the two surfaces.[29] The phenomenon could only be observed if the spring was sufficiently stiff that the outward jump on separation was about 100 nm or less, otherwise the cavity presumably ruptured during the jump apart and could not be observed. Cavity formation has also been noted between surfaces of *N*-(2-hydroxyethyl)-*N*-methyldidodecylammonium bromide adsorbed from solution.[33] In this case, however, cavitation occurred when *bilayer* surfaces were brought into contact, after the second layer was pressed out and hydrophobic contact was reached. The difference in behavior has not been satisfactorily explained.

5. THE HYDROPHOBIC ATTRACTION BETWEEN DEPOSITED MONOLAYERS

5.1. Weakly Charged Monolayers

An initial cause for concern with the surfaces prepared by adsorption from solution was the extent to which the presence of fairly large amounts of surfactant in solution could influence the results. It was suggested by many that the result could simply be due to bridging of the surfaces by surfactant molecules. Admittedly, it was hard to rationalize the observed functional form of the force law in terms of such behavior. Also, dilution experiments were carried out by Pashley and Israelachvili to see whether the actual concentration of surfactant in solution had any effect[26] (it did not).

The obvious way to resolve the matter was found by Claesson *et al.*[34] They prepared hydrophobic ($\theta_a = 94°$) surfaces by Langmuir–Blodgett deposition of dioctadecyldimethylammonium bromide on mica. Such monolayers are strongly bound to the mica surface by electrostatic forces—the positively charged surfactant cations replace the cations present on the mica surface in solution. Furthermore, the area per headgroup on the mica surface closely matches the lattice site area ($\approx 0.5\ nm^2$), and this gives a very close packed monolayer, since each headgroup has two hydrocarbon chains associated with it.

The deposited monolayers were weakly charged, and there was a long-range double-layer repulsion (see Fig. 2). By fitting this to theory at separations beyond 30 nm, they could extract the attractive interaction from the measurements. This time the attraction appeared to show two separate exponential decay lengths and could be described empirically by the equation

$$\frac{F}{R} = C_1 \exp\left(\frac{-D}{D_{01}}\right) + C_2 \exp\left(\frac{-D}{D_{02}}\right) \tag{9}$$

where $C_1 = -0.36\ N\,m^{-1}$, $C_2 = -0.0066\ N\,m^{-1}$, $D_{01} = 1.2$ nm, and $D_{02} = 5.5$ nm. This equation was obtained by determining the onset of instabilities for various spring stiffnesses and subsequently integrating the results (see Section 3.3). This, together with the problems of accurately fitting the double-layer force at long range in water, where the Debye length is large, casts some doubt on the accuracy of the constants in Eq. (9). Even so, the attraction was clearly evident at separations of 25–30 nm! (See Fig. 2.)

The salt dependence of the measured attraction was investigated by adding potassium bromide to 10^{-2} *M*. At this concentration of electrolyte the force was everywhere net attractive (see Fig. 2). By using data obtained in a separate investigation of the forces between one bare mica surface and one hydrophobed mica surface,[35] the authors were able to estimate the potential of the hydrophobed

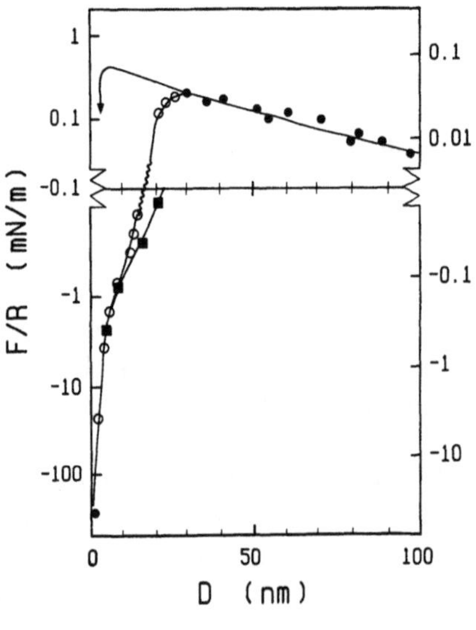

FIGURE 2. Force measured between mica surfaces coated with deposited monolayers of dimethyldioctadecylammonium (DDOA) bromide and immersed in water (circles) and 10^{-2} *M* KBr solution (squares). Note the logarithmic scale for both repulsive and attractive forces. The interaction between these hydrophobic surfaces shows a weak double-layer repulsion at long range and then becomes attractive at about 20 nm due to a very strong and long-range hydrophobic attraction. In 10^{-2} *M* KBr no double-layer repulsion is measurable, but the hydrophobic attraction is of similar magnitude. The right-hand ordinate shows the corresponding free energy per unit area between flat surfaces [Eq. (4)]. (From Ref. 34.)

mica surface to +30–40 mV. On subtraction of the double-layer force, it emerged that the longer of the two decay constants had decreased from 5.5 to 4.5 nm.

That a strong attraction was found even with systems where no surfactant was present in solution lent added weight to the ideas that the hydrophobic attraction was a real effect and not due to some artifact like surfactant bridging. The weak electrolyte dependence suggested that the force was not electrostatic in origin (for example, a double-layer attraction between unequally charged surfaces). Such an attraction was measured by Claesson *et al.* between mica and a hydrophobic surface prepared by Langmuir–Blodgett deposition of the same surfactant.(35) In this system the attraction was well described by double-layer equations for unequally charged surfaces, except at high concentrations of electrolyte, where the short range of the electrostatic force revealed the existence of an additional attraction in this case, too.

5.2. Neutral Monolayers

The discrepancy between the results of Claesson *et al.*(34) and Pashley(29) were nonetheless disconcerting. After all, the advancing contact angle of water on the two different surfaces was the same and, hence, at least according to popular belief, was their hydrophobicity. The surfactants making up the surfaces differed only by two methylene groups in their hydrocarbon chains. The thicknesses of the adsorbed and deposited layers were virtually the same.

About this time Dr. Claesson and the author decided to measure the forces between fluorocarbon surfactant surfaces. The original aim was simply to look at forces between surfaces that were more hydrophobic, using the contact angle criterion. By depositing the double-chain fluorinated quaternary ammonium surfactant *N*-(α-trimethylammonioacetyl)-0-0′-bis-(1*H*, 1*H*, 2*H*, 2*H*-perfluorodecyl)-*L*-glutamate chloride, we obtained surfaces with an advancing contact angle of water of 113°. A series of experiments (initially together with Berg and Herder) was carried out and the results proved startling.(32,36) The measured attraction was now much greater, and at first sight it seemed as if the substantial increase in hydrophobicity as determined by the contact angle measurements did show up in the strength of the attraction. Furthermore, the surfaces were apparently uncharged (no double-layer repulsion was evident), and the results therefore much more accurate. The results will be discussed in more detail below, but first we will return briefly to the hydrocarbon surfaces. We made the fortunate decision to go back and check the dimethyldioctadecylammonium bromide surfaces.(36)

By changing the deposition conditions slightly through a reduction of the surface pressure during deposition from the originally(34) used 35 to 25 mN/m, we managed to obtain hydrocarbon surfaces that were also uncharged. The attraction we measured for these surfaces was of much longer range than that deduced by Claesson *et al.* and, in fact, almost as long range as for the fluorocarbon surfaces.(36,37) Apart from some of the measurements by Pashley and Israelachvili, the hydrophobic attraction had usually been determined by observing the points at which the spring system of the surface force apparatus encountered an instability and, hence, a jump into contact (see Section 3.3). With the uncharged hydrocarbon and fluorocarbon surfaces, however, it turned out to be difficult to establish the

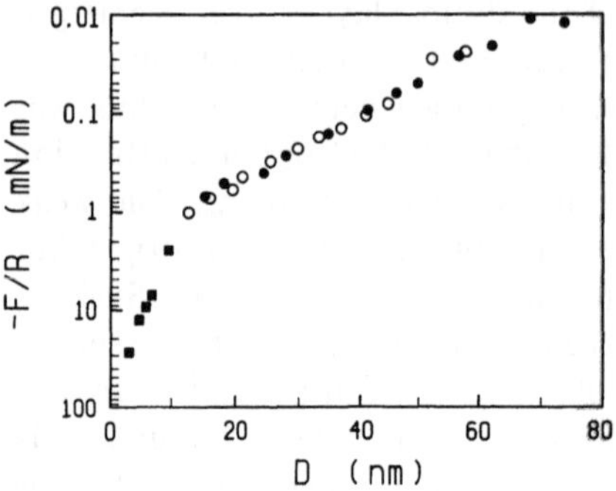

FIGURE 3. Force measured between deposited monolayers of DDOA, showing the range and magnitude of the attraction when neutral surfaces are obtained by a slight change in the deposition conditions (cf. Fig. 2). The attractive force beyond about 15 nm is well described by an exponential function with a decay length of 13 nm. At smaller separations the force turns more attractive. The various symbols are the results of different series of measurements. [Used with permission from the *Journal of Physical Chemistry* **92**, 1650 (1988), copyright 1988, American Chemical Society.]

jump separation (point of instability) with any certainty. The surfaces appeared instead to drift together, slowly at first, and it was impossible to say with any precision at which point there was an actual jump. They would drift over times of ≈ 30 s from 30–40 or even 50 nm apart and into contact. We decided to attempt to measure the force directly, outside the instability, using the traditional method of measuring the spring deflection. This was ultimately more satisfactory and also allowed us to perform measurements at far greater separations, at distances where no jumps were observed even with the weakest possible force-measuring spring. The results displayed the amazing range of the hydrophobic attraction between macroscopic surfaces, with any errors due to subtraction of double-layer forces eliminated. The interaction still appeared to be exponential but with two separate decay lengths, as in the work by Claesson *et al.*[34] For the hydrocarbon surfaces [monolayers of dimethyldioctadecylammonium (DDOA) bromide] the force was measurable out to 70 nm and had a long-range decay of about 13 nm and a short-range decay of 2–3 nm (see Fig. 3). The results are plotted in Fig. 4 on a linear scale, together with the results obtained by Pashley *et al.*[29] with adsorbed monolayers (Section 4.2) and the continuum van der Waals force between two mica surfaces across water calculated from Lifshitz theory (dashed line). The effect of the deposited monolayers would be to slightly decrease the magnitude of the effective van der Waals interaction, especially at short range. The hydrophobic attraction is two orders of magnitude stronger than the van der Waals force over most of the

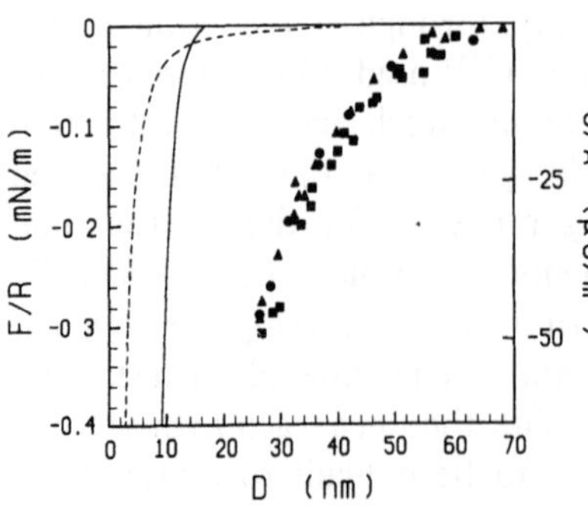

FIGURE 4. Force measured between deposited monolayers of DDOA in water, plotted on a linear scale. The dashed line shows the continuum van der Waals force expected between bare mica surfaces in water, which is an upper limit for the strength of the van der Waals interaction in this system. The solid line is the interaction measured between monolayers of dihexadecyldimethylammonium acetate adsorbed from solution.[29] Over a large part of the range the hydrophobic interaction is two orders of magnitude stronger than the continuum van der Waals force. The right-hand scale shows the corresponding free energy of interaction between flat surfaces. [From Christenson and Claesson, *Science* **239**, 390–392 (1988), copyright 1988 by the AAAS.]

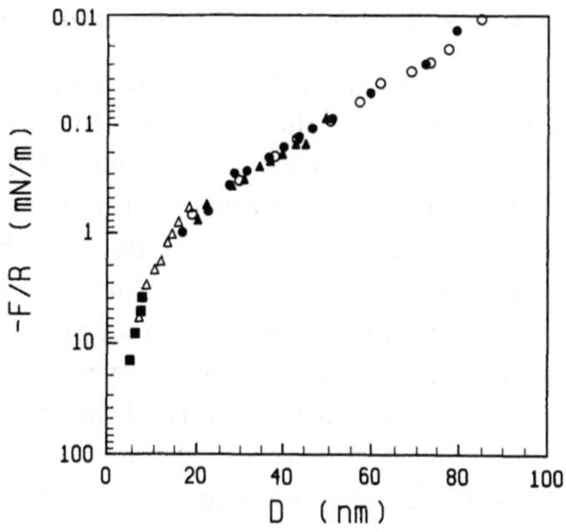

FIGURE 5. Hydrophobic attraction between neutral surfaces prepared by deposition of a double-chain fluorocarbon surfactant on mica. The long-range exponential attraction has a decay length of 16 nm and turns more attractive below about 20 nm. The different symbols represent separate series of measurements. [Used with permission from the *Journal of Physical Chemistry* **92**, 1650 (1988), copyright 1988, American Chemical Society].

measurable range! Moreover, the attraction found between adsorbed monolayers does not seem very remarkable, by comparison.

The fluorocarbon surfaces actually showed a slightly more long-range attraction, measurable out to 90 nm, as seen in Fig. 5, where the force is plotted on a semilog scale. The long-range decay was approximately 16 nm, with a short-range decay similar to that between the hydrocarbon surfaces, namely 2–3 nm. Figure 6 shows some results obtained with the flurocarbon surfaces on a linear scale. The measurements from two different experiments give an idea of the reproducibility of the results.

For both the hydrocarbon and the fluorocarbon surfaces the measured short-range force, when extrapolated to $D = 0$ (contact), was in approximate agreement with the measured pull-off force. Some uncertainity comes from the unresolved question of how exactly to relate the force between deformable surfaces (i.e., the pull-off force) to the force between curved surfaces (i.e., the measurements at finite separations). The surface energy of the hydrocarbon surfaces was usually found to be about 35–50 mJ/m^2, depending on whether Eq. (6) or (7) is used. The fluorocarbon surfaces, however, showed a *smaller* adhesion, contrary to expectation based on the energies of hydrocarbon–water and fluorocarbon–water interfaces.[38] The adhesion was only 20–30 mJ/m^2, and the reason for this remains a mystery. It is possibly related to the numerous and irregularly placed vapor cavities that form around the perimeter of the contact area (see Section 5.3).

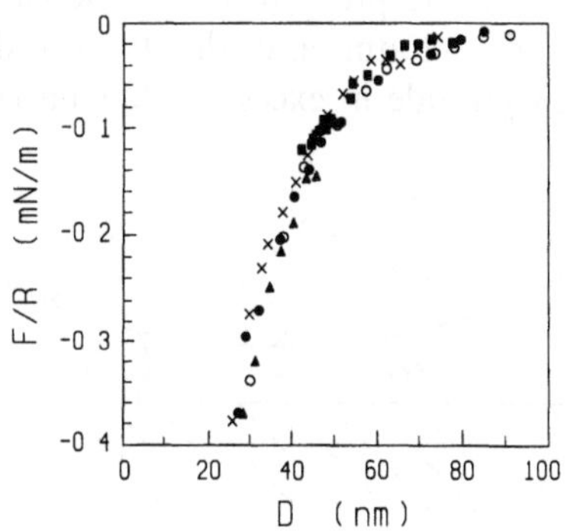

FIGURE 6. Measured force (on a linear scale) between flurocarbon monolayer-coated mica surfaces, showing the reproducibility of different measurements in the same and different experiments. The crosses and filled squares are the results obtained in one experiment. [Used with permission from the *Journal of Physical Chemistry* **92**, 1650 (1988), copyright 1988, American Chemical Society].

5.3. Cavitation between Deposited Monolayers

The predictions of Yushchenko *et al.*[9] regarding cavitation were very nicely confirmed with the deposited hydrocarbon and fluorocarbon surfaces. As in the case of the adsorbed monolayers of dihexadecyldimethylammonium acetate, cavity formation was observed with the hydrocarbon surfaces on separation from molecular contact. With the fluorocarbon monolayers, however, cavitation occurred as the surfaces came into contact. The fact that the cavities contained only vapor was proven unambigously by refractive index measurements. These cavities were directly visible through a microscope eyepiece when the surfaces were viewed from above.[37]

There was a very marked difference in the geometry of the cavities that formed. With the hydrocarbon surfaces a single, large bridge or neck joining the two surfaces formed (see Fig. 8a, b). This cavity remained stable out to separations of typically 0.5–1.0 μm, whereupon it ruptured instantaneously. On bringing the surfaces together again, the same long-range hydrophobic attraction was once more measured and only on separation from contact could the cavity be recreated. If the surfaces were brought into contact with the cavity still joining them, without letting it snap, it vanished as soon as the surfaces came into contact, only to reappear on a second separation.

With the fluorocarbon surfaces numerous, small cavities formed around the perimeter of the contact between the two surfaces (see Fig. 8c, d). On separation these would move about and coalesce gradually, the final, large cavity often remaining stable out to separations of many microns. It was further observed that the size of the cavities seemed to depend on the amount of dissolved air in the water, in particular their size would usually increase slowly during the course of an experiment, presumably as more and more air dissolved in the initially deaerated water.

These observations may be rationalized by considering the constraints that the Laplace equation places on the allowed size of a vapor cavity. The pressure difference ΔP across the vapor–water interface is given by

$$\Delta P = \gamma\left(\frac{1}{r_1}+\frac{1}{r_2}\right) \tag{10}$$

where γ is the interfacial tension (=surface tension of water) and r_1, r_2 are the two principal radii of curvature of the meniscus. This is shown schematically in Fig. 7. Since the pressure inside the cavity is lower than outside, ΔP must be between 0 and -1 atm and the total radius of curvature is negative and must have a magnitude in excess of 700 nm (with $\gamma = 72$ mN/m). An annular cavity around the

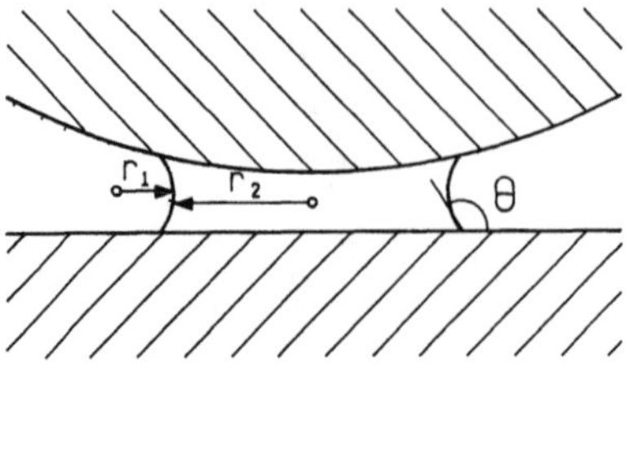

FIGURE 7. Schematic illustration of two hydrophobic surfaces (depicted as a sphere on a flat, which is equivalent to crossed cylinders, to leading order) with a bridging vapor cavity; θ is the contact angle of water on the hydrophobic surface (>90°), and r_1 and r_2 are the principal radii of curvature of the liquid–vapor interface where r_2 is positive and r_1 negative.

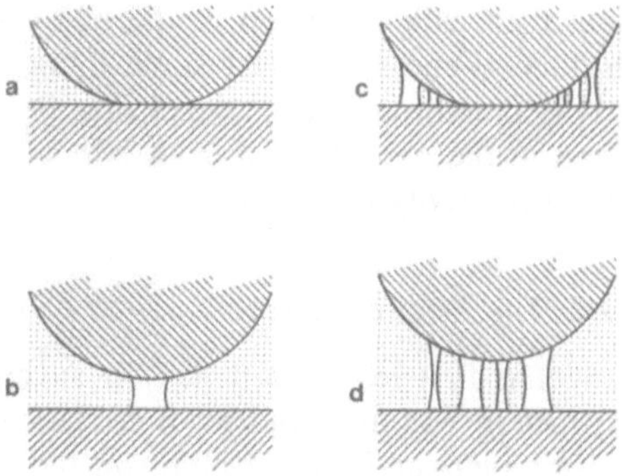

FIGURE 8. Schematic illustration of cavity formation between hydrophobed surfaces in water as deduced from the appearance of the interference fringes (FECO): (a), (b) hydrocarbon monolayer surfaces where a cavity is formed only on separation of the surfaces; a similar situation is observed with surfaces prepared by silanation of plasma etched mica surfaces; (c), (d) fluorocarbon monolayer surfaces, where many small, discrete cavities are formed as the surfaces come into contact.

flattened contact zone would have to have one large, positive radius of curvature and a negative radius of some minimum magnitude (i.e., the curvature cannot be too large). Consequently, such a cavity can only become arbitrarily small for contact angles approaching 90°.

In the case of the hydrocarbon surfaces, where the cavity is only formed after separation from contact, the resulting neck or bridge joining the two surfaces may have a total radius of curvature of the order of 700 nm. It cannot form as a single, annular cavity at contact, because to do so it would necessarily have to pass through some intermediate stage with a much smaller net radius of curvature. With the fluorocarbon surfaces the problem is resolved by the formation of numerous small cavities around the contact area. These have two small radii of curvature of nearly equal magnitude but opposite sign, and the net radius may still be large enough to satisfy Eq. (10).

It should be stressed that the hydrophobic attraction is *not* a direct result of cavitation. The large attractive interaction is reproducibly measured in the absence of cavity formation, and a comparison of the forces measured in the absence and in the presence of a cavity confirms that the state with a cavity is the lower-energy state. This is illustrated in Fig. 9, where the force measured on approach between fluorocarbon surfaces is shown in the presence and the absence of a cavity. At con-

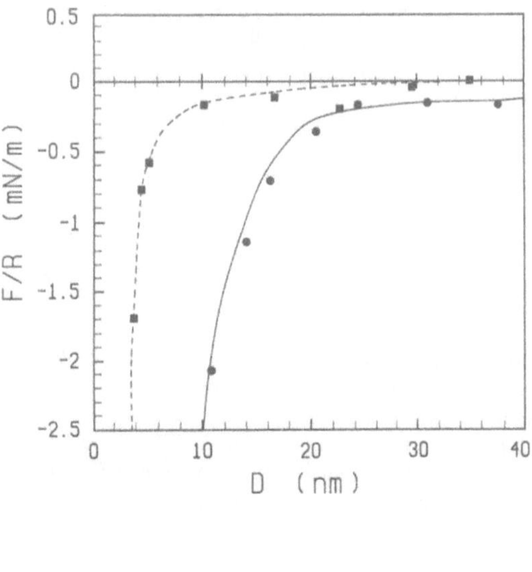

FIGURE 9. Measured force between mica surfaces with a deposited monolayer of fluorocarbon surfactant in the presence (dashed curve) and in the absence (solid curve) of a bridging vapor cavity. [From Christenson and Claesson, *Science* **239**, 390–392 (1988), copyright 1988 by the AAAS.]

tact there is a minimum in the free energy and it is independent of the previous history of the system. At some finite separation the free energy is lower with a cavity present and hence the change in free energy with decreasing separation is less than in the absence of a cavity. In other words $-\partial G/\partial D$, or the force, is greater when no cavity is present.

5.4. The Electrolyte Dependence of the Hydrophobic Attraction

The first results with CTAB suggested that the strength of the hydrophobic attraction was independent of electrolyte concentration.[26] The initial experiments with Langmuir–Blodgett films, on the other hand, did show some, if only a slight, effect on the long-range decay.[34] We started a more systematic investigation of electrolyte effects with the fluorocarbon surfaces.[32] The experiments were hampered by the fact that the deposited layers became unstable in more concentrated ($\geqslant 10^{-3}$ *M*) electrolyte solutions, especially in salts like alkali metal halides. After the first approach and subsequent separation multilayers of surfactant built up on the surfaces, a process aided no doubt by the cavitation that occurred as the surfaces came into contact. The instability of both fluorocarbon and hydrocarbon surfactant monolayers in concentrated alkali halides was confirmed by contact angle measurements. Eventually we found that the fluorocarbon surfactant monolayers were reasonably stable in salts of alkylammonium halides and used tetrapentylammonium bromide to study salt effects up to 10^{-2} *M*.

The results were quite different from those found previously: the range of the attraction was found to decrease substantially with electrolyte concentration, although the contract adhesion suffered only a minor decrease. The analysis of the experimental data was complicated by a very marked charging up of the surfaces; the surface charge density increased with electrolyte concentration, and it became

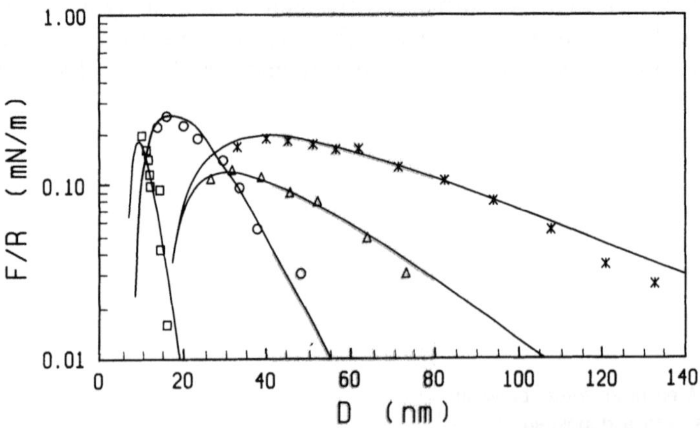

FIGURE 10. Measured force between deposited fluorocarbon surfactant monolayers immersed in solutions of tetrapentylammonium bromide. There is a double-layer repulsion present for all concentrations, and the surface charge increases with electrolyte concentration. The solid lines are fits using an attraction of the form $F/R = C\exp(-D/D_0)$ and a repulsive double-layer interaction calculated from solutions to the nonlinear Poisson–Boltzmann equation. Parameters for the fits are given in Table 1. 5×10^{-5} *M* (crosses); 10^{-4} *M* (triangles); 10^{-1} *M* (circles); 10^{-2} *M* (squares).

TABLE 1. Parameters Used To Fit Hydrophobic Attraction in Electrolyte Solutions
Hydrocarbon Surface + KBr[a]

Conc. M	Debye length κ^{-1} (nm)	Surface potential ψ_0 (mV)	Area per charge (nm^2)	Preexponential C (mN/m)	Decay length D_0 (nm)
$\sim 10^{-4}$	27.5	30	196	−2.0	10.5
$\sim 10^{-3}$	9.0	40	46	−4.0	4.2
$\sim 10^{-2}$	3.0	83	5.4	−20	2.5
$\sim 10^{-3}$	9.6	42	46	−4.0	3.6
$\sim 10^{-4}$	35.0	33	224	−2.7	10.5
Fluorocarbon Surface + Tetrapentylammonium Bromide					
$\sim 5 \times 10^{-5}$	42.5	73	97	−2	14.3
10^{-4}	25	48	103	−2	10
10^{-3}	9	90	14	−8	5.8
10^{-2}	2.6	86	4.5	−60	1.7

[a] The first three entries are the result of an experiment with increasing concentration, the last two are from an experiment with decreasing concentration.

rather difficult to accurately extract the functional form of the extra attraction, given that a substantial double-layer force had to be subtracted out. A reasonable fit over part of the range of the hydrophobic attraction could be obtained on the assumption that the decay length of the long-range part of the attraction decreased with salt concentration, but that the preexponential factor increased. This, however, failed to explain the occurrence of a "secondary minimum" at 10^{-2} M. The results of a series of measurements with fluorocarbon surfaces and tetrapentylammonium bromide concentrations ranging from 5×10^{-5} to 10^{-2} M are shown in Fig. 10. The double-layer parameters and parameters for the hydrophobic attraction [of the form of Eq. (8)] that give a fit to the measured data are given in Table 1.

Similar results were obtained in a series of repeat measurements on DDOA monolayers with potassium bromide.[39] With a deposition pressure of 25 mN/m instead of 35 mN/m, as originally used, stability problems were once again encountered. Qualitatively and semiquantitatively, however, the behavior turned out to be similar to the fluorocarbon surfaces. The surface charge increased with salt concen-

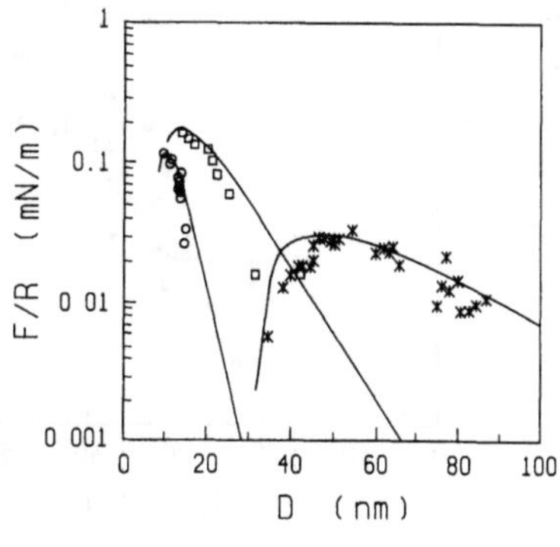

FIGURE 11. Measured force (logarithmic scale) between hydrocarbon (DDOA) surfaces with added electrolyte in the form of KBr at 10^{-4} (crosses), 10^{-3} (squares), and 10^{-2} M (circles). Parameters for the fits are given in Table 1 (first three entries).

tration, and the decay length of the long-range attraction decreased with a concentration dependence similar to that found with the fluorocarbon surfactant. Figure 11 shows some representative force curves with the potassium bromide concentration ranging from 10^{-4} to 10^{-2} *M*. Table 1 shows the double-layer parameters and decay lengths and preexponential factors needed to give fits to the measured curves (first three entries).

It should be emphasised that the hydrophobic attraction extracted from these measurements only gives a reasonable fit over part of the range, specifically at separations from the force maximum to about 2 D lengths further out. The functional form thus deduced does not extrapolate to give the pull-off force at contact, and it does not always fit the results at larger separations.[32] The fitting exercise merely gives a semiquantitative illustration of the changes that occur on addition of electrolyte. An implicit assumption is that the hydrophobic attraction and the eectrostatic double-layer repulsion are independent and additive.

6. *THE HYDROPHOBIC ATTRACTION BETWEEN CHEMICALLY HYDROPHOBED SURFACES*

At the time Dr. Claesson and myself were writing up the results with deposited hydrocarbon and fluorocarbon surfaces, a manuscript from Rabinovich and Derjaguin in Moscow reached our hands. They had been studying the interaction between crossed quartz filaments rendered hydrophobic by methylation, and their results were almost exactly the same as ours. Using a different force-measuring device, with differently prepared surfaces of advancing contact angle $\approx 100°$ and working independently, they had obtained an attractive force with a long-range decay of 12.2 nm, versus 12.9 in our case.[40] Very importantly, too, their system was surfactant-free, and that objection could finally be put to rest. Their measurements are compared with our results for the hydrocarbon surfaces in Fig. 12. As can be seen, their results are virtually identical, apart from more scatter of the data

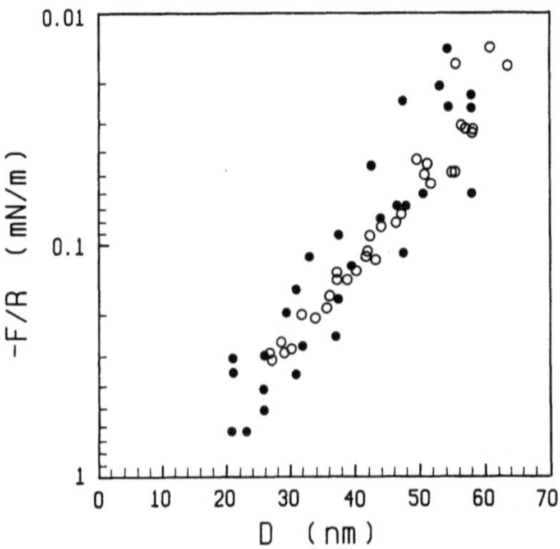

FIGURE 12. A comparison between the hydrophobic attraction measured between deposited monolayers of DDOA[36] (open circles) and methylated silica[40] (filled circles). Both these surfaces are neutral, showing no detectable-layer repulsion.

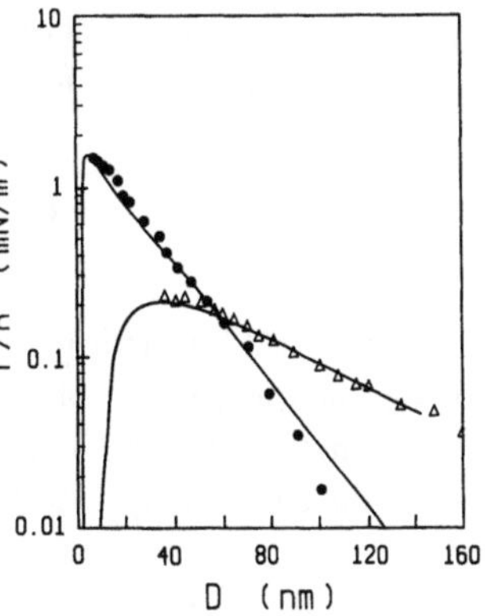

FIGURE 13. Measured force between two fluorocarbon surfaces prepared by plasma-treating and then silanating mica. Triangles are points measured in water and these results are fitted by assuming an exponentially decaying hydrophobic attraction with a decay length of 12 nm. Circles are points measured in 10^{-4} *M* KBr where the attraction needed to fit the data has a decay length of 8 nm.

points. They, too, observed an increased attraction at separations below about 20 nm and from hysteresis of the measured forces deduced that cavity formation occurred at contact between the hydrophobed filaments.

More recently, Parker, Claesson, and co-workers have prepared hydrophobic surfaces by first exposing mica surfaces to water vapor plasma and then chemically modifying the resulting, reactive surface with fluorinated silanes.[17] These fluorinated surfaces show large advancing contact angles of water (up to 105°) and are also stable at high salt concentrations for several days. A drawback is that they carry a substantial surface charge in dilute electrolyte solutions, where one is once more faced with the tricky procedure of subtracting electrostatic forces. The results are very similar to those measured with Langmuir–Blodgett films, vindicating once again their use as model hydrophobic surfaces. An example is shown in Fig. 13, where the force measured between fluorocarbon silane treated mica in water and 10^{-4} *M* potassium bromide is illustrated. The hydrophobic attraction has a decay length of 12 nm (water) and 8 nm (KBr). Cavitation is observed as with the Langmuir–Blodgett films of hydrocarbon surfaces—only on separation from molecular contact.

The results of measurements at electrolyte concentrations of up to 0.1 *M* with such chemically modified surfaces confirm the findings with the less stable Langmuir–Blodgett films. With increasing salt concentration the range of the hydrophobic attraction decreases markedly, and at 0.1 *M* KBr no attraction is measurable beyond about 10 nm,[41] although the force at smaller separations is still much stronger than the continuum van der Waals force.

7. THEORY

The first serious attempt to provide at least a "phenomenological" description of the hydrophobic attraction is due to Eriksson *et al.*[42] Their model is based on the square gradient approximation used by Marcelja and Radic[43] in an earlier treatment of repulsive hydration forces and examines the changes in average number of hydrogen bonds close to a hydrophobic surface. The mathematical form of the force law predicted by this model matches fairly closely the experimental results over the entire range, including the short-range attraction, and is of the form

$$\frac{F}{R} = -B\left[\coth\left(\frac{bD}{2} - 1\right)\right] \quad (11)$$

where both the factor B and the "decay" length $2/b$ are determined by fitting the experimental data to this equation. The theory thus provides no prediction of the magnitude or rate of decay of the attraction, only its general functional form. There is, moreover, no really convincing physical basis for the postulated hydrogen bond structure at a hydrophobic surface.

An alternative explanation has been advanced by Podgornik.[44] He suggests that the attraction is due to electrostatic correlations between fluctuating "non-equilibrium states" on the opposing hydrophobic surfaces. The metastability of the water film between the surfaces could lead to the existence of such nonequilibrium states. The model predicts an attractive interaction with a decay length of $\kappa^{-1}/2$, or half of the Debye screening length.

Attard[45] has proposed a theory where the hydrophobic surfaces are assumed to induce anomalous polarization fluctuations in the adjacent water layers. These fluctuations couple electrostatically between the surfaces, and an attraction with a decay length of $\kappa^{-1}/2$ results. The model does not predict the magnitude of the attraction nor the stronger, short-range attraction. Attard is able to fit reasonably well the long-range results for both hydrocarbon and fluorocarbon monolayer surfaces with added electrolyte as well as the measurements by Rabinovich and Derjaguin with methylated silica surfaces. One less than convincing result is that the electrolyte concentration in distilled water would have to be about 4×10^{-5} M, which is much higher than known conductivity and pH data would predict. Furthermore, the results of Claesson *et al.*[34] on hydrocarbon surfaces with 10^{-2} M KBr are in disagreement with the model.

As long as it is unclear which, if any, of the foregoing suggestions is correct, the hydrophobic attraction will continue to be controversial. It is to be hoped that an increasing number of theoreticians will turn their attention to the problem.

8. *REMAINING PROBLEMS AND FUTURE RESEARCH*

Although the existence of a long-range attractive interaction between hydrophobic surfaces in water is by now well documented, a number of questions remain unanswered.

The discrepancy between the results obtained with surfactant monolayers adsorbed from solution as opposed to deposited ones remains puzzling. Surfactants adsorbed from solution almost invariably give an attraction of much shorter range. The advancing contact angle of water on these surfaces is, however, often the same ($\approx 95°$) as that measured on deposited monolayers of double-chain hydrocarbon surfactants. Obviously, the surface hydrophobicity as determined by the contact angle of a sessile droplet of water is not a good indication of the strength and range of the attraction measured between two such surfaces. The difference is most likely related to the state of the monolayers. With adsorbed monolayers the surfactant molecules are likely to be present in a much more disorganized state and, in particular, additional adsorption of molecules with different orientations (e.g., headgroup facing the aqueous phase) may occur. This would lead to a rougher and more heterogeneous surface, which might conceivably reduce the range of the attraction considerably. This is supported by a comparative study of adsorbed and deposited monolayers[(46)] using contact angle measurements and electron spectroscopy for chemical analysis (ESCA). With the fluorocarbon surfactant it has indeed been shown[(32)] that even the *same* surfactant, when adsorbed from solution, gives a slightly shorter range attraction than when deposited. In this case good evidence for additional, time-dependent adsorption to the hydrophobic monolayers was obtained.

The effect of electrolyte concentration on the range of the interaction is also far from being understood. The theoretical models that predict a $\kappa^{-1}/2$ decay of the force certainly do not agree with the experimental results in the dilute regime. Disregarding theory, it is not clear whether the experimentally observed electrolyte effects are due to the surfaces becoming more and more charged, or whether it is the ionic concentration in the interlayer of water between the surfaces that effects the range of the attraction. A decrease in the range of the attraction with electrolyte concentration has so far invariably been accompanied by an increase in the surface charge.

If the hydrophobic attraction is due to water structure, an interesting question is why the range of the hydrophobic attraction is so much greater (by an order of magnitude, at least) than the range of repulsive "hydration" forces between hydrophilic surfaces.[(47,48)] These surfaces are hydrophilic by virtue of individual sites on the surface, i.e., adsorbed ions, charges, dipoles, etc. Water molecules will thus orient around these sites, which are often randomly distributed over the surface. At a hydrophobic surface on the other hand, the water will orient so as to direct hydrogen bonding *away* from the surface, presumably without much local variation in the degree of orientation. It is consequently easy to imagine that any propagation of structure away from a hydrophobic surface might be more pronounced than at a hydrophilic surface, where structure around one site might actually interfere with the structure at a neighboring site.

A number of features of the hydrophobic attraction are still to receive experimental attention. Among these are the temperature dependence; such a study may give information on the relative importance of enthalpic and entropic contributions to the free energy, as has been done for interactions between the headgroups of nonionic surfactants.[(49)] The effects of divalent ions on the interaction need to be investigated, and this may finally establish whether an electrostatic mechanism is responsible for the hydrophobic attraction. The possible existence of similar, long-range attractive forces in nonaqueous systems should be studied. Much remains to be done in this relatively new area of surface science.

Note Added in Proof. Measurements of the effect of divalent electrolyte on the hydrophobic attraction have unambiguously shown that the decay length is not significantly affected by salt.[(50)] Earlier results with monovalents can be refitted to yield a unified picture of electrolyte effects on the long-range (at surface separations above 20 nm) part of the hydrophobic attraction.[51] With increasing electrolyte concentration the pre-exponential factor decreases but the decay length is unchanged. These results conclusively rule out any electrostatic mechanism[44,45] as the origin of the hydrophobic attraction, and its nature remains a mystery.

REFERENCES

1. C. Tanford, *The Hydrophobic Effect*, 2nd ed., Wiley, New York (1980).
2. F. Franks, in *Water: A Comprehensive Treatise* (F. Franks, ed.), Vol. 4, pp. 1–94, Plenum, New York (1975).
3. C. Pangali, M. Rao, and B. J. Berne, *J. Chem. Phys.* **71**, 2975 (1979).
4. L. R. Pratt and D. Chandler, *J. Chem. Phys.* **67**, 3683 (1977).
5. D. Y. C. Chan, D. J. Mitchell, B. W. Ninham, and B. A. Pailthorpe, in *Water: A Comprehensive Treatise* (F. Franks, ed.), Vol. 6, pp. 239–278 Plenum, New York (1975).
6. C. Y. Lee, J. A. McCammon, and P. J. Rossky, *J. Chem. Phys.* **80**, 4448 (1984).
7. N. I. Christou, J. S. Whitehouse, D. Nicholson, and N. G. Parsonage, *Mol. Phys.* **55**, 397 (1985).
8. F. M. Fowkes, *Ind. Eng. Chem.* **56** (12), 40 (1964).
9. V. S. Yushchenko, V. V. Yaminsky, and E. D. Shchukin, *J. Colloid Interface Sci.* **96**, 307 (1983).
10. Ya. I. Rabinovich, B. V. Derjaguin, and N. V. Churaev, *Adv. Colloid Interface Sci.* **16**, 63 (1982).
11. V. V. Yaminsky, V. S. Yushchenko, E. A. Amelina, and E. D. Shchukin, *J. Colloid Interface Sci.* **96**, 301 (1983).
12. J. Laskowski and J. A. Kitchener, *J. Colloid Interface Sci.* **29**, 670 (1969).
13. T. D. Blake and J. A. Kitchener, *J. Chem. Soc. Faraday Trans. 1* **68**, 1435 (1972).
14. J. N. Israelachvili and G. E. Adams, *J. Chem. Soc. Faraday Trans. 1* **74**, 975 (1978); J. L. Parker, H. K. Christenson, and B. W. Ninham, *Rev. Sci. Instrum.* **60**, 3135 (1989).
15. S. Tolansky, *Multiple-Beam Interferometry of Surfaces and Films*, Oxford University Press (1948).
16. J. N. Israelachvili, *J. Colloid Interface Sci.* **44**, 259 (1973).
17. J. L. Parker, D. L. Cho, and P. M. Claesson, *J. Phys. Chem.* **93**, 6121 (1989).
18. B. V. Derjaguin, *Kolloid-Z.* **69**, 155 (1934).
19. R. G. Horn, J. N. Israelachvili, and F. Pribac, *J. Colloid Interface Sci.* **115**, 480 (1987).
20. K. L. Johnson, K. Kendall, and A. D. Roberts, *Proc. R. Soc. London A* **324**, 301 (1971).
21. V. M. Muller, V. S. Yushchenko, and B. V. Derjaguin, *J. Colloid Interface Sci.* **92**, 92 (1980).
22. J. N. Israelachvili, E. Perez, and R. K. Tandon, *J. Colloid Interface Sci.* **78**, 260 (1980).
23. H. K. Christenson, *J. Colloid Interface Sci.* **121**, 170 (1988).
24. R. M. Pashley and J. N. Israelachvili, *Colloids Surfaces* **2**, 169 (1981).
25. J. N. Israelachvili and R. M. Pashley, *Nature (London)* **300**, 341 (1982).
26. J. N. Israelachvili and R. M. Pashley, *J. Colloid Interface Sci.* **98**, 500 (1984).
27. R. M. Pashley, P. M. McGuiggan, R. G. Horn, and B. W. Ninham, *J. Colloid Interface Sci.* **126**, 569 (1988).
28. P. Kekicheff, H. K. Christenson, and B. W. Ninham, *Colloids Surfaces* **40**, 31 (1989).
29. R. M. Pashley, P. M. McGuiggan, B. W. Ninham, and D. F. Evans, *Science (Washington, DC)* **229**, 1088 (1985).
30. C. E. Herder, P. M. Claesson, and P. C. Herder, *J. Chem. Soc. Faraday Trans. 1* **85**, 1933 (1989).
31. P. C. Herder, *J. Colloid Interface Sci.* **134**, 346 (1990).
32. H. K. Christenson, P. M. Claesson, J. Berg, and P. C. Herder, *J. Phys. Chem.* **93**, 1472 (1989).
33. J. L. Parker, unpublished results.
34. P. M. Claesson, C. E. Blom, P. C. Herder, and B. W. Ninham, *J. Colloid Interface Sci.* **114**, 234 (1986).
35. P. M. Claesson, P. C. Herder, C. E. Blom, and B. W. Ninham, *J. Colloid Interface Sci.* **118**, 68 (1987).
36. P. M. Claesson and H. K. Christenson, *J. Phys. Chem.* **92**, 1650 (1988).
37. H. K. Christenson and P. M. Claesson, *Science (Washington, DC)* **239**, 390 (1988).
38. T. Handa and P. Mukerjee, *J. Phys. Chem.* **85**, 3916 (1981).
39. H. K. Christenson and P. M. Claesson, unpublished results.

40. Ya. I. Rabinovich and B. V. Derjaguin, *Colloids Surfaces* **30**, 243 (1988).
41. J. L. Parker and P. M. Claesson, unpublished results.
42. J. C. Eriksson, S. Ljunggren, and P. M. Claesson, *J. Chem. Soc. Faraday Trans. 2* **85**, 163 (1989).
43. S. Marcelja and N. Radic, *Chem. Phys. Lett.* **42**, 129 (1976).
44. R. Podgornik, *J. Chem. Phys.* **91**, 5840 (1989).
45. P. Attard, *J. Phys. Chem.* **93**, 6441 (1989).
46. P. M. Claesson, P. C. Herder, J. Berg, and H. K. Christenson, *J. Colloid Interface Sci.* **136**, 541 (1990).
47. R. M. Pashley, *J. Colloid Interface Sci.* **80**, 153 (1981).
48. L. J. Lis, M. McAlister, N. Fuller, R. P. Rand, and V. A. Parsegian, *Biophys. J.* **37**, 657 (1982).
49. P. M. Claesson, R. Kjellander, P. Stenius, and H. K. Christenson, *J. Chem. Soc. Faraday Trans. 1* **82**, 2735 (1986).
50. H. K. Christenson, J. Fang, B. W. Ninham and J. L. Parker, *J. Phys. Chem.* **94**, 8004 (1990).
51. H. K. Christenson, P. M. Claesson and J. L. Parker, *J. Phys. Chem.* **95**, (1992), in press.

3

High- and Medium-Energy Surfaces: Ultrahigh Vacuum Approach

Malcolm E. Schrader

1. INTRODUCTION

1.1. High- and Low-Energy Surfaces

Solid surfaces have traditionally been divided into the categories of high energy and low energy.[1,2] This concept has its root in the bulk nature of the solid itself. Metals, glasses, and ceramics exist as materials of high strength because of the chemical bonds that hold their atoms together. A large input of energy is then necessary to fracture these solids, thereby creating two new surfaces of high enthalpy and free energy. In general, these energies[3] are of the order of 10^3 mJ/m^2, ranging from 1000 to 4000 mJ/m^2. By contrast, low-energy surfaces usually derive from soft organic solids, whose molecules are held together by physical, essentially van der Waals, forces. The enthalpy or free energy of these surfaces[3] are of the order of 10^1 mJ/m^2, ranging roughly from 15 to 60 mJ/m^2.

1.2. Contamination

The highly active high-energy surfaces have a strong tendency to pick up contamination. This contamination is of two types: molecular and colloidal. Typical examples of molecular (type a) are oxygen on metals or water on glass. In the ordinary atmosphere these ambients can completely contaminate a surface within nanoseconds. In the colloidal type (type b), hydrocarbonaceous particles in air will contaminate a surface in seconds. Generally, the contamination becomes evident between 30 s and 30 min of initial exposure. To eliminate type a contamination, ultrahigh vacuum (UHV) is necessary. For type b contamination, UHV is not necessary in principle, but is still by far the most reliable approach. When types a and b are both factors, treatment in UHV is the method of choice.

Malcolm E. Schrader • Institute of Chemistry, The Hebrew University of Jerusalem, Jerusalem, Israel.

Modern Approaches to Wettability: Theory and Applications, edited by Malcolm E. Schrader and George Loeb. Plenum Press, New York, 1992.

1.3. Theory of Interfacial Interaction

Girifalco and Good[4] originally proposed a method to predict interfacial tensions of liquid–liquid and liquid–solid systems through use of an interaction term involving the geometric mean of the separate surface tensions of the interacting species. The interaction parameter Φ was calculated by summing the energy of interaction of pairs of molecules across the interface:

$$\gamma_{SL} = \gamma_S + \gamma_L - 2\Phi(\gamma_S \gamma_L)^{1/2} \tag{1}$$

where γ_L is the surface tension of a liquid phase, γ_S is the surface tension of the other phase, which may be liquid or solid, γ_{SL} is the interfacial tension, and Φ is an interaction parameter. This enables the calculation of the right side of the Young equation

$$\gamma_L \cos\theta = \gamma_{SV} - \gamma_{SL} \tag{2}$$

where γ_{SV}, the surface tension of the solid in equilibrium with the vapor of the liquid, is assumed to equal γ_S, the surface tension of the solid in vacuum, so that the contact angle would be predicted from knowledge of the liquid surface tension. Alternatively, the reaction parameters could be calculated from measurement of the contact angle. Fowkes[5] proposed to separate the surface tension of each substance into additive components,

$$\gamma = \gamma^d + \gamma^h + \gamma^m \tag{3}$$

so that, for example, the surface tension of water, γ_L, would be equal to $\gamma^d + \gamma^h$, where γ^d is a component of the surface tension of water resulting from dispersion forces and γ^h is the component resulting from hydrogen bonding forces. Likewise, the surface tension of mercury was described as equal to $\gamma^d + \gamma^m$, where γ^d is the dispersion component and γ^m is the so-called metallic force component. Assuming that the dispersion component γ^d of the surface tension of one substance would interact with only the dispersion component of the surface tension of other substances, the geometric mean of the dispersion components was used as an interaction term:

$$\gamma_{SL} = \gamma_S + \gamma_L - 2(\gamma_S^d \gamma_L^d)^{1/2} \tag{4}$$

By assuming $\gamma = \gamma^d$ for hydrocarbons and that water and mercury each interact with hydrocarbons only via the γ^d component, Fowkes calculated a γ^d for water and for mercury by utilizing the known interfacial tensions of each against one or more hydrocarbons. Utilizing these values of γ^d for water and mercury, Fowkes then calculated an interfacial tension for the mercury–water interface on the basis of the speculative assumption that water and mercury interact with each other in the same manner that each interacts with hydrocarbons. The interfacial tension thus calculated was quite close to the experimentally measured interfacial tension of water and mercury, thus yielding the surprising conclusion that water and mercury interact by means of dispersion forces only.

2. METALS

2.1. Theory of Interaction at the Water–Metal Interface

The implication that interaction at the interface of water and mercury involves only dispersion forces led to extrapolation of this principle to metal surfaces in general.[6] Previous experience with metal surfaces had shown that, in the absence of a contaminating organic layer, all metals are hydrophilic; i.e., water will spontaneously spread on their surfaces with a zero contact angle.[3] However, the "real" metal surfaces heretofore investigated contain combined oxygen (ranging in nature from a monolayer of chemisorbed oxygen to a thick layer of surface oxide) as a result of their exposure to the atmosphere. These surface metal oxides may be capable of strong hydrogen bonding interaction. Fowkes hypothesized that an atomically clean oxygen-free metal surface without this hydrogen bonding capability would interact by means of dispersion forces only, which he expected would be inadequate to yield a zero contact angle.

2.2. Gold

2.2.1. Dispersion Force Hypothesis

The surface of gold seemed to provide a convenient test for this hypothesis. First, it is uniquely inert to oxygen and does not form a stable oxide phase. Second, it was demonstrated[6] that data in the literature could be incorporated into the γ^d system to show that dispersion forces alone would not cause water to spread on gold. Hamaker coefficients[7] for gold had been calculated by Reerink and Overbeek[8] from data on the stability of gold sols in electrolyte solution obtained by Westgren[9] and by Tuorila.[10] The Hamaker coefficient (A) is a proportionality constant for the force exerted by a material based on pairwise summation of the dispersion force of atomic-size surface segments on segments of another surface. Showing that a simple relationship exists between A and γ^d, Fowkes used Reerink and Overbeek's results to calculate a minimum contact angle for water on gold of 60° from the equation

$$\cos\theta = -1 + 0.13(\gamma_S^d)^{1/2} \tag{5}$$

which is obtained by substituting the expression for γ_{SL} in (1) into (2), the value 71.8 mJ/m^2 for γ_L, the surface tension of the water, and 21.8 mJ/m^2 for γ_L^d, the dispersion component of the surface tension of water. γ_S^d is the dispersion component of the surface free energy of gold, which is calculated from the Hamaker constants. This supported the expectation that dispersion forces alone, as calculated for water and gold by the γ^d system, would not be adequate to cause spreading (0° contact angle) of water on gold.

2.2.2. Contact Angles of Water on Gold in Controlled Atmospheres

Previous experience had indicated[11] that in the absence of a contaminating organic layer, gold, like all other metal surfaces, is hydrophilic. In 1964, however, White[12] obtained dropwise condensation of water on gold from air (at atmospheric pressure) saturated with water vapor under conditions of cleanliness during surface preparation and wettability measurement, which were adequate to render base metals hydrophilic, i.e., able to spread water. He ultimately reported the contact angle as $60 \pm 5°$.[13] In 1965, Erb[14] reported continuous dropwise condensation of water vapor in a gold-plated still at atmospheric pressure. Fused quartz and non-noble metals were observed to yield filmwise condensation under these same conditions. Shortly thereafter, Bewig and Zisman[15] reported a water contact angle of 0° on a gold disk that had previously been heated to near melting in a flowing atmosphere of hydrogen (at amospheric pressure) purified of organic contaminants. When the precautions against contamination of the flowing hydrogen were relaxed, the contact angle was no longer zero. The authors concluded that nonzero contact angles reported for water on gold were a result of hydrophobic organic contamination. White,[13] on the other hand, claimed that the zero contact angles observed by Bewig and Zisman resulted from hydrophilic contamination introduced during heating of the sample and by the embedding of alumina abrasive during previous sample polishing. In 1970, Bernett and Zisman[16] reported the results of measurements of the contact angle of water on gold made under conditions designed to eliminate the possibilities of hydrophilic contamination which had been suggested by White and Erb. To avoid surface segregation of metallic impurities, the surface was not heated during preparation. The surface was polished with magnesia followed by leaching with hydrochloric acid, to remove any residual embedded abrasive. A contact angle of 0° was found for water on this gold surface.

2.3. Ultrahigh Vacuum Method of Contact Angle Measurement: Water on Gold

In the late 1960s, a technique was developed for *in situ* measurement of contact angles after surface preparation in an ultrahigh vacuum.[17] It involved admitting a suitably purified vapor to the system (after, of course, valving the system off from the vacuum pumps) to room temperature saturation vapor pressure, condensing the vapor in the sample chamber by means of a cold finger, and depositing a drop on the sample surface through magnetic manipulation. The cold liquid in the finger was removed either immediately after or before deposition of the drop on the sample surface, to avoid maintaining an undesirable low-vapor-pressure region around the finger. The contact angle was then read by means of a goniometer eyepiece mounted on a telescope. In 1970, results were reported on the use of this technique to determine the contact angle of water on gold.[18] Gold surfaces were prepared in the ultrahigh vacuum apparatus by heating polished gold disks to successively higher temperatures in the presence of oxygen followed by vacuum or by evaporating gold films on smooth surfaces *in situ*. The gold disk method was repeated in a conventional vacuum system for comparison. A striking feature of the results was the hysteresis of the water contact angle, which was observed during the various stages of surface activation of the gold disk. When the temperature and time of heating of the gold disk in air were increased, followed by evacuation, the receding angle decreased to zero while the advancing angle decreased to about 20–30°. With more activation (increased heating), the advancing angle decreased further, while the receding angle remained zero. In the conventional vacuum system

TABLE 1. Contact Angles on Gold Disk in an Ultrahigh Vacuum System

Surface activation	Drop no.	Advancing angle (deg)	Drop history after initial reading: time lapse from previous reading or mechanical agitation	Receding angle (deg)
Vacuum, 560°C, 2 h	1	31	—	0
		16	Mechanical agitation	
Air, 1 torr, 550°C, 1.5 h	1	6		
	2	5		
Air, 710°C, 2 h followed by vacuum, 710°C, 1 h	1	2		
		0	1 min	
	2	3		
		0	5 min	
Air, 715°C, 2.5 h followed by vacuum, 715°C, 2.5 h	1	5		
		0	5 min	
	2	6.5		
		2	1 min	
		1	5 min	
		0	30 min	

Reprinted with permission from *J. Phys. Chem.* **74**, 2313 (1970). Copyright 1970 American Chemical Society.

a limit was reached in the decrease of the advancing angle. Further activation either raised the angle or ceased to lower it. In the ultrahigh vacuum system, on the other hand, increased activation decreased the advancing angle to zero degrees (Table 1).

For gold films evaporated in situ, the contact angle of water on a gold film[18] evaporated onto the surface of a polished fused silica disk was zero. The system was evacuated again, and another layer of gold deposited on top of the original. The contact angle was zero once more (Table 2).

In another experiment the gold was deposited on a polished graphite surface.[18] A zero contact angle was again observed a few minutes after deposition of the water drop (Table 2). The contact angle of water on the graphite disk surface which was shielded from the gold vapor flux was approximately 22°.

TABLE 2. Contact Angles on Gold Film Evaporated in an Ultrahigh Vacuum System

Run no.	Substrate	Film no.	Drop	Advancing angle (deg)	Time after deposition of drop (min)
1	Polished silica disk	1	First	Spread with low angle	
			Subsequent	0	1
		2	First	10	1
				0	6
2	Polished silica disk	1	First	Spreading at low angle	1
				0	2
			Subsequent	0	1
3	Polished graphite disk	1	First	Spread at 5	
			Second	8	1
				0	5

Reprinted with permission from *J. Phys. Chem.* **74**, 2313 (1970). Copyright 1970 American Chemical Society.

The hysteresis observed on the surfaces prepared by heating a thick gold disk in conventional or ultrahigh vacuum can be explained by the assumed presence of contamination on the gold surface. The behavior of this hysteresis with heating in the two vacuum systems implies gradual removal of organic hydrophobic contamination from a hydrophilic surface. Now, on the basis of the disk-heating experiments alone, there remains the possibility of surface segregation of inorganic (hydrophilic) impurities under the layer of organic (hydrophobic) contamination. However, the experiments in which gold was evaporated in an ultrahigh vacuum and deposited as a thin film on substrates, followed by measurement of a 0° contact angle, are completely conclusive.[18] There is no possibility of surface segregation of impurities in the deposited film, since it is not heated and, furthermore, is too thin to contain enough impurity to contaminate a surface.[19] Furthermore, the film was deposited on graphite, a relatively hydrophobic substrate, as well as on silica, and when silica was used multilayers of gold were deposited. These latter precautions eliminate the possibility that the substrate has influenced the results. For the case of any metal other than gold, an uncontaminated real surface would contain chemically combined oxygen, either as a built-up surface oxide or as a chemisorbed monolayer. The attainment of a surface which is clean (oxygen free) as well as uncontaminated with organic material would then entail a procedure such as ion bombardment to remove any chemically combined oxygen already present, followed by maintenance of a suitable UHV to prevent oxygen from recombining with the surface before the measurement is completed. Of course, for the case of a film evaporated *in situ*, the ion bombardment is not necessary. In the present work, where water vapor is introduced for the measurement, extensive degassing of the water source is not necessarily sufficient to avoid introduction of a sufficient number of oxygen molecules to chemisorb to a few square centimeters of an active metal surface. For metals in general, therefore, a special gettering technique would have to be devised to ensure measurement on an oxygen-free surface. The surface of gold, however, is unique among all metals in its relative lack of affinity for oxygen. It is the only metal that does not form a bulk oxide that is thermodynamically stable at room temperature. While this does not preclude the existence of a chemisorbed monolayer on the gold surface, the literature[20–22] indicates that such a monolayer will not form at room temperature, even at high oxygen pressure and much longer periods of time than the duration of these wettability experiments on evaporated gold films.

2.4. Ultrahigh Vacuum Contact Angle Measurements on Copper and Silver

Investigation of the contact angle of water on clean metal surfaces was subsequently extended to copper and silver, active metals that form surface oxides, or chemisorbed oxygen monolayers, when exposed to oxygen.[23] In particular, the possibility was considered that clean metals do indeed interact physically with water due to dispersion forces only, and that gold spreads water due to an especially high γ^d resulting from its high molecular weight. The main experimental problem in measuring the wettability of clean copper and silver consisted of admitting water vapor without any accompanying trace of oxygen. The extreme precautions taken to accomplish this consisted of the following: (1) Degassing the liquid water into the sorption pump by momentarily opening a valve and then closing it as the top water layer started to freeze from evaporation. This operation was performed 60 times. (2) The water was exposed to a chamber evacuated to the ultrahigh vacuum region by means of the ion pump. This operation was performed seven times. The water reservoir remaining after treatment according to this and the previous step served as the original source of water vapor throughout all the experiments with copper and silver. For some of the experiments with copper, the following step was added. (3) Prior to admittance to the sample chamber for contact angle measurement, water vapor from the degassed liquid was adsorbed to clean (oxygen-free) germanium powder in an intermediate chamber and allowed to equilibrate for at least 1 h. The germanium had been previously cleaned by heating *in vacuo* at 700°C,[24] and was regenerated after each run. Clean germanium rapidly chemisorbs a monolayer of oxygen. The efficiency of the germanium powder was monitored in a separate experiment by deliberate adsorption of oxygen (after one of the 700°C cleanings), which was measured by pressure difference with a thermistor pressure sensor. The amount of oxygen chemisorbed was approximately equal to that which would have been dissolved in the entire water reservoir if it were in equilibrium with the atmosphere. Since the reservoir was actually thoroughly degassed, and since only a small portion of it vaporizes into the vacuum chamber for contact angle measurement, the capacity of the germanium powder far exceeded that necessary to completely free the water vapor of any possible oxygen residue.

The contact angle of water, purified in this manner, on copper film deposited in UHV was found to be 0°. The water contact angle for silver deposited by evaporation in UHV was also found to be 0°. In the silver experiments, however, the Ge purification step was omitted after analysis with a quadrupole residual gas analyzer did not detect any molecular oxygen in the degassed water. The residual gas analysis also indicated that the hydrogen and carbon monoxide were released from the evaporated silver and copper surfaces after admission of water vapor. As is the case for gold, it is seen that measurements of the contact angle of water on evaporated films of copper and silver in UHV fail to yield any evidence that clean, oxygen-free, metallic surfaces are hydrophobic.

2.5. Surface Segregation

Investigation in the late 1970s on the chemisorption of oxygen to gold highlighted the importance of avoiding the possibility of surface segregation of hydrophilic impurities. The surface of 99.999+% pure gold was ion-bombarded and annealed in UHV, then exposed to molecular oxygen at various temperatures, and oxygen chemisorption measured by Auger electron spectroscopic analysis (AES). It was found that calcium segregated to the surface during annealing in the presence of oxygen,[25] and even in its absence.[26] Furthermore, calcium that segregated to the surface during annealing even in the absence of oxygen was found to promote subsequent chemisorption at low pressures of water vapor.[26] It is clear that the use of evaporated (in UHV) gold films to avoid surface segregation in the wettability experiments[18] was a prudent precaution. Note that the calculated purity of gold films evaporated in UHV receives confirmation from observation of the Auger spectrum of evaporated gold.[27] It can be seen that no peak whatsoever is detectable at 291 eV, the characteristic energy of the calcium peak.

2.6. Surface Structure

Another possible concern in evaluating the significance of the 0° contact angle of water on gold is the question of surface structure. Macroscopically disorganized surfaces are termed "rough" surfaces, and these follow Wenzel's law, where the roughness of a surface as compared to a geometrically smooth surface is equal to the ratio of the cosine of the contact angle of the rough to the smooth one. In general, a highly polished mirrorlike surface is considered to be perfectly smooth. The state of organization of a surface on an atomic scale, however, cannot necessarily be described in terms of roughness and may often be more advantageously viewed as a difference in constitution, resulting from the effect of surface structure on the bonding capability of individual atoms. For example, in principle, measurement of the contact angle of a liquid on different crystal faces of a metal could yield a different result on each face. It is more likely, however, that a difference, if it exists, would be detected when the contact angle on a crystal face is compared with that on an ion-bombarded surface or evaporated film. The results obtained with experiments on graphite,[28] where ion bombardment lowered the contact angle considerably while destroying the low-energy electron diffraction (LEED) patterns, are illustrative of this effect. To be sure, graphite may be considered an extreme case of anisotropy, where a high degree of covalent bonding in the basal plane is in contrast to the dangling bonds available on a disordered surface. Nevertheless, the fact that an effect does exist should serve, at least, to alert us to this possibility on metallic surfaces. If they do follow graphite behavior, then amorphous metallic surfaces such as produced by ion bombardment or film deposition will yield lower contact angles than crystallographic surfaces. The zero contact angles observed on metal surfaces, whose preparation included precautions taken against surface segregation of hydrophilic impurities, were either on films deposited by evaporation[18, 23, 27] or on etched polycrystalline surfaces,[16] all of which were mirror smooth but amorphous. The free energy of the liquid–solid interfacial interaction may be taken to be equal to the sum of the free-energy change due to interaction of the liquid with an equilibrium surface, that is, a crystallographic face, and that due to the disorganization on the surface. It is possible, therefore, that although water has a contact angle of 0° on, for example, an amorphous gold surface, that the contact angle would be greater than zero on a crystal face or even a well-annealed polycrystalline surface. Even if that were the case, however, it is unlikely that the equilibrium surface contact angle would be much greater than zero or, more precisely, that its cosine would be much less than unity. Rather, the surface would still be essentially hydrophilic, with $\cos\theta$ slightly less than 1.

2.7. Theoretical Interpretation of Experimental Results: Impact of New Hamaker Coefficient Calculations

It would seem, therefore, that the experimental results on the wettability of clean metal surfaces disprove the assumption that the interaction of water with solid metal surfaces occurs through dispersion forces only, which can be treated with the γ^d approach of separation of forces. We must remember, however, that this conclusion is based on Fowkes's use of Hamaker coefficients calculated by Derjaguin–Landau–Verwey–Overbeek (DLVO) theory. More recently, developments in the theoretical treatment and calculation of van der Waals (vdW) forces using macroscopic continuum models[(29–32)] have enabled the calculation of Hamaker coefficients[(33–36)] from dielectric spectroscopic data alone. The new Hamaker coefficients were generally one or two orders of magnitude higher than those calculated by the DLVO method, which relies on coagulation rate data. Agreement among the

TABLE 3. Hamaker Coefficients for Gold, Silver, and Copper Calculated by the DLVO Method and by the Lifshitz Method[a]

Metal	Reference	A (erg × 10^{-12})	Medium[b]	γ^d (erg cm^{-2})	θ (deg)
a. DLVO Method					
Au	8	0.1	aq	53	93
Au	8	0.05	aq	43	98
Au	8	0.6	aq	123	64
Au	40	0.087	aq	51	93
b. Lifshitz (macroscopic) method					
Au	38	4.5	vac	313	0
Au	38	3.3	aq	392	0
Au	36	1.9	vac	132	60
Au	37	2.4	aq	309	0
Au	39	2	vac	139	58
Au	39	0.9	aq	158	51
Au	39	4	vac	278	0
Au	39	3	aq	365	0
Au	39	2	vac	139	58
Au	39	1	aq	169	46
Ag	38	4.0	vac	278	0
Ag	36	1.9	vac	132	60
Ag	39	2	vac	139	58
Ag	39	1	aq	169	46
Ag	39	5	vac	347	0
Ag	39	4	aq	455	0
Cu	38	2.8	vac	194	36
Cu	36	1.8	aq	251	0
Cu	36	2.3	vac	160	50
Cu	39	4	vac	278	0
Cu	39	3	aq	365	0
c. Derjaguin (crossed wires) method					
Au	41	4.1	aq	464	0

[a] Also included is a value obtained from the method of crossed polarized wires.
[b] Aq refers to the Hamaker coefficient in an aqueous medium and vac to that in vacuum.

macroscopic Hamaker coefficients calculated from different data sources should improve as more complete and more accurate spectral data become available. In Table 3, DLVO calculations of Hamaker coefficients of gold are compared with macroscopic theory calculations for gold and for copper and silver. Also included are γ^d values derived from the Hamaker coefficients (A) and contact angles θ for water calculated from the γ^d values, using the equations proposed by Fowkes.[6] It can be seen that when the DLVO Hamaker coefficients are used, it is predicted that gold would be quite hydrophobic if its interaction with water were due to dispersion forces alone, the contact angles being in the range of 60° to 90°. On the other hand, when the macroscopic Hamaker coefficients are used, it is predicted that dispersion forces alone may be adequate to render gold, and copper and silver as well, hydrophilic, with about half the predicted contact angles being 0°. Furthermore, refinements in the measurement of spectral data and the availability of more data will most likely yield higher values for the macroscopic Hamaker coefficients, thus shifting more predictions to 0°. Note, however, that a prediction of 0° from dispersion force interaction cannot be regarded as proof that dispersion forces alone operate at the water–clean-metal interface, since one would continue to obtain a 0° contact angle if forces other than dispersion were also operative, for all possible values of these other forces. However, in Tables 1 and 2 a minute (or more) is sometimes required for the drop to advance from a near-zero contact angle to the final value of zero. One among a number of possible explanations for this effect (not usually found on organic-contamination-free oxide-covered metals) is that gold, possibly interacting with water through dispersion forces only, provides barely enough interaction energy to spread the water.

2.8. Conclusions on Metal Wettability by Water

1. Clean solid metal surfaces are hydrophilic.
2. When the metal surfaces are prepared in the amorphous state, such as thin films deposited on substrates under UHV, they yield a 0° contact angle with water.
3. While measurements have not as yet been made on defect-free single-crystal surfaces, it is expected that they will yield either 0° contact angles, i.e., $\cos\theta = 1$, or $\cos\theta$ close to 1.
4. Recent values for the Hamaker coefficients of gold, silver, and copper calculated from Lifshitz theory (in contrast with previous values calculated by the DLVO method) show that these experimental results do not necessarily contradict theoretical speculation that physical interaction at the metal–water interface consists essentially of dispersion forces.

3. GLASSES

3.1. Relative Humidity Experiments at Atmospheric Pressure

Another category of high-energy surface susceptible to molecular contamination is that consisting of the siliceous glasses. Water vapor is the molecular ambient of importance in the study of these surfaces. Shafrin and Zisman[39] undertook an investigation of the wetting of a glass surface by the non-hydrogen-bonding liquid methylene iodide (CH_2I_2) at various relative humidities. They found a contact angle for methylene iodide on soda–lime glass of 36° at 95% RH (relative humidity). This value is nearly the same as the value of 37° found on a duplex water film. On decreasing the RH the contact angle drops until it reaches 13°, sometimes at an RH as high as 40%. Further decrease of RH down to 1% yields no further change in the contact angle. Shafrin and Zisman explain that at high RH a multilayer of adsorbed H_2O is formed that is thick enough to possess the properties of bulk water and mask the force field of the glass surface. As this thickness decreases with lower RH, the masking decreases and the contact angle is lowered. Assuming that this 13° methylene iodide contact angle is characteristic of a silanol surface covered by a single layer of adsorbed molecular water, the properties of the silanol surface after this monolayer of water is removed remain to be determined. The silanol surface may spread methylene iodide, yield a very low contact angle, or act the same as a surface covered by a monolayer of water.

3.2. *Experiments in Conventional Vacuum*

In an attempt to obtain the properties of a silanol surface, measurements in the region of low p/p° of H_2O were made, by Schrader,[17] in a conventional (non-bakable) high-vacuum system. Since there is no possibility of removing traces of water vapor from the ambients in this system, e.g., of amounts equal to partial pressures of roughly 10^{-6} torr, possible chemisorption that increases the number of silanol groups on the sample surface cannot be ruled out. However, it is reasonable to suppose that adequate evacuation of the sample, possibly at room temperature but certainly at elevated temperatures, e.g., 150°C, would remove virtually all the adsorbed molecular H_2O. It can be seen from Table 4 that even after baking at 420°C the residual contact angle for methylene iodide of 8.8° remains. Furthermore, there is no significant change when H_2O is introduced at 1% RH (0.2 torr) or much change even at 3% RH. It seems likely, then, that the residual Shafrin–Zisman angle of 13° occurs on a surface that is essentially silanol, with very little molecular water adsorbed.

3.3. *Ultrahigh Vacuum: Silanol–Siloxane*

Another interesting problem in the very low RH region is the properties of the siloxane surface. Heating of a glass surface in UHV will, after driving off adsorbed

TABLE 4. Contact Angles of Methylene Iodide on Fused Silica Disk in Conventional Vacuum System

Run no.	Vacuum treatment of surface	H_2O introduced	Contact angle (deg)
1	400°C	1%RH	9.0
2	Liquid H_2O wash, room temp. evacuation	None	10.5
3	None additional	3% RH	12.0
4	Liquid H_2O wash, vac. bake at 420°C	None	8.8

Reprinted with permission from *J. Colloid Interface Sci.* **27**, 743 (1968). Copyright 1968 Academic Press.

molecular water, continue to cause emission of water from the surface as a result of condensation of silanol groups to siloxane.[40,41] Evidence clearly indicates that, whereas surface silanol groups are excellent sites for adsorption of H_2O molecules through hydrogen bond formation, the siloxane sites are not. The question of the relative ability of silanol-rich versus siloxane-rich surfaces to spread non-hydrogen-bonding organic liquids, however, remains.

The UHV technique was then used[17] to prepare a silica glass surface not only free of molecular water but, if possible, also depleted in silanol groups. Techniques were devised to introduce CH_2I_2 and measure its contact angle on the silica glass without rehydrating the surface.

3.3.1. Experimental

The vacuum system and experimental method are described in detail in Ref. 17. Pressure was lowered to the 10^{-10} decade by means of an ion pump and bakeout oven with a maximum temperature of 500°C. The methylene iodide was purified in a conventional vacuum system, then transferred to UHV by means of break-seal tubes.

3.3.2. Results

The results of the UHV experiments are listed in Table 5, with the final pressure listed for each run referring to the system pressure after bakeout, sample heating, and final pumpdown before introduction of methylene iodide.

TABLE 5. Contact Angles of CH_2I_2 on Fused Silica Disk in Ultrahigh Vacuum System

Final pressure (torr)	Surface treatment	Contact angle of CH_2I_2 (deg)	
1.0×10^{-9}	Vacuum baked, 350°C max.	5	
	In equilibrium with H_2O vapor at saturation	45	
3.2×10^{-10}	Heated to 350°C max.	Time exposed to CH_2I_2 vapor (min)	Cont. ang. of CH_2I_2
		15	4
		25	8
		40	13
		45	12
		1000	12.5
6.8×10^{-10}	Heated to 500°C max.	Time exp. to H_2O vapor (min)	
		No H_2O	[0]
	H_2O vapor slowly admitted to surface	0.25	10
		0.5–80	20
		165–215	38
7.2×10^{-10}	Heat, 340°C max., in presence of baked silica gel getter, total area $>10^6$ m^2	0, Rapid spreading	

3.4. Discussion

3.4.1. Nature of Surface in UHV

It was pointed out that even the conventional vacuum system will have removed all molecular water from the sample surface. It is not possible, therefore, for any molecular water to be present on the surfaces prepared in UHV (except, of course, when water vapor is subsequently deliberately introduced). Methylene iodide contact angles lower than approximately 10° must therefore result from depletion of silanol sites and their assumed replacement by siloxane sites. Young[40, 41] reports data on removal of silanol groups through condensation when heating silica gel in vacuum. Removal of the first half of the silanol surface is complete at 400°C, while much higher temperatures are required to appreciably deplete the second half. This is due to the fact that only adjacent silanols can condense to a siloxane group plus molecular water. For the case of a single (not powdered) low-area surface in UHV such as our sample, the required temperature should be lower. It is therefore assumed that:

a. In the first three UHV experiments, especially when the final pressure is in the 10^{-10} torr decade, the surface is devoid of at least half the silanol groups before the methylene iodide is introduced.
b. For these first three runs, the methylene iodide vapor may possibly contain enough H_2O impurity to replenish the silanol surface through chemisorption, but its pressure will not be adequate to cause adsorption of molecular water.
c. In the fourth UHV run, where a few grams of silica gel were used in the system, the sample surface is in equilibrium with the gel during bakeout. In this case, the data for removal of silanol groups from silica gel during a vacuum bake would be applicable. Although the system was not heated to 400°C, the total time of vacuum bake in the range 250–350°C was quite long, and the surface may have approached 50% depletion of the silanols. At any rate depletion was substantial, as further indicated by a large drop in pressure as bakeout proceeded. An estimate of 30–50% depletion would seem reasonable.

3.4.2. Nature of Rehydrated Surface

In the conventional high-vacuum experiments, it is very unlikely that any measurements were made on a depleted silanol surface. On the other hand, the presence of adsorbed molecular water is also unlikely unless water vapor is deliberately introduced. The results are therefore interpreted as seen in Table 6. According to this hypothesis, then, the rise in contact angle from 4° to 12° as a function of time (run 2) is due to gradual diffusion of H_2O impurity through the saturated methylene iodide vapor to the sample surface, to complete the silanol monolayer.

Timmons and Zisman[42] have found some evidence of autophobic behavior of methylene iodide on platinum. If this effect takes place on glass, then the 10° angle could result from an oriented methylene iodide layer forming on a silanol (but not

TABLE 6. Interpretation of Contact Angles of CH_2I_2 on Silica in Ultrahigh Vacuum

Contact angle (deg)	Nature of surface
0	70–50% silanol 30–50% siloxane or other
4–5	Mainly silanol, some siloxane
9–10	All silanol
11–20	Formation of adsorbed H_2O monolayer

Reprinted with permission from *J. Colloid Interface Sci.* **27**, 743 (1968). Copyright 1968 Academic Press.

siloxane) surface. In this case, the angle would be only indirectly characteristic of the silanol surface and would be specific to the methylene iodide–glass system. There is, however, no independent evidence of autophobicity of methylene iodide on glass.

3.5. Conclusions: Apparent Nature of Glass Surface Based on Vacuum Experiments Alone

1. It may be concluded from the above results that a silanol surface has a critical surface tension somewhat below 50.7 mJ/m^2, thus yielding a finite contact angle with methylene iodide.
2. When a combined siloxane–silanol surface contains about 40% or more siloxane sites, the average critical surface tension will be above 50.7 mJ/m^2 and methylene iodide will spread.

3.6. Contact Angles on Clay Minerals

3.6.1. Comparison with Glass in Vacuum

Looking at these results in terms of nature and number of sites alone, it might seem that siloxane sites form a relatively energetic surface with respect to dispersion forces that will easily spread methylene iodide. Recent results on the wettability of talclike mineral cleavage surfaces[(43)] imply that the situation is somewhat more complicated than that. The experiments on these cleavage surfaces were performed in the open air, relying on rapid measurements on freshly cleaved surfaces to avoid excessive organic hydrophobic contamination. This technique could not avoid adsorption of a molecular "contaminant" such as water vapor. Rather high advancing contact angles of tetrabromoethane were found, equaling approximately 40° for two of the minerals. One would be immediately tempted to conclude that the mineral surfaces were hydrated, leading to contact angles in that range. However, whereas the deliberately hydrated fused silica in the UHV experiments spreads water, the mineral surfaces appear to be considerably hydrophobic, yielding very high contact angles with water. The mineral siloxane groups consequently behave as true low-energy surfaces.

3.6.2. Glass Surface Structure

The answer to this apparent discrepancy appears to lie in the crystallographic structure of the mineral surfaces as compared with the amorphous fused silica glass. The siloxane groups found on the surface of the siliceous glasses are highly strained, which leads to the ease with which they are hydrolyzed to silanol groups. The siloxane groups in talc are not strained nor easily hydrolyzed. As such they are low in dispersion force as well as hydrogen bonding capability. The strained fused silica surface, when dehydrated, has centers of electron concentration, such as dangling bonds, or lone electron pairs in hybridized orbitals[(43)] that contribute to dispersion-type interaction.

4. GRAPHITE: A MEDIUM-ENERGY SURFACE

4.1. Introduction

4.1.1. Elemental Carbon

Elemental carbon surfaces represent a category that does not seem to fall unambiguously into either the high- or low-energy classification. On the one hand, these surfaces have been regarded as high energy due to their efficacy as adsorbents for gases and vapors. On the other hand, their contact angles are nonzero, which are traditionally associated with poor ability to adsorb the vapor of the liquid; i.e., there is relatively little decrease in free energy on equilibrating the clean solid surface with vapor.

4.1.2. Graphite

Any attempt to systematically categorize the physicochemical nature of elemental carbon surfaces must first specify the type of surface under consideration (e.g., graphite, amorphous carbon). The investigation reviewed here focuses on graphite. It will be seen, however, that, even when considering pure graphite alone, it is important to specify whether the surface under investigation is basal plane or contains a large portion of edge sites.

4.2. Experimental Approach: UHV-AES-LEED

As a precaution, graphite must be treated as a high-energy surface, where the possibility of two types of contamination must be considered: (1) chemisorbed molecular contamination, e.g., by oxygen from the atmosphere: (2) contamination by colloidal hydrocarbons from the atmosphere. The former was monitored by a cylindrical mirror Auger analyzer in the vacuum system used for contact angle measurement.[(44)] The latter, on the other hand, could not conveniently be detected by this method, since the contaminant is a hydrocarbon present on a substrate of carbon, and hydrogen is not detectable by Auger. This type of contamination was therefore monitored by the presence or absence of LEED patterns on the graphite basal plane.[(28)]

4.2.1. Nature of Graphite Sample

The experiments were performed on a disk of grade ZYB Union Carbide "oriented graphite" with a minimum bulk purity of 99.9997%. The disk consisted of graphite crystallites with common orientation, so that the exposed face consisted of 0001 planes. The claimed mosaic spread was $0.8 \pm 0.2°$ for the ZYB grade. Spot checks on spreads with greater or slightly less mosaic spreads yield the same results.

4.2.2. UHV Apparatus: AES, LEED, Contact Angles

The UHV apparatus was a Varian Inc., ion-pumped bell jar with instrumentation for AES and LEED. An XYZ flip-sample manipulator permitted AES and LEED measurements to be followed by *in situ* contact angle measurement all in one vacuum cycle.

4.2.3. Water Vapor: Purification and Equilibration with Surface at p°

A dosing tube with distilled water was degassed many times according to the procedure used in previous investigations. As always, the pump was sealed off from the bell jar, and the doser was allowed to reach vapor equilibrium with the volume in the bell jar. We note that, in general, the Young equation assumes that the solid surface on which the drop stands is in equilibrium with the vapor of the drop at its saturation pressure. When measurements are made in the presence of air, interference by adsorption of other components of air to the solid surface is generally a possibility. In the UHV experiments, though, where liquid–vapor is the only ambient, the condition of the Young equation is correctly fulfilled.

4.3. Results

4.3.1. Measurements in Air

When the contact angle of water on the graphite basal plane is measured in air, contact angles from 50° to 70° were observed.

4.3.2. Electron Bombardment in UHV: In Situ Contact Angles

In UHV, contact angle measurement following bakeout of the vacuum system with sample heating to approximately 300°C yields angles below 40°. These angles do not decrease further with increased heating. When the surface was heated by electron bombardment in vacuum to 800°C, the contact angle on the resulting surface was 38°. AES of the surface showed it to be essentially free of oxygen, and sharp LEED patterns, characteristic of the system of crystallites oriented on the sample plane but rotated with considerable randomness with respect to one another, appeared. When the surface was bombarded with argon ions for short periods, the pattern became fuzzy and the contact angle decreased. Increasing ion bombardment caused the pattern to gradually disappear, with simultaneous reduction of the contact angle. The contact angle reached 0° after the pattern disappeared completely.

4.4. *Discussion*

4.4.1. Nature of Surface with 38° Water Contact Angle

AES of the graphite basal plane indicated virtually no impurities on the surface; i.e., only the carbon peak appeared. The Auger technique, however, does not detect hydrogen (or helium), so that the significance of this analysis is that if there are any impurities on the graphite surface, they must be hydrocarbons. There are then two scenarios for lowering of the water contact angle on graphite (0001) by ion bombardment of the surface. In the first, the ion bombardment is removing hydrocarbon contamination, which is hydrophobic, leaving behind a clean, intact graphite (0001) surface that yields a 0° contact angle. In the second, the ion bombardment disorders an already clean graphite (0001) surface, which in the clean, ordered state (arrived at via electron bombardment heating) has a water contact angle of 38°. Gradual disordering lowers the contact angle until finally a clean, amorphous surface appears that yields a 0° water contact angle. Experience indicates that hydrocarbon layers of contamination can be removed, by ion bombardment, much more quickly than was required to yield the 0° angle on the graphite basal plane. Furthermore, lowering of the contact angle was always accompanied by deterioration of the LEED pattern, dependent on structure. Finally, the actual 0° angle was not obtained until after the LEED pattern had completely disappeared. The results, therefore, conclusively show that the 38° water contact angle is characteristic of the basal plane surfaces of a specularly smooth array of oriented graphite crystals.

4.4.2. Estimated Angle on Perfect Surface

Utilization of different grades of oriented graphite (the grades describe degrees of perfection of orientation) did not produce a substantial difference in angle. It is consequently not likely that this result is greatly influenced by interaction with the edges of the crystals. Taking into account a possible minor effect, however, the water contact angle on a hypothetical perfect crystal surface is estimated as $42° \pm 7°$.

4.4.3. γ^d of Graphite (0001) Surface

The method of separation of forces offers a convenient means of comparing water contact angles with other surface energy determinations and calculations on graphite (0001). The comparisons must be confined to determinations that offer a reasonable probability of not being affected by hydrocarbonaceous contamination of the graphite surface. Table 7 lists a series of surface energies that fulfill this criterion. The results of Brennan[45] and of Girifalco and Lad[46] are theoretical, while Crowell's[47] calculations are based on compressibility measurements that do not involve exposed surfaces. Bryant *et al.*[48] measured the force of cleavage in UHV. Fort and collaborators[49, 50] utilized UHV valves, with two cold traps between their pumps and the manifold, for their measurements of krypton and xenon adsorption on graphitized carbon black. It is reasonable to suppose that the interaction of water with perfectly structured graphite (0001) consists mainly of

TABLE 7. Surface Energy of Graphite (0001). Comparison of Values Calculated from UHV Water Contact Angles with Results Obtained by Other Methods

Group	Method	Type of surface energy reported	Result (mJ/m^2)
1. Bryant *et al.*	Experimental: UHV cleavage	Total	875
2. Brennan	Theoretical	Total	514
			35
3. Girifalco & Lad	Theoretical	Total	165
4. Crowell	Experimental: compressibility	Total	167
			176
5. Fort *et al.*	Experimental: krypton and xenon adsorption	Dispersion	151
			157
6. Schrader	Experimental: water contact angle on ZYB	Dispersion	195
	Experimental: estimated water contact angle on ideal surface	Dispersion	185

dispersion forces, with the possible addition of a hydrogen bonding component arising from interaction of water hydrogen with the π electron cloud of the graphite surface. This would be qualitatively similar to, for example, the situation thought to exist at the benzene–water interface. If one neglects any hydrogen bond contribution to the interaction, the dispersion component of the graphite–surface free energy, γ_S^d, can be obtained from the contact angle of water on graphite by rearranging (5) so that

$$(\gamma_S^d)^{1/2} = \frac{\cos\theta + 1}{0.128} \tag{6}$$

where γ_L is the surface tension of water, θ is the contact angle, and γ_L^d is the dispersion component of the surface tension of water, 21.8 mJ/m^2. From our estimate of $42° \pm 7°$ as the contact angle of water on a perfect (0001) surface, γ_S^d is calculated to be 185 ± 17 mJ/m^2. When one takes each of the total energies reported in Table 5 as equal to γ_S^d, it can be seen that our value, and those of groups 3, 4, and 5, form a cluster from 150 to 200 mJ/m^2. The fact that our value is toward the upper portion of this range may be due to neglect of the hydrogen bonding contribution to the water–graphite (0001) interaction.

5. CONCLUSIONS

1. Presently available data indicate that clean high-energy surfaces spread water and all other ordinary liquids (except Hg).
2. The above conclusion is not inconsistent with separation of force theory.
3. The high-energy surface of the siliceous glass framework is due to the strained condition of this type of surface.
4. In addition to high- and low-energy surfaces (with energies of the order of 10^3 and 10^1 mJ/m^2, respectively) there exists the category of medium energy of the order of 10^2 mJ/m^2.
5. The graphite basal plane, a medium-energy surface, yields a water contact angle of approximately 40°.

REFERENCES

1. W. D. Harkins and A. J. Feldman, *J. Amer. Chem. Soc.* **44**, 2665 (1922).
2. H. W. Fox and W. A. Zisman, *J. Colloid Sci.* **5**, 514 (1950).
3. W. A. Zisman, Adv. Chem. Ser. No. 43, p. 1, American Chemical Society, Washington, DC (1963).
4. L. A. Girifalco and R. J. Good, *J. Phys. Chem.* **61**, 904 (1957).
5. F. M. Fowkes, *J. Phys. Chem.* **67**, 2538 (1963).
6. F. M. Fowkes, *Ind. Eng. Chem.* **56** (12), 40 (1964).
7. A. C. Hamaker, *Physica* **4**, 1058 (1937).
8. H. Reerink and J. Th. Overbeek, *Discuss. Faraday Soc.* **18**, 74 (1954).
9. Westgren, *Ark. Chem. Mineral Geol.* **7** (6) (1918).
10. Tuorila, *Kolloidchem. Beih.* **22**, 191 (1926), **27**, 44 (1928).
11. H. W. Fox, E. F. Hare, and W. A. Zisman, *J. Phys. Chem.* **59**, 1097 (1955).
12. M. L. White, *J. Phys. Chem.* **68**, 3083 (1964).
13. M. L. White and J. Drobek, *J. Phys. Chem.* **70**, 3432 (1966).
14. R. A. Erb, *J. Phys. Chem.* **69**, 1306 (1965).
15. W. A. Zisman and K. W. Bewig, *J. Phys. Chem.* **69**, 4238 (1965).
16. M. K. Bernett and W. A. Zisman, *J. Phys. Chem.* **74**, 2309 (1970).
17. M. E. Schrader, *J. Colloid Interface Sci.* **27**, 743 (1968).
18. M. E. Schrader, *J. Phys. Chem.* **74**, 2313 (1970).
19. P. W. Palmberg and T. N. Rhodin, *Phys. Rev.* **161**, 586 (1967).
20. N. V. Kul'kova and L. L. Levchenko, *Kinet. Katal.* **6**, 765, 688 (1965).
21. B. J. Hopkins, C. H. B. Mee, and D. Parker, *Br. J. Appl. Phys.* **15**, 865 (1964).
22. W. M. H. Sachtler, G. J. H. Dorgelo, and A. A. Holscher, *Surface Sci.* **5**, 221 (1966).
23. M. E. Schrader, *J. Phys. Chem.* **78**, 87 (1974).
24. A. J. Rosenberg, P. H. Robinson, and H. C. Gatos, *J. Appl. Phys.* **29**, 771 (1958).
25. M. A. Chesters and G. A. Somorjai, *Surface Sci.* **52**, 21 (1975).
26. M. E. Schrader, in *Colloid and Interface Science* (M. Kerker, ed.), Vol. 3, p. 105, Academic Press, New York (1976).
27. T. Smith, *J. Colloid Interface Sci.* **75**, 51 (1980).
28. M. E. Schrader, *J. Phys. Chem.* **84**, 2774 (1980).
29. E. M. Lifshitz, *Sov. Phys. JETP* (Engl. Transl.) **2**, 73 (1956).
30. I. E. Dzyaloshinskii, E. M. Lifshitz, and L. P. Pitaevskii, *Adv. Phys.* **10**, 165 (1961).
31. V. A. Parsegian and B. W. Ninham, *Nature (London)* **2224**, 1197 (1969).
32. V. A. Parsegian, in *Physical Chemistry: Enriching Topics from Colloid and Surface Sciences* (H. van Olphen and K. J. Mysels, eds.), pp. 26–72, Theorex, La Jolla, CA (1975).
33. T. Matsunaga and Y. Tamai, *Surface Sci.* **57**, 431 (1976).
34. V. A. Parsegian, G. H. Weiss, and M. E. Schrader, *J. Colloid Interface Sci.* **61**, 356 (1977).
35. H. Krupp, W. Schnabel, and G. Walter, *J. Colloid Interface Sci.* **339**, 421 (1972).
36. V. A. Parsegian and G. H. Weiss, *J. Colloid Interface Sci.* **81**, 285 (1981).
37. S. Demirci, B. V. Enustun, and J. Turkevich, *J. Phys. Chem.* **82**, 2710 (1978).
38. B. V. Derjaguin, V. M. Muller, and Ya. I. Rabinovich, *Kolloid Zh.* **31**, 304 (1969).
39. E. G. Shafrin and W. A. Zisman, *J. Am. Ceram. Soc.* **50**, 478 (1967).
40. G. J. Young, *J. Colloid Sci.* **13**, 67 (1958).
41. M. E. Schrader, in *Surface Characteristics of Fibers and Textiles* (M. J. Schick, ed.), pp. 525–562, Marcel Dekker, New York (1977).
42. C. O. Timmons and W. A. Zisman, *J. Phys. Chem.* **68**, 1336 (1964).
43. M. E. Schrader and S. Yariv, *J. Colloid Interface Sci.* **136**, 85 (1990).
44. M. E. Schrader, *J. Phys. Chem.* **79**, 2508 (1975).
45. R. D. Brennan, *J. Chem. Phys.* **20**, 40 (1952).
46. L. A. Girifalco and R. A. Lad, *J. Chem. Phys.* **25**, 593 (1956).
47. A. D. Crowell, *J. Chem. Phys.* **29**, 446 (1958).
48. P. J. Bryant, P. L. Gutschall, and L. H. Taylor, *Wear* **7**, 118 (1964).
49. F. A. Putnam and T. Fort, Jr., *J. Phys. Chem.* **79**, 459 (1975).
50. T. Fort Jr. and V. P. Toan, presented at the 177th National Meeting of the American Chemical Society, April (1979).

4

Determination of the Surface Energy of Solids by the Two-Liquid-Phase Method

Jacques Schultz and Michel Nardin

1. INTRODUCTION

The surface free energy of solids is a characteristic parameter that determines most of the surface properties such as adsorption, wetting, adhesion, etc. The surface energetics of solids may be characterized by measurement of contact angles of different liquids. Nevertheless, the calculation of surface free energy from contact angle measurements has been the subject of much controversy. Indeed, this characteristic of a solid cannot be measured directly because of elastic and viscous restraints of the bulk phase, which necessitate indirect methods.

For *low-energy solids*, such as polymers, many authors[1–6] have estimated the thermodynamic parameters of solid surfaces, in particular surface free energy, from contact angle measurements. The first approach to the characterization of low-energy solid surfaces was an empirical one developed by Zisman and co-workers.[7–11] They established that a linear relationship often existed between the cosine of contact angle of several liquids and their surface tension γ_{LV}. Zisman introduced the concept of critical surface tension, γ_C, which corresponds to the value of the surface energy of an actual or hypothetical liquid that will just spread on the solid surface, giving a zero contact angle. However, there is no general agreement about the meaning of γ_C, and Zisman has always emphasized that γ_C is not the surface free energy of the solid but only an empirical parameter closely related to this quantity.

Some authors have determined the surface free energy of solids by calculating

---

Jacques Schultz and *Michel Nardin* • Centre de Recherches sur la Physico-Chimie des Surfaces Solides, CNRS F-68200 Mulhouse, France, *and* Laboratoire de Recherches sur la Physico-Chimie des Interfaces de l'École Nationale Supérieure de Chimie de Mulhouse, 68093 Mulhouse Cedex, France.

Modern Approaches to Wettability: Theory and Applications, edited by Malcolm E. Schrader and George Loeb. Plenum Press, New York, 1992.

the energy of adhesion from the contact angle values, θ_{SL}, according to the following equations:

$$W_{SL} = \gamma_S + \gamma_L - \gamma_{SL} \tag{1}$$

$$W_{SL} = \gamma_L(1 + \cos\theta_{SL}) + \Pi_e \tag{2}$$

which are based on both Young[12] and Dupré's[13] analyses. In many cases, it may be assumed that the contribution to W_{SL} from the spreading pressure Π_e, particularly for low-energy solids like polymers, is negligible. Therefore, with the help of the equations of Good and Girifalco,[14] Fowkes,[15] or Neumann *et al.*[16], the surface free energy of solids can be estimated.

For *high-surface energy solids* (metal oxides, glass, metals, etc.), on the one hand, the wetting is generally complete and determination of surface energy from simple contact angle measurements becomes impossible. Nevertheless, Murr[17,18] has described a standard test system for high-energy surfaces involving contact angle measurements using mercury as the liquid and determining the drop shape by means of scanning electron microscopy. On the other hand, high-energy surfaces present the proneness to surface contamination and contact angles have to be measured in a clean environment, i.e., under vacuum. For these reasons, a method consisting of the measurement of contact angle of drops of a liquid, deposited on high-surface energy solids previously immersed in another liquid, was developed. Of course, two immiscible liquids have to be used, and, therefore, systems involving water in hydrocarbon media are usually chosen. This method is generally called *two-liquid-phase contact angle measurement.*

This paper is a review of the work performed at our research center in Mulhouse, concerning the development of this method.

In the first section the principle of the two-liquid phase method will be presented. A special attention will be devoted to the determination of the surface characteristics of mica, used as a model of the solid phase, because by cleavage in a liquid, clean surfaces of large dimensions can be obtained.

Second, the application of the two-liquid phase contact angle measurement to other high-surface energy solids, like metals, glass, carbon, etc., will be examined. Experimental developments of this method in order to characterize the surface properties of fibers (glass, carbon) by tensiometric measurements using two immiscible liquids will also be described. Such a characterization of fiber surfaces is of great importance for the understanding of the mechanical behavior of composite materials.

It is also possible to apply the two-liquid-phase method to low-energy solids and, for instance, to polymeric materials as will be seen in a third section. In particular, orientation phenomena at polymer–liquid interfaces can be directly analyzed. The ability of polymeric surfaces to modify their structure in contact with liquid media, especially water, may be studied *in situ* by contact angle measurements using water as the bulk liquid phase. Finally, an explanation of the hysteresis phenomenon frequently observed for contact angle measurements on polymers will be presented.

2. PRINCIPLE OF THE TWO-LIQUID-PHASE METHOD

Earlier work was performed by Peper and Berch,[19] who measured contact angle in water of organic liquids against polymeric materials, i.e., poly(tetrafluoroethylene) and polyethylene. A few years later, Tamai and co-workers[20] developed the theoretical background of the method. They determined values of the surface energy of different metals (iron, copper, etc.) and some polymers (polyethylene, polyvinylchloride). More recently, Hamilton[21, 22] used this technique on octane–water (bulk liquid phase)–polymer interfaces and determined values of critical surface tension, γ_C, for these low-energy surfaces (polyethylene, polystyrene, polypropylene, etc.). Hamilton's results are in good agreement with those obtained by the classical contact angle method.

Finally, an important contribution for determining surface properties of high-energy solids by means of the two-liquid-phase method was done by ourselves.[23–25] We improved the theoretical approach and performed experiments on a model surface, i.e., sheets of muscovite mica directly cleaved in liquids.

2.1. Principle of the Method

Figure 1 shows a schematic representation of the two-liquid-phase contact angle system. Assuming that Young's equation can be applied to a liquid L_1–liquid L_2 (bulk phase)–solid S system, the following relationship can be obtained:

$$\gamma_{SL_2} = \gamma_{L_1L_2} \cos\theta_{SL_1} + \gamma_{SL_1} \tag{3}$$

where γ_{SL_2}, $\gamma_{L_1L_2}$, and γ_{SL_1} are, respectively, the interfacial free energies of $S-L_2$, L_1-L_2, and $S-L_1$ interfaces, and θ_{SL_1} is the contact angle of a drop of liquid L_1 on solid S. It is seen that γ_{SV}, which could be considered as the driving force of spreading in the one-liquid phase method, is replaced by γ_{SL_2}, which is generally smaller than γ_{SV}. This leads to a finite and thus measurable value of contact angle. According to Fowkes,[15] γ_{SL_1} and γ_{SL_2} are given by

$$\gamma_{SL_1} = \gamma_S + \gamma_{L_1} - 2(\gamma_S^D \gamma_{L_1}^D)^{1/2} - I_{SL_1}^P \tag{4}$$

$$\gamma_{SL_2} = \gamma_S + \gamma_{L_2} - 2(\gamma_S^D \gamma_{L_2}^D)^{1/2} - I_{SL_2}^P \tag{5}$$

where γ and γ^D are the surface energy and its dispersive component, respectively, and I_{SL}^P is a specific (or nondispersive) interaction term that includes all the interactions established between the solid S and the liquid L (dipole–dipole, dipole-induced dipole, hydrogen bonds, Π bonds,...) except London dispersion interactions.

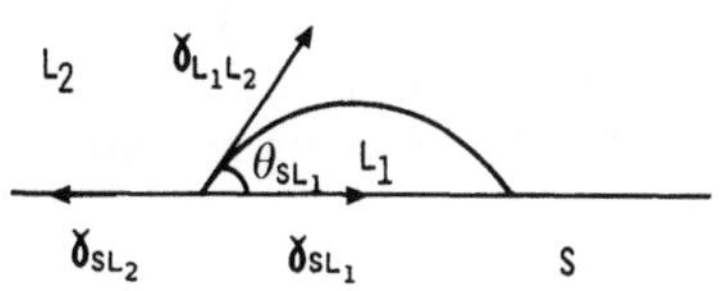

FIGURE 1. Schematic representation of a two-liquid-phase contact angle system (S: solid; L_1: liquid 1; L_2: bulk liquid phase).

Substituting (4) and (5) into (3) leads to

$$\gamma_{L_1} - \gamma_{L_2} + \gamma_{L_1L_2} \cos\theta_{SL_1} = 2(\gamma_S^D)^{1/2} [(\gamma_{L_1}^D)^{1/2} - (\gamma_{L_2}^D)^{1/2}] + I_{SL_1}^P - I_{SL_2}^P \tag{6}$$

If liquid L_1 is water (subscript W) and liquid L_2 is an *n*-alkane (subscript H), the term $I_{SL_2}^P$ may be considered equal to zero since the surface free energy of *n*-alkanes consists of only the London dispersion term. Finally, Eq. (6) can be rewritten as

$$\gamma_W - \gamma_H + \gamma_{WH} \cos\theta_{SW} = 2(\gamma_S^D)^{1/2} [(\gamma_W^D)^{1/2} - (\gamma_H)^{1/2}] + I_{SW}^P \tag{7}$$

If the contact angles of water droplets on the solid surface are measured in several *n*-alkane bulk phases, the plot of $\gamma_W - \gamma_H + \gamma_{WH} \cos\theta_{SW}$ versus $(\gamma_W^D)^{1/2} - (\gamma_H)^{1/2}$ should give a straight line with slope $2(\gamma_S^D)^{1/2}$ and intercept at the origin I_{SW}^P. This method therefore allows the determination of the dispersive component of the surface energy of the solid, γ_S^D, as well as the magnitude of the nondispersive interactions between water and solid surface.

Nevertheless, this principle is based on the assumption that a droplet of water can displace the alkane layer from the solid surface at contact. It is not certain to what extent this will happen and that there will not remain a thin film of the nonpolar liquid between the solid and water, thus preventing direct contact. Therefore, it is necessary to define a criterion for displacement.

2.2. Wetting Criteria

We performed thermodynamic studies in 1982[26] in order to develop criteria to determine whether displacement takes place. In this study, it was first assumed that the spreading pressure at equilibrium, Π_e, of liquids L_1 and L_2 on the solid is negligible so that the free surface energy, γ_S, of the homogeneous and smooth solid surfaces may be taken into account instead of γ_{SV}, the value modified by adsorption.

Two theoretical approaches were then proposed. The first one deals with the free interfacial energy changes involved in a thought experiment when depositing a drop of liquid L_1 onto a flat surface S in the presence of liquid L_2. The second considers the equivalent free-energy changes when the solid crosses the interface between the two liquids. Finally, it was shown that the conclusions of the two derivations are identical, providing in both cases that the free interfacial energy change is always negative so that $S - L_1$ contact will be achieved and then a thin intermediate layer of liquid L_2 will not exist to prevent this contact. The resulting criterion of displacement of L_2 by L_1 can be written as

$$\gamma_{SL_1} - \gamma_{SL_2} - \gamma_{L_1L_2} < 0 \tag{8}$$

In the reverse sense, the above reasoning concerning the displacement of L_1 by L_2 can be used in the same way, interchanging the subscripts referring to the two liquids. The resulting condition is

$$\gamma_{SL_2} - \gamma_{SL_1} - \gamma_{L_1L_2} < 0 \tag{9}$$

Finally, a general criterion necessary for displacement to be mutually reversible can be obtained, considering first that Eqs. (8) and (9) must be satisfied simultaneously and, second, that all free interfacial energy terms must be nonnegative. This criterion corresponds to

$$|\gamma_{SL_2}-\gamma_{SL_1}|<\gamma_{L_1L_2} \tag{10}$$

Therefore, relations (8) to (10), which are quite general, can now be applied to water–n-alkanes–solid systems. In these systems, the bulk liquid phase is nonpolar. Substituting in Eq. (8), the expression of the three free interfacial energies calculated according to Fowkes's approach[15] [see, for example, (4) and (5)] leads to

$$I^{P}_{SW}>2[(\gamma^{D}_{S})^{1/2}-(\gamma_{H})^{1/2}][(\gamma_{H})^{1/2}-(\gamma^{D}_{W})^{1/2}] \tag{11}$$

This shows that, provided the polar interaction term between the solid and water satisfies relation (11), water is capable of displacing the n-alkane.

2.3. Application to a Model Surface: Mica

Muscovite mica can be considered as a model for the solid phase, since by cleavage, smooth and clean surfaces of large area are obtained.

In our work[22–25] a sheet of mica was directly cleaved in n-alkanes to avoid contamination in air. The new surface in the n-alkane bulk phase was then directly used to perform measurements of the contact angle of water droplets, deposited by means of a microsyringe. Contact angles of at least 10 droplets of water for each n-alkane are measured with a goniometer telescope.

Very reproducible results were then obtained. First, it was shown that the contact angle of water on mica under n-alkanes remains independent of the contact time. This indicates that both diffusion between liquid phases and adsorption of diffused liquid at the area adjacent to the droplet can be neglected. Moreover, the same experiments done with the saturated liquids led to the same results.

By plotting $\gamma_W-\gamma_H+\gamma_{WH}\cos\theta_{SW}$ against $(\gamma^{D}_{S})^{1/2}-(\gamma_{H})^{1/2}$, as shown in Fig. 2,

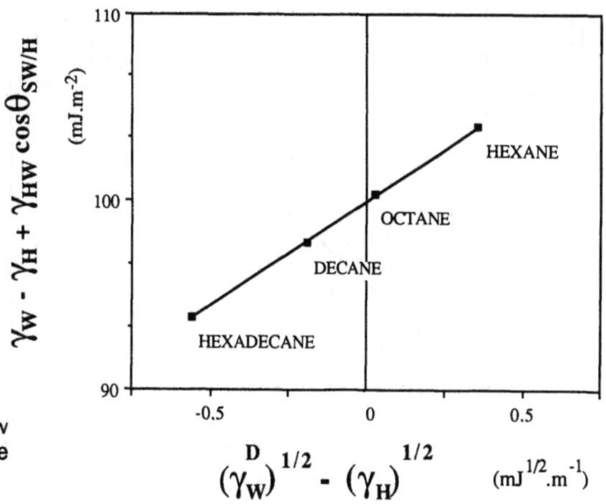

FIGURE 2. Plot of $\gamma_W-\gamma_H+\gamma_{WH}\cos\theta_{SW}$ versus $(\gamma^{D}_{W})^{1/2}-(\gamma_{H})^{1/2}$ for muscovite mica.

a straight line is obtained according to Eq. (7). The slope and the intercept at the origin of this linear relationship leads to the values $\gamma_S^D = 30 \pm 2.5\ \text{mJ/m}^2$ and $I_{SW}^P = 100 \pm 2\ \text{mJ/m}^2$, respectively. The value of γ_S^D is in good agreement with those (21 to 38 mJ/m^2) obtained[24] by measurement of contact angles of high-energy liquids that have to a first approximation no polar interaction with mica, such as mercury, tricresylphosphate, and α-bromonaphthalene. Moreover, this agreement is confirmed by comparing the energy of adhesion at mica–liquid interfaces calculated from surface energies of both liquid and solid, and the interfacial free energy of mica measured by cleavage in various liquid media according to Bailey *et al.*[27,28]

The measurement of the contact angle of water on mica immersed in chloroalkanes, nitroalkanes, aromatics, or alcohols were also performed[25] in order to quantitatively determine the polar part I_{SL}^P of the adhesion energy of these liquids to mica. A linear relationship (Fig. 3) between I_{SL}^P and the square root of the polar component of the surface energy of most of the liquids was established. These results suggest that all nondispersive interactions (dipole–dipole, dipole-induced dipole, Π bonds, hydrogen bonds) can be considered together as a polar interaction. Moreover, this polar interaction may be represented by the geometric mean of the polar component of the surface free energy of liquid and solid according to the following expression, previously proposed by Owens and Wendt[1]:

$$I_{SL}^P = 2(\gamma_S^P \gamma_L^P)^{1/2} \tag{12}$$

Nevertheless, in agreement with Fowkes,[29] there are no theoretical reasons to represent all the nondispersive interactions by this type of expression.

According to Eq. (12), the polar component of the surface energy γ_S^P of muscovite mica was found to be about 90 mJ/m^2. Considering that the surface free energy γ_S is given by[25]

$$\gamma_S = \gamma_S^D + \gamma_S^P \tag{13}$$

one obtains a value of γ_S for mica of about 120 mJ/m^2. This value is exactly the

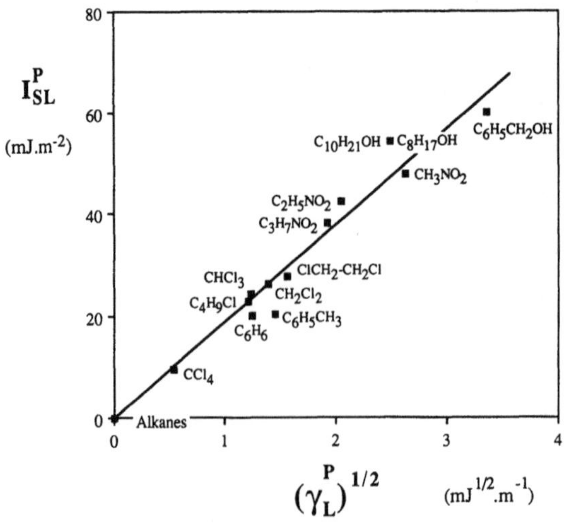

FIGURE 3. Plot of the nondispersive interaction between mica and liquid versus the square root of the polar component of the surface free energy of the liquids.

same as that measured by Bailey *et al.*[28] by cleavage of nonmatching mica sheets. This fact confirms the validity of the two-liquid-phase method applied to an uncontaminated model surface, directly created by cleavage in *n*-alkanes.

3. CHARACTERIZATION OF HIGH-ENERGY SURFACES

Since our early work in 1975,[23–25] the two-liquid-phase method has been applied at our laboratory to numerous high-energy solids in order to characterize their surface properties. Thus, knowing the surface characteristics of these solids, it is possible to estimate their adhesion to other materials, in particular, to polymers. Indeed, the knowledge of the magnitude of the physicochemical interactions, which may be established at solid–polymer interfaces, constitutes the main purpose of such studies. In particular, the estimation of metal, glass, or carbon–polymer adhesion is of great importance. Therefore, the determination by means of the two-liquid-phase method of the surface energy of metal, glass, and carbon in relation to their surface treatment will first be presented in this section.

It is well known that the structure and properties of the fiber–matrix interface play a major role on the mechanical and physical properties of composite materials. The quality of this interface is particularly dependent on any modification affecting the fiber surface, such as fiber surface treatments, fiber sizing, etc., and it is necessary, prior to other investigations, to know precisely the surface characteristics of the fibers, in particular their surface free energy. However, measurements of contact angle by methods that work well with flat surfaces do not give very accurate and reproducible results when applied to fibers. Then, in the last part of this section, the development of the tensiometric method by using two liquid phases for determining the wettability of glass and carbon fibers will be examined. A comparison will also be made between the results obtained by this modified tensiometric method and those obtained, for the same fibers, by a recently developed technique, i.e., the inverse gas–solid chromatography at infinite dilution. In addition to surface energies, this method allows the determination of the electron acceptor and electron donor (acid–base) characters of the solid surfaces, according to Lewis's concept.

3.1. Aluminum

The purpose of our work[30,31] was to study the effect of surface treatments of aluminum on the surface energetics and work of adhesion to acrylonitrile-butadiene (NBR) and styrene-butadiene (SBR) elastomers. The surface free energy of aluminum surfaces, which have been preliminarily submitted to different treatments, is determined by measuring the contact angle between the surface and two immiscible liquid phases: water in *n*-alkanes.

Aluminum is generally covered with a natural oxide layer, contaminated by various compounds difficult to extract. In our study, this contaminated layer, which usually confers to aluminum poor adhesive qualities, is removed or modified by the following surface treatments:

- Solvent degreasing (hexane or dimethylformamide): this treatment eliminates, partly at least, impurities from the oxide layer.

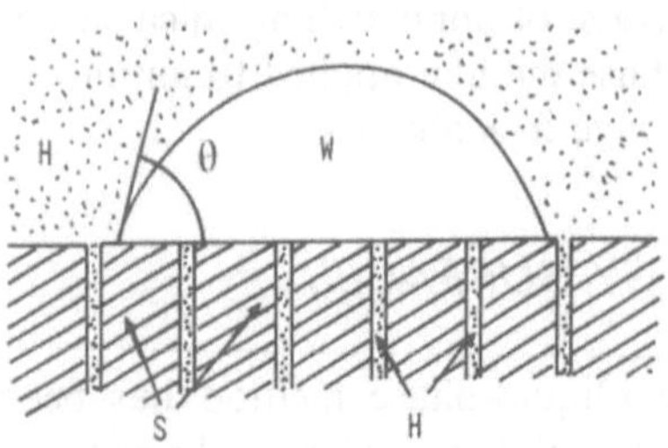

FIGURE 4. Schematic representation of the water (W) contact angle on a porous aluminum oxide layer in a hydrocarbon (H) environment.

- Conversion by amorphous phosphatization: aluminum is treated with a solution containing phosphoric acid, chromic acid, and fluorides.
- Anodization: the surface layer, about 20 μm thick, formed by anodic oxidation in a sulfuric acid medium is a compact system of hexagonal cells, each containing a pore of radius equal to about 10 nm.
- Natural or salt solution sealing of anodized surfaces.

However, these surface treatments can also modify the roughness of aluminum by affecting the number of surface defects. Moreover, particularly in the case of anodized aluminum, the effects of porosity on contact angle cannot be neglected, since pores can be filled by liquids due to capillary phenomena. This is schematically illustrated in Fig. 4. Consequently, the effects of roughness and porosity are taken into account for the determination of the surface free energy of treated aluminum, according to respective Wenzel[32] and Cassie-Baxter[33] analyses concerning contact angle values in such configurations.

Figure 5 presents the straight lines of linear regression corresponding to Eq. (7), whereas Table 1 gathers the calculated values of dispersive γ_S^D and polar γ_S^P

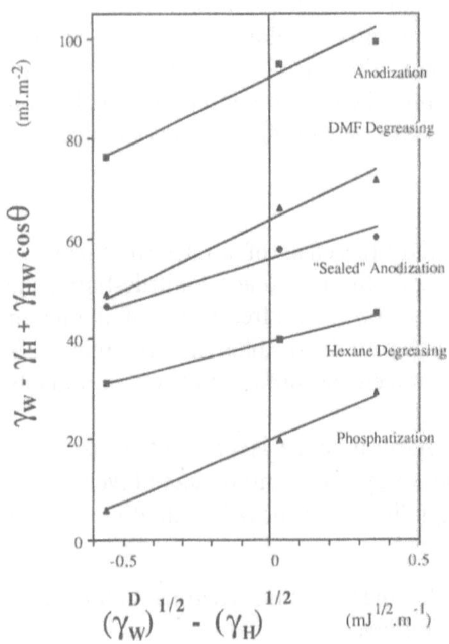

FIGURE 5. Determination of the surface energy of treated aluminum.

TABLE 1. Surface Energies of Surface Treated Aluminum and Elastomers (in mJ m^{-2})

Solid	γ_S^D	γ_S^P	γ_S
Hexane extracted Al	42	7.5	49.5
DMF extracted Al	135	19	154
Phosphated Al	150	1.5	151.5
Anodized Al	125	44	169
Sealed anodized Al	41	15	56
SBR	29.5	0.5	30
NBR	26.5	9.5	36

components of the surface energy for each treated aluminum sample. All these results clearly show to what extent the surface characteristics of aluminum can be modified by surface treatment. The values of the surface energy of SBR and NBR elastomers, determined by classical contact angle measurements, are also given in Table 1. The reversible work of adhesion of phosphated, anodized, and sealed anodized aluminum to NBR and SBR can be calculated and compared to the adhesive strength measured by means of a peel test at different peel rates.(31)

Finally, all the results obtained in this study, on the adhesive strength of model elastomer surface-treated aluminum assemblies, allowed us(31) to propose a new model of the adhesion of viscoelastic materials. According to this model, the energy of separation of an assembly is described by the product of the three following terms: (i) the reversible energy of adhesion or cohesion, (ii) a macroscopic dissipation factor due to viscoelastic losses, (iii) a molecular dissipation factor related to the degree of cross-linking of the elastomer in the vicinity of the interface. This new model constitutes an improvement of the one previously developed in other studies.(34,35)

3.2. Glass

Another example of application of the two-liquid-phase method (n-alkane–water) concerns the determination of the surface characteristics of flat surfaces of float glass.(36,37) These surfaces are preliminarly cleaned by one of the following treatments:

- A slight corrosion in alkaline medium according to Tichane's procedure:(38) glass plates are dipped consecutively in solutions of sodium hydroxide, hydrochloric acid, and then rinsed in water at room temperature. This treatment increases the surface density of silanol groups.
- A solvent cleaning: glass plates are washed with pure hexane in a soxhlet at the boiling temperature of the solvent.

The surface properties of treated glasses are then determined directly or after drying at 350°C under vacuum, in order to remove adsorbed water. Heat treatment leads to an increase of the surface concentration of siloxane groups.

Figure 6 presents the straight lines obtained according to Eq. (7) for float glass surfaces after drying at high temperature. It immediately appears that both the

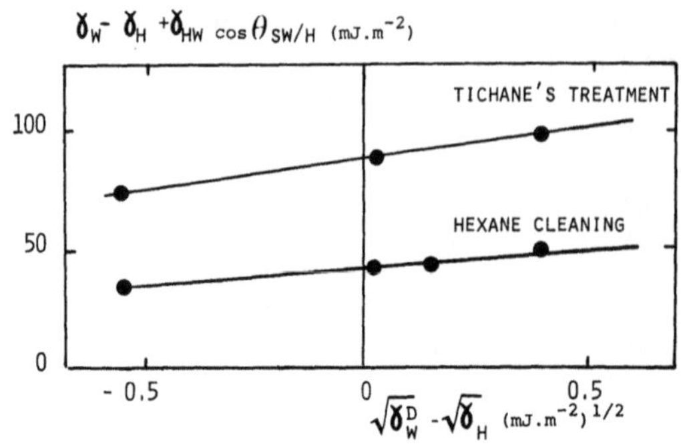

FIGURE 6. Surface characteristics of cleaned float glass plates, after drying at 350°C.

dispersive component γ_S^D of the surface energy and the water–glass polar interaction term I_{SW}^P are strongly affected by the surface preparation. The value of these quantities are gathered in Table 2. The cleaning procedure according to Tichane leads to very high values of γ_S^D and I_{SW}^P compared with those obtained by a simple solvent washing. However, these values are quite different from those measured before heat treatment. Therefore, the influence of the temperature on the surface characteristics of glass has to be determined.

Figure 7 shows the results obtained on a corroded float glass, dried under vacuum at 20°C or at 350°C. The experimental values of γ_S^D and I_{SW}^P, after drying at 20°C, are also given in Table 2. The low value of γ_S^D for the glass sample dried at 20°C can be explained by the presence on the glass surface of adsorbed layers of water.[10] It is clear that the measured value of γ_S^D, in that case, is close to that of pure water, equal to 21.6 mJ m^{-2}.[36] This assumption is confirmed by the value of I_{SW}^P in the following way. If the glass surface can be effectively considered as a pure "water surface," this term must be equal, to a first approximation, to $I_{WW}^P = 2\gamma_W^P$ according to (12). Since $\gamma_W^P \sim 51$ mJ m^{-2}, the numerical value of this term, about 102 mJ m^{-2}, is very close to the measured value of I_{SW}^P (97 mJ m^{-2}). It can be concluded that, after drying at 20°C, water layers adsorbed on the glass surface totally mask the actual surface of the float glass.[10]

The thermal treatment at 350°C removes adsorbed water molecules and leads to a decrease of the surface density of polar chemical groups, by reducing the number of silanol functions. This decrease of the polar character of the glass surface is evidenced by the decrease of the value of the I_{SW}^P term from 97 at 20°C to 80 mJ m^{-2} at 350°C. Finally, the removal of the adsorbed water by heat treatment allows the determination of the actual surface energy of the glass.

Qualitatively, the results obtained[36,37] are in good agreement with the theoretical approach of Dietzel,[39] which allows the calculation of the surface

TABLE 2. Surface Energies of Treated Glasses after Drying at Different Temperatures

T (°C)	Treatment	γ_S^D (mJ m^{-2})	I_{SW}^P (mJ m^{-2})
350	Hexane	36 ± 10	42 ± 1
	Tichane	125 ± 40	80 ± 2
20	Tichane	25 ± 5	97 ± 1

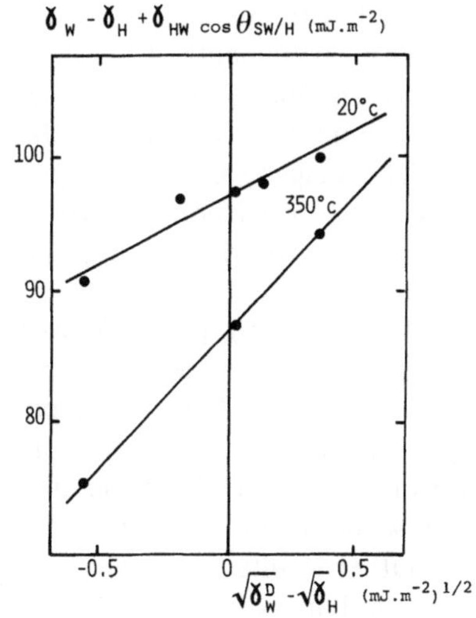

FIGURE 7. Influence of the thermal treatment on the surface characteristics of float glass plates, cleaned according to Tichane.

energy of melted glasses, taking into account their chemical composition. They are also in rather good agreement with Fowkes's approach,[40] in which the dispersive component of the surface energy of a solid is related to the polarizability, the ionization potential, and the number and the radius of the atoms located at the solid surface. For both calculations, it is necessary to know precisely the chemical composition of the treated glass surfaces. This was done, in particular, by Scherrer and Naudin,[41] using secondary ion mass spectrometry.

Developments of the two-liquid-phase method in the case of glass fibers by means of tensiometric measurements will be examined later. The results will then be compared with those concerning glass plates.

3.3. Carbon

In composite materials, it is the fiber–matrix interface that controls the magnitude of the stress transfer and thus strongly influences the mechanical performance of the composite. Therefore, it is of prime importance to determine the surface characteristics of carbon fibers. In a first approach, however, the flat carbon surfaces were examined. The study concerning fibers will be presented in the following section.

Since graphite is a typical anisotropic material, its properties are particularly dependent on the crystalline orientation. This anisotropy must be reflected in the surface properties of such a material. In particular, the surface energy of the basal planes is certainly different from that of the prismatic edge surfaces.

Three types of flat carbon have been studied in our work.[42] These carbons, listed below, are considered as being representative of different carbon structures:

1. A vitreous carbon, with a nonoriented "amorphous" structure, prepared by controlled carbonization of polymer
2. A pyrolytic carbon with a partially oriented structure
3. A highly oriented pyrolytic graphite (HOPG).

Both vitreous and pyrolytic carbons were cleaned before undertaking surface energy measurements, while in the case of HOPG new surfaces were easily formed by cleavage for each measurement. Contact angle determination of water droplets on the solid surfaces under *n*-alkanes were then performed.

The dispersive γ_S^D and polar γ_S^P components of the surface energy of the three types of carbon studied, as well as their polar interaction term I_{SW}^P, were determined according to Eqs. (7) and (12). The experimental values are given in Table 3.

Vitreous and pyrolytic carbons have very similar values of γ_S^D, of about 30 mJ m^{-2}, and a negligible polar component γ_S^P. Both values are identical to those of apolar polymers, like polethylene or polypropylene for example. These two types of carbon can be considered, therefore, as low-surface-energy solids. On the contrary, HOPG clearly exhibits a higher dispersive component of its surface energy and also a polar interaction term that is low but cannot be neglected.

Other results[43] were obtained on two types of carbon (vitreous and pyrolytic) after applying different treatments, such as polishing in the presence of water, heat treatment at 800°C under argon, or plasma treatment under low partial pressure of air. From this study, it appears that large variations exist in the surface energy of prismatic planes for partially or nonoriented structure, depending upon the polishing and cleaning methods used. The highest values of both γ_S^D ($\sim$400 mJ m^{-2}) and γ_S^P (30 mJ m^{-2}) are obtained on plasma-treated samples.

3.4. Fibers

Continuous or chopped fibers of carbon, glass, or polymer, such as aramid, have increasingly important applications in highperformance thermosetting and thermoplastic composites. Good wetting of fibers by the matrix and controlled fiber–matrix interactions is requisite for obtaining adequate characteristics of the fiber–matrix interface and achieving optimal composite mechanical properties.

A method for measuring the contact angle on fibers was described by Carroll.[44] Symetrical drops of liquid are directly applied onto a fiber and analytical expressions relating drop length, drop radius, and fiber radius to the contact angle allow the determination of the surface free energy of fibers. However, it is difficult to place droplets on the thinnest fibers (for instance, carbon fibers commonly have a 7 μm diameter); consequently, only a few contact angle measurements using this technique have been reported in the literature.[45–47]

One of the most appropriate methods for determining the wettability of fibers is the tensiometric method (or Wilhelmy method). If a thin fiber is hanging below the pan of an electromicrobalance, the force acting on the fiber, when it touches the

TABLE 3. Surface Energies of Different Types of Flat Carbon (in mJ m^{-2})

Type of Carbon	γ_S^D	I_{SW}^P	γ_S^P
Vitreous	32 ± 2	1.5 ± 0.1	$\sim$0
Pyrolytic	30 ± 3	1.2 ± 0.2	$\sim$0
Highly oriented pyrolytic (HOPG)	150 ± 40	10 ± 1	$\sim$0.5

surface of a liquid and, further, penetrates this liquid, can be related to the morphological parameters of the fiber, the density of the liquid, the depth of immersion, and the contact angle of the liquid on the solid surface. Nevertheless, most of the fibers used in composite materials are high-surface-energy solids. Therefore, a modified tensiometric method using two immiscible liquid phases was developed and will be described first.

With inverse gas chromatography (IGC), which is a more recent technique, it is also possible to determine the surface energy of fibers and powders. Moreover, this method was also used to determine the acceptor–donor or acid–base characteristics of the surface of such divided solids. In a second section, the results obtained by IGC on glass and carbon fibers will be compared with those previously determined by tensiometric measurements.

3.4.1. Tensiometric Method with Two Liquid Phases

Principle. The two immiscible liquids, which are generally used for determining the surface properties of fibers, are water (or formamide) and *n*-alkanes.[48] Formamide is chosen rather than water because of the experimental difficulty of forcing the fiber across the water–hydrocarbon interface. For two-liquid-phase systems, Eq. (7) can be rewritten as follows, the subscript F referring now to formamide:

$$\gamma_F - \gamma_H + \gamma_{HF} \cos \theta_{SF/H} = 2(\gamma_S^D)^{1/2} [(\gamma_F^D)^{1/2} - (\gamma_H)^{1/2}] \tag{14}$$

The fiber is suspended from the pan of an electrobalance above a beaker containing the two liquid phases. By means of a motorized support, the beaker can be raised or lowered. During immersion and emersion cycles, the change in force, ΔF, due to menisci of liquid raised or lowered at the two interfaces is recorded. When the fiber is dipped in both liquids, at equilibrium, three contributions are responsible for this force difference, ΔF (Fig. 8):[48] the apparent weight, F_{HA}, of liquid raised at the air–hydrocarbon–fiber interface, the apparent weight, F_{HF}, of liquid raised at the hydrocarbon–formamide–fiber interface, and the effect of buoyancy, F_P, acting on the immersion length of the fiber. In the case of small-diameter fibers, the term F_P can be neglected.

The term F_{HA} is given by

$$F_{HA} = C\gamma_H \cos \theta_{SH/A} \tag{15}$$

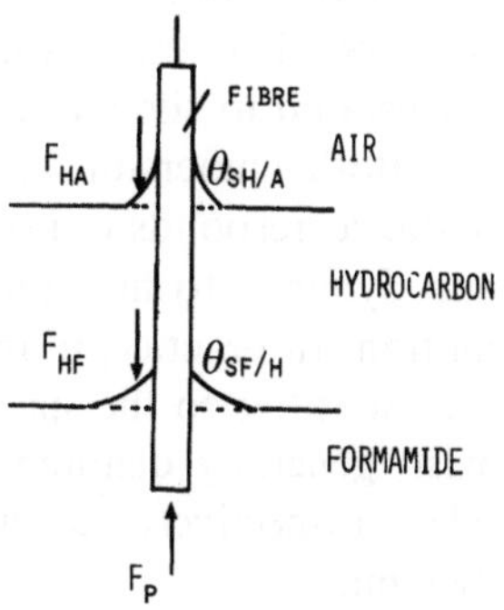

FIGURE 8. Schematic representation of fiber in two-liquid-phase system.

where C is the perimeter of the fiber and $\theta_{SH/A}$ is the contact angle of the hydrocarbon on the fiber in air. As seen above, $\theta_{SH/A}$ is equal to zero, since the hydrocarbon perfectly wets the high-energy surface of the fiber; thus

$$F_{HA} = C\gamma_H \tag{16}$$

As a consequence, knowing γ_H, the circumference of the fiber can be determined by immersion only in the n-alkane phase. The term F_{HF} is given by an expression analogous to Eq. (15):

$$F_{HF} = C\gamma_{HF} \cos \theta_{SF/H} \tag{17}$$

where $\theta_{SF/H}$ is the contact angle of formamide on the fiber in the presence of hydrocarbon.

Writing $\gamma_{HF} \cos \theta_{SF/H} = \tau$, where τ is defined as the tension of the adhesion,[48] Eq. (17) becomes

$$F_{HF} = C\tau \tag{18}$$

Note that τ is the major variable on the left side of (14). Combining (16) and (18) leads to

$$\tau = \frac{F_{HF}}{F_{HA}} \gamma_H \tag{19}$$

Consequently, according to (19), the calculation of $\theta_{SF/H}$ is unnecessary when using the present tensiometric method. Nevertheless, the problem of displacement on the solid surface of one liquid by the other one, in immersion and emersion, respectively, has to be studied very carefully. As described in detail above, the criteria to be fulfilled for the formamide to replace the hydrocarbon in immersion, and reciprocally for the hydrocarbon to displace the formamide in emersion, are respectively given by the following inequalities:

$$I_{SF}^{P} > 2[(\gamma_S^D)^{1/2} - (\gamma_H)^{1/2}][(\gamma_H)^{1/2} - (\gamma_F^D)^{1/2}] \tag{20}$$

$$I_{SF}^{P} < 2\{\gamma_F^P - [(\gamma_S^D)^{1/2} - (\gamma_F^D)^{1/2}][(\gamma_F^D)^{1/2} - (\gamma_H)^{1/2}]\} \tag{21}$$

With respect to formamide, values of τ obtained from immersion measurements correspond to the advancing value of the contact angle, and values on emersion correspond to the receding contact angle.

Both immersion and emersion can be studied in static and dynamic modes. Static corresponds to no relative movement between the liquid system and the fiber, and dynamic to measurements taken while the beaker containing the liquids is in motion. In practice, it appears that, in the second case, the values recorded are not very sensitive to the speed of motion. Therefore, in principle, four values of τ could exist, generally denoted τ_{AD}, τ_{AS}, τ_{RS}, τ_{RD}, where the subscripts A, R, S, and D refer, respectively, to advancing (immersion), receding (emersion), static, and dynamic.

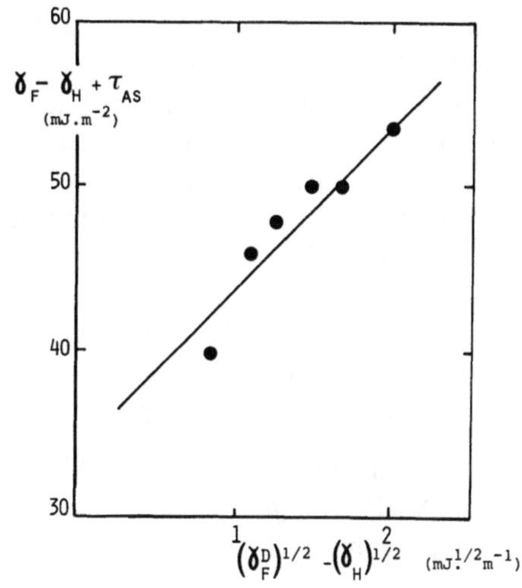

FIGURE 9. Surface characteristics of the carbon fiber T300.

Application to Carbon Fibers. Results are now presented for two types of polymer-coated fibers[48,49] from Toray: a high-strength fiber T300 and a high-modulus fiber M40. On the one hand, in the case of fiber T300, reproducible values of τ are obtained in both static and dynamic modes. Moreover, τ_{AS} and τ_{RS} values (typically 15 ± 0.1 mJ m^{-2} using hexadecane as the hydrocarbon) are similar, indicating that static advancing and receding contact angles are close. It is also observed, as expected, that values corresponding to dynamic measurements are equivalent to static values, although the scatter is larger.

On the other hand, in the case of fiber M40 the situation is totally different. Large differences between values of τ_{AS} and τ_{RS} are observed. Moreover, with dynamic measurements, an important hysteresis phenomenon is clearly evidenced, since, for instance, immersion values correspond to a contact angle greater than 90° and emersion values to a contact angle less than 90°. This suggests that the surface of the fiber M40 is greatly nonhomogeneous. Therefore, results concerning this fiber are unsuitable for further analysis.

On the contrary, for the fiber T300, static values of τ can be used to calculate surface energy data, since they are closer to equilibrium than are the dynamic data. As shown in Fig. 9, a linear relationship between $\gamma_F - \gamma_H + \tau_{AS}$ and $(\gamma_F^D)^{1/2} - (\gamma_H)^{1/2}$ is obtained in agreement with Eq. (14). This figure refers to the immersion mode, but emersion data are very similar. Finally, the calculated values of the surface energy components for the fiber T300 are given in Table 4. It can be seen that advancing and receding contact angle results are in good agreement. Substitution

TABLE 4. Surface Free Energy Data for the Carbon Fiber T300 (in mJ m^{-2})

	γ_S^D	I_{SF}^P	γ_S^P
Immersion	24 ± 11	34 ± 3	15 ± 9
Emersion	25 ± 10	34 ± 3	15 ± 9

of the obtained values of γ_S^D and I_{SF}^P into (20) and (21) shows that the present system is reversible and the liquids employed are capable of mutual displacements.

In conclusion, this tensiometric method, in which a two-liquid-phase system is used, leads to reproducible and coherent values for the components of surface free energy and polar interaction with formamide, for carbon fibers. Nevertheless, surface inhomogeneities, promoting, in particular, preferential adsorption and surface energy gradients, constitute a limitation of the validity of this method.

Application to Glass Fibers. Using the same experimental conditions as previously, the surface energy data of untreated or treated glass fibers were determined.[50] The three types of untreated glass fibers were

- An alkali-resistant glass (AR)
- An R-type glass
- A common E-type glass

In addition, the surface energy of the alkali-resistant glass fibers treated by chemical etching according to Tichane's procedure previously described or modified by trimethylchlorosilane (TMCS) in toluene solution or by α-aminopropylethoxysilane (A1100 from Union Carbide) have been also studied.

In each case, it was carefully checked that Eq. (14) was verified and that the systems were reversible.

The values of γ_S^D and I_{SF}^P are gathered in Table 5. Although glasses R and E exhibit higher values of γ_S^D than AR glass fibers due to different chemical compositions, it appears that the three types of untreated glass fibers present γ_S^D values close to that of water ($21.6\,\mathrm{mJ\,m^{-2}}$). This result confirms previous conclusions[36] obtained on glass plates; i.e., glass is totally covered by water layers that mask its actual surface properties.

The surface energy data of treated AR glass fibers are not very different from the previous ones, except for TMCS-modified fibers. In this last case, the γ_S^D value is close to that of a polymer, such as polyolefin for example, and the term I_{SF}^P is very low. These results are in agreement with the presence on the glass surface of methyl groups, which are more hydrophobic than metallic oxides.

3.4.2. Inverse Gas Chromatography

Inverse gas chromatography (IGC) is a valuable method for the surface characterization of fillers for polymers, like glass and carbon fibers and powders. In

TABLE 5. Surface Energy Data of Untreated and Treated Glass Fibers

Glass fibers	Treatment	γ_S^D ($\mathrm{mJ\,m^{-2}}$)	I_{SF}^P ($\mathrm{mJ\,m^{-2}}$)
R type	None	31.7	35.6
E type	None	28.9	36.8
AR type	None	21.7	37.4
	Tichane	20.1	38.1
	TMCS	33.7	7.6
	A 1100	21.6	36.2

particular, IGC at infinite dilution, i.e., near-zero surface coverage adsorption, in the absence of interaction between the probe molecules themselves, allows the determination of the surface free energy and the acid–base surface characteristics of such divided solids.

Principle of IGC at Infinite Dilution. With the IGC technique, the surface characteristics of the solid are analyzed by injecting probes of known properties into a column containing the solid. The retention time or the net retention volume, V_N, measured at infinite dilution allows the evaluation of the interactions between the probes and the solid and, thus, the characterization of the solid surface.

Measurements are carried out with a chromatograph equipped with a flame ionization detector of high sensitivity. Stainless steel columns are first filled with the solid and, then, usually conditioned at 150°C for about 48 hr in the oven of the chromatograph under helium gas flow. Afterwards, the retention data, determined at different temperatures, are recorded and analyzed.

By simple thermodynamic considerations, the following relationship can be established:[51,52]

$$-\Delta G^\circ = RT \,\mathrm{Ln}\, V_N + K \tag{22}$$

where ΔG° is the free energy of adsorption of the probes and K is a constant depending on the reference states. Assuming that the free energy is proportional to the heat of adsorption,[53] ΔG° can be related to the work of adhesion W_A, between the probe and the solid surface, per unit surface area:

$$-\Delta G^\circ = NaW_A \tag{23}$$

where N is Avogadro's number and a is the surface area of one adsorbed molecule of the probe. Taking into account, in the case of nonpolar probes (n-alkanes), the relation of Fowkes,[15] stating that the energy of interaction through London forces is the geometrical mean of the dispersive components of the surface energy of the interacting bodies, Eq. (23) becomes[54–56]

$$-\Delta G^\circ = 2N\sqrt{\gamma_S^D}\,(a\sqrt{\gamma_L^D}) \tag{24}$$

Therefore, considering a series of nonpolar probes, like n-alkanes, the plot of $(-\Delta G^\circ)$ or RT Ln V_N against the quantity $a(\gamma_L^D)^{1/2}$ must be a straight line. The slope of this line allows the determination of γ_S^D according to Eq. (24).[54–56] Recall that Dorris and Gray[57] use an equivalent method for the determination of γ_S^D by considering the increment of ΔG° per methylene group in the n-alkanes series.

Through injection of polar probes, nondispersive interactions (specific interactions) are now established between these probes and the solid surface. In order to determine quantitatively these specific interactions, it can be considered, to a first approximation, that the specific interactions are simply added to the dispersive interactions previously defined. Therefore, the experimental point corresponding to a polar probe should lie above the reference straight line of $-\Delta G^\circ$ versus $a(\gamma_L^D)^{1/2}$, corresponding to the *n*-alkanes. Finally, Eq. (24) can be rewritten as[55,56]

$$-\Delta G^\circ = 2N\sqrt{\gamma_S^D}\,(a\sqrt{\gamma_L^D}) + \Delta G_{SP} \tag{25}$$

where $\Delta G_{SP} = \Delta H_{SP} - T\,\Delta S_{SP}$ is the free energy of desorption corresponding to specific interactions. Specific interactions have usually been referred to in terms of polar interactions. Recently, in the work of Drago *et al.*,[58,59] Gutmann,[60] and Fowkes,[61] and Fowkes and Maruchi,[62] these interactions are essentially considered as Lewis acid–base interactions or electron acceptor–donor interactions. According to this concept, strong interactions develop only between an acid and a base.

Generally, for each probe, Gutmann's semiempirical scale[60] is used in terms of acceptor (AN) and donor (DN) numbers which, respectively, describe the ability of a molecule to attract or to release electrons. According to Saint Flour and Papirer's approach,[51] the specific enthalpy of adsorption ($-\Delta H_{SP}$) can be related to AN and DN as follows:

$$-\frac{\Delta H_{SP}}{AN} = K_A \frac{DN}{AN} + K_D \tag{26}$$

where K_A and K_D are the acid and base coefficients of the solid surface, respectively.[55,56] Experimentally, K_A and K_D are, respectively, the slope and the intercept at the origin of the linear variation of $-\Delta H_{SP}/AN$ versus DN/AN.

Application to Carbon and Glass Fibers. Recently, a comparison between the values of dispersive component, γ_S^D, of the surface energy of carbon fibers T300 (Soficar, France) measured by the two-liquid-phase tensiometric method and IGC at infinite dilution has been systematically made.[54–56] Three types of high-strength carbon fibers corresponding to different stages of manufacturing were considered: (i) untreated, (ii) oxidized, and (iii) sized fibers. The comparison between the results obtained using the two methods can be seen in Table 6. It appears that, in each case, excellent agreement exists between the two sets of values. Moreover, the values obtained for coated carbon fibers are close to those we obtained previously[48] on similar fibers.

TABLE 6. Comparison between Dispersive Component of the Surface Energy of Carbon Fibers Determined by Tensiometric Method or by IGC

	γ_S^D (mJ m^{-2})	
Carbon fibers	Tensiometric method	IGC
Untreated	50 ± 8	50 ± 4
Oxidized	48 ± 10	49 ± 2
Coated	34 ± 6	36 ± 3

TABLE 7. Electron Acceptor–donor (Lewis Acid–Base) Characters of the Carbon Fibers, in Arbitrary Units

Carbon fibers	K_A	K_D
Untreated	6.5	1.5
Oxidized	10.0	3.2
Sized	8.6	13.0

Finally, it can be concluded that IGC and the two-liquid phase method are well adapted for characterizing the surface properties of fibers.

In addition, another advantage of IGC is the easy determination of the acid–base characteristics of these fibers. Table 7 gives the values of K_A and K_D found for the three carbon fibers,[55,56] in arbitrary units. Both untreated and oxidized fibers exhibit a strong acid character (high value of K_A and low value of K_D), while a coated fiber surface seems to be more amphoteric (high values of K_A and K_D).

The knowledge of the acid–base characters of a fiber surface is of great importance in the domain of composite materials. It is well known that the quality of the fiber–matrix interface is one of the most relevant factors in controlling, to a large extent, the final performance of a composite. We have proposed that, knowing the K_A and K_D values for the carbon fibers and the matrix (epoxy resin, for instance), and by analogy with Eq. (26), it is possible to define a specific interaction parameter A, describing the acid–base interactions at the fiber (f)–matrix (m) interface:[55,56]

$$A = K_A^f K_D^m + K_A^m K_D^f \tag{27}$$

We have then shown that a linear relationship exists between the interfacial shear strength at the carbon fiber–epoxy resin interface, measured by means of a fragmentation test on single-fiber composites, and the specific interaction parameter A. Thus, this concept of acid–base or acceptor–donor interactions constitutes an interesting approach to the understanding of the interfacial properties of composite materials.

At present, fewer IGC measurements have been performed on glass fibers than on carbon fibers. Nevertheless, Papirer *et al.*[51,52] have found a value of about 50 mJ m^{-2} for γ_S^D of glass fibers. This value is about twice that found by ourselves.[50] This is due to the heat treatment performed by Papirer *et al.* under helium gas flow, applied to the column in the chromatograph, prior to IGC measurements. This treatment can remove most of the water layer adsorbed on the glass fiber surface.

4. CHARACTERIZATION OF LOW-ENERGY SURFACES

As seen in the preceding section, the two-liquid-phase method is an appropriate technique for determining the surface characteristics of high-surface-energy solids. However, this method can be successfully applied to low-surface-energy materials and, in particular, to polymers. Recall that all earlier work[19–22] on this method was performed on polymeric surfaces. Since 1978, we have systematically compared the surface energy of different polymers, determined either by standard contact angle measurements (one liquid phase) or by the two-liquid-phase method.[63,64] Whatever the nature of the polymer, the results are identical, within the experimental scatter.

In most cases, because of their inherently poor adhesion properties, polymers and, more particularly, polyethylene (PE) and polypropylene are subjected to surface modification, prior to bonding, for instance to metal or other polymers. One interesting route of modification of polyolefins is to graft onto the polymer small quantities of polar species such as acrylic acid (about 1% by weight). This is currently done on PE in particular. Thus, we have observed[65] that the adhesion of PE and of grafted PE to aluminium differed considerably, although the surface energies of the two polymers seemed to be identical. The enhanced adhesion with grafted PE was attributed to the reorientation of polar groups (acrylic acid groups) contained in the bulk polymer, when this polymer is in contact with a polar or a high-surface-energy substrate, in agreement with previous observations.[66–69] Moreover, this orientation can lead to the formation of chemical linkages between the metal oxide and the polymer at the polymer–metal oxide interface.

The variation with time of the surface energy of the grafted polymer in contact with an "orienting" polar liquid (water, for example) can be directly studied by means of the two-liquid-phase method in the following way. A plate of modified polymer is put into contact with water. At different contact times, droplets of *n*-alkanes are placed at the water–polymer interface and the surface characteristics of the polymer are then determined as seen above [Eq. (7)].

Consequently, the first part of this study will be devoted to the orientation ability of acrylic-acid-modified polyethylene studied *in situ* by the two-liquid-phase method.

The second part will concern the hysteresis phenomenon experimentally observed in most wettability measurements. It is well known that, in practice, it is very common to obtain a range of experimental values of contact angles, between advancing and receding contact angles, on polymeric surfaces. Such a hysteresis phenomenon can be also explained by the reorientation of polar groups of the polymer chains at the polymer–liquid interfaces, and this can be easily studied, as previously, by using the two-liquid-phase method.

4.1. Orientation Phenomenon

In order to explain the behavior of grafted PE (PEg), we have determined directly, as described above, the dispersive γ_S^D and polar γ_S^P components of the surface energy of PEg in contact with model orientation environments (water or formamide).[65] Figure 10 shows the variation of γ_S^D and γ_S^P of PE and PEg as a function of the contact time on water at 20°C. The dispersive components of PE and PEg are practically the same and about 39 ± 2 mJ m^{-2}. The polar component γ_S^P of PE remains negligible, whatever the contact time, while γ_S^P of PEg increases considerably to reach a maximum value of 22 mJ m^{-2} after 13 days of contact

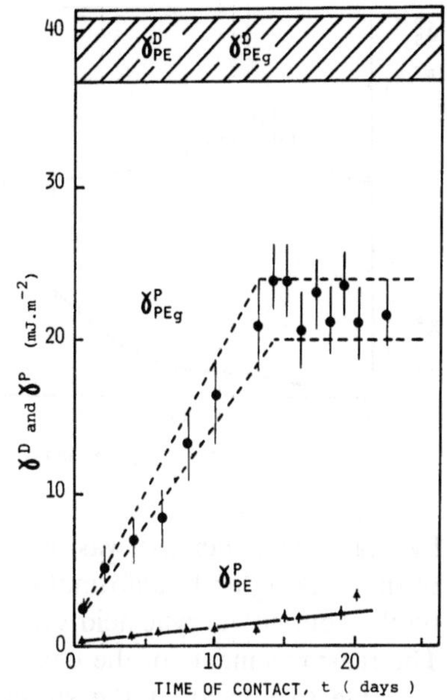

FIGURE 10. Variations of γ^D_S and γ^P_S of PE and PEg versus contact time on water at 20°C.

with water. With formamide, the maximum value of γ^P_S of PEg is 8.8 mJ m^{-2}. Consequently, it seems that the maximum value of γ^P_S of PEg is proportional to the polarity, γ^P_L, of the orientation liquid medium; the ratio between γ^P_S and γ^P_L is quite constant and equal to about 0.45 ± 00.5.

In addition, we have pointed out the following observations:[65]

- The kinetics of orientation is increased by increasing temperature.
- The orientation of PEg corresponds to a decrease of the interfacial energy, γ_{SL}, between the PEg and the polar liquid, in agreement with the thermodynamics of systems tending toward an equilibrium state.
- The results are different on polymeric films obtained by extrusion and blowing; the differences are attributed to a change of morphology or crystallinity.

It can be concluded that, in contact with a polar orientation medium, the polymer chains move and the conformation of the chains in a surface layer is therefore modified. As a consequence, the surface energy of a polymer can no longer be considered as an intrinsic property of this polymer, but should be replaced by the new concept of potential surface energy.[65] Thus, we have explained the enhanced adhesion observed between PEg and aluminum by using this new concept.

More recently,[70,71] we have reexamined this problem of orientation phenomenon concerning films of PEg in contact with water. Complementary techniques have been used in order to follow the evolution and to determine the molecular and macroscopic properties of the reoriented surface: two-liquid-phase contact angle measurements, coloration tests, water sorption measurements, IGC, and x-ray photoelectron spectroscopy (XPS). It was shown that the polymer film surface is reconstructed in two steps, as illustrated in Fig. 11 by the variation of both γ^D_S (upper curve) and γ^P_S (bottom curve) of PEg versus contact time with water at room temperature. Initially, γ^D_S increases from 35 to 55 mJ m^{-2} in about

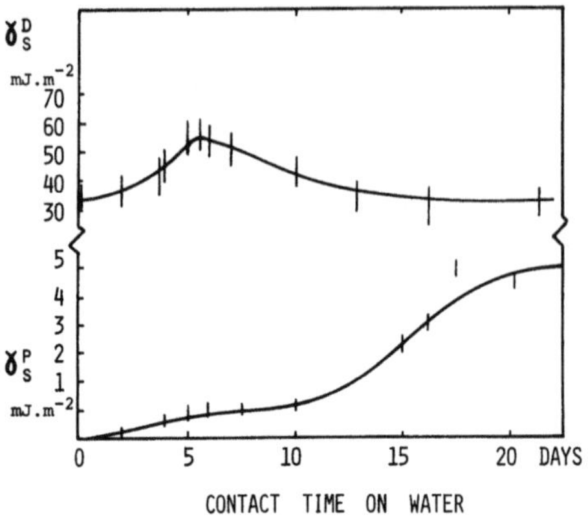

FIGURE 11. Variations of γ_S^D and γ_S^P of PEg films versus contact time with water.

five days. This increase was tentatively attributed to a change, in the surface layer, of the packing and conformation of the polymer chains under the pressure exerted by the "buried" acrylic acid groups, which tend to orient outside the bulk polymer. The rearrangements of the chains are necessarily required so that the polar groups are able to appear at the surface. Water is both adsorbed and absorbed in the polymer during this reconstruction phenomenon. XPS measurements confirm that the acrylic groups, initially not present in the surface layer, begin to appear on the surface after about five days. The second step, corresponding to the decrease of γ_S^D and the increase of γ_S^P after five days, is essentially due to the gradual appearance of acrylic groups on the PEg surface.

Analogous results[72] were also obtained on other grafted polymers and copolymers, and it seems that this orientation phenomenon is general for polar polymers.

This phenomenon at the polymer–water interface has been also analyzed through the thermodynamics of irreversible processes.[71] The formalism employed by Sanfeld and Steinchen[73] was used to describe the orientation at the interface, considering that the variation of the surface free energy for the polymer surface in equilibrium with water depends on water adsorption and reorientation of the polar groups. It was also assumed that the behavior of the polymer surface is comparable to that of a fluid surface. Finally, it has been shown that the surface can be characterized by a phenomenological coefficient related to the reorientation at the interface. This coefficient gives an indication of the ability of different polar groups, having different lengths of grafted chains, to reorientate at the surface for a given ratio of grafting. It may be calculated from the dipole moment of the polar groups, the contact time, and the interfacial tension variation. In conclusion, the application of the concepts of irreversible thermodynamic processes in this domain leads to a semiquantitative characterization of the time-dependent process of surface reorientation at the polymer–water interface.

4.2. *Hysteresis Phenomenon*

According to the classical theory of wetting, it is assumed that the measured contact angle is unique provided that the phases in contact (solid, liquid 1, liquid 2 or air) are strictly homogeneous and smooth. Unfortunately, in practice, a whole range of contact angles is usually obtained. On the one hand, the largest angle measurable corresponds to the advancing angle θ_A, observed just after a drop of liquid has advanced on the solid surface. On the other hand, the smallest angle is obtained just after the arrest of a receding liquid drop and corresponds to the receding angle θ_R. In the past, this hysteresis has been attributed to different causes, such as composition heterogeneity and roughness of the solid surface, diffusion, swelling, etc. Recently,[74,75] we tentatively attributed this phenomenon to the polar character of the polymer surface, and more precisely, to the reorientation of polar groups on the surface in contact with a polar liquid such as water.

Consider the special system polymer–drop of water–*n*-octane (system 1). For this system, the dispersive components γ_W^D and γ_H^D $(=\gamma_H)$ of the surface energy of water and octane, respectively, are almost equal. Consequently, Eq. (7) simplifies to

$$\gamma_W - \gamma_H + \gamma_{WH} \cos\theta_1 = I \tag{28}$$

where $I = I_{SW}^P$ and $\theta_1 = \theta_{SW}$. A direct estimation of the polar interaction I is therefore accessible. By inversion of the system (system 2), the contact angle of *n*-octane on the polymer in water, θ_2, can be measured, and the corresponding equation is

$$\gamma_W - \gamma_H - \gamma_{WH} \cos\theta_2 = I \tag{29}$$

where $\theta_2 = \theta_{SH}$. Equations (28) and (29) lead to the supplementary relationship between θ_1 and θ_2:

$$\cos\theta_1 + \cos\theta_2 = 0 \qquad \text{or} \qquad \theta_1 + \theta_2 = \pi \tag{30}$$

In each case, both advancing, θ_A, and receding, θ_R, contact angles can be measured. Indeed, advancing angle in system 1 should be related to receding angle in system 2, and, therefore, the following equalities can be written:

$$\theta_{1A} = \pi - \theta_{2R} \qquad \text{and} \qquad \theta_{1R} = \pi - \theta_{2A} \tag{31}$$

If the hysteresis phenomenon exists, two terms I and I^*, referring respectively to the polar interactions using θ_{1A} and θ_{1R}, are now defined:

$$\gamma_W - \gamma_H + \gamma_{WH} \cos\theta_{1A} = I \tag{32}$$

$$\gamma_W - \gamma_H + \gamma_{WH} \cos\theta_{1R} = I^* \tag{33}$$

Subtracting (32) from (33), the hysteresis of wetting, H, may be defined by

$$H = \cos\theta_{1R} - \cos\theta_{1A} = (\gamma_{WH})^{-1}(I^* - I) \tag{34}$$

A similar expression is obtained for system 2. Experimentally, using systems 1

TABLE 8. Values of the Hysteresis H in Relation to the Polar Interactions I and I^*

Polymer	θ	θ^*	$H = (\cos\theta^* - \cos\theta)$	I (mJ m^{-2})	I^* (mJ m^{-2})
PTFE	177	167	0.02	0.4	1.6
PE	176	161	0.05	0.4	3.1
PP	176	157	0.08	0.4	4.4
PS	174	142	0.21	0.6	11.1
PVDF	122	93	0.48	24.3	48.6
PA 11	123	87	0.60	23.5	54.0
PET	120	82	0.64	25.8	58.4
PMMA	124	76	0.80	22.8	63.6

and 2, it is possible to define the mean values, θ and θ^*, of θ_{1A} and θ_{1R}, respectively, according to (31), as follows:

$$\theta = \frac{1}{2}(\theta_{1A} + (\pi - \theta_{2R})) \quad \text{and} \quad \theta^* = \frac{1}{2}(\theta_{1R} + (\pi - \theta_{2A})) \tag{35}$$

Finally, the hysteresis H is

$$H = \cos\theta^* - \cos\theta \tag{36}$$

Measurements of advancing and receding contact angles were performed[74,75] on various apolar and polar polymers for systems 1 and 2. It was checked that conditions (31) were satisfied, whatever the nature of the polymers. Table 8 gives the values of θ, θ^*, the hysteresis H, and the polar interaction terms, I and I^*, as calculated from Eqs. (32), (33), (35), and (36). It is clear that H increases with polarity defined by either I or I^*. If hysteresis is directly related to the orientation of polar groups near the polymer surface in contact with water, it can be considered that the values of I and I^* correspond to two distinct states of the polymer surface: I and I^* represent a random distribution and a more or less oriented distribution, respectively, of the polar groups at the solid–liquid interface. Therefore, it can be assumed that two types of surface exist: S and S^*, respectively, as schematically shown in Fig. 12. Nevertheless, the kinetics of orientation or disorientation of the polar groups near the polymer surface is certainly slower than that associated with the rate of flow of a liquid drop on the solid during contact angle measurements. This means that an advancing drop of water will be in contact with the polymer surface

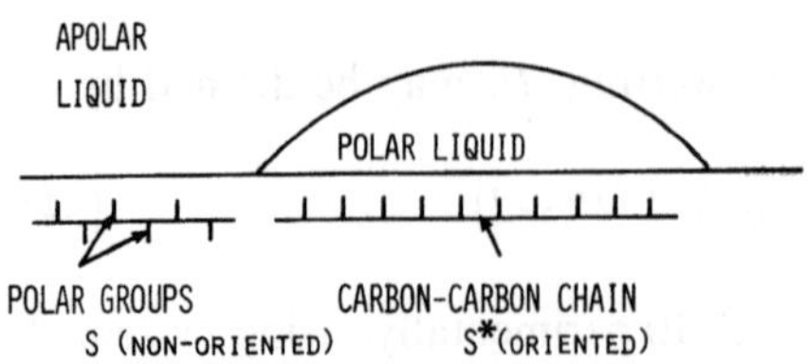

FIGURE 12. Schematic representation of polar group orientation on the polymer surface due to polar liquid contact.

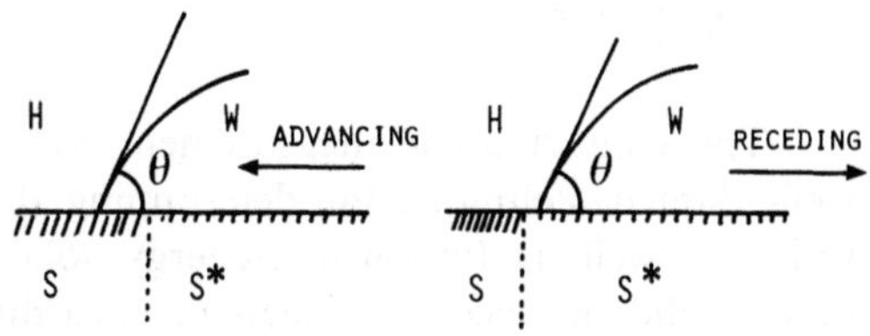

FIGURE 13. Schematic representation of the surface characteristics of the polymer during advancing and receding contact angle measurements.

that was just in contact with *n*-octane and, thus, equivalent to solid *S*. Inversely, when the drop of water recedes, contact angle measurement will be performed on solid S^*. This procedure is schematically represented in Fig. 13.

To observe a hysteresis phenomenon on a polymer surface, it is necessary that the following factors differ from zero: (i) the intrinsic polarity of the polymer, (ii) the mobility *m* of either the macromolecular chains or the polar groups attached to these chains. It is concluded that the hysteresis *H* is equal to zero when the following conditions are fullfilled:

$$I = I^* = 0 \qquad \text{but } m \neq 0$$

or

$$I = I^* \qquad \text{but } m = 0$$

Assuming a simple relationship of proportionality between *H* and polarity and mobility leads to the fact that *H* is directly proportional to the oriented polar interaction term I^*. This assumption seems to be experimentally verified, as shown in Fig. 14 for various polymers.

Finally, by analogy with electrostatic interactions and taking into account a potential of orientation of the polar groups around the macromolecular chains, a simple model has been proposed.[75] This model suggests that the polar interactions, which are responsible for the hysteresis phenomena, are essentially those established between the liquid molecules and polar groups of the polymer closest to the solid–liquid interface, i.e., at a distance of about 0.3 nm.

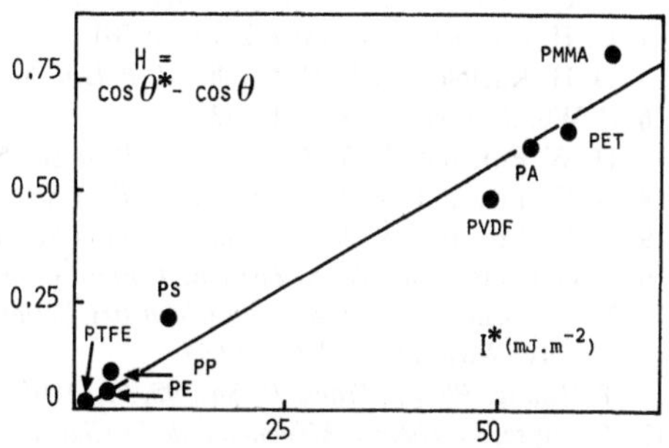

FIGURE 14. Hysteresis *H* versus polarity I^* for different polymers.

5. CONCLUSION

The contact angle measurement by using two immiscible liquid phases is a well-adapted technique for determining the surface energy of high-surface-energy solids as well as low-surface-energy solids. This method, initially developed on model solid surfaces, i.e., sheets of mica directly cleaved in hydrocarbon media, is very fruitful for the surface characterization of air-exposed highenergy solids such as metals, oxides, glass, and carbon. Its application to tensiometric measurements leads to accurate determination of surface energy of fibers, in particular, glass and carbon fibers. Such a determination is of great importance for a better understanding of the mechanical behavior of fiber–matrix interfaces in composite materials. The results obtained by the tensiometric method with two liquid phases are in good agreement with those reached by other techniques, particularly, inverse gas chromatography at infinite dilution.

Finally, the most important fact is that the two-liquid-phase method has constituted the basis of the discovery and the explanation of fundamental phenomena in wettability and adhesion science. In particular, the three following phenomena have to be mentioned:

1. The improvement of our model concerning the adhesion of viscoelastic materials proposed in 1971,[(34)] by taking into account a molecular dissipation factor depending on the degree of cross-linking of the polymer near the interface, in addition to the reversible adhesion energy and the dissipation factor due to viscoelastic losses.[(31)]
2. The orientation phenomenon, near the interface, of polar groups contained in the bulk of polar polymers, when these polymers are in contact with a polar medium. Such a phenomenon has led to the new concept of potential surface energy, pointing out that the surface energy cannot be considered as an intrinsic property of materials.[(65)]
3. A new complementary explanation of the hysteresis of wetting of polymers, considering that this phenomenon is also due to the polarity of materials and, in particular, to the reorientation of polar groups at the solid–liquid interface.[(74,75)]

REFERENCES

1. D. K. Owens and R. C. Wendt, *J. Appl. Polym. Sci.* **13**, 1740 (1969).
2. J. R. Dann, *J. Colloid Interface Sci.* **32**, 302 (1970).
3. J. R. Dann, *J. Colloid Interface Sci.* **32**, 321 (1970).
4. D. H. Kaelble, *J. Adhesion* **2**, 66 (1970).
5. D. H. Kaelble and E. H. Circlin, *J. Polym. Sci. A2* **9**, 363 (1971).
6. S. Wu, *J. Adhesion* **5**, 39 (1973).
7. H. W. Fox and W. A. Zisman, *J. Colloid Sci.* **5**, 514 (1950).
8. H. W. Fox and W. A. Zisman, *J. Colloid Sci.* **7**, 109 (1952).
9. H. W. Fox and W. A. Zisman, *J. Colloid Sci.* **7**, 428 (1952).
10. E. G. Shafrin and W. A. Zisman, *J. Phys. Chem.* **64**, 519 (1960); *J. Am. Ceram. Soc.* **50**, 478 (1967).
11. W. A. Zisman, in *Advances in Chemistry Series* (R. F. Gould, ed.), Vol. 43, p. 1, American Chemical Society, Washington, DC (1964).
12. T. Young, *Philos. Trans. R. Soc.* **95**, 65 (1805).
13. A Dupré, in *Théorie Mécanique de la Chaleur*, p. 369, Gauthier-Villars, Paris (1869).

14. L. A. Girifalco and R. J. Good, *J. Phys. Chem.* **61**, 904 (1957).
15. F. M. Fowkes, *Ind. Eng. Chem.* **56**, 40 (1964).
16. O. Drieger, A. W. Neumann, and P. J. Sell, *Kolloid-Z. Z. Polym.* **201**, 52 (1965).
17. L. E. Murr, in *Adhesion Measurement of Thin Films, Thick Films and Bulk Coatings* (K. L. Mittal, ed.), ASTM STP 640, p. 82, American Society for Testing and Materials, Philadelphia (1978).
18. L. E. Murr, *Mater. Res. Sci.* **14**, 107 (1981).
19. H. Peper and J. Berch, *J. Phys. Chem.* **68**, 1586 (1964).
20. Y. Tamai, K. Makuuchi, and M. Suzuki, *J. Phys. Chem.* **71**, 4176 (1967).
21. W. C. Hamilton, *J. Colloid Interface Sci.* **40**, 219 (1972).
22. W. C. Hamilton, *J. Colloid Interface Sci.* **47**, 672 (1974).
23. J. Schultz and K. Tsutsumi, *Proc. 4ème Conf. Thermodynamique Chimique*, Montpellier (France), VII, 126–134 (1975).
24. J. Schultz, K. Tsutsumi, and J. B. Donnet, *J. Colloid Interface Sci.* **59**, 272 (1977).
25. J. Schultz, K. Tsutsumi, and J. B. Donnet, *J. Colloid Interface Sci.* **59**, 277 (1977).
26. M. E. R. Shanahan, C. Cazeneuve, A. Carré, and J. Schultz, *J. Chem. Phys.* **79**, 241 (1982).
27. A. I. Bailey and S. M. Kay, *Proc. R. Soc. London A* **301**, 47 (1967).
28. A. I. Bailey and A. G. Price, *Discuss Faraday* **Soc.**, 118 (1970).
29. F. M. Fowkes, *162nd ACS Meeting, Div. Org. Coat. Plast. Chem.* **31**, 11 (1971).
30. A. Carré and J. Schultz, *J. Adhesion* **15**, 151 (1983).
31. A. Carré and J. Schultz, *J. Adhesion* **17**, 135 (1984).
32. R. N. Wenzel, *Ind. Eng. Chem.* **28**, 988 (1936).
33. A. B. D. Cassie and S. Baxter, *Trans. Faraday Soc.* **40**, 546 (1944).
34. A. N. Gent and J. Schultz, *Proc. 162nd ACS Meeting* **31**(2), 113 (1971).
35. A. N. Gent and J. Schultz, *J. Adhesion* **3**, 281 (1972).
36. J. Schultz and H. Simon, *Verres Réfractaires* **34**, 23 (1980).
37. J. Schultz and H. Simon, *Verres Réfractaires* **34**, 192 (1980).
38. R. M. Tichane, *Am. Ceram. Soc. Bull.* **42**, 441 (1963).
39. A. Dietzel, *Sprechsaal* **75**, 82 (1942).
40. F. M. Fowkes, in *Chemistry and Physics of Interfaces*, pp. 1–12, American Chemical Society, Washington, DC (1965).
41. S. Scherrer and F. Naudin, *11th Int. Congress of Glass*, Vol. III, pp. 301–310, Prague (1977).
42. J. B. Donnet, J. Schultz, and M. E. R. Shanahan, *Int. Symp. on Carbon, New Processing and New Applications*, pp. 57–60, Toyohashi, Japan (1982).
43. J. B. Donnet, M. Brendle, and J. Schultz, *16th Biennial Conf. on Carbon*, pp. 379–380 (1983).
44. B. J. Carroll, *J. Colloid Interface Sci.* **57**, 488 (1976).
45. J. I. Yamaki and Y. Katayama, *J. Appl. Polym. Sci.* **19**, 2897 (1975).
46. B. J. Carroll, *Langmuir* **2**, 248 (1986).
47. M. Nardin and I. M. Ward, *Mater. Sci. Technol.* **3**, 814 (1987).
48. J. Schultz, C. Cazeneuve, M. E. R. Shanahan, and J. B. Donnet, *J. Adhesion* **12**, 221 (1981).
49. C. Cazeneuve, J. B. Donnet, J. Schultz, and M. E. R. Shanahan, *Int. Conf. Adhesion and Adhesives: Science, Technology and Applications*, pp. 19.1–19.7, Durham, U.K., The Plastics and Rubber Institute, London (1980).
50. H. Simon, Ph. D. Thesis, Université de Haute-Alsace, Mulhouse (France), (1984).
51. C. Saint Flour and E. Papirer, *Ind. Eng. Chem.—Prod. Res. Dev.* **21**, 666 (1982).
52. E. Papirer, H. Balard, and A. Vidal, *Eur. Polym. J.* **24**, 783 (1988).
53. G. M. Dorris and D. G. Gray, *J. Chem. Soc.—Faraday Trans. I* **77**, 713 (1981).
54. J. Schultz, L. Lavielle and H. Simon, *Int. Symp. on Science and New Applications of Carbon Fibers*, Toyohashi, Japan, 125–139 (1984).
55. J. Schultz, L. Lavielle, and C. Martin, *J. Adhesion* **23**, 45 (1987).
56. J. Schultz, L. Lavielle, and C. Martin, in *Adhesion International 1987* (L. H. Sharpe, ed.), Gordon and Breach, pp. 513–528, New York (1988).
57. G. M. Dorris and D. G. Gray, *J. Colloid Interface Sci.* **77**, 353 (1980).
58. R. S. Drago, G. C. Vogel, and T. E. Needham, *J. Am. Chem. Soc.* **93**, 6014 (1971).
59. R. S. Drago, L. B. Parr, and C. S. Chamberlain, *J. Am. Chem. Soc.* **99**, 3203 (1977).
60. V. Gutmann, *The Donor–Acceptor Approach to Molecular Interactions*, Plenum Press, New York (1983).

61. F. M. Fowkes, *Rubber Chem. Technol.* **57**, 328 (1978).
62. F. M. Fowkes and S. Maruchi, *Organic Coat. Plastics Chem.* **37**, 605 (1977).
63. A. Carré, J. Schultz, and H. Simon, *5ème Conf. Européenne des Plastiques et des Caoutchoucs*, pp. 1–6, Vol. 1, C-17 (1978).
64. J. Schultz, A. Carré, and H. Simon, *Double Liaison-Chim. Peintures* **322**, 263 (1982).
65. J. Schultz, A Carré, and C. Mazeau, *Int. J. Adhesion Adhesives* **4**, 163 (1984).
66. A. Baszkin, M. Nishimo, and L. Ter-Minassian-Saraga, *J. Colloid Interface Sci.* **54**, 317 (1976).
67. J. D. Andrade, S. M. Ma, R. King, and D. E. Gregonis, *J. Colloid Interface Sci.* **72**, 488 (1979).
68. Y. C. Ko, B. D. Ratner, and A. S. Hoffman, *J. Colloid Interface Sci.* **82**, 25 (1981).
69. A. Carré and H. P. Schreiber, *J. Coat. Technol.* **54**, 31 (1982).
70. L. Lavielle and J. Schultz, *J. Colloid Interface Sci.* **106**, 438 (1985).
71. L. Lavielle, J. Schultz, and A. Sanfeld, *J. Colloid Interface Sci.* **106**, 446 (1985).
72. J. Schultz and L. Lavielle, *Macromol. Chem., Macromol. Symp.* **23**, 343–353 (1989).
73. A. Sanfeld and A. Steinchen, *Introduction to the Thermodynamics of Charged and Polarized Layers*, Wiley, London (1968).
74. M. E. R. Shanahan, A. Carré, S. Moll, and J. Schultz, *J. Chem. Phys.* **83**, 351 (1986).
75. A. Carré, S. Moll, J. Schultz, and M. E. R. Shanahan, in *Adhesion 11* (K. W. Allen, ed.), pp. 82–96, Elsevier, New York (1987).

5

The Equation of State Approach to Interfacial Tensions

J. K. Spelt, D. Li, and A. W. Neumann

1. INTRODUCTION

The determination of solid–vapor (γ_{sv}) and solid–liquid (γ_{sl}) interfacial tensions is of importance in a wide range of problems in pure and applied science. Because of the complexities involved in measuring directly surface tensions involving a solid phase, there exist, at present, a wide range of indirect approaches for obtaining these values. The attractiveness of using contact angles to estimate γ_{sv} and γ_{sl} results from the relative ease with which contact angles can be measured on suitably prepared solid surfaces.

In order to calculate γ_{sv} and γ_{sl}, the Young equation,

$$\gamma_{sl} = \gamma_{sv} - \gamma_{lv} \cos \theta \tag{1}$$

must be combined with a relation of the form

$$\gamma_{sl} = f(\gamma_{sv}, \gamma_{lv}) \tag{2}$$

to provide two equations in terms of the two unknowns γ_{sv} and γ_{sl}. For a liquid of known surface tension, γ_{lv}, a single measurement of the contact angle θ on the solid of interest permits the determination of γ_{sv} and γ_{sl} from Eqs. (1) and (2). Equation (2) is called an *equation of state for interfacial tensions*, and the first part of this chapter outlines two proofs of its existence. The remainder of the chapter explains the empirical formulation of an explicit equation of state and the experimental evidence supporting the validity of both the equation of state concept and the accuracy of the explicit formulation.

J. K. Spelt, D. Li, and A. W. Neumann • Department of Mechanical Engineering, University of Toronto, Toronto, Ontario, Canada M5S 1A4.

Modern Approaches to Wettability: Theory and Applications, edited by Malcolm E. Schrader and George Loeb. Plenum Press, New York, 1992.

Historically, we see our equation of state approach for interfacial tensions as a development based on the pioneering work of W. A. Zisman and his co-workers.[1] While contact angles and contact angle measurements prior to the work of Zisman were somewhat suspect, particularly among physical chemists, they have since gained respectability to the extent that whole symposia are dedicated to contact angle phenomena.[2, 3]

Zisman *et al.* conducted many studies of contact angles on low-energy solid surfaces, such as Teflon, with liquids of relatively high surface tension (we know that this corresponds to $\gamma_{lv} > \gamma_{sv}$). The key observation made by Zisman *et al.* was that for a given solid the measured contact angles did not vary randomly as the liquid was varied; rather, they found that for a homologous series of liquids, say alkanes, and a given solid, say Teflon, $\cos\theta$ changed smoothly with γ_{lv} in a fashion that suggested a straight-line relationship. The extrapolation of this straight line to the point where $\cos\theta = 1$ yielded "the critical surface tension γ_c"; i.e., the surface tension of a liquid that would just wet the solid surface completely. While Zisman stated that γ_c behaved as one would expect the surface tension γ_{sv} of the solid to behave, he took great care not to identify γ_c with γ_{sv}. When other types of liquids, say a homologous series of alcohols, were used instead, the contact angles changed with the liquid surface tension in a similar manner but did not superimpose completely on the alkane data. Such observations have been discussed in terms of a band, in which all experimental points fall, or, alternatively in terms of different straight-line fits for different homologous series and hence different γ_c for one and the same solid depending on the types of liquids.[1]

Our equation of state approach started out in a somewhat different way: We asked whether it might be possible that the $\cos\theta$ versus γ_{lv} relation might not be more or less universal and whether the deviations of the experimental points from a smooth curve could not have causes that differ from a lack of uniqueness of γ_c. To test this hypothesis, we give, in Fig. 1, a plot of contact angle data for Teflon, $\cos\theta$ versus γ_{lv}. In this plot, we have drawn a solid line calculated from our equation of state (see below) for $\gamma_{sv} = 20\ \mathrm{mJ/m^2}$. We have also drawn two lines, corresponding to $\gamma_{sv} = 19\ \mathrm{mJ/m^2}$ and $\gamma_{sv} = 21\ \mathrm{mJ/m^2}$. We can readily see that essentially all experimental points fall within the band defined by these two lines. Thus, it appears that, at least in this case, the deviations from a unique value for γ_{sv} are less than $1\ \mathrm{mJ/m^2}$, and the question we wish to consider is whether such deviations could have reasons different from lack of universality of Eq. (2).

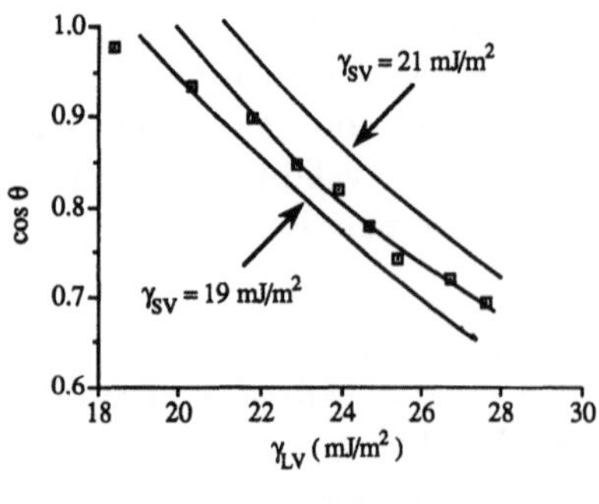

FIGURE 1. Contact angle data for Teflon.[14]

Several factors could account for the scatter of the points in Fig. 1. Some of the more important ones are given here.

1. The equilibrium spreading pressure π is given by the equation

$$\pi = \gamma_s - \gamma_{sv} \tag{3}$$

where γ_s is the solid surface tension in a vacuum. This issue has been studied recently by Spelt *et al.*[4] In Fig. 1, the first two data points are off the fitted solid line. However, note that for these two points, the surface tensions of the liquids are near that of the solid substrate. Hence, it may be expected that the adsorption of the liquid vapor on the solid surface may become more pronounced in such situations and thus change the apparent γ_{sv}. This is discussed more fully in Section 3.1. Although there is no general concensus on the magnitude of the equilibrium spreading pressure, it would seem that a value of the order of 1 mJ/m^2 is not too high an estimate. Since the equilibrium spreading pressure will depend *inter alia* on vapor pressure, it will vary from liquid to liquid within a homologous series as well as from one type of liquid to the next. Thus a variety of patterns of experimental points is possible, without conflicting with the idea of a unique equation of state. Scatter due to this cause would not interfere with the possibility of calculating γ_{sv} from individual pairs of (γ_{lv}, θ) data: Changes of γ_{sv} with changes of liquid and liquid surface tension would simply reflect changes in the equilibrium spreading pressure and might be used to determine the latter.

2. Contact angle measurement. In Fox and Zisman's contact angle measurements, the sessile drop was formed by depositing the liquid from above onto the solid surface, and the contact angle was measured by a goniometer. In this procedure, certain vibrations or oscillations of the drop are inevitable. This may produce a value of contact angle θ between the true advancing contact angle θ_a and the receding contact angle θ_r. Furthermore, the error of the contact angles measured by a goniometer may be as large as $\pm 3°$. But a 3° difference in contact angle at $\gamma_{lv} = 27$ mJ/m^2 and $\theta = 50°$ leads to, approximately, a 1-mJ/m^2 difference in γ_{sv}.

3. Another possible error involved in contact angle measurements can arise from the drop size dependence of contact angles, a possibility that has not been considered adequately yet. The dependence of contact angles on drop size may be caused by the line tension.[5] There are a number of studies of contact angle drop size dependence. As examples, several investigations[6–8] have reported that the contact angle changed by approximately 3–5°, while the radius of the three-phase contact circle increased from 1 to 5 mm.

Only after clarification of these points will be know whether, experimentally, Eq. (2) is unique or universal. While it may not be possible to anticipate the answer to all these questions, we note that, in our experience, the more careful the experimentation, including the preparation of the solid surface, the closer do the experimental points fall to a smooth curve. An example is given in Fig. 2, where we reproduce two $\cos\theta$ versus γ_{lv} plots for hexatriacontane[9] and cholesteryl acetate[10] surfaces. Both surfaces were so smooth and homogeneous that contact angle hysteresis with water was zero. The measurements were made dynamically with the method of capillary rise at a vertical plate at very low rate of advance of

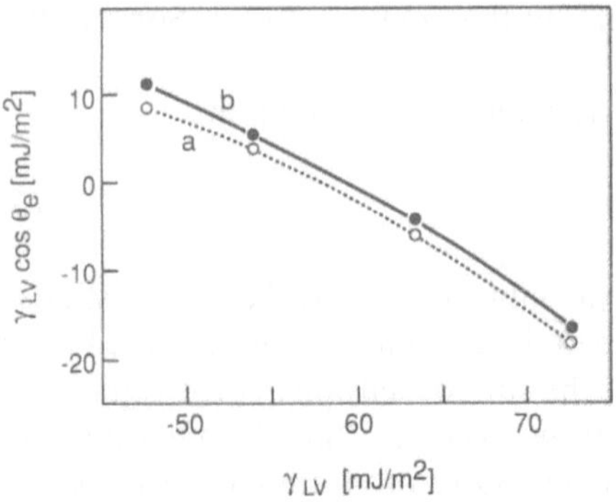

FIGURE 2. Test for the absence of adsorption: curve a, hexatriacontane; curve b, cholesteryl acetate.

the three-phase line. Figure 2 suggests that the equation of state is indeed universal, that the contact angles are meaningful, and that the equilibrium spreading pressure is negligible, in these cases. Clearly, more high-quality contact angle data are sorely needed.

On the theoretical side, the first question that must be asked is whether a relation of the form of Eq. (2) exists.

2. EXISTENCE OF AN EQUATION OF STATE

The ability to determine γ_{sv} and γ_{sl} from a single contact angle measurement depends on having an equation of state of the form of Eq. (2). The explicit formulation of an empirical equation of state is discussed in the following section. The fact that such an equation must exist can be demonstrated in two ways.

2.1. Interfacial Gibbs–Duhem Equations

Following the approach of Ward and Neumann,[11] consider the system shown in Fig. 3, where three phases are in equilibrium under the following conditions.

1. The surface of the solid is smooth and homogeneous.
2. There is no dissolution of the solid, nor is there any absorption by the solid of any of the components of the liquid or gaseous phases.
3. The solid is assumed sufficiently rigid so that its state of strain is unaffected by movements of the three-phase line.

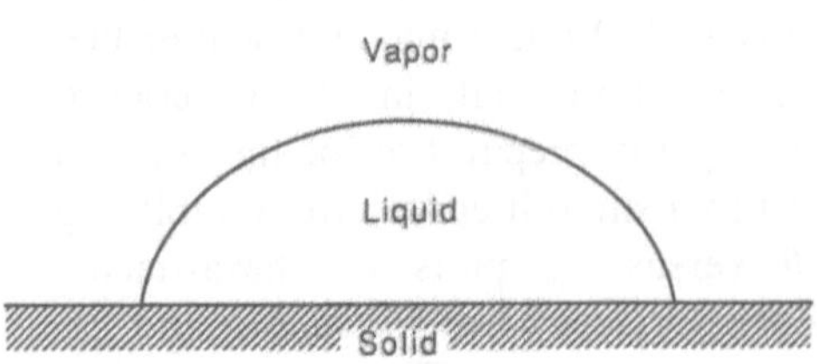

FIGURE 3. Ideal solid–liquid–vapor system.

For simplicity consider the case in which the liquid is pure and the gas is just the vapor phase of the liquid. The three interfacial Gibbs–Duhem equations are then

$$d\gamma_{sv} = -s^{sv}_{(1)}\, dT - \Gamma^{sv}_{2(1)}\, d\mu_2 \tag{4}$$

$$d\gamma_{sl} = -s^{sl}_{(1)}\, dT - \Gamma^{sl}_{2(1)}\, d\mu_2 \tag{5}$$

$$d\gamma_{lv} = -s^{lv}\, dT - \Gamma^{lv}_{2}\, d\mu_2 \tag{6}$$

where the subscript 2 indicates the liquid component and the subscript (1) refers to the definition of the Gibbs dividing surface chosen to eliminate adsorption of the solid component at the particular interface. The surface entropy of the solid–vapor interface is $s^{sv}_{(1)}$, T is the absolute temperature, $\Gamma^{sv}_{2(1)}$ is the surface excess concentration of component 2 (the liquid) at the solid–vapor interface, and μ_2 is the chemical potential of the liquid component. Similarly, $\Gamma^{sl}_{2(1)}$ is the surface excess concentration of component 2 at the solid–liquid interface, and Γ^{lv}_{2} is the surface concentration of component 2 at the liquid–vapor interface. The surface entropies at the solid–liquid and liquid–vapor interfaces are, respectively, $s^{sl}_{(1)}$ and s^{lv}.

Equations (4)–(6) indicate that each of the surface tensions is a function of T and μ_2; i.e.,

$$\gamma_{sv} = \gamma_{sv}(T, \mu_2) \tag{4a}$$

$$\gamma_{sl} = \gamma_{sl}(T, \mu_2) \tag{5a}$$

$$\gamma_{lv} = \gamma_{lv}(T, \mu_2) \tag{6a}$$

Thus, there are three equations in terms of the two variables T and μ_2, implying that any one of Eqs. (4a)–(6a) may be expressed as a linear combination of the other two. In other words, there must exist an equation of the form of Eq. (2). The existence of an equation of state for interfacial tensions, Eq. (2), is thereby proven.

2.2. *Phase Rule for Interfacial Systems*

The existence of an equation of state for interfacial tensions has also been proven[12] by determining the number of degrees of freedom in the equilibrium state of the system shown in Fig. 3.

The Gibbs phase rule

$$f = r + 2 - M \tag{7}$$

gives the number of degrees of freedom, f, in a system of r independent chemical components and M phases under the following restrictions:

1. The system must have negligible boundary effects and all boundaries between phases must be thermally conducting, deformable, and permeable to all components.
2. No chemical reactions occur.
3. Volume is the only work coordinate; i.e., $P\,dV$ work is the only mode of work.

These conditions are not satisfied by the system of Fig. 3, and a different form of the phase rule must be used to determine the number of independent intensive variables or degrees of freedom.

The required phase rule for surface systems may be derived by subtracting the number of equilibrium constraint equations from the number of variables required to describe the system. For a surface system of M phases (bulk and surface) each with r independent components, each bulk phase, α, may be described by the variables T^α, P^α, $x_1^\alpha, x_2^\alpha, \ldots, x_{r-1}^\alpha$, where T^α and P^α are, respectively, the temperature and pressure of phase α, and x_i^α $(i = 1, 2, \ldots, r-1)$ is the mole fraction of the ith component in phase α. The surface phases in the system may be described by a similar set of independent variables, only replacing P^α by $\gamma_{\alpha\beta}$, the interfacial tension between adjacent bulk phases α and β. Thus, the total number of intensive variables describing the surface system is $M(r+1)$.

Considering the number of constraint equations, the equilibrium of the surface system is defined as follows:

thermal equilibrium conditions:

$$T^\alpha = T^\beta = \cdots = T^M \qquad M-1 \text{ equations} \tag{8}$$

chemical equilibrium conditions:

$$\mu_i^\alpha = \mu_i^\beta = \cdots = \mu_i^M \qquad r(M-1) \text{ equations} \tag{9}$$

where $i = 1, 2, \ldots, r$.

Mechanical equilibrium conditions of three possible types:

(i) Laplace equations:

$$P^\alpha - P^\beta = \gamma_{\alpha\beta} J^{\alpha\beta} \tag{10}$$

where α and β represent adjacent bulk phases separated by a curved liquid–fluid interface and $J^{\alpha\beta}$ is the mean curvature of the $\alpha\beta$ interface. If this interface is planar, then $J^{\alpha\beta}$ equals zero and Eq. (10) reduces to $P^\alpha = P^\beta$, which is the mechanical equilibrium condition used in the derivation of the Gibbs phase rule, Eq. (7).

(ii) Young's equation, Eq. (1).

(iii) Neumann triangle relations,

$$2\gamma_{12}\gamma_{23} \cos\theta = (\gamma_{12})^2 - (\gamma_{23})^2 - (\gamma_{13})^2 \tag{11}$$

where the phases are defined as in Fig. 4a and θ is the angle within phase 3 between the 1–3 and 2–3 interfaces.

It should be emphasized that Eqs. (1), (10), and (11) do not, in general, serve as constraint equations, since neither $J^{\alpha\beta}$ nor θ are members of the set of intensive variables describing the state of the surface system. For example, Eq. (10) can always be satisfied by adjusting $J^{\alpha\beta}$, while the values of the pressures and the

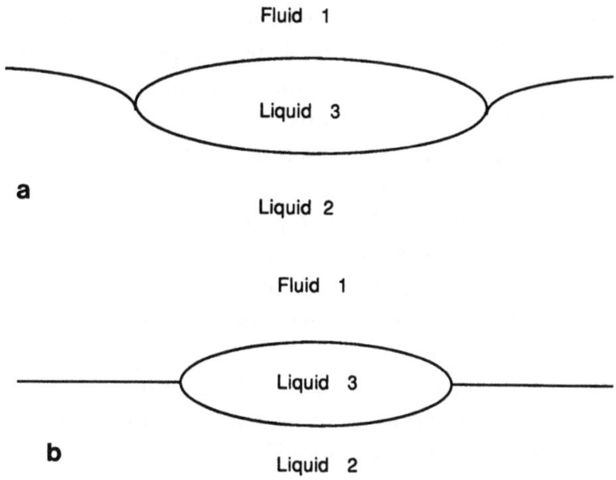

FIGURE 4. (a) Liquid–liquid–fluid system. (b) Liquid–liquid–fluid system with a planar interface.

surface tension in Eq. (10) can be assigned freely. Also, in Young's equation and the Neumann triangle relation the presence of the contact angle θ introduces a new unknown variable. Therefore, it is not possible to use these equations to calculate one of the interfacial tensions from a knowledge of the other two; i.e., these are not constraint equations for the set of independent intensive variables. It is only when $J^{\alpha\beta}=0$ (a planar interface) and, hence, $P^{\alpha}=P^{\beta}$ that a mechanical constraint is imposed by any of the Eqs. (1), (10), or (11). With this is in mind, let N be the number of distinct $P^{\alpha}=P^{\beta}$-type relations among the mechanical equilibrium conditions; then for a surface system with M phases and r independent chemical components the total number of constraint equations is

$$\underset{\substack{\text{thermal}\\ \text{equilibrium}}}{(M-1)} + \underset{\substack{\text{chemical}\\ \text{equilibrium}}}{r(M-1)} + \underset{\substack{\text{mechanical}\\ \text{equilibrium}}}{N}$$

Remembering that the number of intensive variables of the system is $M(r+1)$, the degrees of freedom f are given by

$$f=M(r+1)-[(M-1)+r(M-1)+N]=r+1-N \tag{12}$$

where r is the number of independent chemical components in each phase of the system, and N is the number of $P^{\alpha}=P^{\beta}$ relations among the mechanical equilibrium conditions, i.e., the number of distinct planar interfaces between adjacent bulk phases. Equation (12) is the phase rule for surface systems.

Consider now the application of Eq. (12) to the two-component (solid and liquid–vapor) surface system shown in Fig. 3, which has three bulk phases and three surface phases. Note that such a system is typically called a two-component, three-phase system. Assuming that the solid phase is isotropic and may be characterized by a single hydrostatic pressure, P^{S} (see Ref. 13 for a discussion of other cases), one of the mechanical equilibrium conditions is then $P^{S}=P^{V}$, and Eq. (12) gives $f=2$, indicating that any two of the intensive variables describing the system may be independently varied. Any of the other variables is then a function of these two arbitrarily chosen independent ones. Thus, if we choose γ_{sv} and γ_{lv} as the two independent variables out of the complete set of $M(r+1)$ variables, then γ_{sl} may be expressed as a function of these variables, i.e., Eq. (2). The existence of an equation of state for interfacial tensions is, therefore, established by consideration of the phase rule for surface systems, subject to the aforementioned restrictions on the solid, liquid, and vapor.

The inclusion of linear phases in the development of the phase rule for surface systems does not affect this result.[12]

It is also important to note that for a system composed of three bulk fluid phases, since, in general, all of the interfaces are curved, there are no mechanical equilibrium constraints of the type $P^{\alpha} = P^{\beta}$ and, therefore, $N=0$. For a two-component liquid lens system as shown in Fig. 4a, Eq. (12) predicts 3 degrees of freedom, and no equation of state relation can exist among the three interfacial tensions. Such a relation can be formed only if one of the interfaces is planar (Fig. 4b) so that $N=1$ and Eq. (12) gives $f=2$.

We have thus demonstrated thermodynamically that an equation of state must exist relating γ_{sl}, γ_{sv}, and γ_{lv} in the system of Fig. 3 comprising a pure liquid, its vapor, and a rigid, insoluble solid on which there is no liquid or vapor absorption or adsorption, and which is smooth and homogeneous. Assuming that the presence of air may be ignored, the explicit form of such a relation has been determined empirically by curve-fitting large quantities of contact angle data.[14] The present equation of state is, therefore, not so much a single equation as it is a computer program[14] or a set of tables[15] that give the empirical relation.

3. FORMULATION OF THE EQUATION OF STATE

In the previous section, the existence of an equation of state of the form of Eq. (2) was proven in two independent ways. The explicit formulation of Eq. (2) can be done empirically through the interpretation and curve fitting of contact angle data[14] or, in principle, through statistical mechanics, although at present this remains beyond our capabilities. This section will examine two equivalent empirical methods that have been used to formulate explicitly the equation of state for interfacial tensions.[14]

3.1. Role of Adsorption

In order to obtain an explicit formulation of $\gamma_{sl} = f(\gamma_{lv}, \gamma_{sv})$, it is desirable to keep one of the three variables γ_{sl}, γ_{sv}, and γ_{lv} constant, to subject the second one to a known change, and to register the effect of this change on the third variable. Of these three quantities only γ_{lv} can be readily measured and serve as the independent variable; γ_{sl} will change with γ_{lv}; and γ_{sv} remains unchanged if the adsorption of the liquid's vapor on the solid–vapor interface can be neglected. The validity of this assumption is demonstrated as follows:

Consider a solid of surface tension γ_s against vacuum and a liquid of surface tension γ_{lv}. Let us assume that the vapor of the liquid is initially prevented from contacting the solid surface so that the vapor pressure near the solid surface is equal to zero; then the vapor of vapor pressure P_v is allowed to contact the solid surface. In order to obtain an expression for the resulting solid–vapor interfacial tension γ_{sv}, we perform a Taylor series expansion and retain only the first-order term:

$$\gamma_{sv} = \gamma_s + \left(\frac{\partial \gamma_{sv}}{\partial P_v}\right)_T \Delta P_v \tag{13}$$

which, since the original vapor pressure was equal to zero, can be written as

$$\gamma_{sv} = \gamma_s + \left(\frac{\partial \gamma_{sv}}{\partial P_v}\right)_T P_v \tag{14}$$

From the Gibbs-Duhem equation for the solid–liquid interface, Eq. (4), we have

$$-\Gamma^{sv}_{2(1)} = \left(\frac{\partial \gamma_{sv}}{\partial \mu_2}\right)_T \tag{15}$$

and

$$\left(\frac{\partial \gamma_{sv}}{\partial \mu_2}\right)_T = \left(\frac{\partial \gamma_{sv}}{\partial P_v}\right)_T \left(\frac{\partial P_v}{\partial \mu_2}\right)_T \tag{16}$$

Assuming for simplicity that the vapor is an ideal gas, we have

$$\mu_2 = \mu_2^\circ(T) + RT \ln P_v \tag{17}$$

so that

$$\left(\frac{\partial \mu_2}{\partial P_v}\right)_T = \frac{RT}{P_v} \tag{18}$$

It follows that

$$\left(\frac{\partial \gamma_{sv}}{\partial P_v}\right)_T = -\Gamma^{sv}_{2(1)} \left(\frac{RT}{P_v}\right) \tag{19}$$

and

$$\gamma_{sv} = \gamma_s - RT\Gamma^{sv}_{2(1)} \tag{20}$$

Since $\Gamma^{sv}_{2(1)} > 0$, we infer from Eq. (20) that adsorption will decrease γ_s. On the other hand, if we only consider the systems for which $\gamma_s < \gamma_{lv}$ the adsorption of the vapor of the liquid would increase the solid surface tension. Since this would contradict Eq. (20), it may be expected that adsorption will not play a major role for systems having $\gamma_s < \gamma_{lv}$.

Adsorption from the vapor has also been investigated experimentally. The few experimental data available at present[16] seem to indicate that the so-called equilibrium spreading pressure is normally less than approximately 1 mJ/m^2, if the contact angle is not too low, for instance, greater than 20 or 30°. Based on the above arguments, the solid–vapor surface tension γ_{sv} will be considered as a constant, independent of the wetting liquid.

3.2. Equation of State—Original Formulation

An equation of state relation, $\gamma_{sl} = f(\gamma_{lv}, \gamma_{sv})$, can be formulated from contact angle data on low-energy solids. Such a formulation was first attempted in the 1960s[17, 18] and completed in the 1970s, using extensive contact angle data on eight polymeric solids.[14]

In Fig. 5, the data for the eight solids are plotted in terms of $\gamma_{lv} \cos\theta$ versus γ_{lv}. As seen in these diagrams, all experimental points fall reasonably close to smooth curves, which have, moreover, the same general shape in all cases. In view of Young's equation, Eq. (1), these continuous curves are consistent with the hypothesis that for any constant γ_{sv}, γ_{sl} is a unique function of γ_{lv}. The contact angles of Fig. 5 are advancing contact angles that were measured on carefully prepared solid surfaces with pure liquids.[19] Such contact angles are denoted "Young contact angles," θ_Y, because they are thermodynamically significant and

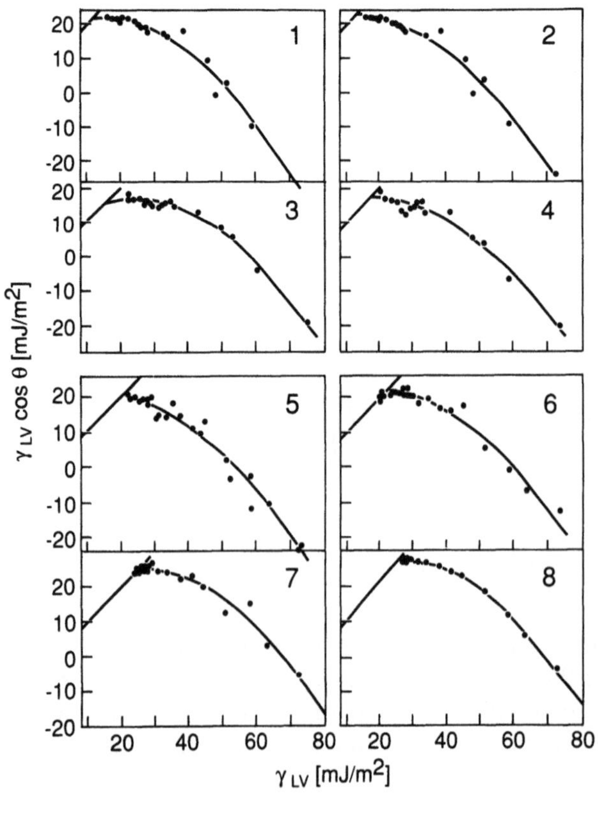

FIGURE 5. $\gamma_{lv} \cos\theta$ as a function of the surface tension γ_{lv} of various liquids: 1. MMM methacrylic polymer A with fluorinated side chain. 2. MMM methacrylic polymer S with fluorinated side chain. 3. 17-(perfluoropropyl)-heptadecanoic acid. 4. 17-(perfluoroethyl)-heptadecanoic acid. 5. Polytetrafluoroethylene. 6. 80-20 copolymer of tetrafluoroethylene and chlorotrifluoroethylene. 7. 60-40 copolymer of tetrafluoroethylene and chlorotrifluoroethylene. 8. 50-50 copolymer of tetrafluoroethylene and polyethylene.

satisfy Young's equation for a given solid and liquid. As mentioned in the introduction, the measurement of a Young contact angle is complicated by the influence of surface roughness, vapor adsorption, and liquid impurities.

Two general conclusions can be drawn from the plots in Fig. 5.

1. As γ_{lv} decreases, $\gamma_{lv}\cos\theta$ increases and, by Young's equation, since γ_{sv} is assumed constant, γ_{sl} decreases.
2. The slope $d(\gamma_{lv}\cos\theta_Y)/d\gamma_{lv}$ is equal to zero at $\theta_Y = 0$. This was demonstrated quantitatively by computer-curve-fitting the experimental data to a second-order polynomial in each case.

$$\gamma_{lv}\cos\theta_Y = a\gamma_{lv}^2 + b\gamma_{lv} + c \tag{21}$$

The 45° line

$$\gamma_{lv}\cos\theta_Y = \gamma_{lv} \tag{22}$$

i.e., the limiting condition $\theta_Y = 0$, is also shown in each case. The intercept of the computed curve [Eq. (21)] with the 45° line is given by

$$a\gamma_{lv}^2 + (b-1)\gamma_{lv} + c = 0 \tag{23}$$

The intercepts and the limiting slopes at the intercepts are given in Table 1. The average limiting angle of inclination, calculated from the average limiting slope given, is $0.1° \pm 4.2°$. The standard deviation includes zero, so that the hypothesis that the true value of the limiting slope is zero is acceptable. Thus, it follows that

$$\lim_{\theta_Y \to 0} \frac{d(\gamma_{lv}\cos\theta_Y)}{d\gamma_{lv}} = 0 \tag{24}$$

Since γ_{sv} is assumed constant, it follows that

$$\frac{d(\gamma_{lv}\cos\theta_Y)}{d\gamma_{lv}} = \frac{d(\gamma_{sv} - \gamma_{sl})}{d\gamma_{lv}} = -\frac{d\gamma_{sl}}{d\gamma_{lv}} \tag{25}$$

TABLE 1. Limiting Slopes and Intercepts for the Eight Systems in Figure 5

No.	Solid	Slope	Intercept
1	MMM methacrylic polymer with fluorinated side chain	0.0611	11.66
2	MMM methacrylic polymer S with fluorinated side chain	0.0089	12.44
3	17-(Perfluoropropyl)-heptadecanoic acid	0.1714	15.87
4	17-(Perfluorethyl)-heptadecanoic acid	−0.0520	17.38
5	Polytetrafluoroethylene	−0.1759	20.23
6	80-20 copolymer of tetrafluoroethylene and chlorotrifluoroethylene	−0.0207	20.93
7	60-40 copolymer of tetrafluoroethylene and chlorotrifluoroethylene	0.0379	24.63
8	50-50 copolymer of tetrafluoroethylene and ethylene	−0.0443	26.90

Equations (23) and (24) then imply that

$$\lim_{\theta_Y \to 0} \frac{d\gamma_{sl}}{d\gamma_{lv}} = 0 \tag{26}$$

From the above fact that γ_{sl} decreases as γ_{lv} decreases, and from Eq. (26), we conclude that γ_{sl} has its minimum value when $\theta_Y = 0$.

From our knowledge of liquid–liquid interfaces, zero is the lower limit for the interfacial tension between the two liquid phases at equilibrium. It would then be very difficult to understand if arbitrarily small solid–liquid interfacial free energies were not possible. Therefore, the minimum value of the solid–liquid interfacial tension is zero as the contact angle approaches zero; i.e.,

$$\lim_{\theta_Y \to 0} \gamma_{sl} = \gamma_{sl}^* = 0 \tag{27}$$

This will be discussed further in the next section where the possibility of negative interfacial tensions will be considered.

The formulation of the equation of state is essentially an empirical curve fit to contact angle data. As such there are a variety of ways of proceeding, and in this instance it was decided to correlate the data in terms of Good's interaction parameter:[14]

$$\Phi = \frac{\gamma_{sv} + \gamma_{lv} - \gamma_{sl}}{2\sqrt{\gamma_{lv}\gamma_{sv}}} \tag{28}$$

This can be done as follows:

1. Assuming γ_{sv} constant and $\gamma_{sl}^* = 0$, determine γ_{sv} graphically from Fig. 5 using

$$\gamma_{sv} = \lim_{\theta_Y \to 0} \gamma_{lv} = \gamma_{lv}^* \tag{29}$$

2. Using this constant γ_{sv} and experimental values of γ_{lv} and $\cos\theta_Y$, obtain γ_{sl} as a function of θ_Y from Young's equation, Eq. (1).
3. Using the values of γ_{sv} and γ_{sl} as obtained in steps 1 and 2, compute Φ using Eq. (28).

Figure 6 shows graphs of Φ versus γ_{sl}, for the eight systems given in Fig. 5. Clearly, the fit of the data to straight lines is satisfactory, and we conclude that

1. As a good approximation, Φ is a linear function of γ_{sl}, for a particular solid with a series of liquids. The straight lines shown in Fig. 6 are least-square fits.
2. All data points of all eight systems can be fitted to a single straight line

$$\Phi = -\alpha\gamma_{sl} + \beta \tag{30}$$

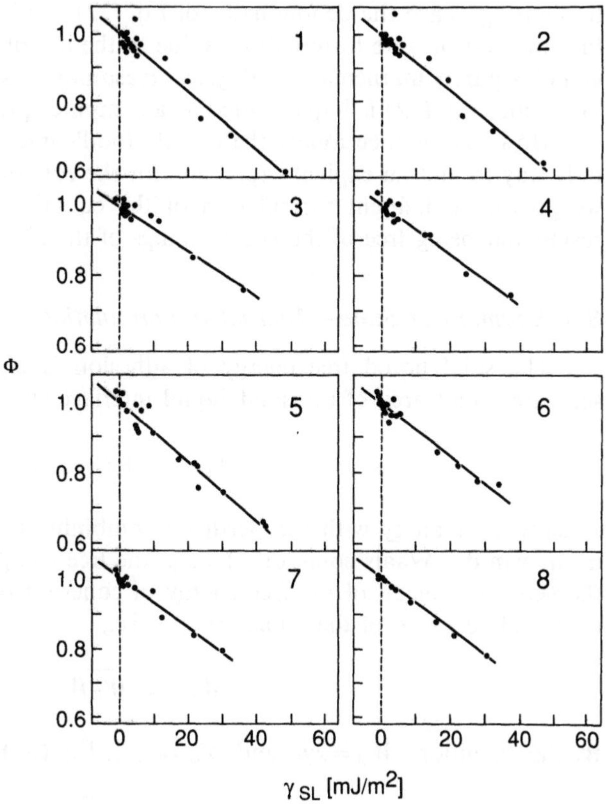

FIGURE 6. Interaction parameter Φ as a function of γ_{sl} for the eight systems given in Figure 5.

From this, we may conclude that the relationship $\Phi = f(\gamma_{sl})$ holds for all low-energy surfaces.

The values of the two constants, α and β, in Eq. (30) were found by the best fit of Eq. (30) to the experimental data:[14]

$$\alpha = \frac{d\Phi}{d\gamma_{sl}} = -0.0075 \text{ m}^2/\text{mJ} \qquad \beta = 1.000 \tag{31}$$

Combining Eq. (28) with Eqs. (30) and (31), we can obtain an explicit form of the equation of state:

$$\gamma_{sl} = \frac{(\sqrt{\gamma_{lv}} - \sqrt{\gamma_{sv}})^2}{1 - 0.015\sqrt{\gamma_{lv}\gamma_{sv}}} \tag{32}$$

Combining this equation of state with Young's equation, we have

$$\cos\theta_Y = \frac{0.0015\gamma_{sv} - 2.00)\sqrt{\gamma_{lv}\gamma_{sv}} + \gamma_{lv}}{\gamma_{lv}(0.015\sqrt{\gamma_{lv}\gamma_{sv}} - 1)} \tag{33}$$

Notice that difficulties may arise for liquids with relatively large surface tensions γ_{lv}, since the denominator of Eq. (32) can become zero. This limitation of the equation of state formulation is due to the use of Good's interaction parameter Φ and is purely mathematical. Physical reasoning was used to "mend" Eq. (32), and in practice Eq. (32) is implemented as a computer program[14] or a set of tables.[15]

However, as mentioned, the use of Good's interaction parameter Φ is not the only way to find an explicit expression for the equation of state. In the next section we present a different formulation of the equation of state,[20] giving the same results but being free of the shortcomings of the above development.

3.3. Equation of State—Alternative Formulation

The solid–liquid free energy of adhesion is equal to the work required to separate a unit area of the solid–liquid interface;[21] that is,

$$W_{sl} = \gamma_{lv} + \gamma_{sv} - \gamma_{sl} \tag{34}$$

Usually, in analogy with the Berthelot combining rule for the attractive constants in the van der Waals equation of state, the free energy of adhesion W_{sl} is taken as the geometric mean of the free energy of cohesion of the solid, W_{ss}, and the free energy of cohesion of the liquid, W_{ll};[22] i.e.,

$$W_{sl} = \sqrt{W_{ll} W_{ss}} \tag{35}$$

By the definitions $W_{ll} = 2\gamma_{lv}$ and $W_{ss} = 2\gamma_{sv}$, Eq. (35) becomes

$$W_{sl} = 2\sqrt{\gamma_{lv}\gamma_{sv}} \tag{35a}$$

Therefore, combining Eq. (34) with Eq. (35a), the solid–liquid interfacial tension γ_{sl} can be written as

$$\gamma_{sl} = \gamma_{lv} + \gamma_{sv} - 2\sqrt{\gamma_{lv}\gamma_{sv}} = (\sqrt{\gamma_{lv}} - \sqrt{\gamma_{sv}})^2 \tag{36}$$

However, it has been found that this simple equation of state, Eq. (36), works only for situations where γ_{lv} values are close to the values of γ_{sv}. This is because, as we will show below, the functional form of Eq. (36) is the result of using the geometric mean combining rule, Eq. (35), which is valid only for $W_{ll} \approx W_{ss}$.

To modify the geometric mean combining rule, Girifalco and Good[22] introduced the interaction parameter Φ as the ratio of the free energy of adhesion between two phases to the square root of the product of the free energies of cohesion of these two phases, leading to

$$W_{sl} = \Phi\sqrt{W_{ll} W_{ss}} \tag{37}$$

An alternative form of Φ is given in Eq. (28), which was derived by combining the definition of interaction parameter Φ, Eq. (37), with Eq. (34). It was found that[14]

$$\Phi \leqslant 1 \tag{38}$$

This implies that the geometric mean combining rule, Eq. (35), generally overestimates the value of W_{sl}.

Actually, the above pattern holds generally for bulk systems too. In the theory of intermolecular interactions and the theory of mixtures, the combining rule is used to evaluate the parameters of unlike-pair interactions in terms of those of the like interactions. As for many other combining rules, the Berthelot rule, i.e., the geometric mean combining rule

$$\varepsilon_{ij} = \sqrt{\varepsilon_{ii}\varepsilon_{jj}} \tag{39}$$

where ε_{ij} is the energy parameter for unlike-pair interactions and ε_{ii}, ε_{jj} are the energy parameters for like-pair interactions, is only a useful approximation and does not provide a secure basis for understanding the unlike-pair interactions. Finding better combining rules to characterize the unlike-pair interactions in terms of like ones has been the subject of much research related to equations of state of liquid mixtures. Reviews on this subject can be found elsewhere.[23, 24] An important question is how far the interactions of unlike molecules can be expressed in terms of the two like interactions, as in the Berthelot rule Eq. (39). By London's theory of dispersion forces, it has been shown[23] that the geometric mean combining rule Eq. (39) is applicable only for similar molecules, because implicit in this rule is the condition that the two energy parameters of like-pair interactions must be very close to each other; i.e., $\varepsilon_{ii} \approx \varepsilon_{jj}$. However, for the interactions between two very dissimilar molecules or materials, where there is an apparent difference between ε_{ii} and ε_{jj}, it has been demonstrated[25, 26] that the geometric mean combining rule generally overestimates the strength of the unlike-pair interactions. This is the case even for the interactions between similar molecules,[23] although the extent of the overestimation is smaller than that between dissimilar molecules. Clearly, this is why the geometric mean combining rule does not work for situations of large differences $|W_{ll} - W_{ss}|$ or $|\gamma_{lv} - \gamma_{sv}|$.

In the study of mixtures, it has become common practice to introduce a factor $1 - K_{ij}$ to the geometric mean combining rule,

$$\varepsilon_{ij} = (1 - K_{ij})\sqrt{\varepsilon_{ii}\varepsilon_{jj}} \tag{40}$$

where K_{ij} is an empirical parameter quantifying deviations from the geometric mean combining rule. Since the geometric mean combining rule overestimates the strength of the unlike-pair interactions, the modifying factor $1 - K_{ij}$ should decrease with the difference $\varepsilon_{ii} - \varepsilon_{jj}$ and be equal to unity when the difference $\varepsilon_{ii} - \varepsilon_{jj}$ is zero. Based on this thought, we may consider a modified combining rule of the form

$$\varepsilon_{ij} = \sqrt{\varepsilon_{ii}\varepsilon_{jj}}\, e^{-\alpha(\varepsilon_{ii} - \varepsilon_{jj})^2} \tag{41}$$

where α is an empirical constant; the square of the difference $\varepsilon_{ii} - \varepsilon_{jj}$, rather than the difference itself, reflects the symmetry of this combining rule and, hence, the anticipated symmetry of the equation of state.[27]

Correspondingly, for the cases of large differences $|W_{ll} - W_{ss}|$ or $|\gamma_{lv} - \gamma_{sv}|$, the combining rule for the free energy of adhesion of a solid–liquid pair can be written as

$$W_{sl} = \sqrt{W_{ll} W_{ss}}\, e^{-\alpha(W_{ll} - W_{ss})^2} \tag{42}$$

or, more explicitly, by using $W_{ll} = 2\gamma_{lv}$ and $W_{ss} = 2\gamma_{sv}$,

$$W_{sl} = 2\sqrt{\gamma_{lv}\gamma_{sv}}\, e^{-\beta(\gamma_{lv} - \gamma_{sv})^2} \tag{43}$$

In the above, α and β are as yet unknown constants. Clearly, when the values of γ_{lv} and γ_{sv} are close to each other, Eq. (43) will revert to Eq. (35a), the geometric mean combining rule. Coupling Eq. (43) with Eq. (34), we can write an equation of state of interfacial tensions as

$$\gamma_{sl} = \gamma_{lv} + \gamma_{sv} - 2\sqrt{\gamma_{lv}\gamma_{sv}}\, e^{-\beta(\gamma_{lv} - \gamma_{sv})^2} \tag{44}$$

Obviously, Eq. (44) will not have the difficulty of a discontinuity as Eq. (32). Combining Eq. (44) with Young's equation, Eq. (1), will yield

$$\cos\theta = -1 + 2\sqrt{\frac{\gamma_{sv}}{\gamma_{lv}}}\, e^{-\beta(\gamma_{lv} - \gamma_{sv})^2} \tag{45}$$

where θ is understood to be the Young contact angle, θ_Y. For simplicity we will omit the subscript Y in the following derivations.

By fitting Eq. (45) to the experimental data[(14)] used for deriving the equation of state, Eq. (32), the constant β in Eqs. (44) and (45) is obtained as $\beta = 0.000115$ $(\mathrm{m^2/mJ})^2$. Details of the purely mathematical data handling procedure are given elsewhere.[(28)] It is apparent that Eq. (45) has three variables, γ_{lv}, θ, and γ_{sv}, and thus will enable us to determine the solid surface tension, γ_{sv}, when we have the experimental data for the liquid surface tension, γ_{lv}, and the contact angle θ. For this purpose, Eq. (45) may be rearranged as

$$\frac{\cos\theta + 1}{2} = \sqrt{\frac{\gamma_{sv}}{\gamma_{lv}}} \exp\left[-\beta\gamma_{lv}^2\left(1 - \frac{\gamma_{sv}}{\gamma_{lv}}\right)^2\right] \tag{46}$$

Let

$$x^2 = \frac{\gamma_{sv}}{\gamma_{lv}} \tag{47}$$

and

$$a = \frac{\cos\theta + 1}{2} \qquad b = \beta\gamma_{lv}^2 \tag{48}$$

Equation (45) can then be written as

$$a = x e^{-b(1 - x^2)^2}$$

or

$$x = ae^{b(1-x^2)^2} \tag{49}$$

For given γ_{lv} and θ, a and b in Eq. (49) can be readily determined by Eq. (48); the only unknown is x. Since the contact angle data are obtained from the solid–liquid–vapor systems with $\gamma_{sv} < \gamma_{lv}$, by Eq. (47), x must have a value between 0 and 1; i.e., $0 \leqslant x \leqslant 1$. Therefore, for a given pair of γ_{lv} and θ, finding x from Eq. (49) is equivalent to finding the root of the following equation:

$$f(x) = x - ae^{b(1-x^2)^2} = 0 \qquad (0 \leqslant x \leqslant 1) \tag{50}$$

Mathematically, Eq. (50) can be easily solved, for example, by using the Newton method.[29] Once the x value is obtained, the solid–vapor surface tension γ_{sv} can be determined from Eq. (47). The solid–liquid interfacial tension γ_{sl} can then be calculated either from the equation of state (44) or from Young's equation (1). A FORTRAN computer program that does this is provided in the appendix. Alternatively, based on the standard Newton method, such calculations can also be performed easily by using an ordinary calculator. This is illustrated below.

Newton's method[29] is one of the most popular techniques for finding the root of an equation $f(x) = 0$. According to the Newton method, the real root of Eq. (50) can be approached by the successive approximations

$$x_{n+1} = x_n - \frac{f(x_n)}{f'(x_n)} \qquad (n = 1, 2, ...) \tag{51}$$

This approximation process is continued until the desired degree of accuracy is obtained. To use a handheld calculator to find the root of Eq. (50) by Newton's method, the explicit form of Eq. (51) can be written as

$$x_{n+1} = x_n - \frac{x_n - ae^{b(1-x_n^2)^2}}{1 + 4abx_n(1 - x_n^2)e^{b(1-x_n^2)^2}} \tag{51a}$$

If we are given that $\gamma_{lv} = 72.8\ \text{mJ/m}^2$ and $\theta = 60°$, by Eq. (48) the values of a and b in Eq. (51) are determined as

$$a = \frac{\cos 60° + 1}{2} = 0.750 \qquad b = 0.000115(72.8)^2 = 0.6094816$$

Remembering that $0 \leqslant x \leqslant 1$ in Eq. (47), we may start searching for the real root by choosing

$$x_1 = 1.0$$

as the first approximation. By Eq. (51a), we have

$$x_2 = 1 - \frac{1 - ae^0}{1 + 0} = a = 0.750$$

Using this x_2 value in Eq. (47), we can calculate the value of the solid–vapor surface tension as

$$\gamma_{sv,2} = x^2\gamma_{lv} = (0.750)^2 \times 72.8\ \mathrm{mJ/m^2} = 40.95000\ \mathrm{mJ/m^2}$$

Next, using $x_2 = 0.75$ in Eq. (51a), we find

$$x_3 = 0.805430$$

and, by Eq. (47),

$$\gamma_{sv,3} = 47.22663\ \mathrm{mJ/m^2}$$

Similarly, we have

$$x_4 = 0.807453 \qquad \gamma_{sv,4} = 47.46417\ \mathrm{mJ/m^2}$$

$$x_5 = 0.807456 \qquad \gamma_{sv,5} = 47.46521\ \mathrm{mJ/m^2}$$

TABLE 2. Comparison between the Equations of State (33) and (45)

		γ_{sv} (mJ/m^2)	
γ_{lv} (mJ/m^2)	θ (deg)	Eq. (33)	Eq. (45)
70.0	20.0	66.0	66.1
70.0	30.0	61.7	61.9
70.0	40.0	56.5	56.8
70.0	50.0	50.8	51.2
70.0	60.0	45.5	45.3
70.0	70.0	39.1	39.2
70.0	80.0	33.2	33.0
70.0	90.0	27.1	26.9
70.0	100.0	21.1	20.8
70.0	110.0	15.2	15.1
50.0	20.0	47.1	47.1
50.0	30.0	44.0	43.9
50.0	40.0	40.0	39.9
50.0	50.0	35.5	35.4
50.0	60.0	30.7	30.7
50.0	70.0	25.7	25.8
50.0	80.0	20.9	20.9
50.0	90.0	16.2	16.2
50.0	100.0	11.9	11.9
30.0	20.0	28.2	28.2
30.0	30.0	26.2	26.2
30.0	40.0	23.6	23.6
30.0	50.0	20.6	20.7
30.0	60.0	17.5	17.5
30.0	70.0	14.3	14.3
30.0	80.0	11.3	11.2

The accuracy at this point is adequate for most purposes. Therefore, if we take

$$\gamma_{sv} = 47.5 \text{ mJ/m}^2$$

the corresponding solid–liquid interfacial tension can then be determined by Young's equation to be

$$\gamma_{sl} = \gamma_{sv} - \gamma_{lv} \cos\theta = 47.5 - 72.8 \cos 60° = 11.1 \text{ mJ/m}^2$$

Equation (33) was compared with Eq. (45) by calculating the solid surface tension γ_{sv} with hypothetical values of liquid surface tension γ_{lv} and contact angle θ. Examples of these results are given in Table 2. It is evident that Eq. (45) yields essentially the same results as that of Eq. (33). The larger discrepancies at high values of γ_{lv} occur due to the linear interpolation used to overcome the mathematical difficulty associated with Eq. (33). We suspect that, if extreme accuracy matters, the results of Eq. (45) are more reliable than those of Eq. (33).

4. EXPERIMENTAL SUPPORT FOR THE EQUATION OF STATE

For many years a persistent problem in surface science has been the direct measurement of interfacial tensions involving a solid phase. The many uncertainties associated with such measurements have dissuaded most authors from seeking independent, direct experimental support for their predictions of γ_{sv} and γ_{sl} obtained indirectly from contact angles, for example. In this section we consider a number of experimental observations that verify the γ_{sv} and γ_{sl} predictions of the equation of state from contact angles.

4.1. Direct Force Measurements

The surface force apparatus of Israelachvili is capable of measuring directly the intermolecular forces between solid substrates separated by gases or liquids.[30] If such forces are recorded as a function of the separation distance, it is possible to integrate to obtain the energy of surface interaction, which is just twice the surface tension.

Claesson *et al.*[31, 32] have measured the interaction force in air and water between two hydrophobic surfaces of mica coated with dimethyldioctadecylammonium ($DDOA^+$). These surfaces had an advancing water contact angle of 94°. Values of γ_{sl} for the $DDOA^+$–mica surface against water were obtained in two ways:

1. Measurements of force and separation were fitted with an exponential function to give a force law that was extrapolated to zero separation. This adhesion force F_a is then related to γ_{sl} by

$$\gamma_{sl} = \frac{F_a}{4\pi R} \tag{52}$$

where R is the radius of the crossed mica cylinders used in the experiment.[31]

2. The adhesion force was measured directly as the crossed mica cylinders were pulled apart while immersed in water.

The two methods gave γ_{sl} values of 29 mJ/m^2 and 34 mJ/m^2, respectively.

Method 2 was also used to obtain γ_{sv}, the solid surface tension of the DDOA$^+$-coated mica. In this case the crossed cylinders were separated in air, and the measured adhesion force yielded $\gamma_{sv} = 27$ mJ/m^2. This value together with $\gamma_{lv} = 72$ mJ/m^2 and $\theta = 94°$ in Young's equation, Eq. (1), yields $\gamma_{sl} = 32$ mJ/m^2, which is consistent with 29 mJ/m^2 and 34 mJ/m^2 from the direct measurements in water.

Using the advancing water contact angle of 94° and $\gamma_{lv} = 72$ mJ/m^2, the equation of state gives $\gamma_{sv} = 26$ mJ/m^2 and $\gamma_{sl} = 31$ mJ/m^2.[32] Table 3 summarizes the comparison between the equation of state values and the direct measurements.

The relatively good agreement among the values of Table 3 is an important confirmation of the accuracy of the equation of state. It will be most interesting to see whether direct force measurements with other solids and liquids continue to support the equation of state.

4.2. Solidification Front Experiments

The behavior of microscopic solid particles, at advancing liquid solidification fronts, can be explained in terms of interfacial free energy changes.[33] This fact was used by Omenyi and Neumann[34, 35] to provide an independent test of predictions of solid and solid–liquid interfacial tensions determined by the equation of state from contact angle measurements.

When a microscopic particle, initially embedded in the liquid phase of a matrix material, is approached by the solid–liquid interface of the solidifying matrix, its subsequent behavior is governed largely by the free energy of adhesion of the particle and the contacting interface. Particle engulfment must be preceded by adhesion, and vice versa for particle rejection. Thus, the initial rejection or engulfment of the particle may be predicted by the free energy of adhesion given by

$$\Delta F^{adh} = \gamma_{ps} - \gamma_{pl} - \gamma_{sl} \tag{53}$$

where γ_{ps}, γ_{pl}, and γ_{sl} are, respectively, the interfacial tensions between the particle and the matrix solid phase, the particle and the matrix liquid phase, and the matrix solid and liquid. If ΔF^{adh} is positive, the adhesion of the particle is thermodynamically unfavorable because of the predicted increase of the system's free energy.

TABLE 3. Comparison of γ_{SV} and γ_{SL} mJ/m^2 Values Obtained by Direct Force Measurements and by the Contact Angle–Equation of State Approach

	Method direct force measurement	Contact angle–equation of state
γ_{SV}	27	26
γ_{SL}	29, 34	31

Particle adhesion is favored if ΔF^{adh} is negative, thereby resulting in a decrease in the overall system free energy. Therefore, a knowledge of the three relevant interfacial tensions can be used to calculate ΔF^{adh} and predict the outcome of a particle engulfing experiment. The accuracy of such predictions can in turn be used to judge the accuracy of the method used to obtain γ_{ps}, γ_{pl}, and γ_{sl}. This is the basis of the present test of the equation of state.

The experimental apparatus consists of a copper cell with external dimensions of 80 × 30 × 3 mm and a groove 30 mm long, 5 mm wide, and 1 mm deep, machined centrally near one end of the cell. The temperature at each end of the cell is regulated by thermoelectric devices that are programmable in order to establish the desired cooling rate and speed of the solidification front. Observations of particle rejection or engulfing are made with a reflected light microscope at magnifications between 64 × and 128 × for particle diameters between 10 and 200 μm.

In order to test the predictions of the equation of state, very slow solidification rates of less than 2 μm/s were chosen. In case of doubt, the solidification front was made to recede, melting again the matrix material surrounding a particle that was just engulfed and reinitiating the process of solidification, possibly at a slower speed. This procedure is of importance because all particles, irrespective of their properties, will be engulfed at sufficiently high rates of solidification due to liquid drag forces.[35]

Table 4 lists the materials that have been used for particles and matrices.[35] In both cases, solid surface tensions γ_{sv} and γ_{pv} were determined from contact angles with liquids of known surface tension using the equation of state. The data were obtained at different temperatures, thereby allowing the calculation of $d\gamma_{sv}/dT$ and $d\gamma_{pv}/dT$, which then could be used to find γ_{sv} and γ_{pv} at the melting points of the various matrix materials. The surface tension of each matrix liquid phase was measured with the Wilhelmy plate technique. Finally, in order to calculate the solid–solid interfacial tension γ_{ps}, it was necessary to treat the equation of state as a "generic" equation of the form

$$\gamma_{12} = f(\gamma_{1v}, \gamma_{2v}) \tag{54}$$

This represents an extrapolation to two solid phases, because the explicit form of the equation of state was derived empirically from contact angle data. Although this lacks a rigorous justification, the fact that this approach does, in fact, accurately predict particle engulfing behavior, lends confidence to this generic usage of the equation of state.

Table 4 lists the predictions of ΔF^{adh} calculated using the equation of state (ΔF_{I}^{adh}) and shows the corresponding experimental observation: particle engulfment or rejection. In the vast majority of cases the equation of state predictions of ΔF^{adh} are substantiated by the experimental observation; i.e., $\Delta F^{adh} < 0$ leads to engulfing (E) and $\Delta F^{adh} > 0$ leads to rejection (R). In some cases, where ΔF^{adh} is near zero, the experimental error in the determination of the interfacial tensions makes the prediction unclear.[35] This is compounded by the transitional behavior of particles in these cases, whereby the particle is reoriented by the solidification front but not clearly pushed. Such observations are denoted by the letter E in parentheses in

TABLE 4. Theoretical Predictions and Microscopic Observations of Particle Behavior at Solidification Fronts

		Naphthalene	Biphenyl	Thymol	*o*-Terphenyl	Salol	2-Phenylphenol	Pinacol
Acetal	ΔF_{I}^{adh}	2.58	1.99	0.198	1.12	0.204	0.013	0.047
	ΔF_{II}^{adh}	1.56	−0.181	−5.66	−1.90	−5.41	−5.47	−5.33
	Observation	R	R	R	R	R	R	R
Nylon 6	ΔF_{I}^{adh}	1.98	1.76	0.186	1.00	0.189	0.011	0.045
	ΔF_{II}^{adh}	1.42	−0.168	−5.39	−1.77	−5.14	−5.06	−5.12
	Observation	R	R	R	R	R	—	E
Nylon 6,6	ΔF_{I}^{adh}	2.12	1.67	0.177	0.922	0.174	0.009	0.043
	ΔF_{II}^{adh}	1.29	−0.154	−5.02	−1.62	−4.74	−4.57	−4.81
	Observation	R	R	R	R	E	—	E
Nylon 12	ΔF_{I}^{adh}	1.57	1.28	0.147	0.631	0.125	0.003	0.037
	ΔF_{II}^{adh}	1.12	−0.135	−4.50	−1.41	−4.19	−3.86	−4.36
	Observation	R	R	R	R	R	E	R
Nylon 6,10	ΔF_{I}^{adh}	0.720	0.735	0.109	0.263	0.070	−0.003	0.029
	ΔF_{II}^{adh}	0.967	−0.118	−4.05	−1.22	−3.69	−3.27	−3.97
	Observation	R	R	R	R	R	E	R
Nylon 6,12	ΔF_{I}^{adh}	−0.279	0.038	0.056	−0.236	−0.013	−0.014	0.018
	ΔF_{II}^{adh}	0.667	−0.084	−3.06	−0.822	−2.56	−1.97	−3.06
	Observation	R	R	R	R	E	E	E
Polystyrene	ΔF_{I}^{adh}	−1.42	−0.757	−0.002	−0.797	−0.105	−0.025	0.006
	ΔF_{II}^{adh}	0.187	−0.031	−1.67	−0.250	−1.02	−0.082	−1.88
	Observation	E	E	R	E	E	E	R
Teflon	ΔF_{I}^{adh}	−5.10	−3.21	−0.186	−2.51	−0.380	−0.059	−0.030
	ΔF_{II}^{adh}	−0.962	0.098	1.91	1.20	2.99	4.66	1.28
	Observation	E	E	E	E	E	E	R
Siliconized glass	ΔF_{I}^{adh}	−6.58	−4.05	−0.231	−3.00	−0.437	−0.068	−0.038
	ΔF_{II}^{adh}	−1.47	0.145	2.86	1.64	3.85	6.06	1.98
	Observation	E	E	E	E	E	E	E

ΔF_{I}^{adh} = free energy of adhesion in mJ/m^2 calculated from equation of state; ΔF_{II}^{adh} = free energy of adhesion in mJ/m^2 calculated from Fowkes' theory; R = rejection; E = engulfment.
Teflon in benzophenone: $\Delta F_{I}^{adh} = -3.2$ mJ/m^2; $\Delta F_{II}^{adh} = 2.6$ mJ/m^2; observation: E.
Teflon in bibenzyl: $\Delta F_{I}^{adh} = -0.96$ mJ/m^2; $\Delta F_{II}^{adh} = 0.03$ mJ/m^2; observation: E.

Table 4, and, as expected, they correspond to small absolute values of ΔF^{adh}. There are, however, two cases (nylon 6,12 in naphthalene and *o*-terphenyl) in which the absolute value of ΔF^{adh} is relatively large but which produced discrepancies between the equation of state prediction and the experimental observation. The reason for this is unknown. Table 4 also lists the predictions of the Fowkes equation (ΔF_{II}^{adh}), which will be discussed in Section 4.6.

The solidification front experiment has since been developed into a technique for measuring the surface tensions of small particles. In this case, we measure the interface speed at which rejected particles first begin to be engulfed.[36]

In conclusion, predictions of particle engulfing or rejection were observed to be accurate in almost all cases. For example, if we consider only those experiments in which $|\Delta F_{I}^{adh}| \geqslant 0.2$ to be representative of the most clear-cut observations (unambiguous pushing or rejection), then the equation of state produces agreement 29 out of 31 times (94%).

4.3. Contact Angles for Different Liquids

The equation of state concept holds that γ_{sl} is a function only of γ_{sv} and γ_{lv}. Therefore, in the absence of vapor adsorption on the solid, contact angle measurements on a single substrate with a range of liquids should all give the same value of γ_{sv}. In this case, θ and γ_{lv} of the measuring liquid are the only two inputs into the equation of state and γ_{sv} and γ_{sl} are the outputs.

Before we consider a series of contact angle experiments, it is appropriate to note the considerable care that must be taken to obtain good-quality contact angles. The measurement of contact angles is subject to many sources of error that are often overlooked. This lack of appreciation of the subtleties involved has caused many investigators to be misled by spurious data, thereby adding great confusion to the field of surface science. Particularly important are the effects of surface roughness and the necessity of establishing a truly advancing contact angle. Scratches less than 0.5 μm deep can cause an advancing liquid drop to "hinge" at the scratch, rendering the contact angle essentially meaningless.[37, 38] Surface roughness and homogeneity should always be assessed by measurement of both the advancing and receding contact angles to give the contact angle hysteresis. Also note that the common practice of adding liquid to an existing drop by touching it momentarily with a pendant drop suspended from a platinum wire or the tip of a pipet may be incorrect. Although the additional liquid indeed causes the drop to grow, the contact angle is often not the maximal advancing angle because of the vibration caused by the sudden absorption of the pendant drop. It has been observed many times that this effect can produce intermediate, metastable contact angles as much as 7° below the advancing angle.[38] If a pipet or syringe is used to add liquid to a sessile drop, it must penetrate the interface and not be withdrawn prior to observation.

Returning to the equation of state, consider the contact angles reported in Table 5. The solid substrate in this case was hexatriacontane deposited by vacuum sublimation on clean glass.[9] These surfaces were so smooth and homogeneous that no contact angle hysteresis (difference between advancing and receding angles) was observed even though the measurement technique was capillary rise at a vertical

TABLE 5. Solid Surface Tension γ_{SV} (mJ/m^2) of *n*-Hexatriacontane at 20°C Calculated with the Equation of State. Contact Angle Data (degrees)[9].

Liquid	γ_{LV}	θ	γ_{SV}
Water	72.8	104.6	19.8
Glycerol	63.4	95.4	20.0
Thiodiglycol	54.0	86.3	19.8
Ethylene glycol	47.7	79.2	19.8
Hexadecane	27.6	46.0	20.1
Tetradecane	26.7	41.0	20.7
Dodecane	25.4	38.0	20.4
Decane	23.9	28.0	21.2
Nonane	22.9	25.0	20.8

plate, which has a precision of about 0.1° (better than this if used with digital image analysis; see Section 7). As discussed in Section 3.1 vapor adsorption of the test liquids should be negligible on such a homogeneous, low-energy material as hexatriacontane,[39] and this is reflected in the consistency of the γ_{sv} values calculated by the equation of state. It is significant that, regardless of the relative magnitudes of the various intermolecular forces within the different liquids, the equation of state, using only the measured contact angle and the total liquid surface tension, predicts correctly a constant solid surface tension γ_{sv}.

The ability of the equation of state to correlate a wide range of measured $\theta-\gamma_{lv}$ data with a single γ_{sv} is a fundamental test of both the equation of state concept [that $\gamma_{sl}=f(\gamma_{sv},\gamma_{lv})$] and the explicit formulation of the equation.

4.4. Sedimentation Volumes

Sedimentation experiments are a well-established technique to study the stability of powder dispersions in liquids. The behavior of such systems is governed largely by van der Waals and electrostatic interactions, although a complete model has yet to be developed. In this section we consider sedimentation volume data that indicates the role of van der Waals forces as reflected by the relevant interfacial tensions of the solid particles suspended in a liquid.

The relationship between van der Waals interactions and surface thermodynamics is evident when one realizes that the free energy of adhesion between two solids of surface tension γ_{pv} and γ_{sv}, in a liquid of surface tension γ_{lv}, is just the integral of the van der Waals forces from infinity to the equilibrium separation distance and is given by Eq. (53). The surface tensions of Eq. (53) result from the interactions of all possible intermolecular forces. If only van der Waals forces are considered, the free energy of adhesion of two solid half-spaces in a liquid at separation d_0 can also be written in terms of Hamaker coefficients as

$$\Delta F^{adh}_{pls}=\frac{-A_{pls}}{12\pi d^2_{0_{pls}}} \tag{55}$$

where d_0 is the equilibrium separation distance between the solids. If we assume that d_0 is so small that there is in essence contact between the two solid phases, then substitution of Eq. (55) into Eq. (53) gives

$$A_{pls}=-12\pi d^2_{0_{pls}}(\gamma_{ps}-\gamma_{pl}-\gamma_{sl}) \tag{56}$$

which makes explicit the relationship between interfacial tensions and Hamaker coefficients. If $A_{pls}<0$, the two solids will repel each other ($\Delta F^{adh}_{pls}>0$), and if $A_{pls}>0$ the solids will be attracted ($\Delta F^{adh}_{pls}<0$).

Consider now, the case of like particles suspended in a liquid. Instead of the free energy of adhesion we now have the free energy of cohesion

$$\Delta F^{coh}_{plp}=-2\gamma_{pl} \tag{57}$$

which is maximum when $\gamma_{pl}=0$ corresponding to $\gamma_{pv}=\gamma_{lv}$. From Eq. (56) it is clear that the Hamaker coefficient, A_{plp}, will then be positive or zero, depending on γ_{lv} of the liquid. In other words, for a given level of electrostatic repulsion, the degree of particle attraction and, hence, sedimentation volume will be a function of the surface tension of the suspending liquid, γ_{lv}, reaching a minimum when $\gamma_{\text{lv}} = \gamma_{\text{pv}}$. This provides us with another method of independently measuring γ_{pv} and comparing it with the value obtained from contact angles and the equation of state.(40–42)

Before considering the experimental results, it is useful to recognize two possible sedimentation mechanisms, depending on whether particle agglomeration occurs.

1. If there is no agglomeration at finite values of van der Waals attraction, then for zero van der Waals attraction ($\gamma_{\text{lv}} = \gamma_{\text{pv}}$; $\gamma_{\text{pl}} = 0$) the repulsive electrostatic forces are unopposed and the sediment will exhibit a maximum volume.
2. If there is agglomeration at finite values of van der Waals attraction, then this agglomeration will cease when the van der Waals attraction is zero ($\gamma_{\text{lv}} = \gamma_{\text{pv}}$). The sedimentation volume is then expected to be a minimum when $\gamma_{\text{lv}} = \gamma_{\text{pv}}$ because the irregularly shaped aggregates produced by agglomeration do not pack as well as individual particles.

The condition $\gamma_{\text{lv}} = \gamma_{\text{pv}}$ may, therefore, result in either a maximum or a minimum in the sedimentation volume.

Sedimentation volumes have been recorded for a number of polymer powders in both pure liquids and binary liquid mixtures of various surface tensions.(40–42) Tables 6 and 7 list the polymers and the liquids used in Ref. 40 along with certain physical properties. The liquids were selected to have the following characteristics: (1) they are chemically inert with respect to the solid particles; (2) the boiling temperature is relatively high to minimize evaporation; (3) the density is less than that of the particles; (4) they cover a wide range of dipole moments; (5) the liquid com-

TABLE 6. Physical Properties of Polymers Used for Sedimentation

Polymer	Density ρ (kg/m^3)	Surface tension γ_{SV} (mJ/m^2)	Particle size range D (μm)
Polytetrafluoroethylene PTFE Grade 1 / No. 6	2000	20.0	500–600 / 200–400
Polyvinylidenefluoride PVDF	1740	25.5	3–30
Polyvinylfluoride PVF	1380	29.4	2–35
Polyethylene high density HDPE	965	30.3	30–120
Polyhexamethylene adipamide, nylon 6,6 PA 66	1090	41.1	60–250
Polysulfone PSF	1240	43.1	10–100

ponents of the binary mixtures are miscible in all ratios; and (6) the surface tension range of the liquid mixtures includes that of the selected particles. The liquids of Table 7 were combined as shown in Table 8 in the ratios required to produce the desired γ_{lv} ranges.

Powder dispersions in these mixtures were placed in 1-mL microtubes, and the sedimentation volumes V_{sed} were recorded over a period of three to seven days until V_{sed} reached a steady value.

The use of pure liquids instead of liquid mixtures was studied in Ref. 41. It was observed that the sedimentation behavior remained unchanged, and possible preferential adsorption of one liquid over another did not seem to influence the results.

Figures 7 and 8 show the pattern of V_{sed} as a function of γ_{LV} for polytetrafluoroethylene (PTFE) in two liquid mixtures. As expected, the extrema in V_{sed} occur at approximately the same value of $\gamma_{LV} = 20.1$ mJ/m^2 for both liquid mixtures. We hypothesize that Fig. 7 represents a situation with particle agglomeration at finite van der Waals attraction and that Fig. 8 shows the effect of no agglomeration. This has been verified with photomicrographs[41] and is consistent with the observation that it is far easier to resuspend those systems with a maximum than those with a minimum. The data of Figs. 7 and 8 are plotted in Fig. 9 together with the data for the mixture diethyl ether and tetralin. According to our model of the mechanisms of particle sedimentation, the locations of the extrema show that the surface tension of PTFE is very close to 20.1 mJ/m^2.

The values of γ_{lv} corresponding to the indicated extrema in V_{sed} are summarized in Table 8. Extrema are minima for nonpolar or slightly polar liquids and maxima for polar mixtures.

In the context of the equation of state, it is most interesting to compare the values of particle surface tension predicted by the γ_{lv} extrema in Table 8 with the values obtained from contact angle measurements and the equation of state. This is done in Table 9 for advancing water contact angles on sheets or films of the particle materials. The equation of state–contact angle predictions of γ_{PV} are in

TABLE 7. Physical Properties of Liquids Used for Sedimentation

Liquids		Boiling point (°C)	Density at 20°C ρ_L (kg m^{-3})	Dipole moment at 20°C $\mu \cdot 10^{-30}$ (cm)	Surface tension at 20°C γ_{LV} (mJ m^{-2}) Literature	Measured
n-Hexane	C_6H_{14}	68.95	660.3	0	18.40	18.3
n-Octane	C_8H_{18}	125.66	702.5	0	21.62	21.8
n-Hexadecane	$C_{16}H_{34}$	286.80	773.3	0	27.47	27.0
Tetralin	$C_{10}H_{12}$	207.57	970.2	1.63	35.60	36.5
Diethyl ether	$C_4H_{10}O$	34.51	713.8	3.84	17.10	17.4
n-Propanol	C_3H_8O	97.40	803.5	5.60	23.71	23.8
n-Hexanol	$C_6H_{14}O$	157.85	813.6	5.17	26.21	25.2
Ethylene glycol	$C_2H_6O_2$	198.93	1108.8	7.61	48.43	48.9
Cyclohexanone	$C_6H_{10}O$	155.65	947.8	9.01	35.19	34.3
2,2′-Thiodiethanol	$C_4H_{10}O_2S$	165/20 nm	1182.0	not available	54.00	52.5

TABLE 8. Extrema of Sedimentation Volumes for Polymers in Different Liquid Combinations

Liquid combinations \ Polymers	Surface tension range γ_{LV} (mJ/m²)	Dipole moment range $\mu \times 10^{-30}$ (cm)	γ_{LV} of Liquid mixtures at 20°C at indicated V_{sed} extrema (mJ/m²)[a] PTFE Grade 1	PTFE No. 6	PVDF	PVF	HDPE	PA 66	PSF
n-Hexane / *n*-Hexadecane	18.3 / 27.0	0 / 0	20.0 min	20.0 min					
n-Octane / Tetralin	21.8 / 36.5	0 / 1.63			27.1 min	29.0 min	29.2 min		
Diethyl ether / Tetralin	17.4 / 36.5	3.84 / 1.63	20.2 min	19.6 min			28.4 min		
Diethyl ether / *n*-Hexanol	17.4 / 25.2	3.84 / 5.17	20.2 max	19.4 max					
n-Propanol / Ethylene glycol	23.8 / 48.9	5.60 / 7.61			27.8 max	28.6 max		38.5 max	42.2 max
n-Propanol / Cyclohexanone	23.8 / 34.3	5.60 / 9.01					30.5 max		
n-Propanol / Thiodiethanol	23.8 / 52.5	5.60 / ?						39.6 max	44.0 max

[a] min and max denote a minimum and maximum, respectively, in V_{sed}.

good agreement (average difference <3%) with those inferred independently from the thermodynamic model of sedimentation. Because the latter did not, in any way, involve the explicit form of the equation of state and depended only on liquid surface tension measurements, these data constitute another piece of evidence that the equation of state approach is correct.

TABLE 9. Comparison of the Surface Tension of Polymers Obtained from Contact Angles and from Sedimentation Volumes V_{sed}

Polymer	θ_{H_2O} [a] (deg)	Polymer surface tension γ_{SV} (mJ/m²) from Contact angle	Sedimentation volume
PTFE Grade 1 / No. 6	104.0 ± 2.0	20.0 ± 1.3	20.1 / 19.7
PVDF	94.8 ± 2.0	25.5 ± 1.3	27.4
PVF	88.6 ± 2.5	29.4 ± 1.7	28.8
HDPE	87.1 ± 2.0	30.3 ± 1.3	29.4
PA 66	70.0 ± 3.0	41.4 ± 1.8	39.0
PSF	66.0 ± 2.0	43.1 ± 1.1	43.1

[a] $\gamma_{LV} = 72.0$ mJ/m² (26°C) except with PTFE and PA 66, where $\gamma_{LV} = 72.4$ mJ/m² (23°C).

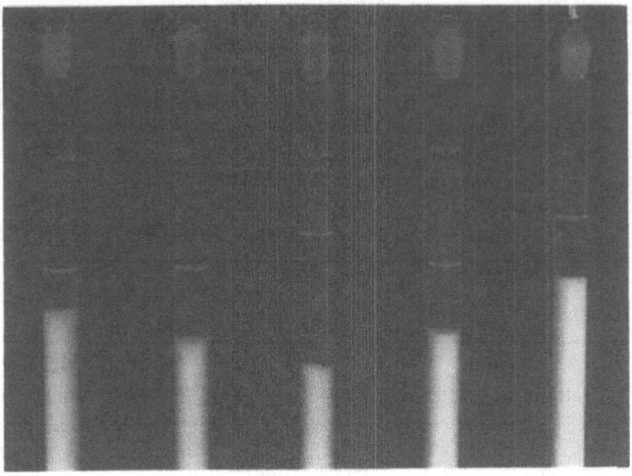

FIGURE 7. Sedimentation volumes of powdered PTFE in mixtures of *n*-hexane and *n*-hexadecane.

4.5. Particle Suspension Layer Stability

When a dilute suspension of small particles in a liquid is carefully layered on a dense liquid as, for example, in zone electrophoresis, centrifugation or isoelectric focusing, the suspension often forms a zone or layer of finite thickness with a well-defined interface between the suspension layer and the supporting liquid. For a stable suspension layer, the suspension layer–liquid cushion interface is planar and "sharp." When the suspension layer becomes unstable, droplets of suspension form at the interface and fall through into the liquid cushion. This instability is generally referred to as *streaming* or *droplet sedimentation*, and is affected by the initial particle concentration, the diffusion of solutes, particle charge, and van der Waals attractive forces.(43) The latter two factors were the focus of a study of suspension layer stability of fixed erythrocytes on a D_2O cushion, which is described below.(43)

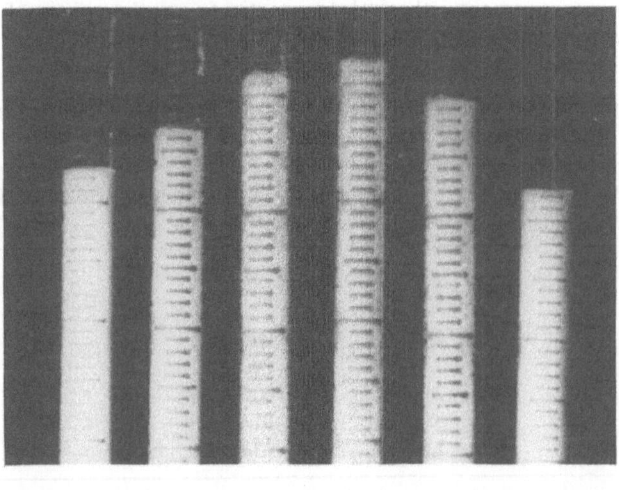

FIGURE 8. Sedimentation volumes of powdered PTFE in mixtures of diethyl ether and *n*-hexanol.

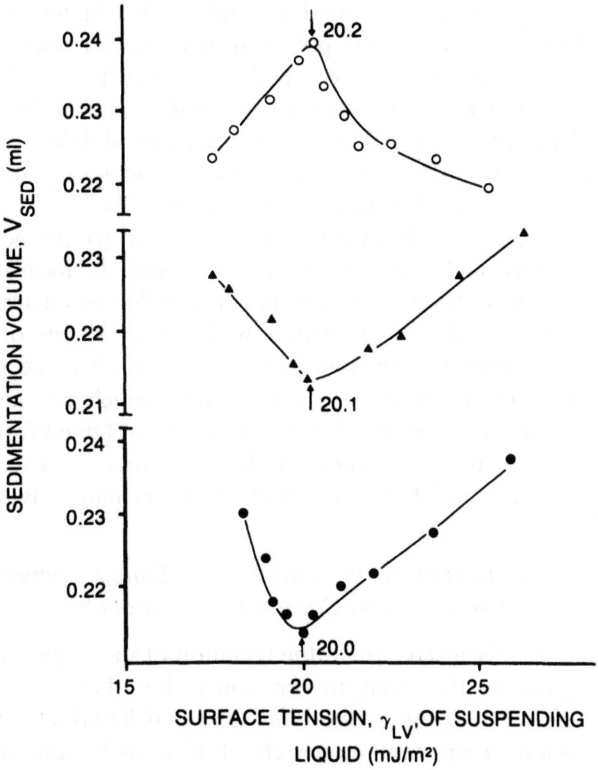

FIGURE 9. Sedimentation volume of PTFE (Grade 1) as a function of suspending liquid surface tension: (●) *n*-Hexane/*n*-Hexadecane, (▲) Diethyl ether/Tetralin, (○) Diethyl ether/*n*-Hexanol.

Under the hypothesis that colloidal stability theory was applicable to suspensions of biological cells, the total energy of interaction between like cells was modeled as the sum of a repulsive component due to electrostatic charge and an attractive component due to van der Waals forces. It has been shown that the Hamaker coefficients, which give the van der Waals potential energy of attraction, may be expressed in terms of the surface tensions of the solid particle (the cell) and the suspending liquid.[44, 45] The Hamaker coefficient for the cells in the liquid reaches its minimum value of zero when the cell surface tension equals that of the suspending liquid. At this point the attractive van der Waals forces are reduced to zero, and the repulsive action of the electrostatic charge is maximized. This serves to prevent cell agglomeration and subsequent droplet sedimentation. In other words, the model predicts that the cell suspension is most stable when γ_{cv}, the cell surface tension, equals the liquid surface tension γ_{lv}. Thus, we have yet another way of measuring a solid surface tension, this time by determining the liquid surface tension that gives maximum particle suspension stability.

The model was tested using five species of glutaraldehyde fixed erythrocytes (human, horse, chicken, canine, turkey) which are red blood cells, treated to render the cell wall rigid. Because of their uniformity of size and shape, fixed erythrocytes make an excellent model particle in many physical studies. The cells were suspended in a saline solution and then layered onto a cushion of D_2O. Stability was recorded as the time required for the onset of droplet sedimentation as indicated by the distortion of the D_2O suspension interface. The surface tension of the saline was varied by adding dimethyl sulfoxide (DMSO), which had a relatively small effect on the cell surface charge potential as measured electrophoretically. It was, therefore, possible to regulate the van der Waals attraction forces while leaving the electrostatic repulsive forces relatively constant. Table 10 lists the surface tensions of the suspending liquids that produced greatest stability, as measured by the elapsed time before the interface became distorted. According to the model, this liquid surface tension should be equal to that of the cells.

So far we have not made reference to the equation of state for interfacial tensions. Unfortunately, it is not possible to measure contact angles on layers of fixed erythrocytes as it is with, say, platelets or bacteria.[46] However, the cell surface tensions could be compared with earlier values obtained with the solidification front technique (Section 4.2), which is itself based on the equation of state.[36, 47] These values of cell surface tension, which are also listed in Table 10, are in remarkably good agreement with those obtained from suspension stabilities. Thus, we have another reassurance that the equation of state, which was used to develop and "calibrate" the solidification front technique, is indeed reliable.

4.6. Verification of the Equation of State Concept and Comparison with the Surface Tension Component Approach

As discussed earlier, the equation of state approach is based on thermodynamic arguments that lead to the conclusion that the solid–liquid interfacial tension is only a function of the *total* solid and liquid surface tensions. Unlike the surface tension component approach of Fowkes[48] and others, the types and relative magnitudes of the intermolecular forces in either phase are not directly relevant. The interfacial tension is completely defined by the total surface tensions of the separate phases. This section summarizes a number of experimental and theoretical observations that reinforce the validity of the equation of state concept, and demonstrates the inadequacy of the present surface tension component approaches. The reason for doing this is that these two methods are mutually exclusive from a theoretical perspective and often yield results in considerable disagreement.

In both the equation of state approach and the theory of surface tension components, solid surface tensions are evaluated using contact angle measurements. From Young's equation, Eq. (1), it is seen that if γ_{lv} and γ_{sv}, the liquid–vapor and solid–vapor surface tensions, respectively, are fixed for a series of solids and liquids,

TABLE 10. Comparison of the Surface Tension of Erythrocytes Obtained via Two Independent Methods

		Technique	
		Droplet sedimentation (mJ/m^2)	Freezing front (mJ/m^2)
Erythrocyte species	Turkey	65.7	65.4
	Chicken	65.2	65.1
	Canine	64.4	64.2
	Horse	65.4	64.5
	Human	64.3	64.1

then the contact angle is directly related to γ_{sl}, the solid–liquid interfacial tension. This concept forms the basis for a direct experimental comparison of the predictions of the two theories.

Consider first two different pure liquids that are chosen to have equal overall surface tensions, as measured by, for example, the Wilhelmy plate technique. These same liquids are, however, also selected to have widely disparate compositions of intermolecular forces. In other words, one liquid may be an alkane (a liquid that has only dispersion forces), whereas the other may be characterized by a large dipole moment and perhaps significant hydrogen bonding. According to the theory of surface tension components, the contact angles of these two liquids on a single solid surface should differ in proportion to the differences in the makeup of the intermolecular forces. In contrast, the equation of state approach predicts that the contact angles will be equal since the total liquid and solid surface tensions are constant. This simple experiment provides a direct test of the basic premise of each of the two theories, and, moreover, it is independent of the specific form of any Fowkes-type equation or of any particular equation of state.

The details of this experiment have been reported elsewhere.[49, 50] In summary, it was observed that for five pairs of liquids the contact angle was essentially identical on solid substrates of Teflon FEP. This is fully consistent with the expectations of the equation of state, but is very difficult to explain with the theory of surface tension components.

Section 4.2 discussed the use of the solidification front experiment to verify the explicit form of the equation of state of Neumann *et al.* Table 4 contains the equation of state prediction of ΔF^{adh} ($\Delta F_{\text{I}}^{\text{adh}}$) and the corresponding observation of particle engulfment or rejection. Also shown in Table 4 are the predictions,[51] $\Delta F_{\text{II}}^{\text{adh}}$, based on the Fowkes equation for systems that interact only by dispersion forces,

$$\gamma_{sl} = \gamma_s + \gamma_l - 2\sqrt{\gamma_s^d \gamma_l^d} \tag{58}$$

where γ_s, γ_l, γ_s^d, and γ_l^d are, respectively, the solid and liquid surface tensions and the corresponding "dispersion components." In the vast majority of cases the Fowkes approach does not correctly predict particle behavior at solidification fronts. This is generally true, even for those matrix and particle materials that are completely or overwhelmingly dispersive, namely polystyrene, Teflon, siliconized glass in naphthalene, biphenyl, and *o*-terphenyl. As in Section 4.2, consideration of only those experiments in which $|\Delta F^{\text{adh}}| \geqslant 0.2$ shows that Eq. (58) produces a prediction that agrees with 32% of the observations as compared with the equation of state that produced agreement in 94% of cases.

It was mentioned at the beginning of this section that, from a theoretical standpoint, the equation of state and surface tension component approaches are mutually exclusive. In Sections 2.1 and 2.2, two thermodynamic proofs were given to support the contention that an equation of state, Eq. (2), exists. This has far-reaching consequences, the most important of which is that an equation such as Eq. (58) would also have to satisfy Eq. (2), but this is apparently impossible, since the dispersion components of the surface tensions would then have to be functions of the total surface tensions. This would preclude the possibility of γ_d being a unique function of intermolecular forces, as stipulated by the Fowkes approach. This impossibility carries over into contact angles. Combining Eqs. (1) and (2) yields

$$\cos\theta = \frac{f(\gamma_s, \gamma_l) - \gamma_s}{\gamma_l} \tag{59}$$

so that

$$\cos\theta = g(\gamma_s, \gamma_l), \tag{60}$$

indicating that the contact angle is completely determined by the total surface tensions γ_s and γ_l. This implies that surface tension components cannot be determined from contact angles. Clearly, any arbitrary solid–liquid–vapor system with the same total surface tensions γ_s and γ_l must have the same contact angle θ. This is true regardless of the relative magnitudes and types of intermolecular forces, including hydrogen bonding.

Overall, it is apparent that any attempt to develop a theory of surface tension components would have to come to terms with Eq. (2). Particularly, there is no apparent way of using contact angles to measure surface tension components. This does not imply that the existence of surface tension components is theoretically impossible, only that contact angles cannot be used to measure them should they exist.

It must, therefore, be concluded that the surface tension components reported in the literature are incorrect because they are all based on Eq. (58). Such components cannot be regarded as physically real quantities.[52] Moreover, since γ_{sl} is a known function of the total solid and liquid surface tensions, in many instances there is little incentive to attempt an evaluation and treatment of surface tension components. Total solid surface tensions can in principle be determined from a single contact angle measurement using the equation of state approach.

5. ALTERNATIVE EQUATIONS OF STATE

As shown in the previous sections, the equation of state approach has a substantial thermodynamic basis and is in excellent agreement with various experimental results. However, as distinct from the equation of state, Eqs. (32) or (44), there exist some other equation-of-state-type relations; two typical ones are the unmodified Good equation, Eq. (36), and Antonow's rule

$$\gamma_{12} = \gamma_{1v} - \gamma_{2v} \tag{61}$$

Equation (36) represents the standard Good equation, Eq. (28), with a value of unity for the interaction parameter Φ. Equation (61), although never having been derived in any fashion, appears in the literature from time to time. The question may then arise as to what will be the differences when these equation of state relations are used to determine solid surface tensions. A comparative study of this question has been done by employing five approaches to the determination of solid surface tensions.[53] Three of these approaches are sedimentation volumes, solidification front, and contact angles. In the following, we compare the results obtained from Eqs. (44), (36), and (61) with the results of the three approaches.

5.1. Sedimentation Volumes

The sedimentation experiments have been discussed in Section 4.4. Figure 10 shows the free energy of cohesion ΔF^{coh} versus the liquid surface tension γ_{lv} for hypothetical particles having a surface tension $\gamma_{pv} = 20\ \mathrm{mJ/m^2}$. All three equations produce a minimum at $\gamma_{lv} = \gamma_{pv}$, suggesting that in a sedimentation experiment one should observe an extremum (presumably a minimum) in the sedimentation volumes at that point.

As shown, Eqs. (44), (36), and (61) all yield identical results for the prediction of sedimentation volumes, suggesting that all three equations have common features. By inspection we find that for all three equations

$$\gamma_{12} = f(\gamma_{1v}, \gamma_{2v}) = f(\gamma_{2v}, \gamma_{1v}) \tag{62}$$

and

$$\gamma_{12} = f(\gamma_{1v}, \gamma_{2v}) = 0 \qquad \text{when} \quad \gamma_{1v} = \gamma_{2v} \tag{63}$$

Here f denotes an equation of state, and γ_{12}, γ_{1v}, and γ_{2v} have the same meaning as in Eqs. (44), (36), and (61). Equation (62) is a condition of symmetry and Eq. (63) states that the interfacial tension between two phases of equal surface tension is zero.

The sedimentation experiments may be interpreted by referring only to the basic characteristics that are common to the three equations of state, Eqs. (44), (36), and (61). By Eq. (63), when $\gamma_{lv} = \gamma_{pv}$ then $\gamma_{pl} = 0$ and ΔF^{coh} is a maximum, which will generate the minima in Fig. 9. It becomes apparent that the conditions (62) and (63) are crucial for the correct interpretations of the sedimentation volumes by Eqs. (44), (36), and (61). However, condition (63) may still be relaxed slightly. It can be shown that as long as γ_{12} is a minimum when $\gamma_{1v} = \gamma_{2v}$, and the minimum may not necessarily be zero, the minima for the free energy of cohesion

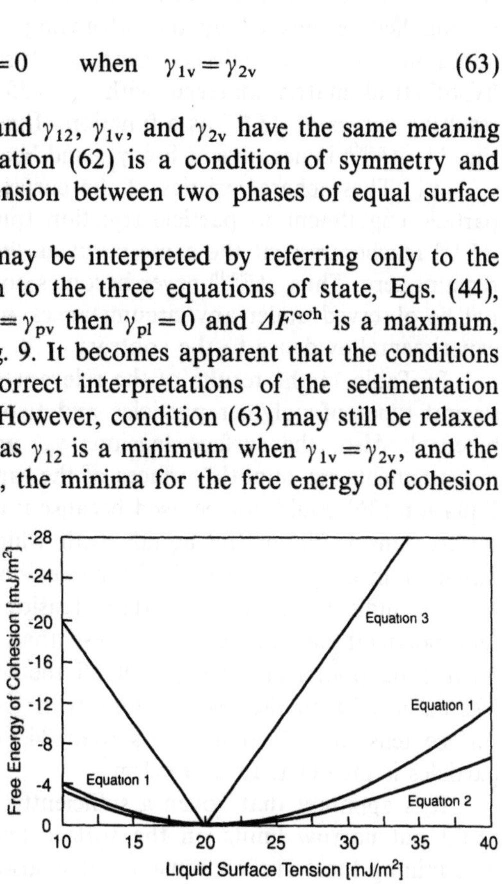

FIGURE 10. Free energy of cohesion calculated using three equation-of-state type relations: Eq. (44) as Eq. (1), Eq. (36) as Eq. (2), Eq. (61) as Eq. (3).

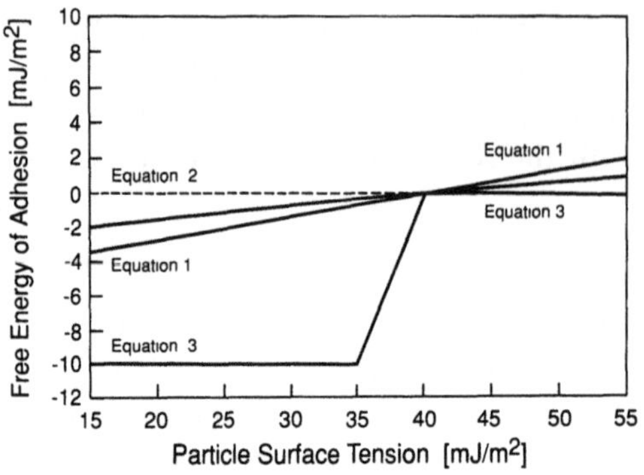

FIGURE 11. Free energy of adhesion calculated using three equation-of-state type relations: Eq. (44) as Eq. (1), Eq. (36) as Eq. (2), Eq. (61) as Eq. (3).

ΔF^{coh} in Fig. 10 and the extrema of the sedimentation volumes in Fig. 9 will move only vertically. Hence, the predictions of the above three equations of state will remain the same.

5.2. *Solidification Front Technique*

The solidification front technique has been described in Section 4.2. Recall that the sign of the free energy of adhesion, ΔF^{adh}, will indicate whether a particle will be engulfed or pushed by the advancing solidification front. The free energy of adhesion can be calculated using, in turn, Eqs. (44), (36), and (61). For a hypothetical matrix material with $\gamma_{sv} = 35$ mJ/m^2 and $\gamma_{lv} = 40$ mJ/m^2, the three resulting curves of ΔF^{adh} as a function of particle surface tension γ_{pv} are plotted in Fig. 11: ΔF^{adh} is negative at low γ_{pv} and becomes positive with increasing γ_{pv} when $\gamma_{pv} = \gamma_{lv}$. These characteristics of the curves imply that there is a transition from particle engulfment to particle rejection (pushing) when $\gamma_{pv} = \gamma_{lv}$. With Eq. (61), ΔF^{adh} reaches zero at the same point as the curves from Eqs. (44) and (36), but remains zero. Thus, ΔF^{adh} never becomes positive so pushing of the particles would not be observed under any circumstances, in contradiction with the overwhelming experimental evidence to the contrary.

In Table 11 the results of the relevant microscopic observations are listed for several types of polymer particles and two matrix materials, benzophenone and bibenzyl. Also, the surface tensions γ_{pv} are calculated from the contact angle measurements on smooth surfaces of the same polymeric materials using Eq. (44). Equation (36) could not be used because it does not yield a γ_{pv} independent of the surface tension γ_{lv} of the liquids with which the contact angles were measured. Table 11 in conjunction with the curves obtained from Eq. (44) and Eq. (36) in Fig. 11 indicates that the surface tensions of Teflon, nylon 6,12, nylon 6,10, and polymethylmethacrylate are less than 39.9 mJ/m^2 (γ_{lv} of benzophenone at its melting point) and the γ_{pv} of all the other polymer particles is greater than 39.9 mJ/m^2. From the observations with bibenzyl we conclude similarly that the surface tension of Teflon is less than 24.9 mJ/m^2 and that of all other polymer particles is greater than 24.9 mJ/m^2.

It is apparent that, given a sufficiently large number of matrix materials, one could put narrow limits on the surface tension of each particle, thus effectively determining these surface tensions. It is also clear that these results would not be

TABLE 11. Observation of Particle Pushing or Engulfment in the Melt Materials Benzophenone and Bibenzyl

	Benzophenone $\gamma_{lv} = 39.9$ mJ/m²		Bibenzyl $\gamma_{lv} = 24.9$ mJ/m²	
Particle material	γ_{pv} (mJ/m²)	Observation	γ_{pv} (mJ/m²)	Observation
Acetal	44.5	Pushed	44.3	Pushed
Nylon 6	43.7	Pushed	43.4	Pushed
Nylon 6,6	43.1	Pushed	42.8	Pushed
Nylon 12	40.8	Pushed	40.6	Pushed
Nylon 6, 10	38.2	Engulfed	37.8	Pushed
PMMA	37.2	Engulfed	36.8	Pushed
Nylon 6,12	34.3	Engulfed	34.0	Pushed
Teflon	18.1	Engulfed	17.8	Engulfed

The surface tensions refer to the respective melting points of the matrix materials, 48°C for benzophenone and 52°C for bibenzyl.

a unique consequence of Eq. (44). While Eq. (61) is at variance with experimental observations of particle rejection, the above inferences can be made on the basis of Eq. (36) as well as Eq. (44). The surface tensions γ_{pv}, as calculated from contact angles using Eq. (44), are in complete agreement with the implications of both Eq. (44) and Eq. (36) with respect to the observations of "engulfment" and "rejection." Therefore, unlike the previous technique, where all three equations gave identical and presumably correct values for the surface tension γ_{pv}, Eq. (61) does not predict correctly experimental observations of particles at solidification fronts.

5.3. Contact Angles

Any of Eqs. (44), (36), and (61) can be used to calculate values for the solid surface tension γ_{sv} from contact angles in conjunction with Young's equation, Eq. (1). There is one immediate criterion that the results obtained with any of the three equations (or, for that matter, any other equations) must satisfy. When

TABLE 12. Liquid Surface Tensions, Contact Angles, and Solid Surface Tensions of Hexatriacontane at 20°C

			γ_{sv} (mJ/m²)		
Liquid	γ_{lv} (mJ/m²)	θ (deg)	Eq. (44)	Eq. (36)	Eq. (61)
Water	72.8	104.6	19.8	10.2	27.2
Glycerol	63.4	95.4	20.0	13.0	28.7
Thiodiglycol	54.0	86.3	19.8	15.3	28.7
Ethylene glycol	47.7	79.2	19.7	16.8	28.3
Hexadecane	27.6	46	20.0	19.8	23.3
Tetradecane	26.7	41	20.7	20.6	23.4
Dodecane	25.4	38	20.4	20.3	22.7
Decane	23.9	28	21.2	19.1	21.3
Nonane	22.9	25	20.8	20.8	21.8

measuring contact angles with a number of liquids on a low-energy solid, the surface tension γ_{sv} is expected to be constant, independent of the liquid surface tension γ_{lv}. This expectation is based on the fact that the equilibrium spreading pressure in such situations is very low, about 1 mJ/m^2.[16] For the reasons discussed in the introduction, contact angle data obtained on hexatriacontane are used here.[54] The results obtained with Eqs. (44), (36), and (61) are given in Table 12. We infer that only Eq. (44) gives results that are essentially independent of γ_{lv}, suggesting that Eqs. (36) and (61) are deficient. The fact that all three equations predict γ_{sv} values near 20 mJ/m^2 as the contact angle decreases reflects the fact that all three equations imply $\gamma_{sl} \rightarrow 0$ as $\theta \rightarrow 0$.

We conclude that only Eq. (44) is consistent throughout. Equation (36), while correctly interpreting sedimentation and freezing front data, interprets contact angles in a way that is not only unreasonable physically but also at variance with its own predictions on the basis of the other techniques. Equation (61) has even more deficiencies. Nevertheless, most of the direct techniques are not sensitive to the exact form of the equation of state used.

6. *THE POSSIBILITY OF NEGATIVE SOLID–LIQUID INTERFACIAL TENSIONS*

As shown in Section 3, the formulation of the equation of state, Eqs. (32) or (44), is based on the assumption that the minimum value of solid–liquid interfacial tension, γ_{sl}, is zero. In this section, the possibility of negative solid–liquid interfacial tensions will be discussed. For this purpose, the experimental data obtained from a variety of essentially independent methodologies are examined in conjunction

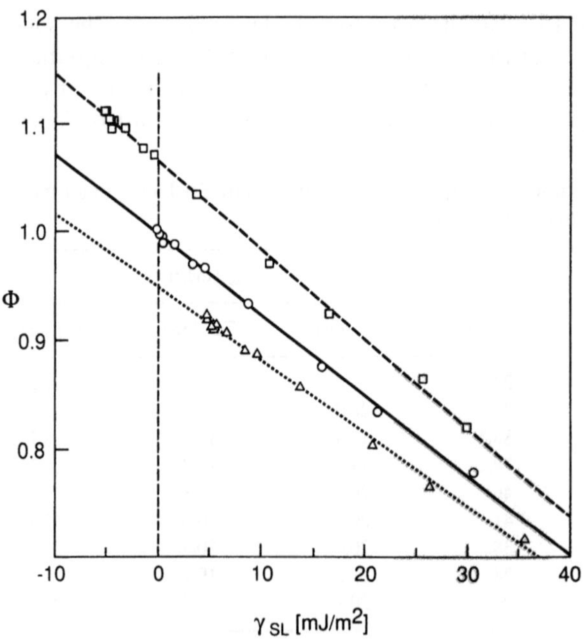

FIGURE 12. Copolymer (50-50) of tetrafluoroethylene and ethylene; Φ as a function of γ_{sl} for different estimates of γ_{sv}: □ γ_{sv} = 21.9 mJ/m^2; ○ γ_{sv} = 26.9 mJ/m^2; △ γ_{sv} = 31.9 mJ/m^2.

with the equation of state approach for calculating interfacial tensions. Contact angle measurements, liquid–liquid interfacial tensions, and advancing solidification front–particle interactions are employed to demonstrate that zero is the lower limit of solid–liquid interfacial tensions.[55]

In order to study this question in more detail, let us consider the plots in Fig. 12. For a given set of contact angle data on a given solid, three hypothetical γ_{sv} values were selected. For each of these, the corresponding hypothetical γ_{sl} and the Good interaction parameter Φ were calculated from Young's equation and the definition of Φ, Eq. (28), respectively. It was found that, in each case, there was a linear relation between the hypothetical Φ and γ_{sl} values, as shown in Fig. 12. The question of determining the correct γ_{sv} value was thus reduced to determining the correct straight line from such a family of curves. It has been argued[14] that the only possible choice was the straight line that intersects the Φ axis at $\Phi = 1$ when $\gamma_{sl} = 0$. This argument was, however, partially based on the assumption that γ_{sl} could not be negative, and the approach taken only considered situations of finite contact angles; i.e., it excluded spreading situations throughout.

Regarding this aspect of the problem, let us consider pairs of polymer melts that are all mutually insoluble to a degree comparable with that of polymeric solids and low molecular weight liquids. We reproduce in Fig. 13 a plot of Φ versus γ_{12}, the interfacial tension between pairs of polymer melts. In some cases, the free energy of spreading is negative, while in others it is positive, so that conditions conducive to negative interfacial tensions should exist in some of these cases. However, it is clear from Fig. 13 that, independent of spreading or nonspreading, all points fall close to a straight line, giving Φ as a function of γ_{12} with a limiting value of $\Phi = 1.0$ at $\gamma_{12} = 0$, in excellent agreement with our choice for the solid–liquid case in Fig. 12.

It is worth repeating that the use of Φ in the explicit (empirical) formulation of the equation of state is essentially arbitrary. As noted in Section 3, since Φ was used as a correlating parameter in the original equation of state formulation, it is convenient to refer to it in the present context.

As a second test of the validity of a particular Φ versus γ_{sl} relation in Fig. 12, consider again the interaction of small particles, initially embedded in a liquid, with

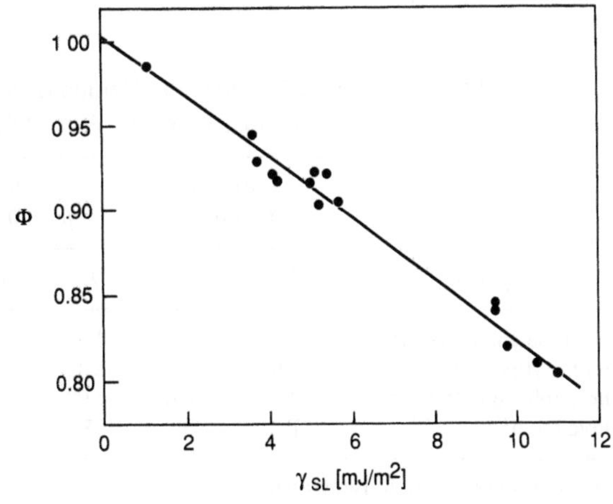

FIGURE 13. Polyethylene melt in contact with various polymer melts; Φ as a function of the polymer–polymer interfacial tension, γ_{12}. All interfacial tensions in this case are readily measurable.

an advancing solidification front. Whether a particle, when encountered by the solidification front, is engulfed or swept along by the solidification front is expected to depend on the sign of the free energy of adhesion, Eq. (53) or, alternatively, the free energy of engulfment

$$\Delta F^{\text{eng}} = \gamma_{\text{ps}} - \gamma_{\text{pl}} \tag{64}$$

where γ_{ps} is the particle–solid interfacial tension and γ_{pl} the particle–liquid interfacial tension. For $\Delta F^{\text{adh}} < 0$ or $\Delta F^{\text{eng}} < 0$, engulfment is predicted, whereas at low rates of solidification, $\Delta F^{\text{adh}} > 0$ or $\Delta F^{\text{eng}} > 0$ indicate that particle rejection should occur. Thus, the equation of state approach and its choice of Φ that allows the prediction of solid–liquid interfacial tensions can be verified experimentally through observations of particle behavior at solidification fronts.[35] To do so, the surface tension of the melt, γ_{lv}, and the contact angles on the solid matrix materials as well as the contact angles on the particle materials were measured. From the contact angle data, the surface tension of the solid matrix, γ_{sv}, was calculated by the equation of state. In a second step, the required interfacial tensions in Eqs. (53) and (64) were calculated.

The various straight lines in Fig. 12 may be represented by

$$\Phi = \beta - \alpha\gamma_{\text{sl}} \tag{65}$$

In the equation of state approach,[14] the constants were chosen to be $\beta = 1.00$ and $\alpha = 0.0075\ \text{m}^2/\text{mJ}$. Solidification front observations can then be used to test the validity of choices other than the straight line with the limiting value $\gamma_{\text{sl}} = 0$ at $\Phi = 1.00$. The results of comparisons of such calculations with experimental observations for several matrix–polymer systems are listed in Tables 13 and 14.[35] Overall, it can be seen that for $\beta > 1.00$, particle rejection is predicted in a number of cases where the microscopic observation is engulfment; for $\beta < 1.00$, particle engulfment is predicted where the microscopic observation is rejection; only when $\alpha = 0.0075\ \text{m}^2/\text{mJ}$ and $\beta = 1.00$ are the predictions confirmed by the experimental observations. For over 60 cases, $\beta = 1.00$ has yielded agreement between experimental observations and thermodynamic predictions, with very few exceptions when

TABLE 13. Testing for Possible Values of $\alpha \geqslant 0.0075\ \text{m}^2/\text{mJ}$ and $\beta \geqslant 1.00$ against Freezing Front Observations

	$\alpha = 0.0075$, $\beta = 1.00$		$\alpha = 0.0075$, $\beta = 1.05$		$\alpha = 0.0085$, $\beta = 1.05$		
System	ΔF^{adh}	ΔF^{eng}	ΔF^{adh}	ΔF^{eng}	ΔF^{adh}	ΔF^{eng}	Observation[a]
Naphthalene/polystyrene	−1.46	−0.02	1.36	0.30	2.83	0.77	E
Biphenyl/polystyrene	−0.68	−0.10	2.45	0.15	3.94	0.57	E
Benzophenone/PMMA	−0.39	−0.01	5.08	0.55	7.01	1.07	E
Prediction[a]	E	E	R	R	R	R	

[a] E, particle engulfment; R, particle rejection.

TABLE 14. Testing for Possible Values of $\alpha \leqslant 0.0075\ m^2/mJ$ and $\beta \leqslant 1.00$ Against Freezing Front Observations

System	$\alpha = 0.0075$ $\beta = 1.00$		$\alpha = 0.0065$ $\beta = 0.95$		Observation[a]
	ΔF^{adh}	ΔF^{eng}	ΔF^{adh}	ΔF^{eng}	
Thymol/Nylon 6,10	0.088	0.11	−0.98	−7.08	R
Prediction[a]	R	R	E	E	

[a] E, particle engulfment; R, particle rejection.

ΔF^{adh} was very close to zero so that experimental error could not be ruled out.[56] Thus $\beta = 1.00$ corresponding to a minimum $\gamma_{sl} = 0$ produces the best agreement with these experiments.

Overall, it can be concluded that, considering a variety of systems and situations (including situations of spreading and nonspreading), there is no evidence for negative solid–liquid interfacial tensions. Instead, there is considerable evidence that zero is the lower limit of all solid–liquid interfacial tensions.

7. FUTURE DEVELOPMENT OF THE EQUATION OF STATE

The previous sections have demonstrated the existence of an equation of state for interfacial tensions and the validity of the explicit form of such an equation of state, i.e., Eqs. (32) or (44).

Although the mathematical difficulty associated with Eq. (32) can now be avoided by using the new equation of state, Eq. (44), there are still some other improvements that can be made in the future. For instance, the contact angle data used to derive Eqs. (32) or (44) were obtained over 30 years ago. Today we have a better understanding of the thermodynamic status of contact angles, as well as more convenient and accurate measurement techniques, such as axisymmetric drop shape analysis—profile (ADSA-P),[57] axisymmetric drop shape analysis—contact diameter (ADSA-CD)[58, 59] and the improved method of capillary rise at a vertical plate.[60] Higher-accuracy contact angle data with a wider variety of solids and liquids will permit an improved formulation of the equation of state. These important new measurement techniques are summarized below.

The ADSA-P technique was developed to determine liquid–fluid interfacial tensions and contact angles from the shape of axisymmetric menisci, i.e., from sessile as well as pendent drops. The strategy employed is to construct an objective function that expresses the error between the physically observed drop profile and a theoretical Laplacian curve (i.e., a curve satisfying the Laplace equation of capillarity). The objective function is minimized numerically using the method of incremental loading in conjunction with the Newton–Raphson method in such a fashion as to find the optimal value for surface tension and contact angle. Apart from local gravity and the densities of the liquid and fluid phases, the only data required are several arbitrary coordinate points selected from the drop profile. The experimental system consists of a video camera, mounted on a microscope, which is connected to a computer. Images of drops are automatically digitized and processed giving an accuracy of ±0.05 mJ/m^2 for interfacial tensions and ±0.2° for contact angles.[57]

For the case of very low contact angles, say below 20°, it becomes increasingly difficult to measure contact angles with most techniques; the precision of ADSA-P also decreases since it becomes more difficult to acquire accurate coordinate points along the edge of the drop profile. ADSA-CD is useful in these situations because the drop is viewed from above rather than in profile. The contact angle is determined by minimizing the difference between the experimentally determined drop volume and the drop volume calculated from the contact diameter in conjunction with the Laplace equation of capillarity. The input for ADSA-CD consists of the surface tension of the liquid, the density of the liquid, the volume of the drop, and the contact diameter of the drop. As with ADSA-P, the contact diameter is determined by using digital image acquisition and analysis. The accuracy of the technique for small contact angles on smooth and homogeneous surfaces is found to be better than ±0.3°.[58, 59]

The capillary rise method remains the most accurate technique for measuring contact angles. In this method, the contact angle is determined indirectly by measuring the height of the capillary rise at a vertical plate. Using a cathetometer, this can be done to within ±4 μm, corresponding to an uncertainty in the contact angle value of approximately ±0.1°. Recently, the method has been automated, and height measurements are now made using digital image analysis. This has substantially reduced the errors associated with a manual setup, and the accuracy in contact angles is now about ±0.01°.[60]

Finally, the earlier contact angle data were obtained usually on solid surfaces having surface tensions γ_{sv} near 20 mJ/m^2. It would be desirable to produce new contact angle data on a series of high-quality solid surfaces with γ_{sv} ranging from 10 to 50 mJ/m^2, or wider if possible.

APPENDIX

This FORTRAN program is to calculate the solid–liquid surface tension γ_{sl}, the solid–vapor surface tension γ_{sv}, and $\cos\theta$ by the equation of state (44) using Newton's method. The required input data are the liquid surface tension γ_{lv} and the contact angle θ.

```
   implicit real*8(a-h,o-z)
10 write(6,*)‘‘input gamlv, theta’’
   read(5,*)glv,theta
   write(6,*)‘‘glv=’’,glv, ‘‘theta=’’,theta

   const=3.141592654• 180.0
   a=0.5d0*(dcos(theta*const)+1.0d0)
   b=0.000115d0*glv*glv
   y=a
```

```
20  x=y
    e=dexp(b*(1.0d0-x*x)**2)
    u=x-a*e
    v=1.0d0+4.0d0*a*b*x*(1.0d0-x*x)*e
    y=x-u/v
    if (y .lt. 1.0d0) go to 30
    y=a*dexp(b)
    go to 20
30  if (dabs(y-x) .lt. 0.00001d0) go to 40
    go to 20
40  gsv=x*x*glv
    gsl=gsv-glv*dcos(theta*const)
    cosin=dcos(theta*const)
    write(6,*)''gsv='',gsv,''gsl='',gsl,''costheta='',cosin
    write(6,*)''Do you want to run again? Yes=1''
    read(5,*)n
    if (n .eq. 1) go to 10
    stop
    end
```

REFERENCES

1. W. A. Zisman, *Contact Angle, Wettability and Adhesion*, Advances in Chemistry Series, Vol. 43, American Chemical Society, Washington, DC (1964).
2. *Wetting*, S.C.I. Monograph No. 25, Society of Chemical Industry, London, S.W.I. (1967).
3. J. F. Padday, ed., *Wetting, Spreading and Adhesion*, Academic Press, London (1978).
4. J. K. Spelt, D. R. Absolom, and A. W. Neumann, *Langmuir* **2**, 620 (1986).
5. L. Boruvka and A. W. Neumann, *J. Chem. Phys.* **66**, 5464 (1977).
6. J. Gaydos and A. W. Neumann, *J. Colloid Interface Sci.* **120**, 76 (1987).
7. D. Li and A. W. Neumann, *Colloids Surfaces* **43**, 195 (1990).
8. M. Yekta-Fard and A. B. Ponter, *J. Colloid Interface Sci.* **126**, 134 (1988).
9. G. E. H. Hellwig and A. W. Neumann, *5th Int. Congr. on Surface Activity*, Section B, p. 687 (1968).
10. G. E. H. Hellwig and A. W. Neumann, *Kolloid-Z. Z. Polym.* **40**, 229 (1969).
11. C. A. Ward and A. W. Neumann, *J. Colloid Interface Sci.* **49**, 286 (1974).
12. D. Li, J. Gaydos, and A. W. Neumann, *Langmuir* **5**, 1133 (1989).
13. A. Münster, *Classical Thermodynamics* (E. S. Halberstadt, trans.), Wiley, New York (1970).
14. A. W. Neumann, R. J. Good, C. J. Hope, and M. Sejpal, *J. Colloid Interface Sci.* **49**, 291 (1974).
15. A. W. Neumann, D. R. Absolom, D. W. Francis, and C. J. van Oss, *Separation Purification Methods* **9**, 69 (1980).
16. J. W. Whalen and W. H. Wade, *J. Colloid Interface Sci.* **24**, 372 (1967).
17. O. Driedger, A. W. Neumann, and P. J. Sell, *Colloid-Z. Z. Polym.* **201**, 52 (1965).
18. O. Driedger, A. W. Neumann, and P. J. Sell, *Colloid-Z. Z. Polym.* **204**, 101 (1965).
19. A. W. Neumann and R. J. Good, *J. Colloid Interface Sci.* **38**, 341 (1972).
20. D. Li and A. W. Neumann, *J. Colloid Interface Sci.* **137**, 304 (1990).
21. A. Dupré, Théorie Mécanique de la Chaleur, p. 369, Gauthier-Villars, Paris (1969).
22. L. A. Girifalco and R. J. Good, *J. Phys. Chem.* **61**, 904 (1957).
23. G. C. Maitland, M. Rigby, E. B. Smith, and W. A. Wakeham, *Intermolecular Forces: Their Origin and Determination*, Clarendon Press, Oxford (1981).
24. K. C. Chao and R. L. Robinson, Jr., *Equations of State: Theories and Applications*, American Chemical Society, Washington, DC (1986).

25. J. Kestin and E. A. Mason, *AIP Conf. Proc.* **11**, 137 (1973).
26. J. N. Israelachvili, *Proc. R. Soc. London A* **331**, 39 (1972).
27. R. P. Smith, D. R. Absolom, J. K. Spelt, and A. W. Neumann, *J. Colloid Interface Sci.* **110**, 521 (1986).
28. D. Li, Ph.D. Thesis, University of Toronto.
29. E. W. Swokowski, *Calculus with Analytic Geometry*, 2nd ed., Prindle, Weber & Schmidt, Boston (1979).
30. J. N. Israelachvili and G. E. Adams, *J. Chem. Soc. Faraday Trans. 1*, **74**, 975 (1978).
31. P. M. Claesson, C. E. Blom, P. C. Harder, and B. W. Ninham, *J. Colloid Interface Sci.* **114**, 234 (1986).
32. E. Moy and A. W. Neumann, *J. Colloid Interface Sci.* **119**, 296 (1987).
33. D. R. Uhlmann, B. Chalmers, and K. A. Jackson, *J. Appl. Phys.* **35**, 2986 (1964).
34. S. N. Omenyi and A. W. Neumann, *J. Appl. Phys.* **47**, 3956 (1976).
35. S. N. Omenyi, A. W. Neumann, and C. J. van Oss, *J. Appl. Phys.* **52**, 789 (1981).
36. J. K. Spelt, D. R. Absolom, W. Zingg, C. J. van Oss, and A. W. Neumann, *Cell Biophys.* **4**, 117 (1982).
37. J. F. Oliver, C. Huh, and S. G. Mason, *Colloids Surfaces* **1**, 79 (1980).
38. J. K. Spelt, Ph.D. dissertation, Univ. of Toronto (1985).
39. R. J. Good, A.C.S. Symp. Ser. #8, *Adsorption at Interfaces* (K. L. Mittal, ed.), American Chemical Society, Washington, DC (1975).
40. E. I. Vargha-Butler, T. K. Zubovits, H. A. Hamza, and A. W. Neumann, *J. Dispersion Sci. Technol.* **6**, 357 (1985).
41. E. I. Vargha-Butler, E. Moy, and A. W. Neumann, *Colloids Surfaces* **24**, 315 (1987).
42. E. I. Vargha-Butler, T. K. Zubovits, M. K. Weibel, D. R. Absolom, and A. W. Neumann, *Colloids Surfaces* **15**, 233 (1985).
43. S. N. Omenyi, R. S. Snyder, D. R. Absolom, C. J. van Oss, and A. W. Neumann, *J. Dispersion Sci. Technol.* **3**, 307 (1982).
44. A. W. Neumann, S. N. Omenyi, and C. J. van Oss, *J. Phys. Chem.* **86**, 1267 (1982).
45. A. W. Neumann, S. N. Omenyi, and C. J. van Oss, *J. Colloid Polym. Sci.* **257**, 413 (1979).
46. A. W. Neumann, D. R. Absolom, D. W. Francis, S. N. Omenyi, J. K. Spelt, Z. Policova, C. Thomson, W. Zingg, and C. J. van Oss, *Surface Phenomena in Hemorheology* (A. L. Copley and G. V. F. Seaman, eds.), The New York Academy of Sciences, New York (1983).
47. D. R. Absolom, Z. Policova, E. Moy, W. Zingg, and A. W. Neumann, *Cell Biophys.* **7**, 267 (1985).
48. F. M. Fowkes, *Ind. Eng. Chem.* **40** (1964).
49. J. K. Spelt, D. R. Absolom, and A. W. Neumann, *Langmuir* **2**, 620 (1986).
50. J. K. Spelt and A. W. Neumann, *Progr. Colloids Polym. Sci.* **77**, 26 (1988).
51. J. K. Spelt, R. P. Smith, and A. W. Neumann, *Colloids Surfaces* **28**, 85 (1987).
52. J. K. Spelt and A. W. Neumann, *Langmuir* **3**, 588 (1987).
53. R. P. Smith, D. R. Absolom, J. K. Spelt, and A. W. Neumann, *J. Colloid Interface Sci.* **110**, 521 (1986).
54. A. W. Neumann, *Adv. Colloid Interface Sci.* **4**, 105 (1974).
55. A. W. Neumann, J. K. Spelt, R. P. Smith, D. W. Francis, Y. Rotenberg, and D. R. Absolom, *J. Colloid Interface Sci.* **102**, 278 (1987).
56. S. N. Omenyi, A. W. Neumann, W. W. Martin, G. M. Lespinard and R. P. Smith, *J. Appl. Phys.* **52**, 796 (1981).
57. P. Cheng, D. Li, L. Boruvka, Y. Rotenberg, and A. W. Neumann, *Colloids Surfaces* **43**, 151 (1990).
58. F. K. Skinner, Y. Rotenberg, and A. W. Neumann, *J. Colloid Interface Sci.* **130**, 25 (1989).
59. W. C. Duncan-Hewitt, Z. Policova, P. Cheng, E. I. Vargha-Butler, and A. W. Neumann, *Colloids Surfaces* **42**, 391 (1989).
60. C. Budziak and A. W. Neumann, to appear.

6

Thermal Reconstruction of the Functionalized Interface of Polyethylene Carboxylic Acid and Its Derivatives

Gregory S. Ferguson and George M. Whitesides

1. INTRODUCTION

Chemical modification of the surface of a polymer changes its free energy of interaction with contacting solids or liquids while leaving its bulk physical properties largely unchanged. The ability to control interfacial interactions by chemical modification of a surface is important for applications involving properties, such as adhesion,[1-6]* biocompatability,[7-9] printability,[10] static discharge,[11] that depend on wetting or on hydrophilicity.[12,13] We[14-20] and others[21-29] have studied the relations between the structures of the surfaces of surface-modified polymers and their wettability. The meaning of the phrase *structure of the surface* in this context is not sharply defined, but includes those characteristics of the solid–liquid interface—functional group types, orientation, and distribution; surface topology; bulk polymer properties—that influence wetting. These studies have led to a useful understanding of the factors that control wetting, phrased in the qualitative terms of physical organic chemistry.

Our research has explored synthetic methods for introducing organic functional groups at the surface of polyethylene and has examined the thermal stability of the resulting functionalized interfaces in detail.[30] We chose polyethylene for a number of reasons: low dielectric constant, absence of functional groups, trans-

* For an example of work with surface-modified, low-density polyethylene, see Ref. 4.

Gregory S. Ferguson • Department of Chemistry, Lehigh University, Bethlehem, Pennsylvania 18015. *George M. Whitesides* • Department of Chemistry, Harvard University, Cambridge, Massachusetts 02138.

Modern Approaches to Wettability: Theory and Applications, edited by Malcolm E. Schrader and George Loeb. Plenum Press, New York, 1992.

parency in many regions of spectroscopic utility, ready availability. This review summarizes our work in this area and related work from other groups.[27,31–45]

1.1. Surfaces, Interfaces, and Interphases

The field of surface chemistry encompasses a wide range of disciplines having interfacial phenomena as a common interest. The terminology of the field must therefore accommodate a wide range of systems varying in their degree of ideality (e.g., in flatness or in uniformity).

In the context of this review, the term *surface* refers to the face of a solid (or of a liquid) directly exposed to the environment, without reference to any contacting phase. The term *interface* refers to the contact between a surface and a second phase (solid, liquid, or gas). The term *interphase* refers to a surface or interface and some portion of the underlying solid.

Organic surfaces, for example, of polymers or of thin films, present difficulties in nomenclature not often encountered in inorganic systems. Some liquids penetrate or "swell" organic surfaces, creating a state of matter intermediate between solid and liquid. Since this situation defies simple and precise description, we use general terms such as *interfacial region* to emphasize the complex, environmentally determined nature of the interface.

We have used a range of analytical techniques to characterize the interface of surface-modified polyethylene. Measurement of solid–liquid contact angles (θ),[15–20,46–50] X-ray photoelectron spectroscopy (XPS),[15–20,51] attenuated total reflectance infrared spectroscopy (ATR-IR),[20,30,52] and fluorescence spectroscopy[15–17] have been especially useful. Each of these techniques provides information about the constitution of a layer of different thickness at the polymer surface.

For clarity, it is convenient to define different "interphases" associated with the depth sensed by the various analytical tools available for probing the functionalized surfaces (Fig. 1). The "θ interphase" refers to the outer 5–10 Å of a solid—the portion relevant to wetting *(vide infra)*. Likewise, the "XPS interphase" refers to approximately the outer 50 Å—the portion sensed by X-ray photoelectron spectroscopy.[51]* The "ATR-IR interphase," the deepest interphase of interest, refers to the outer micron or more.[52] We were particularly interested in comparing the information provided by contact angles and XPS, since these techniques are the most sensitive to the composition of the *outermost* part of a surface.

In some cases, functional groups too deep to influence wetting—that is, outside the θ interphase—are nonetheless accessible to reagents in solution. We refer to the portion of the solid where this type of interaction is possible as the "sub-θ interphase." The depth of the sub-θ interphase is determined by permeability, pore structure, and liquid–solid interactions, and is generally less well defined than those associated with analytical techniques. The depth of the sub-θ interphase is intermediate between those of the XPS and ATR-IR interphases.[20]

* *XPS interphase* is an approximate term; the intensity of the signal for a given atom type is usually assumed to drop off exponentially as a function of depth: $I_d = I_0 \exp(-d/\lambda)$, where d is the distance of the atom of interest from the surface, λ is the inelastic mean free path of the photoelectron in the medium of interest, and I_d and I_0 refer to the same experimental geometry.

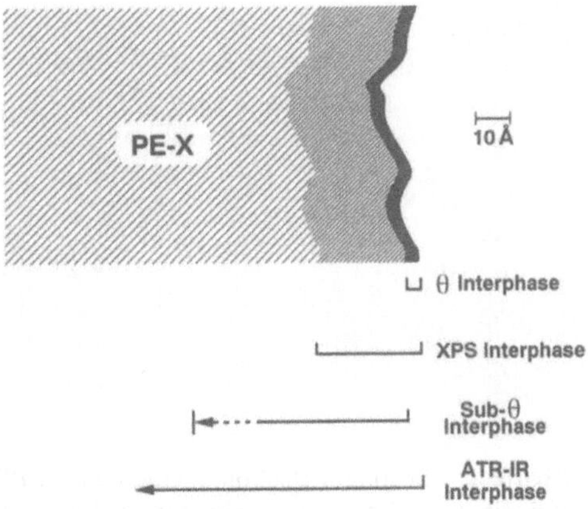

FIGURE 1. Schematic illustration of the various "interphases," labeled with the associated analytical techniques, making up the interfacial region of PE-CO_2H and its derivatives.

1.2. Synthesis of Surface-Modified Polyethylene and Nomenclature Used in Describing These Systems

Modification of the surface of polyethylene (PE-H) is straightforward, albeit not always easily controlled. Several reagents and methods have been reported for the oxidative derivatization of the surface of PE-H, including chromic acid, oxygen (or other gaseous) plasmas, flames, and graft polymerization.[53] Most of our work has focused on PE-H oxidized with aqueous chromic acid.

Treatment of low-density PE-H ($\rho = 0.919$ g/cm^3; melt index = 2) with aqueous chromic acid introduces a thin (10–20 Å) interfacial region composed of carboxylic acids (30%), ketones and aldehydes (20%), and unreacted methylene groups (50%).[15,17,18,20] The presence of oxidized functional groups, particularly the carboxylic acid moieties, greatly modify the interfacial properties of this material. The carboxylic acid groups are also useful as the starting point for further chemical elaboration (Scheme 1).

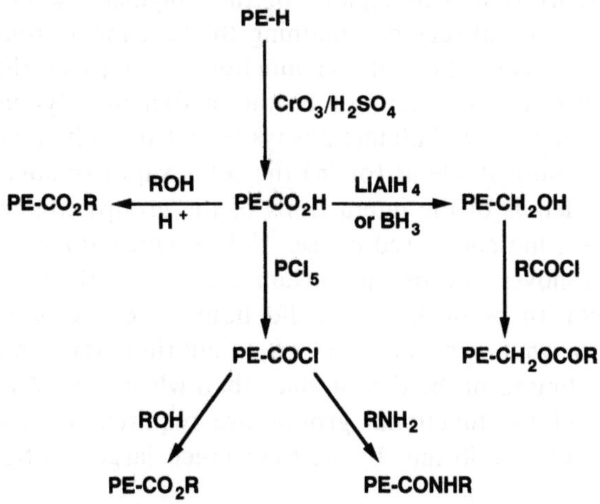

SCHEME 1. Representative reactions used to convert the surface of polyethylene (PE-H) to "polyethylene carboxylic acid" (PE-CO_2H) and derivatives. Reprinted with permission from reference 14. Copyright © 1988 Data Trace Chemistry Publishers.

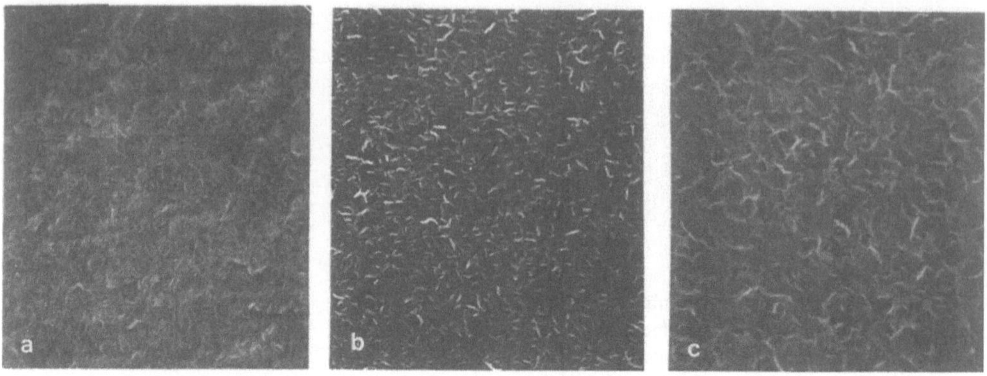

FIGURE 2. Scanning electron micrographs of PE-H (a), and PE-CO_2H oxidized for 1 min (b) or for 6 min (c) with aqueous chromic acid at 72°C. Reprinted with permission from reference 18. Copyright © 1985 American Chemical Society.

To emphasize the central role of the carboxylic acid groups of polyethylene oxidized with chromic acid, we call the oxidized material *polyethylene carboxylic acid* (PE-CO_2H). By analogy, we designate *derivatives* of this material as PE-X, where X specifies the surface-bound functional groups. Oxidation with chromic acid etches the polymer, leaving the surfaces of PE-CO_2H and its derivatives rougher than that of PE-H. Figure 2 shows scanning electron micrographs of PE-H and of PE-CO_2H (oxidized with chromic acid at 72°C, for 1 min and for 6 min) for comparison.

1.3. The Role of Thermodynamics in Determining Interfacial Properties and the Kinetics of Their Approach to Equilibrium

In any application of a modified polymer surface, a critical concern is the stability of the surface. The phrase *modified polymer surface* implies the existence of two distinct, spatially separated regions of the polymeric solid: the bulk of the material and the interfacial region containing the functional groups introduced by the modification. Concentration of the functional groups at the solid–vapor or solid–liquid interface may or may not be thermodynamically stable. Diffusion of these groups into the bulk will almost always be entropically favorable.* Enthalpically, they may be more stable at (or in) the solid–vapor or solid–liquid interface, depending on the details of the interactions of the groups comprising bulk, interface, and any contacting condensed phase.[54] The *apparent rate* at which a system not at equilibrium moves toward equilibrium depends on the probes being used to examine the system. In particular, the solid–liquid interface of a surface-modified polymer will reach what appears to be an equilibrium state more quickly when observed using wetting to probe the interface than when using ATR-IR. Only small motions (<10 Å) of the functional groups are required to bring a solid–liquid interface to apparent equilibrium by wetting; much larger motions ($\approx 10^6$ Å) are

* We note, however, that the combinatorial (ideal) entropy of mixing is usually very small for mixtures of polymers.

required to achieve a spatial distribution that appears to be close to equilibrium by ATR-IR.[52]

The thermal reconstruction of surface-modified polyethylene provides a convenient (although complex) system for kinetic investigation, since the rates at which the interfaces reconstruct are slow at room temperature but rapid at temperatures approaching 100°C (the melting point of polyethylene is about 115°C). Our studies of the kinetics and mechanism of reconstruction focused on several questions: Is the instability of the interfaces due to chemical reactions of the oxidized functionality, to loss of short chains bearing this functionality (by volatilization or solubilization), to diffusion of this functionality into the bulk of the polymer, or to small conformational changes of the functional groups at the interface? What are the driving forces that lead to thermal reconstruction in these systems?

At the outset of this work, we recognized three possible driving forces for diffusion of polar functionality into the bulk of the polymer: minimization of interfacial free energy, entropy of mixing (passive diffusion), and relaxation of mechanical stresses introduced during manufacture of the films.

The central conclusions from our work are that (1) reconstruction initially involves conformational changes of the polymer chains in the interfacial region, driven by the tendency to minimize interfacial free energy; and (2) in a slower process, interfacial functional groups passively diffuse into the bulk of the polymer. As expected, the initial composition of the interface (the surface of the polymer and any contacting phase) plays a critical role in determining the composition of the interface at equilibrium and the rate at which the system reaches equilibrium.

2. WETTABILITY AS A PROBE OF SURFACE STRUCTURE

The sensitivity of studies of wetting to the composition of the outermost part of a surface make it, in principle, a valuable method for monitoring reconstruction. In practice, however, detailed interpretation of contact angles, especially on polymer surfaces, is difficult because the factors that determine γ_{sl} and γ_{sv} [Eq. (1)] are only partially defined, since the surfaces are heterogeneous and rough. To aid in interpreting the relation between the structure of an interface and its wetting behavior, we have explored some structurally well-defined model systems, namely ordered monolayers of *n*-alkanethiols on gold.

2.1. Contact Angles: Definitions and Background

A solid–liquid contact angle is the angle formed at the three-phase contact of a drop of liquid with a surface.[46–50] An advancing contact angle (θ_a) is the static, kinetically stationary angle formed after a drop has advanced across a surface, for instance, after liquid has been added to the drop. A receding contact angle (θ_r) is the angle formed after a drop has receded across the surface, for instance, after liquid has been removed from the drop. The contact angles presented in this chapter are *advancing* contact angles. The contact angles on surface-modified polyethylene, presented in this chapter, were measured by the sessile drop technique.[47]

The contact angles on monolayers on gold were measured either by the sessile drop or a related technique.[55] Although the different experimental techniques used to measure θ_a give different values, the qualitative trends of most interest in our work are largely independent of the technique used.

Hystereses in the wetting experiments [$\theta_a - \theta_r$, or correspondingly, $\Delta(\cos\theta) = \cos\theta_a - \cos\theta_r$] on surface-modified polyethylene were high (40–90°)[56] in all cases, indicating that the systems were not at equilibrium. We have not, however, studied this aspect of the system in detail. According to most descriptions, both theoretical and empirical, the causes of hysteresis include heterogeneity and roughness of the surface,[48,57] and reorganization of functional groups in the interfacial region.[36,40–45,58–60]

Young's equation [Eq. (1)] relates cos θ to the local balance of forces operating at the three phase line (Fig. 3):[61]

$$\cos\theta = \frac{\gamma_{sv} - \gamma_{sl}}{\gamma_{lv}} \tag{1}$$

where

θ = contact angle

γ = interfacial free energy, where the interface is specified by subscripts: s = solid, l = liquid, v = vapor

The quantity cos θ is proportional to the difference in interfacial free energies ($\gamma_{sv} - \gamma_{sl}$); we therefore report values of cos θ, rather than θ, in discussions of surface free energies. Since contact angles are related to interfacial free energies, measurements of the wettability of polymer surfaces can provide qualitative information about the identity, distribution, and orientation of interfacial functionality. The utility of these measurements as probes of functional group identity and position is limited, however, by the importance of additional factors that affect contact angles: roughness and morphology of the surface, swelling of the solid by the contacting liquid, and heterogeneity in the chemical environment of functional groups in the interfacial region. Inferences relating wetting to surface composition are most reliable when surfaces of comparable topology are compared.

The ways in which roughness can affect contact angles are many and complex; important parameters include the scale of length and the topological details of the roughness.[57,62–64] Approaches to describing the relation between roughness and contact angles have, to this point, been empirical and/or approximate. The most common approach to describing the effect of roughness is to treat the observed (i.e.,

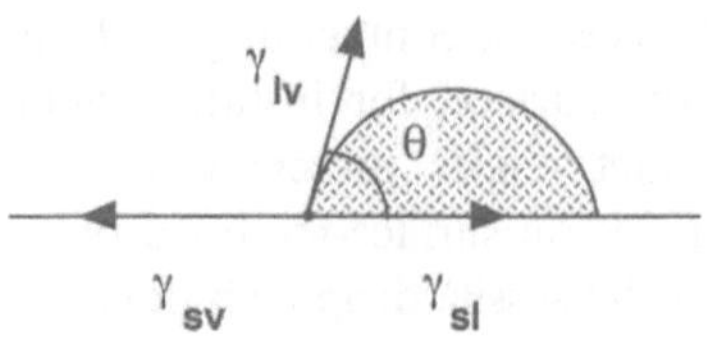

FIGURE 3. Schematic illustration of a sessile drop of liquid in contact with a solid surface, showing important parameters: the contact angle θ; the solid–liquid interfacial free energy γ_{sl}; the liquid–vapor interfacial free energy γ_{lv}; and the solid–vapor interfacial free energy γ_{sv}.

apparent) value of $\cos\theta_{obs}$ as the product of an *intrinsic* $\cos\theta_i$ and a "roughness factor" r [Eq. (2)]:[65–68]

$$\cos\theta_{obs} = r\cos\theta_i \tag{2}$$

The quantity r is defined as the ratio of the actual surface area to the area projected from the surface onto a flat, parallel surface. In addition to the effect of roughness described by Eq. (2), a drop of liquid in contact with a sufficiently rough surface can trap pockets of air to form a "composite" [(solid + air)–liquid] interface, thereby raising the observed contact angle.

For polymers, the morphology of a surface can play a significant role in determining its wetting behavior. Schonhorn and Ryan[27,69] showed that the critical surface tension of wetting (γ_c),* and hence θ_a, on polyethylene depends strongly on the degree of crystallinity in the interfacial region. Although we have not measured the rates of etching by chromic acid of amorphous and of crystalline PE-H, we expect that they are different. Preferential etching of one of these regions (presumably the amorphous) may therefore modify the interface—in its roughness and crystallinity— of the polymer in its wetting behavior.

Many hydrocarbon and halocarbon solvents penetrate or "swell" the interfacial region of PE-CO_2H and its derivatives. Swelling of the interfacial region of a solid by a contacting liquid necessarily changes the composition of the interface, and makes the interpretation of wetting data difficult. For this reason, we limited our studies of wetting on surface-derivatized polyethylene to wetting *by water*, since water does not swell polyethylene. Swelling of polyethylene by organic solvents also limits the solvents that can be used successfully in the preparation of these surfaces.

Chemical heterogeneity at a surface leads to contact angles *averaged* over the interfacial functionality.[66–68] Cassie's equation [Eq. (3)] describes this situation explicitly: $\cos\theta$ on a surface composed of regions of different composition, each having a normalized area fraction (A_i), is a weighted average of the values of $\cos\theta$ that would be observed on the pure surfaces ($\cos\theta_i$):

$$\cos\theta = \sum A_i \cos\theta_i \tag{3}$$

The lack of detailed knowledge about the spatial distribution of functional groups on the surfaces of PE-CO_2H and its derivatives complicates interpretations of wetting experiments.

2.2. Reconstruction of the Surface of PE-CO_2H: Initial Observations

The surface of unoxidized polyethylene (PE-H) is hydrophobic ($\theta_a = 103°$ for water); the surface of PE-CO_2H is relatively hydrophilic ($\theta_a = 55°$ for water at pH 1). At room temperature, the hydrophilicity of the surface of PE-CO_2H is retained almost indefinitely (months to years). At elevated temperatures ($T = 35$–$110°C$), however, the surface of PE-CO_2H becomes hydrophobic; its final

* The critical surface tension of wetting (γ_c) is an experimentally convenient, albeit *qualitative*, measure of γ_{sv}.[70]

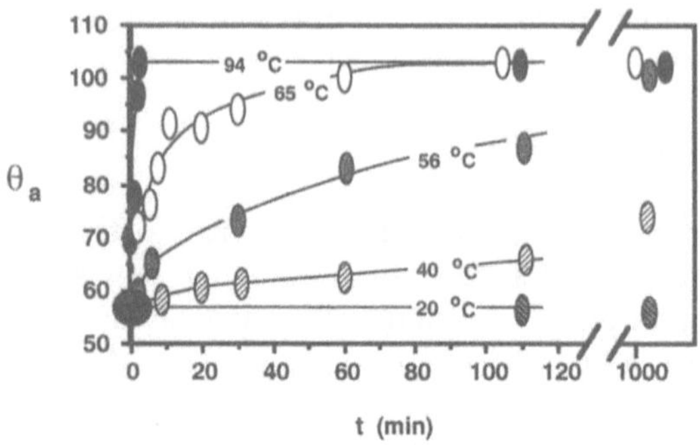

FIGURE 4. Advancing contact angles of water (pH 1) on PE-CO_2H as a function of the time the film had been heated at various temperatures under vacuum. Reprinted with permission from reference 30. Copyright © 1987 American Chemical Society.

state is indistinguishable from unoxidized polyethylene in its wettability by water (Fig. 4).* In earlier papers, Ter-Minassian-Saraga and co-workers had reported similar behavior for a related material, produced by oxidizing PE-H with a mixture of sulfuric acid and potassium chlorate.[28]

These observations led us to examine *how* the surface of PE-CO_2H (and its derivatives) reconstructs on heating. We use the term *reconstruct* broadly to include (1) "passive" diffusion of functional groups into the bulk of the polymer, and (2) conformational changes at the surface that affect its wettability.

2.3. *Model Systems for Studies of Wetting: Self-Assembled Monolayers (SAMs) of Alkanethiols on Gold*

The surface of PE-CO_2H is structurally complex, both in chemical composition and in morphological detail. The interactions between its surface and a contacting liquid phase are, consequently, also complex. For this reason, we have examined structurally well-defined—that is, chemically and morphologically homogeneous (compared to PE-CO_2H)—surfaces as models. These studies provided a basis for understanding the character, and the scales of length, of the interactions relevant to the study of reconstruction of polymer surfaces by measuring their wettability by water.

Long-chain alkanethiols adsorb from solution onto gold surfaces and form monolayers (called self-assembled monolayers, SAMs) in which the sulfur atoms are coordinated to gold and the hydrocarbon chains are densely packed, all-*trans*, and tilted about 30° from the normal to the surface (Fig. 5).[55,70–81]† As a result of this geometry, the organic–air interface is populated predominantly by the functional groups at the termini of the alkyl chains.

These SAMS are useful in studying the relations between microscopic structures (molecular identity and orientation) and macroscopic properties (wetting and adhesion). Synthetic organic chemistry provides techniques for systematically varying the physical properties of these SAMs by modifying the molecular precursors

* For comparison, we treated ultrahigh molecular weight polyethylene with aqueous chromic acid and followed the reconstruction of the resulting surface by measuring contact angles of water. As expected, its surface reconstructed more slowly than did that of PE-CO_2H.[30]

† For adsorption of a sulfur-containing protein onto gold, see Ref. 75; for adsorption of sulfur-containing polystyrene onto gold, see Ref. 76.

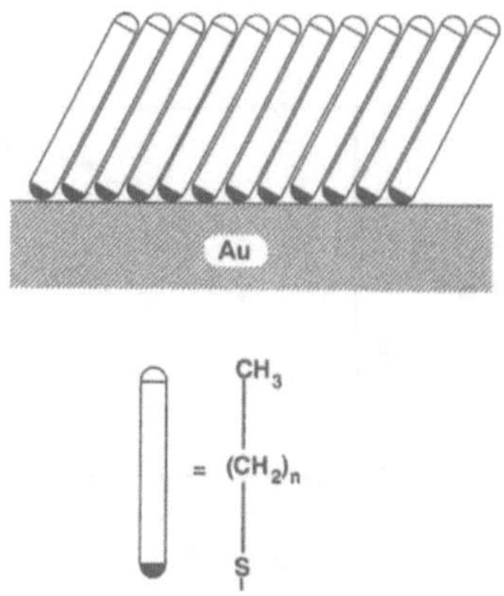

FIGURE 5. Schematic illustration of a self-assembled monolayer formed by adsorption of *n*-alkanethiols on gold.

from which they are assembled. This approach allows control over the identity and orientation of interfacial functional groups, along with their distance from the gold substrate and their depth from the organic–air interface.

2.3.1. Sensitivity of Wetting to the Depth of Interfacial Functional Groups

We believe that the measurement of contact angles constitutes the most surface-sensitive (if not the most easily interpretable) method now available for characterizing organic interfaces, especially solid–liquid interfaces. In order to probe the sensitivity of wetting experiments to the depth of interfacial functional groups, we have examined the variation in the contact angle of water (and other liquids) on SAMs formed by adsorption of mercaptoethers, $HS(CH_2)_{16}OC_nH_{2n+1}$, on gold, as a function of the length of the *n*-alkyl group C_nH_{2n+1}.[82] Figure 6 relates $\cos\theta_a$ of water to the length of the terminal alkyl chain. Contact angles on two additional surfaces are added for reference: (1) a SAM of an unsubstituted *n*-alkanethiol [$HS(CH_2)_{21}CH_3$] on gold, a model for an oriented hydrocarbon monolayer surface in which there are no ether functionalities, and (2) polyethylene glycol [$-(CH_2CH_2O)_n-$, PEG], a substance containing a high concentration of exposed ether groups.*

In the SAM obtained from $HS(CH_2)_{16}OCH_3$, the ethereal oxygen atom is close to the solid–liquid interface, and the surface is relatively hydrophilic ($\theta_a = 75°$). Contact angles of water on SAMs from $HS(CH_2)_{16}OR$, R = *n*-butyl or longer, in contrast, approach that on a SAM from $HS(CH_2)_{21}CH_3$ ($\theta_a = 116°$). Terminal *n*-alkyl groups having lengths of approximately 5 Å or more appear to shield the ethereal oxygen atom from contact with water. These results indicate that the θ interphase, for water, constitutes only approximately the outermost 5 Å of the surface.† This value (5 Å) probably reflects both through-space dipolar interactions and hydrogen bonding between the interfacial functional groups and water that has diffused into the topmost part of the monolayer.

* Polyethylene glycol is soluble in water, so contact angles were measured immediately (~1 s) after application of the drops.

† The depth of the "θ interphase" depends on the choice of liquid used in the wetting studies. On SAMs with structure $Au/AuS(CH_2)_{16}OR$, hexadecane apparently senses only about the top 2 Å of the surface.

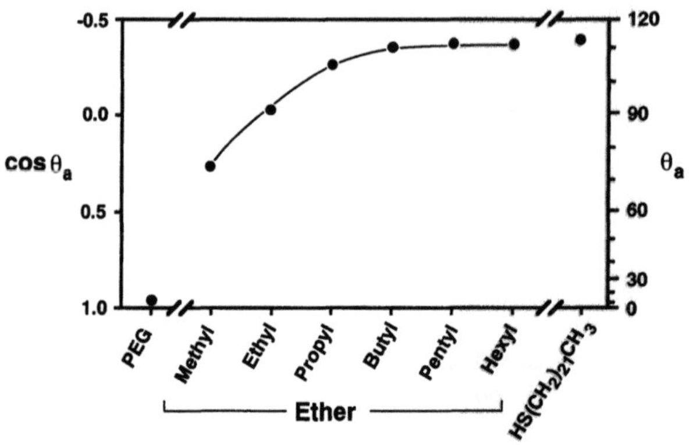

FIGURE 6. Advancing contact angles of water on SAMs of $AuS(CH_2)_{16}OR$ on gold, as a function of the length of the alkyl chain R. The contact of water on polyethylene glycol (a model for a surface in which ether groups are exposed) and on a SAM of docosanethiol on gold (a model for a surface in which ether groups are completely buried) are added for comparison. Reprinted with permission from reference 82. Copyright © 1988 American Chemical Society.

We have obtained similar results in other systems:

1. The wetting behavior of SAMs obtained from mercaptoamides $HS(CH_2)_{11}CONHC_nH_{2n+1}$ varies with the length of the terminal n-alkyl chain, much as does that of the SAMs from mercaptoethers.[83]
2. The wetting behavior of SAMs obtained from simple n-alkanethiols $[HS(CH_2)_nCH_3]$ on gold also varies with the number of carbons in the alkane chain. The contact angle on these SAMs varies smoothly from about 77° for ethanethiol to about 110° for hexanethiol; the angles for longer chains are consistently between 110° and 114°. We infer that alkyl chains of approximately six carbons in length are sufficient to shield the contacting drop of water from interaction (whether van der Waals or dipole-induced dipole) with the underlying gold.* The value of six carbons is actually an upper limit, since SAMs from smaller thiols ($n < 5$) may be more disordered *(vide infra)* than their longer homologues ($n > 5$).[55,73]
3. The highly disordered surfaces of PE-CONHR and $PE\text{-}CO_2R$ show variations in contact angle as a function of the length of the n-alkyl group (R) that are surprisingly similar to those observed for the ordered SAMs.[85] We infer that the size of the pendant alkyl group, and not the degree of order in the arrangement of these groups, is crucially important in masking the underlying polar functionality.[70]

2.3.2. Interfacial Order and Wetting

Since SAMs of organosulfur compounds on gold are structurally well defined, they provide useful model systems for probing the effect of interfacial structure in studies of wetting. SAMs on gold derived from mixtures of two ω-hydroxyalkanethiols, $HS(CH_2)_{19}OH$ and $HS(CH_2)_{11}OH$, differing in the length of their polymethylene chains, show a variation in their contact angles with water as a function of the ratio of concentrations of the two components on the surface.[86] Figure 7 relates two interfacial properties ($\cos\theta_a$ and ellipsometric thickness of the adsorbed monolayer) to the ratio of concentrations of the two components in the

* Clean, bare gold is wet by water.[84]

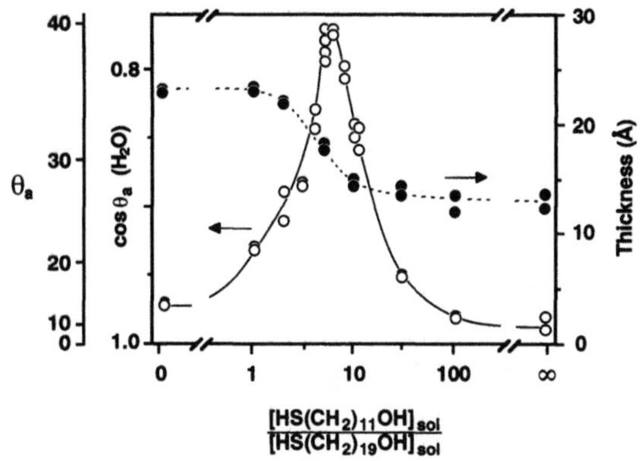

FIGURE 7. Advancing contact angles of water on, and ellipsometric thicknesses of, SAMs derived from mixtures of $HS(CH_2)_{19}OH$ and $HS(CH_2)_{11}OH$ on gold, as a function of the ratio of concentrations of the two thiols in the solutions from which they were adsorbed. Adapted with permission from reference 86b. Copyright © 1989 American Chemical Society.

solutions from which the SAMs were formed. The ellipsometric data allow us to relate the ratio of concentrations of the two components in solution to the ratio of their concentrations on the surface. We infer from comparison of wetting, XPS, and ellipsometry that the midpoint in thickness corresponds to the SAM containing equal concentrations of the two components.* For the molecules used as components of these SAMs, ellipsometry is sensitive primarily to the average thickness of the organic film. This thickness is directly related to the average composition in monolayers composed of mixtures of molecules of different lengths (Fig. 8).

The contact angles of water on these surfaces show a marked dependence on the composition of the organic thin films. SAMs composed of the pure ω-hydroxythiols give very low contact angles ($<15°$), consistent with close-packed, crystalline-like arrays,† exposing hydrophilic hydroxyl groups at the surface. SAMs composed of mixtures of the thiols give contact angles (15°–40°) higher than those on the pure SAMs, and consistent with mixtures of hydroxyl and methylene groups in the θ interphase. We infer from the wetting behavior, and from infrared data,[87] that the outermost part of the mixed SAMs are disordered and liquidlike (Fig. 8). Polarized reflectance infrared spectroscopy provides a valuable method of determining the orientation of the CH_2 groups of the polymethylene chains with respect to the plane of the gold surface. We hypothesize that the disorder in the outer parts of the monolayers causes an increase in the contact angle of water, relative to that on the pure SAMs, by exposing hydrophobic methylene groups. The maximum contact angle for this system is observed on the surface nearing roughly equal concentrations of the two thiols. We infer that this composition also gives rise to the highest degree of disorder for this system at the interface. This system is important because it provides an example of the successful correlation of molecular-scale structure with a macroscopic physical property.

* Compositions of the SAMs are not the same as the compositions of the solutions from which they were formed. The preference for $HS(CH_2)_{19}OH$ at the surface is not surprising, since the longer alkyl chain provides a greater molecular surface to participate in energetically favorable van der Waals interactions with neighboring chains.

† Polarized external reflectance FTIR spectra of a SAM of $HS(CH_2)_{16}OH$ on gold showed CH_2 symmetric and asymmetric stretching vibrations at $2851\ cm^{-1}$ and $2919\ cm^{-1}$, frequencies that are indicative of crystalline packing of polymethylene chains.[73]

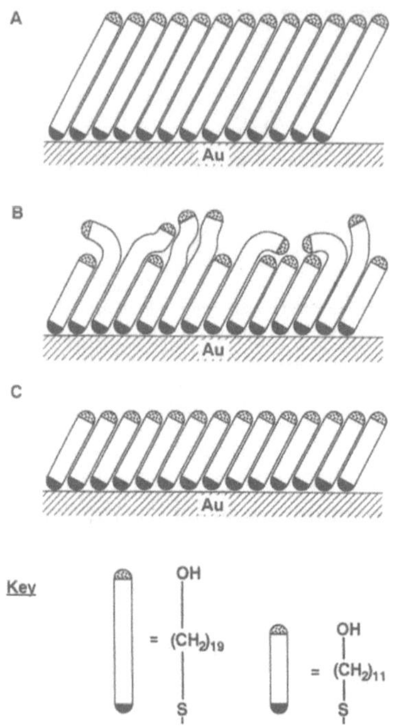

FIGURE 8. Schematic illustration of SAMs of hydroxy-terminated thiols on gold: (A) SAM derived from pure $HS(CH_2)_{19}OH$; (B) SAM derived from equal amounts of $HS(CH_2)_{19}OH$ and $HS(CH_2)_{11}OH$; (C) SAM derived from pure $HS(CH_2)_{11}OH$.

2.4. Wetting Experiments on PE-X

Unlike SAMs of alkanethiols on gold, the surfaces of PE-CO_2H and its derivatives are chemically and morphologically complex. Although interpreting studies of wetting at the surface of derivatized polyethylene is complicated, we have developed experiments that give qualitative (but useful) information about the θ interphase. An example is the variation of the contact angle of water on PE-CO_2H with the pH of the contacting drop, an experiment we call *contact angle titration* (Fig. 9).[18,19] The surface of PE-CO_2H is hydrophilic ($\theta_a = 55°$) relative to that of PE-H. Ionization of the acid groups leaves charged carboxylate ions that are more hydrophilic than neutral CO_2H groups; hence, the contact angle of a drop of water having a high pH ($\geqslant 11$) is low ($\theta \approx 20°$).

Given that SAMs formed by adsorption of ω-mercaptocarboxylic acids on gold are wet by water at all values of pH, it is curious that PE-CO_2H reaches a limiting value of θ (20°) at low pH. In fact, we know of no derivative of PE-CO_2H whose contact angle of water is less than 20°. This fact may reflect the heterogeneity of these modified surfaces: methylene groups are probably always present in the θ interphase of these surface-modified polymers. Although the number of carboxylic acid groups per unit area in the interfacial region of the SAMs on gold are comparable to the same of PE-CO_2H,[17,80] the numbers and arrangements of these groups in the θ *interphases* are probably different.

At the surface of PE-X, sufficiently hydrophilic functional groups (for instance, COO^-) probably adsorb water of hydration at high values of relative humidity at which contact angles are measured. If the polar functional groups (or clusters

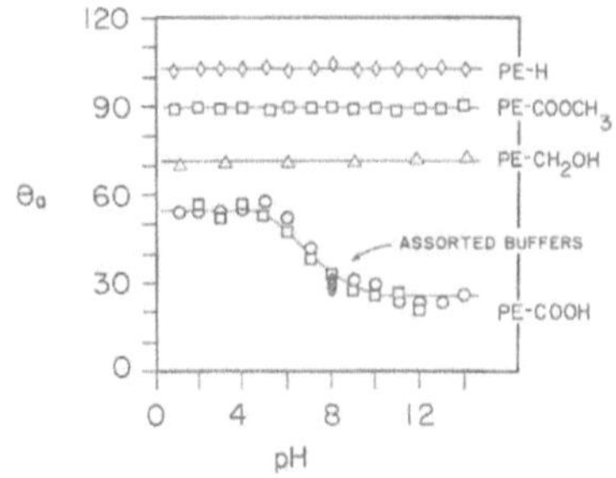

FIGURE 9. Variation in the advancing contact angle of water on PE-CO_2H as a function of pH. Data for PE-H and two nonionizable derivatives (PE-CO_2CH_3 and PE-CH_2OH) are included for reference. Reprinted with permission from reference 18. Copyright © 1985 American Chemical Society.

of them) are well enough separated by methylene groups, the water hydrating them may not condense; as a result, the hydrophilicity of the surface would reach a limiting value, characteristic of a mixture of "puddles" of water dispersed on a background of hydrophobic methylene groups (Fig. 10).

2.5. Sensitivity of Wetting to Small Conformational Changes Within the θ Interphase

It has long been recognized that the surfaces of polymers bearing polar functional groups can reorient, when in contact with polar condensed phases, to expose the polar functionality. This behavior has been reported for the surfaces of several polymers, including hydrogels,[40,41] oxidized polyethylene, other surface-modified polymers,[42–44] polyethylene grafted with acrylic acid[45] or other polar molecules,[60] and several unmodified polymers having polar substituents.[58] Most studies have focussed on the relation between the mobility of interfacial polar functional groups and the hysteresis in the contact angles of polar liquids.

The *ortho*-anthranilate amide (**1**) of PE-CO_2H provides a remarkable example of this type of behavior: it shows a surprisingly large change with pH in its wettability by water (Fig. 11).[88] At low pH ($\leqslant 4$), the advancing contact angle of water is about 110° (i.e., the surface is more hydrophobic than unfunctionalized

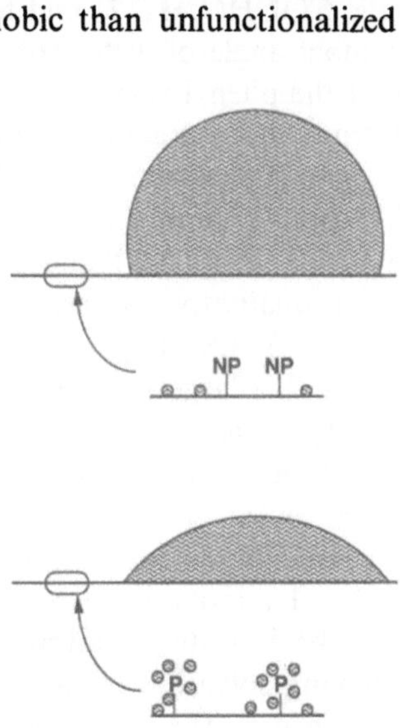

FIGURE 10. Stylized illustration of our model for the surface of PE-X near a contacting drop of water, where X is a polar functional group (P) and where it is a nonpolar functional group (NP). The small filled circles represent adsorbed water molecules. The polar groups are fully hydrated, but well enough separated so that the water of hydration does not condense.

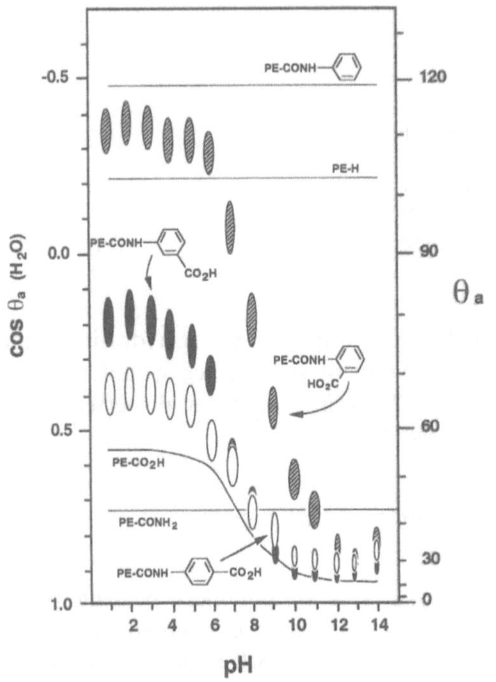

FIGURE 11. Variation in the advancing contact angle of water on PE-CO_2H and several of its anilide derivatives as a function of pH. Data for PE-H and PE-$CONH_2$ are included for reference. Reprinted with permission from reference 88. Copyright © 1988 American Chemical Society.

polyethylene); at high pH ($\geqslant 12$), it is ca. 33°. The *meta*- and *para*-isomers show less pronounced variations, though larger than that of PE-CO_2H.

The difference in the contact angles of water (low pH) between **1** and unfunctionalized polyethylene (PE-H) cannot be interpreted solely in terms of hydrophobicity, since the surface of **1** is rougher than that of PE-H (see Section 2.1). Figure 11 includes data for surfaces bearing the unsubstituted anilide amide (PE-$CONHC_6H_5$, **2**) and for the parent amide (PE-$CONH_2$) for comparison. The contact angle of water (pH $\leqslant 4$) on **2** is only 9° higher than that on **1**, suggesting that the phenyl ring efficiently shields *both* the amide and carboxylic acid groups of **1** from the contacting drop of water. The contact angle of water on PE-$CONH_2$ ($\theta_a = 43°$; invariant with pH) provides an indication of the degree to which the hydrophilic amide group is shielded by the phenyl ring in **1** and **2**. The high hydrophobicity of **1** and **2**, despite the presence of polar amide and acid groups, is in qualitative agreement with the data from SAMs on gold derived from $HS(CH_2)_{16}OC_nH_{2n+1}$: a nonpolar organic group having about six carbon atoms seems able largely to shield a polar group from a contacting drop of water.

We believe that the large decrease in the contact angle of water on **1** (as the pH of the contacting drop is varied from 1 to 14) is due to a conformational change of the amide groups concomitant with ionization of the carboxylic acid moieties at the solid–liquid interface (Scheme 2). We suggest that when in contact with acidic water, **1** adopts conformation **3**, burying the polar functionality; when in contact with basic water, **1** adopts conformation **4**, exposing the polar functionality and allowing solvation of the carboxylate ions. We attribute the preference for conformation **4** to the favorable heat of solvation of the carboxylate ion. Rationalizing

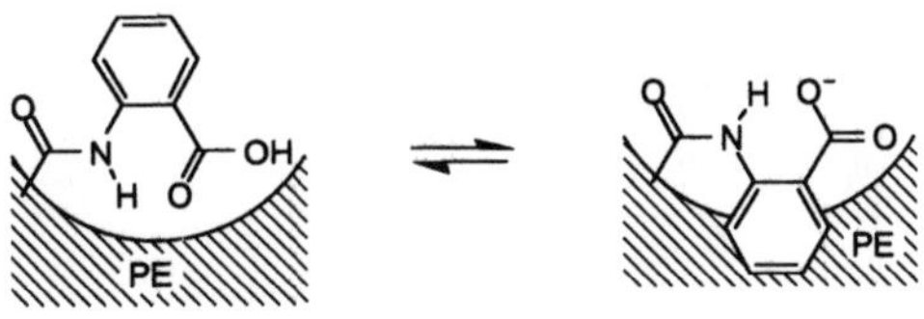

3 4

SCHEME 2. Proposed conformational changes by **1** when in contact with water at acidic or neutral (**3**) and at basic (**4**) pH's. Reprinted with permission from reference 88. Copyright © 1988 American Chemical Society.

conformation **3** is more difficult: why should **1** adopt a conformation that minimizes energetically favorable polar interactions and hydrogen bonding between the amide–acid groups and the contacting water? One possibility is that strong hydrogen bonding (amide–amide, acid–amide, and acid–acid) between the functional groups present at the surface of **1** make it possible for this material to present a low-energy surface composed predominantly of aromatic C–H bonds. The relatively low contact angle of water (pH 1, $\theta_a = 84°$) on the *N*-methyl derivative of **1**, which cannot take part in hydrogen bonding involving N–H moieties, may reflect the importance of such interactions.

3. RECONSTRUCTION OF THE INTERFACE OF PE-CO_2H AND DERIVATIVES ON HEATING

With the background in Sections 1 and 2, the remainder of this chapter describes experiments with which we and others have examined the thermal mobility of functional groups in polyethylene, especially in the interfacial region. The reactions presented in Scheme 1 provided versatility in controlling some of the parameters in the experiments, particularly the polarity and size of the surface-bound functional groups. Several characteristics of the surfaces, however, could not be controlled: microscopic roughness, and chemical and morphological heterogeneity.

3.1. Comparison of Results from Wetting and XPS

Figure 12 shows the contact angles of water (measured at both pH 1 and 13) on PE-CO_2H as a function of the time of heating (at 100°C) under argon and in vacuum; XPS data for the oxygen 1*s* (O_{1s}) signal are included to provide complementary information about the evolution in composition of the interfacial region on heating at approximately the same temperature (106°C).

Several points deserve comment. Water at pH 1 ceases to sense the hydrophilic functionality of PE-CO_2H after it has been heated at 100°C for 5 min; samples heated under argon and in vacuum are indistinguishable in their wetting behavior. We attribute the change in wetting behavior on heating to diffusion of the polar functionality into the bulk of the polymer and out of the θ interphase. Water at pH 13 ceases to sense the polar functionality only after about 15 min on heating at 100°C. The greater apparent depth sensitivity of wetting by basic water (relative to that by acidic water) may reflect long-range electrostatic interactions between the

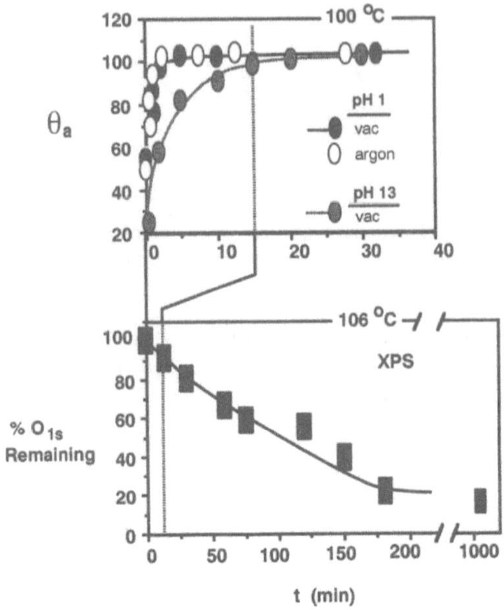

FIGURE 12. (Top) Advancing contact angles of water (pH 1 and 13) on PE-CO_2H as a function the time the film had been heated at 100°C. Data are included for samples heated in vacuum and heated under argon. (Bottom) The normalized intensity of the O_{1s} XPS signal from PE-CO_2H as a function of the time the film had been heated at 106°C. Reprinted with permission from reference 30. Copyright © 1987 American Chemical Society.

polar aqueous phase and the carboxylate ions, or reconstruction of the interface driven by solvation of the carboxylate ions.

The oxygen content of the interfacial region of PE-CO_2H, as measured by XPS, and in contrast to that inferred from wetting, diminishes only slightly (by about 5%) during the first 5 min of heating. We attribute this difference in the apparent rates of reconstruction measured by wetting and by XPS to the different depth sensitivities of these techniques. After 5 min of heating, the polar oxygen-containing groups have diffused out of the θ interphase but still remain (for the most part) within the XPS interphase. The decrease in intensity of the O_{1s} signal is due to attenuation of the signal (scattering of photoelectrons) by methylene groups between the oxygen atoms and the detector. We attribute the residual O_{1s} signal remaining after 1000 min of heating to contaminants in the film blooming to the surface.*

3.2. *Results from ATR-IR*

Attenuated total reflectance infrared spectroscopy samples a thick portion of the polymer interphase (> 10,000 Å), and allows quantitative analysis of functional groups within that region. Results from ATR-IR spectroscopy suggest that the mechanism of the disappearance of polar functionality is diffusion into the bulk of the polymer; we found no evidence for volatilization or for chemical reactions of the polar functionality. Figure 13 shows ATR-IR data for polyethylene films treated under a variety of conditions. Before extraction with methylene chloride, the film has several peaks between 2000 and 1500 cm^{-1}; we attribute these peaks to film

* Additives introduced during manufacture of the film include antioxidants, slip agents, and antistatic agents.

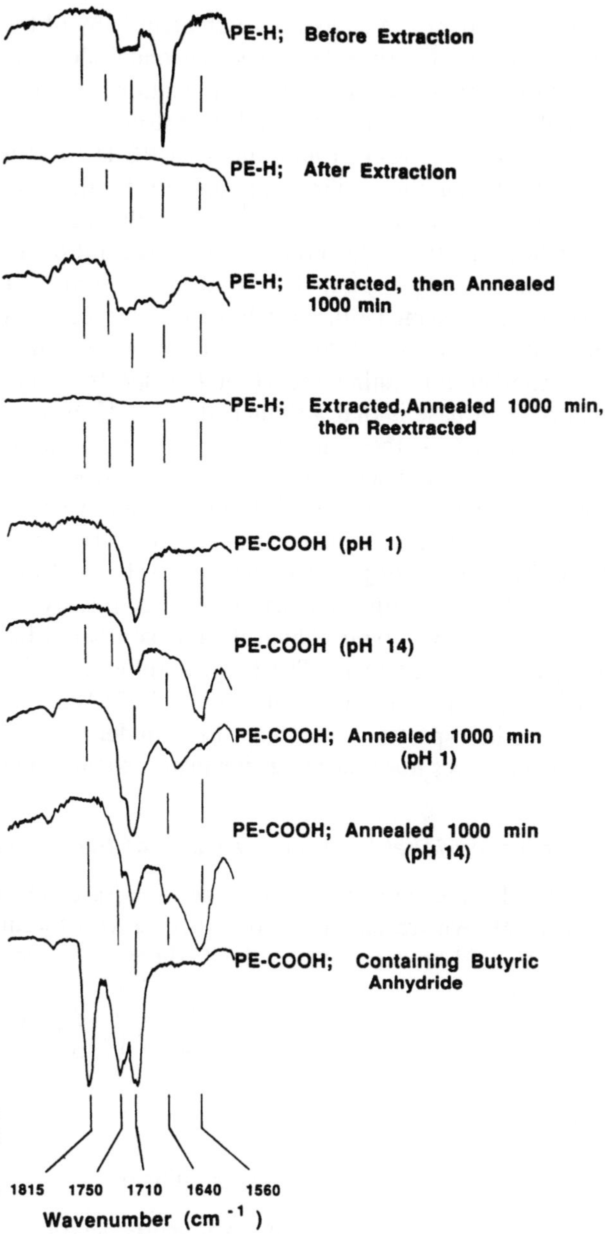

FIGURE 13. The 2000–1500 cm^{-1} regions of ATR-IR spectra of PE-H and PE-CO_2H, treated under a variety of conditions. Reprinted with permission from reference 30. Copyright © 1987 American Chemical Society.

additives. Extraction with methylene chloride removes the contaminants giving rise to these peaks. Heating the extracted film at 100°C for 1000 min causes the peaks to reappear; reextraction again causes them to disappear. These results are consistent with the idea that contaminants bloom to the surface of these films on heating. The ATR-IR spectrum of fully protonated PE-CO_2H shows a broad absorption centered at 1710 cm^{-1} that we attribute to the C=O stretching modes of the carboxylic acids and ketones and aldehydes. Deprotonation of PE-CO_2H with water at pH 14 shifts approximately 75% of the C=O peak to lower energy, characteristic of carboxylate ions; the remainder, assigned to the ketone–aldehyde moieties, remains unchanged. After heating PE-CO_2H at 100°C for 1000 min, its ATR-IR spectra (protonated and deprotonated) are approximate superpositions of the spectra of PE-CO_2H and annealed PE-H (blooming film contaminants).

The blooming of film contaminants to the surface of PE-CO_2H on heating prevents straightforward quantitation of the amount of carboxylic acid before and after heating. We note, however, that the ratio of the integrated absorbance at 1560 cm^{-1} (CO_2^- groups) to that at 1710 cm^{-1} (ketone–aldehyde and CO_2H groups) is the same (~0.75) in heated and unheated samples. These results indicate that the CO_2H groups do not volatilize on heating to 100°C.

Another interesting aspect of the spectrum of the sample heated and then treated with base is that although the CO_2H groups are well outside of the θ and XPS interphases, these groups are still accessible to aqueous hydroxide ion. We have studied this phenomenon in detail, and we designate the portion of the film where it is relevant as the *sub–θ interphase* (Section 1.1).[20]

An alternative explanation for the changes in the wetting behavior of PE-CO_2H on heating is that the carboxylic acid moieties in the interfacial region can condense to form anhydrides. To test this hypothesis, we treated a sample of PE-CO_2H with butyric anhydride and compared its ATR-IR spectrum to that of reconstructed PE-CO_2H. The butyric anhydride diffused into the polymer, giving rise to well-separated peaks at 1815 and 1750 cm^{-1}. These peaks are absent in all of the other spectra, indicating that condensation of carboxylic acid groups to anhydrides does not occur to a significant extent under our experimental conditions.

3.3. *Kinetic Model for Reconstruction of the Interface of PE-CO_2H*

We have examined in detail the kinetics of reconstruction of the interface of PE-CO_2H. We treated the reconstruction as a thermally activated process, whose kinetics could be described by the Arrhenius equation

$$D = A_D \exp\left(\frac{-E_{a,D}}{RT}\right) \tag{4}$$

where

$$D = \text{rate of diffusion}$$

$$A_D = \text{preexponential factor}$$

$$E_{a,D} = \text{energy of activation for diffusion}$$

This approach had important advantages. First, it was the simplest and most straightforward approach available; the reconstruction of the interfacial region of PE-CO_2H is complex, and describing its kinetics by even this simple model required several approximations. Our limited control over the parameters of the experiment (for example, the molecular weight of the diffusing species) prevented us from using more sophisticated theories.[89–92] Second, it allowed useful comparisons

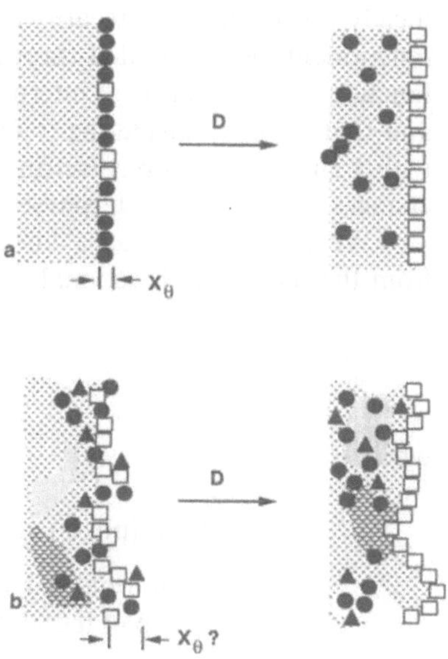

FIGURE 14. Schematic illustration of our model for the interfacial region of $PE\text{-}CO_2H$, and of the actual interfacial region. The filled symbols represent polar functional groups; the open symbols represent nonpolar ones. X_0 denotes the thickness of the θ interphase. The lower figure depicts the heterogeneity in morphology and in chemical composition, characteristic of the actual interface of $PE\text{-}CO_2H$. Reprinted with permission from reference 30. Copyright © 1987 American Chemical Society.

to previous work describing diffusion in bulk polyethylene.[93] We were interested, for instance, in how the energy of activation for diffusion ($E_{a,D}$) from a surface differed from that in the bulk.

We followed the diffusion of the polar functional groups away from the surface by measuring the advancing contact angle of water on the films as a function of the amount of time they had been heated at various temperatures. Since the θ interphase extends only a short distance into the polymer, these measurements selectively monitored diffusion in the interfacial region close to the surface (within 5–10 Å).

Deriving diffusion constants (D) from contact angles required several approximations. Figure 14 schematically compares the actual interface of $PE\text{-}CO_2H$ to our model for the interface. First, we treated the system as if the surface were flat and composed of homogeneously distributed, noninteracting functional groups. The actual interface is rough, at least on the scale of 1000 Å, and there is no reason to believe that the functional groups are evenly distributed on the microscopic level.* The carboxylic acid groups may interact strongly via hydrogen bonding.† Second, we treated the θ interphase as a mixture of only two types of functionality: polar (P) groups and nonpolar (NP) groups. The actual interface, however, is a mixture of many types of polar (carboxylic acids and aldehydes–ketones) and nonpolar (predominantly methylene) groups. Third, we assumed that the bulk polymer was a homogeneous, isotropic medium. The bulk polymer is actually heterogenous and

* An added complication is that heating may change the roughness of the surface. There are no gross morphological changes on heating (as judged by SEM), but changes on a much smaller scale of length than is visible via SEM may be important. We note that the contact angle of water on reconstructed $PE\text{-}CO_2H$ is the same as that on PE-H, a value lower than that on *unreconstructed* $PE\text{-}CO_2C_8H_{17}$.

† The ATR-IR spectra of these films were consistent with the presence of hydrogen-bonded dimers of the carboxylic acid groups.

locally anisotropic: there are crystalline and amorphous regions, and some polar functionality is distributed throughout the film. Finally, we assumed that all of the functional groups in the θ interphase affect the wetting behavior to the same extent, and that groups outside the θ interphase do not affect wetting. As the experiments with **1** showed, however, wetting can be affected strongly by subtle conformational changes within the θ interphase.

We treated $\cos\theta$ as a linear combination of molecular components, arising from the individual functional groups that make up the surface:[19]*

$$\cos\theta = \sum N_i \cos\theta_i \tag{5}$$

where

$i =$ P (polar) or NP (nonpolar) functional group

$N_i =$ normalized, fractional area of the interface made up of the ith functional group

This assumption is not correct, at least for interfaces composed of carboxylic acid and methyl groups;[94] it is, however, a useful first approximation and greatly simplifies the kinetic analysis.

By defining a normalized value of $\cos\theta$, $f_{\cos\theta}$ [Eq. 6], expressing it in terms of N_i, and combining it with the equation describing diffusion from a planar surface† [Eq. 7], we extracted diffusion constants (D) for the interfacial reconstruction of PE-CO_2H at several temperatures:

$$f_{\cos\theta} = \frac{\cos\theta_t - \cos\theta_\infty}{\cos\theta_{t_0} - \cos\theta_\infty} \tag{6}$$

$$\frac{N(x,t)}{N(x=0,t=0)} = (\pi D t)^{-1/2} \exp\left(\frac{-x^2}{4Dt}\right) \tag{7}$$

Figure 15 shows the data plotted according to the Arrhenius equation from 20 to 100°C: they do not fit a single straight line. A least-squares fit to the data over the entire range of temperature gives an activation energy for diffusion, $E_{a,D}$, of 37 kcal/mol. A least-squares fit to the data in only the low-temperature regime gives a value for $E_{a,D}$ of 50 kcal/mol. The data at high temperatures may be suspect since significant reconstruction may have occurred in the time required for the films to equilibrate at these temperatures.

Other groups, who had previously found similar results, suggested alternative explanations associated with the onset of melting of the polymer. The lowest temperature at which Ter-Minassian-Saraga and co-workers observed reconstruction of

* This equation is based upon Cassie's model for heterogeneous interfaces [Eq. (3)].[67]

† The number of ith functional groups within the θ interphase is given by $\int_0^{x_\theta} N(x,t)\,dx$, where x_θ equals the depth of the θ interphase. For the calculations of diffusion constants, we approximated this value as 5 Å.[82]

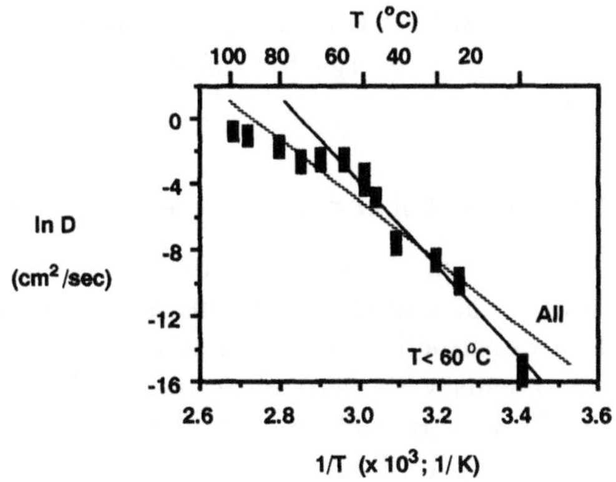

FIGURE 15. Arrhenius plot for the reconstruction of PE-CO_2H in the temperature range of 20–100°C. The solid line is a least-squares fit to the low-temperature ($\leqslant$60°C) data; the dashed line is a least-squares fit to all of the data. Reprinted with permission from reference 30. Copyright © 1987 American Chemical Society.

the surface of their oxidized polyethylene (Section 2.2) coincided with the onset of the melting transition given by differential scanning calorimetry (DSC).[33]

Pennings and Bosman measured the rates of thermal "relaxation" of the surface free energy of polyethylene and other polymers that had been compression-molded against gold foil.[39] Their plot of the rate of relaxation for polyethylene, derived from the measurement of contact angles, as a function of $1/T$, showed a discontinuity at 52°C; the data above this temperature fell on a straight line, and the data below it fell on a different straight line. The discontinuity occurred at approximately the same temperature as the onset of the melting peak given by DSC. Although the process that they were studying—the conversion of crystalline to amorphous polymer—is formally different from the reconstruction of PE-CO_2H, both processes involve conformational changes and diffusion of polymer chains in the interfacial region.

Klein and Briscoe reported a discontinuity in the plot of log D versus $1/T$ for diffusants with long alkyl chains in the bulk of branched, low-density polyethylene; the discontinuity occurred at $1/T_m$ (T_m = melting point of the polymer $\approx$ 107°C).[93] Above the melting point, the data fit a single straight line; below the melting point however, the data did not fit a straight line. The authors attributed the anomalous low temperature behavior to "complex morphological changes" associated with the onset of melting. They also cited dimerization of diffusants (stearamide in one case) as a potential complication at low temperature.

The values for $E_{a,D}$ derived from the plot in Fig. 15 are significantly higher than those reported by Klein and Briscoe for diffusion of molecules with long alkyl chains through the interior of polyethylene. The activation energy for diffusion of behenyl behenate, $CH_3(CH_2)_{20}CO_2(CH_2)_{21}CH_3$, through semicrystalline, low-density polyethylene is 23 kcal/mol.[93] The values of activation energy from the annealing experiments reported by Pennings and Bosman were 8 kcal/mol above 52°C and 31 kcal/mol below it. Baszkin and Ter-Minassian-Saraga did not report an activation energy for the thermal reconstruction of PE-H oxidized with sulfuric acid–potassium chlorate.

These differences in activation energy may be due, at least in part, to hydrogen bonding between carboxylic acid groups at the surface of PE-CO_2H. The enthalpy of dissociation for the H-bonded dimer of stearic acid in a solution of paraffin is

13.4 kcal/mol.[95] The contribution of the surface free energy (γ_{sv}) of the polymer to the activation energy for diffusion away from the solid–vapor interface is small ($\leqslant$5%).*

3.4. Reconstruction of the Interface of Derivatives of PE-CO_2H

Although the contribution of γ_{SV} to $E_{a,D}$ is small *for PE-CO_2H*, the minimization of interfacial free energy is, we presume, an important driving force in the reconstruction of the interfaces of PE-CO_2H and its derivatives. A series of experiments with derivatives of PE-CO_2H that varied in their degree of hydrophobicity supported this hypothesis.

The hydrophobicity of a surface is a function of its interfacial free energy. The value of $\cos\theta$ is proportional to the difference in interfacial free energies, $\gamma_{sv} - \gamma_{sl}$ [Eq. (1)]. These two terms are not separable using wetting data, but at extreme values of $\cos\theta$ (large positive or negative), one of the two components is dominant. For contact angles of water on these surfaces, large positive values of $\cos\theta$ (small values of θ) indicate polar surfaces with high surface free energies (mainly γ_{sv}); large negative values of $\cos\theta$ (values of $\theta \gtrsim 90°$) indicate nonpolar surfaces with low solid–vapor interfacial free energies (γ_{sv}).

Part a of Fig. 16 shows the advancing contact angle of water on surfaces with interfacial free energies lower than that of PE-H as a function of the amount of time the samples had been heated. Part b of Fig. 16 shows the advancing contact angle of water on surfaces with interfacial free energies higher than that of PE-H as a function of the amount of time the samples had been heated. These data show the expected results: (1) all of the surfaces approach the wetting behavior of PE-H on extended heating; and (2) surfaces with interfacial free energies lower than that of PE-H reconstruct more slowly than those with interfacial free energies higher than PE-H. A plot of D versus the cosine of the initial (i.e., before reconstruction) contact angle of water, $(\cos\theta_a)_i$, for PE-CO_2H and several of its derivatives illustrates this relation (Fig. 17).

The surface of PE[$CH_2OCO(CF_2)_6CF_3$] reconstructed very slowly relative to the other surfaces. To determine whether the slow rate was due, in part, to steric problems associated with diffusion of a large functional group, we monitored the reconstruction of several *high* energy surfaces [poly(ethylene glycol) esters of PE-CO_2H] as a function of the size of the interfacial functional groups. Figure 18 shows the rate of reconstruction for these interfaces, as measured by two quantities ($t_{1/2}$ and t_{100}), as a function of the number of monomer units in the ester. The quantity $t_{1/2}$ is the time required, at a given temperature, for $f_{\cos\theta}$ to reach 0.5 (i.e., "half reconstructed"); t_{100} is the time required for the advancing contact angle of water to reach 100°, that is, the point at which it is experimentally indistin-

* The critical surface tension of PE-H is 31 ergs/cm^2;[96] estimating that there are about 10^{15} "molecules"/cm^2 occupying the surface sites, γ_c is about 0.5 kcal/mol. Unfortunately, we do not know γ_c for PE-CO_2H; it is certainly higher than that of PE-H, but probably not by more than a factor of 2 or 3. The value of γ_c reported for a 20.7:79.3 mol% copolymer of acrylic acid and ethylene is 59 erg/cm^2.[97] The value of γ_c reported by Baszkin and Ter-Minassian-Saraga for surface-oxidized polyethylene is 39 erg/cm^2.[28]

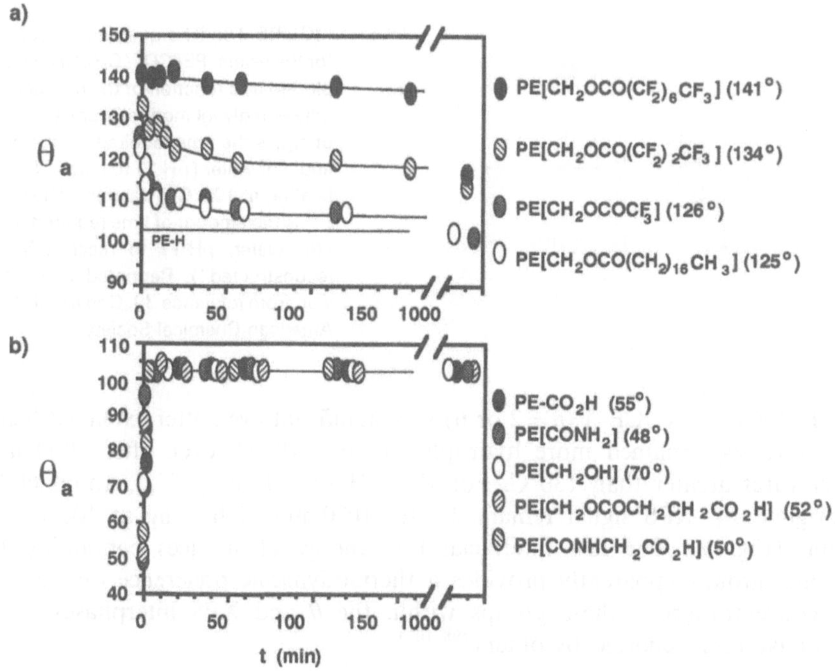

FIGURE 16. Thermal reconstruction of derivatives of PE-CO_2H at 100°C in vacuum: (a) variation in contact angles of water (pH 1) for samples whose interfacial free energies are lower than PE-H as a function of the amount of time of heating; (b) variation in contact angles of water for samples whose interfacial free energies are higher than PE-H as a function of the amount of time of heating. The contact angles of water on these surfaces before reconstruction are given for reference. Reprinted with permission from reference 30. Copyright © 1987 American Chemical Society.

guishable from that on PE-H. The rate of reconstruction of longer esters ($n = 7$, 10, and 14) was significantly slower than that of the shorter esters ($n = 1$–3). The differences, however, are not large enough to explain the slow rate observed for PE[$CH_2OCO(CF_2)_6CF_3$]. These data are consistent with the idea that the low interfacial free energy of the fluorinated surfaces impedes their reconstruction.

The concentration of the fluorinated groups in the interfacial region of

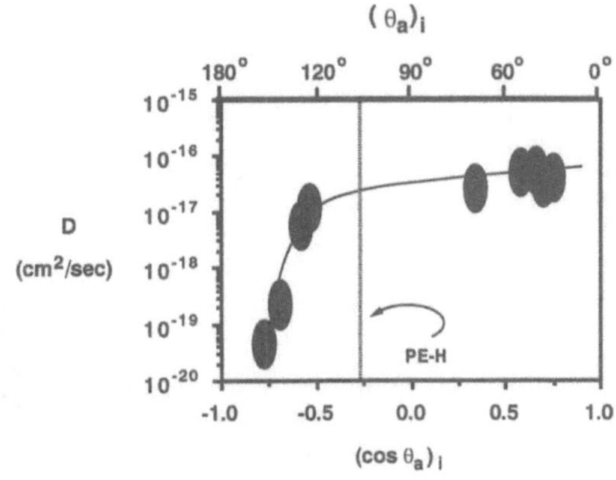

FIGURE 17. Diffusion constants (D) describing the reconstruction of PE-CO_2H and derivatives plotted as a function of $(\cos \theta_a)_i$, the cosine of the initial (before reconstruction) contact angle of water (pH 1). Reprinted with permission from reference 30. Copyright © 1987 American Chemical Society.

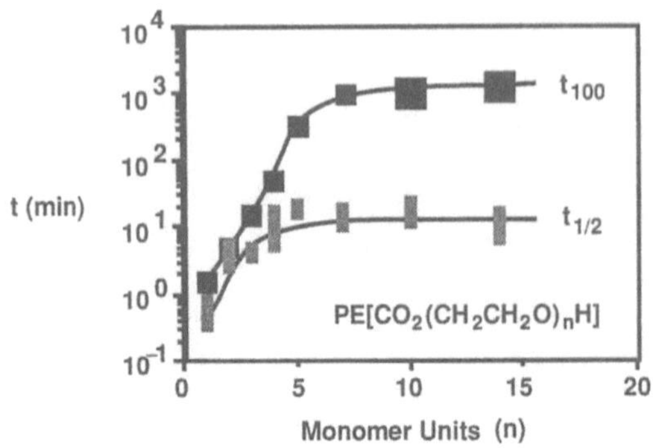

FIGURE 18. The quantities t_{100} and $t_{1/2}$ for the esters, $PE[CO_2(CH_2CH_2O)_n)H]$, plotted as a function of the number (n) of ethylene glycol monomer units. The value of t_{100} is the time required for the contact angle of water (pH 1) to reach 100° upon heating at 100°C in vacuum. The value of $t^{1/2}$ is the amount of time required for $f_{\cos\theta}$ (for water, pH 1) to reach 0.5 ("half reconstructed"). Reprinted with permission from reference 30. Copyright © 1987 American Chemical Society.

$PE[CH_2OCO(CF_2)_nCF_3]$ ($n=2$ or 6) was significant even after extended heating. These surfaces remained more hydrophobic than PE-H, even after 1000 min at temperatures greater than 130°C. For $PE[CH_2OCO(CF_2)_2CF_3]$, about 60% of the original F_{1s} XPS signal remained after 1000 min of heating at 100°C under vacuum (Fig. 19). The low interfacial free energy of surfaces containing these functional groups apparently provides a thermodynamic preference for keeping a large concentration of these groups within the θ and XPS interphases. Similar results have been reported by others.[98–102]

3.5. Reconstruction of the Interface of PE-CO_2H and Derivatives on Heating in Contact with Liquids

We were interested in the influence of a contacting liquid on the thermal reconstruction of the surface of PE-CO_2H and its derivatives, because the introduction of a solid–liquid interface allowed a systematic variation in interfacial free energy. This question also had practical significance, since the surfaces of these films reconstruct at room temperature when in contact with many organic solvents. Some solvents (e.g., methylene chloride, hexane) swell the interfacial region of polyethylene and its derivatives at room temperature; other solvents (e.g., toluene or hexadecane) dissolve the films at elevated temperatures. The advancing contact angle of water on the surface of PE-CO_2H, for example, changes from its initial

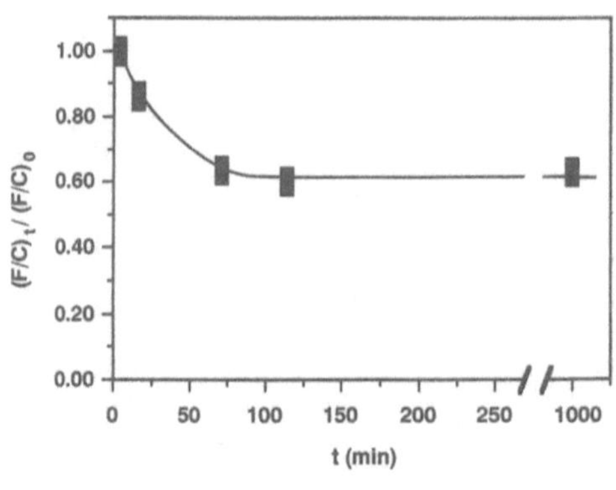

FIGURE 19. Surface concentration of fluorine, relative to that of carbon, by XPS for $PE[CH_2OCO(CF_2)_2CF_3]$ as a function of the amount of time the sample had been heated at 100°C in vacuum. Reprinted with permission from reference 30. Copyright © 1987 American Chemical Society.

value of 55° to 80° after soaking in methylene chloride at room temperature for 4 h. This behavior restricts the number of solvents useful for the synthesis of derivatives of PE-CO_2H; we have not, however, studied the process of swelling in detail.

The previous section described how solid–vapor interfacial free energy (γ_{sv}) can influence the rate at which a surface reconstructs; this section focuses on the influence of γ_{sl}. Surfaces bearing polar functional groups (e.g., PE-CO_2H) interact strongly with polar liquids (e.g., water) by dipolar and hydrogen bonding interactions in addition to dispersive forces. As a result, we expected that polar, hydrophilic surfaces would reconstruct slowly or not at all when in contact with a polar liquid. In the absence of these favorable polar interactions—that is, those when the film is in contact with a nonpolar liquid—we expected that PE-CO_2H and its polar derivatives would reconstruct relatively quickly.

To test these hypotheses, we monitored the reconstruction of PE-CO_2H, a nonpolar derivative (PE[$CO_2C_8H_{17}$]), and PE-H (as a control) with the films in contact with a polar or with a nonpolar liquid. The two liquids chosen for comparison in these experiments, water and perfluorodecalin, do not appear to swell the interfacial region of the polymer.

Figure 20a shows the dependence of the advancing contract angle of water on these surfaces as a function of the amount of time they had been heated in water (pH 6–7) at 100°C. The interfaces of PE-CO_2H and PE-H showed essentially no change in their wettability by water over the course of these experiments. We attribute the thermal stability of the PE-CO_2H–water interface to the favorable

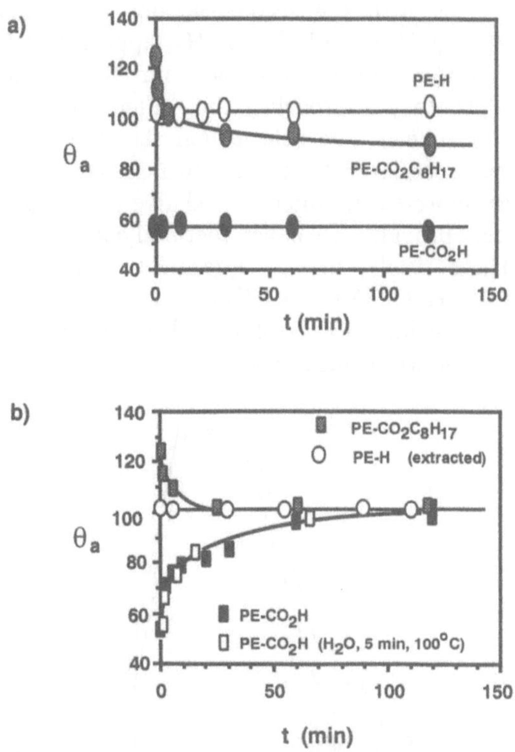

FIGURE 20. Advancing contact angles of water (pH 1) on PE-H, PE-CO_2H, an PE-$CO_2C_8H_{17}$ as a function of the time the films had been heated at various temperatures (a) under water (pH 6–7) and (b) under perfluorodecalin. Plot (b) includes data for PE-CO_2H that had been heated in water for 5 min, then dried, prior to heating in perfluorodecalin. Reprinted with permission from reference 30. Copyright © 1987 American Chemical Society.

interactions between the hydrophilic functionality at the surface and the contacting aqueous phase.

The surface of $PE[CO_2C_8H_{17}]$ behaves differently: initially it is more hydrophobic than PE-H, but on heating in water it becomes more hydrophilic than PE-H. After 2 h of heating in water at 100°C, its contact angle had decreased from the initial (unreconstructed) value of 124° to about 90°. The contact angle dropped from 124° to about 93° in the first 30 min; it fell only slightly in the next hour. We infer from this behavior that, on heating in water, the interface of $PE[CO_2C_8H_{17}]$ reconstructs, exposes polar ester groups to the aqueous phase, and buries hydrophobic alkyl chains. This reconstruction minimizes the interfacial free energy of the system. These results are similar to the wetting behavior of the *o*-anthranilate amide of $PE\text{-}CO_2H$ (**1**), discussed in Section 2.5, and to work reported by others.[40–45, 58–60]

An alternative explanation for this behavior is that heating in water hydrolyzes some of the interfacial ester groups, leaving (more polar) carboxylic acid groups. We feel that this explanation is not correct, however, since the hydrolysis of $PE[CO_2C_8H_{17}]$ in 1 *N* NaOH proceeds to only approximately 30% completion (by ATR-IR) after a week at room temperature.[20]

We heated the same samples (PE-H, $PE\text{-}CO_2H$, and $PE\text{-}CO_2C_8H_{17}$) in the nonpolar solvent, perfluorodecalin. Figure 20b shows the contact angle of water on these surfaces as a function of the amount of time that they had been heated at 100°C in perfluorodecalin. The reconstructions under these conditions were qualitatively similar to those vacuum and under argon.

As with the reconstructions in vacuum or under argon, the surfaces of $PE\text{-}CO_2H$ and $PE\text{-}CO_2C_8H_{17}$ approached the hydrophobicity of PE-H; after 2 h, they were indistinguishable from PE-H in their wettability by water. The *rates* at which they approached this value, however, were slower than those under inert atmospheres. We believe that the slower rates reflect the difference in interfacial free energies between the two systems. The lower free energy of the polymer–perfluorodecalin interface, relative to that of the polymer–vacuum (or argon) interface, provides a smaller thermodynamic driving force to reconstruction (Fig. 21). These results are in agreement with the slow rate of reconstruction of the $PE[CH_2OCO(CF_2)_nCF_3]$–vacuum interface, relative to that of the $PE\text{-}CO_2H$–vacuum interface. As expected, the contact angle on PE-H did not change during the course of these experiments.

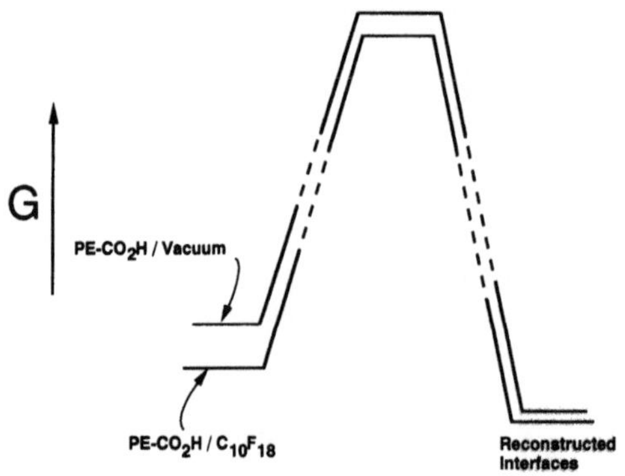

FIGURE 21. Hypothetical plot of free energy versus "reaction coordinate" for the thermal reconstruction of $PE\text{-}CO_2H$ in contact with vacuum and with perfluorodecalin.

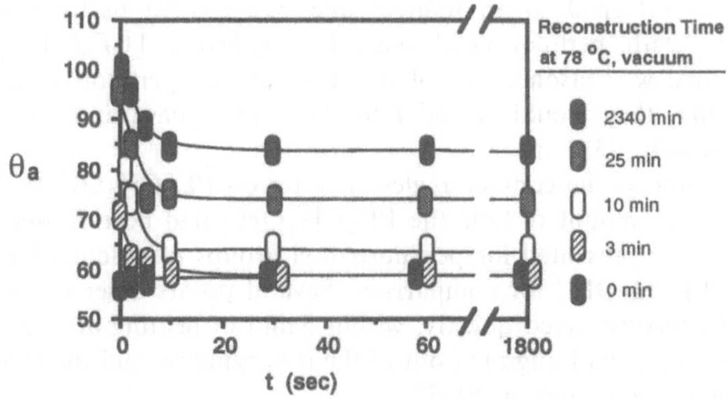

FIGURE 22. Advancing contact angles of water (pH 1) on PE-CO_2H that had been heated at 78°C in vacuum for various amounts of time, as a function of the amount of time the samples were then heated at 100°C in water (pH 6–7). Reprinted with permission from reference 30. Copyright © 1987 American Chemical Society.

3.6. *Recovery of Polar Functional Groups from the Sub–θ Interphase*

In the previous section, we attributed the stability of the interface of PE-CO_2H in water to a thermodynamic preference for having polar, hydrophilic functional groups in contact with the aqueous phase. To test this hypothesis further and to determine the degree of reversibility of the thermal reconstruction of these interfaces, we heated samples of *reconstructed* PE-CO_2H in water to try to "recover" polar functionality that had diffused out of the θ interphase.

Figure 22 shows the contact angle of water as a function of the amount of time samples had been heated in distilled water (pH 6–7) at 100°C. The PE-CO_2H had been reconstructed by heating at 78°C in vacuum for various periods of time. The reconstruction was only partially reversible for samples heated in vacuum for more than 3 min. In all cases however, the surfaces reached limiting contact angles within 30 s on heating in water; no further decrease had occurred after 30 min.

3.7. *Depth Profiling of PE-CO_2H and Derivatives During the Thermal Reconstruction of Their Interfaces*

The accessibility of functional groups in the sub–θ interphase to reagents in solution allowed us to follow the movement of polar functionality out of the θ interphase during reconstruction. Since the thickness of the θ interphase is small, direct measurement of the concentration of polar functionality as a function of depth into the polymer would require resolution on the scale of a few angstroms. Although collecting XPS data at various takeoff angles* can, in principle, be used for depth profiling of this sort,[51] the roughness of the surface of these films prevented us from collecting interpretable data. We followed the reconstruction of PE-$CONHCH_2CO_2H$ by XPS, but found little difference between intensities using takeoff angles of 20° and 75°.

In the absence of a direct method for determining the depth profile of polar

* A takeoff angle is the angle between the plane of the surface of the sample and the detector.

functionality, we turned to an indirect one. We treated partially reconstructed PE[CH_2OH] with perfluorinated *n*-alkyl anhydrides, [$CF_3(CF_2)_nCO]_2O$, of different lengths; we reasoned that if the hydrophobic perfluoroalkyl chains were sufficiently long, they would extend into the θ interphase and raise the contact angle of water (Fig. 23).

Figure 24 shows the contact angles of water on PE[$CH_2OCO(CF_2)_nCF_3$] as a function of the amount of time the PE[CH_2OH] had been heated at 100°C in vacuum. Data are presented for perfluoroalkyl groups of different lengths and for unacylated PE[CH_2OH] for comparison. Several points deserve comment. First, PE[CH_2OH] reconstructed quickly; within 5 min of heating at 100°C in vacuum the hydroxyl groups had migrated out of the θ interphase, and the contact angle of water was the same as that on PE-H.

Second, the perfluoro*methyl* ($n = 0$) ester derived from PE[CH_2OH] that had been reconstructed for 5 min was almost as hydrophobic as that derived from *unreconstructed* PE[CH_2OH]. Even the perfluoromethyl ester of PE[CH_2OH] that had been reconstructed for 20 min was more hydrophobic than PE-H. These results imply that the initial reconstruction (about the first 5 min) involves subtle conformational changes of the hydroxyl-containing groups, reminiscent of the pH-dependent behavior of **1**. After 5 min of heating, the hydroxyls had migrated

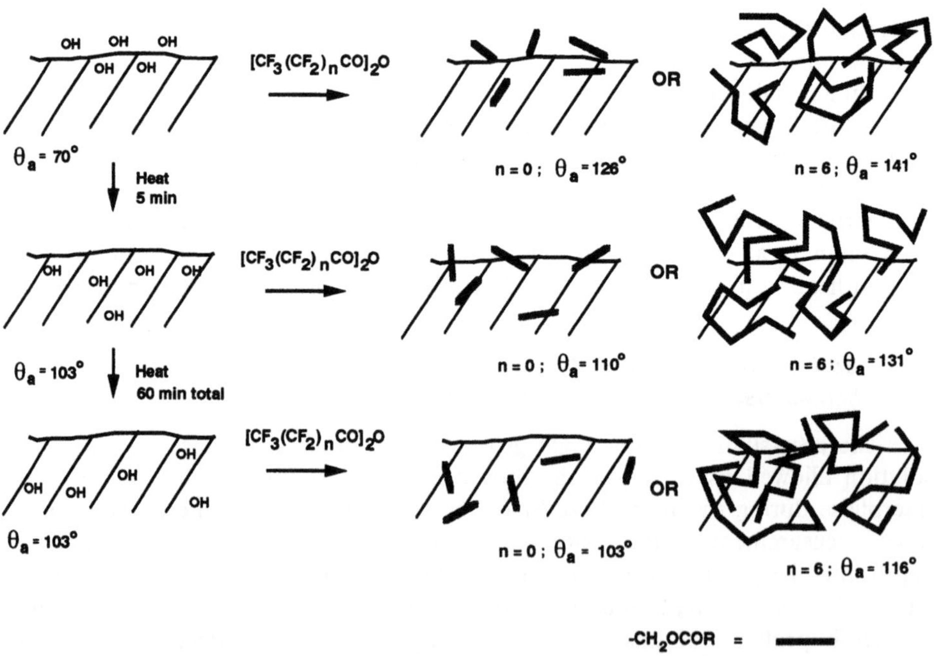

FIGURE 23. Schematic illustration of the interfacial region of PE[CH_2OH] and its derivatives, PE[$CH_2OCO(CF_2)_nCF_3$], before and after initial reconstruction of the PE[CH_2OH] at 100°C in vacuum. Reaction of perfluoroanhydrides with the hydroxyl groups in the sub-θ interphase produced surfaces whose wetting behavior provided information about the depth of the θ interphase. Contact angles of water (pH 1) are shown at each step of the reconstruction and functionalization. Reprinted with permission from reference 30. Copyright © 1987 American Chemical Society.

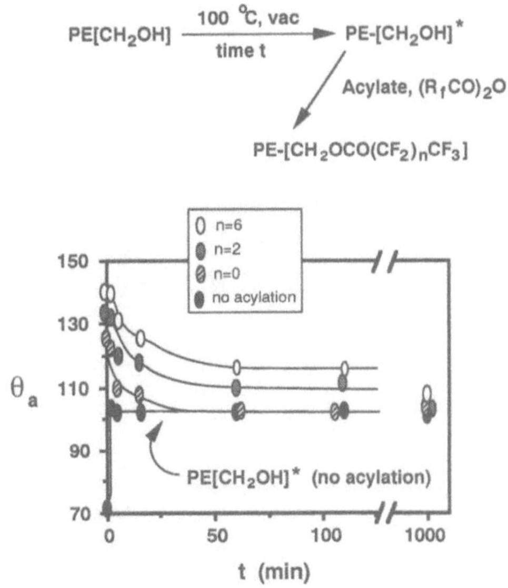

FIGURE 24. Advancing contact angles of water (pH 1) on $PE[CH_2OCO(CF_2)_nCF_3]$ as a function of the amount of time that the $PE\text{-}CH_2OH$ had been heated at 100°C in vacuum prior to acylation. Data for unacylated $PE[CH_2OH]$ are shown for comparison. Reprinted with permission from reference 30. Copyright © 1987 American Chemical Society.

out of the θ interphase, but they remained close enough to the surface that the perfluoromethyl groups of the esters protruded into the θ interphase. These results also confirmed that the θ interphase is very thin ($\sim$5–10 Å).

Third, as the hydroxyl groups diffused away from the interface, the length of the perfluoroalkyl chains of the esters became important for determining the wettability of the interface. For samples at each point in the reconstruction, the hydrophobicity of the esters increased with increasing length of the perfluoroalkyl chains. After further heating ($\sim$1 h), the hydroxyl groups had diffused more than a few angstroms away from the interface, and the perfluoromethyl ester no longer influenced wetting. After extended heating (1000 min), the hydroxyl groups had diffused sufficiently far into the polymer that *none* of the esters influenced wetting. In each case, ATR-IR spectra confirmed the existence of esters outside of the θ interphase.

Treatment of the esters with 1 *N* NaOH regenerated *reconstructed* $PE[CH_2OH]$; that is, the wetting behavior of the product was indistinguishable from that of PE-H. Retreatment with the anhydride produced an esterified surface whose contact angle was the *same* as its original value. The reversibility exemplified in these experiments is important, because it rules out the possibility that the esterification or saponification reactions, *themselves*, reconstruct the surfaces.

3.8. Is Interfacial Strain a Driving Force for the Reconstruction of the Surface of $PE\text{-}CO_2H$?

By comparing the rates of reconstruction for samples annealed either in vacuum before oxidation or in water after oxidation to unannealed samples, we showed that mechanical stress in the films was not an important factor in these processes. Figure 25 compares annealed and unannealed samples in their rates of reconstruction at 65°C.

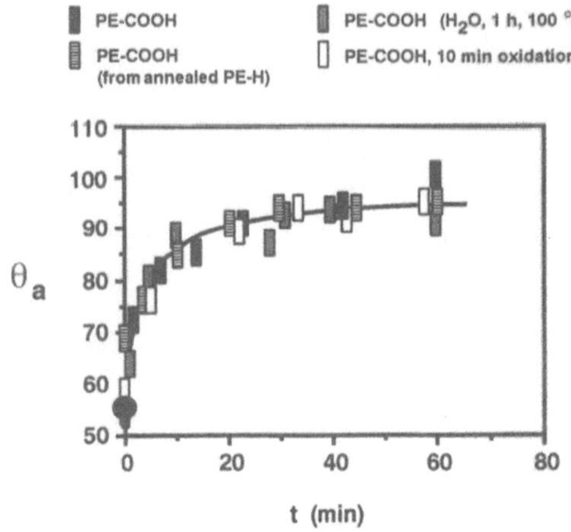

FIGURE 25. Advancing contact angles of water (pH 1) on various types of PE-CO_2H as a function of the amount of time the samples had been heated at 65°C in vacuum. The samples were PE-CO_2H prepared by oxidation of PE-H (solid box); PE-CO_2H prepared by oxidation of PE-H that had been annealed at 100°C under vacuum for five days (box with lines); PE-CO_2H prepared by oxidation of PE-H, then heated at 100°C for 1 h in water (pH 6–7) prior to reconstruction (stippled box); and PE-CO_2H prepared by oxidation of PE-H for 10 min instead of the usual 1 min (unfilled box). All of the samples had θ_a(water) = 54–57° before reconstruction. Reprinted with permission from reference 30. Copyright © 1987 American Chemical Society.

We heated a sample of PE-H in vacuum at 100°C for five days, a treatment longer by a factor of 10^3 than that required for all of the polar groups of PE-CO_2H to diffuse out of the θ interphase. The PE-CO_2H derived from this film reconstructed at the same rate as PE-CO_2H derived from unannealed PE-H. We heated another sample of PE-CO_2H in distilled water at 100°C for 1 h; this treatment allowed extensive annealing of the interface while keeping the hydrophilic CO_2H groups within the θ interphase. As expected, the contact angle of water (pH 1) on this sample was the same before and after treatment. The rate of reconstruction of the annealed sample was indistinguishable from that of unannealed PE-CO_2H.

It is possible that the oxidation itself has an important effect on the thermal reconstruction of these films: treatment of PE-H with hot chromic acid may reveal underlying mechanical strain or *introduce* strain in the interfacial region of the polymer films. Oxidation with chromic acid could produce strain by inducing reactions (i.e., of radicals)[103] that cross-link the methylene chains at the surface. In addition, oxidation could lower the density of the interfacial region of the polymer, relative to that of the bulk, by selectively etching amorphous portions of the interface. The initial reconstruction might then be driven by relaxation to an "equilibrium density" in the interfacial region.[27,29]

To test these hypotheses, we oxidized a sample of PE-H with chromic acid at 72°C for 10 min, instead of the usual 1 min, to produce PE-CO_2H with a deeply etched surface (Fig. 2). The interface of this sample reconstructed at the same rate as PE-CO_2H produced by a 1 min oxidation, indicating that the oxidation neither reveals nor introduces strain into the polymer films. This result does not, however, rule out the possibility that the oxidation introduces strain into the interfacial region in a quickly established ($\leqslant$1 min) steady-state process.

CONCLUSIONS

The unique feature that chemistry offers materials and surface science is the ability to manipulate systems of interest at the molecular level. Synthetic chemistry provides versatile methods for the modification of existing organic surfaces and for the creation of new ones. This versatility makes it possible to address the following question: what are the relations between molecular identity and orientation at a surface and the macroscopic, interfacial (especially liquid–solid) behavior of that surface? Initial oxidation of PE-H with chromic acid introduces carboxylic acid moieties into the solid–liquid interface that serve as starting points for further chemical elaboration. This approach allows access to a wide range of functional group types. The relative orientation of these groups can be systematically varied; syntheses of surfaces bearing amides derived from each isomer (*o*, *m*, and *p*) of anthranilic acid provide examples.

Most of our studies have focused on wetting of solids by liquids. Wetting measurements probe the chemical composition, structure (orientation and order), and dynamics at the surface of solids. An example of the last of these areas—the dynamics of polymer-bound functional groups at or near the surface of PE-X—is the focus of this review.

Surface-modified polyethylene provides a challenging "real" system for studying the relations between interfacial structure and wetting: its surface is rough, chemically and morphologically heterogeneous, and penetrable by many liquids and reagents. Since the surface of PE-X is nonideal, the molecular details of its reconstruction are not amenable to study by conventional spectroscopic techniques alone. For instance, changing the takeoff angle in XPS experiments does not significantly change the surface sensitivity of the measurements. Despite these complexities, convenient experimental protocols can give useful information about the location and environment of the surface-bound functional groups.

The work described in this review illustrates the usefulness of combining spectroscopic and nonspectroscopic experiments to infer interfacial structure. This combination allows resolution on the scale of a few angstroms in experiments designed to locate functional groups at or near the surfaces of the polymer. The analytical techniques discussed—measurement of contact angles, XPS, and ATR-IR—each probe a different portion of the polymeric interphase. These techniques give complementary information about the thermal reconstruction of the interfacial region.

Our ability to interpret data from contact angles and from XPS was aided by reference to model systems. Zisman's method for producing ordered monolayers by molecular self-assembly at the surface of a solid,[71] applied in the system comprising alkanethiols and gold, allows production of structurally well-defined organic surfaces. Modification of the molecular precursors of the SAMs—an activity entirely within the field of synthetic organic chemistry—provides routes to a range of model organic surfaces.

These model systems make it possible to address specific questions about the relations between interfacial structure and wettability. One conclusion from our work with these systems is that wetting is the most surface-sensitive technique available to us: wetting by water senses only the outermost 5–10 Å of a solid. Since wetting is surface sensitive, the degree of order in the interfacial region of systems composed of terminally functionalized alkyl chains is important in determining contact angles; that is, the *orientation* of interfacial functional groups can profoundly affect contact angles on the surfaces bearing these groups.

One of the most generally useful methods for examining the surface of PE-X is contact angle titration, the measurement of the variation in the contact angle of water on a surface with the pH of the contacting drop. The contact angle titration of the *o*-anthranilate amide of PE-CO_2H illustrates the relationship between wetting and microreconstruction of a polymer surface. The results of this study show that (1) subtle conformational changes within the θ interphase of a polymer can strongly influence the contact angle of water on its surface, and (2) even in this disordered system, a small nonpolar group (containing six carbon atoms) is sufficient to shield a polar group(s) from a contacting drop of water. We use the word "shield" *only* within the context of the measurement of contact angles for solid–liquid interfaces at quasi-equilibrium. This "shielding" does not apply to *kinetic* phenomena; for instance, polar groups in the sub–θ interphase are still accessible to reagents in aqueous solution.

We studied in detail the thermal reconstruction of PE-CO_2H and several of its derivatives. The main conclusions from these studies are as follows:

1. The initial reconstruction involves conformational changes close to the surface, akin to those inferred in the PE-anthranilate system, driven by the tendency to minimize the interfacial free energy of the system.
2. In a slower process, functional groups passively diffuse into the bulk of the polymer. This process is presumably driven by entropy.
3. The rate of reconstruction and the composition of the interface at equilibrium depend on the initial composition of the interface. Surfaces with interfacial free energies higher than PE-CO_2H reconstruct quickly to give interfaces indistinguishable from PE-H in their wettability by water. Surfaces with interfacial free energies lower than PE-CO_2H reconstruct more slowly than does PE-CO_2H; surfaces bearing long perfluoroalkyl groups (even PE[$CH_2OCO(CF_2)_2CF_3$]) retain significant concentrations of fluorine in XPS interphase (and even in the θ interphase) after extended heating.
4. Interfacial strain, introduced either in the manufacture of the films or in their surface functionalization, does not seem to be an important factor in determining the kinetics and thermodynamics of reconstruction.

Our experimental value for the activation energy of reconstruction of PE-CO_2H is higher than those measured for related processes (diffusion of large monomers in bulk polyethylene, or breakdown of the crystalline lamellae at the surface of low-density polyethylene). These differences add emphasis to the claim that the interfacial reconstruction of PE-CO_2H is complex, and we believe that several factors are probably important in determining the rate of this process. Aside from the inherent difference between a value of $E_{a,D}$ derived directly from the measurement of contact angles on a complex solid and those derived from infrared spectroscopy or from measurements of critical surface tension, these factors may include hydrogen bonding to form dimers of the carboxylic acid moieties in the interfacial region of PE-CO_2H and preferential etching of amorphous regions of the polyethylene interphase in its synthesis.

ACKNOWLEDGMENTS

This work was supported in part by the Office of Naval Research and the Defense Advanced Research Projects Agency, and by the NSF through grants to GMW (CHE 88-12709) and to the Harvard Materials Research Laboratory (DMR 86-14003).

We wish to acknowledge the creativity and effort of our colleagues without whom the work described in this review could not have been done. These people include Randy Holmes-Farley, Mark Wilson, Tom McCarthy, Ralph Nuzzo, Colin Bain, Barry Troughton, and Lou Scarmoutzos.

REFERENCES

1. J. R. Huntsberger, *Adv. Chem. Ser.* **43**, 180 (1964) L. H. Sharpe and H. Schonhorn, *Adv. Chem. Ser.* **43**, 189 (1964).
2. D. T. Clark and W. J. Feast, *Polymer Surfaces*, Wiley, New York 1978; S. Wu, *Polymer Interface and Adhesion*, Marcel Dekker, New York (1982); B. W. Cherry, *Polymer Surfaces*, Cambridge University Press, Cambridge (1981).
3. W. A. Zisman, in *Handbook of Adhesives*, 2nd ed. (I. Skeist, ed.) Chap. 3, Van Nostrand Reinhold, New York (1977).
4. D. Briggs and C. R. Kendall, *Int. J. Adhesion Adhesives* **2**, 13 (1982).
5. S. L. Kaplan and P. W. Rose, *Plastics Eng.* **44**, 77 (1988).
6. D. H. Kaelble, *Physical Chemistry of Adhesion*, Wiley-Interscience. New York (1971).
7. B. D. Ratner, in *Biomaterials: Interfacial Phenomena and Applications* (S. L. Cooper and N. A. Peppas, eds.) Chap. 2, p. 9, *Advances in Chemistry*, Vol. 199, American Chemical Society, Washington, DC (1982).
8. A. A. Durrani and D. Chapman, in *Polymer Surfaces and Interfaces* (W. J. Feast and H. S. Munro, eds.) Chap. 10, Wiley, New York (1987).
9. I. Lundstroem, B. Ivarsson, U. Joensson, and H. Elwing, in *Polymer Surfaces and Interfaces* (W. J. Feast and H. S. Munro, eds.) Chap. 11, Wiley, New York (1987).
10. D. A. Markgraf, *TAPPI Proc.* 333 (1987); K. Rossmann, *J. Polym. Sci.* **19**, 141 (1956).
11. N. Wilson, in *Polymer Surfaces* (D. T. Clark and W. J. Feast, Chap. 7, Wiley, New York (1978).
12. J. D. Swalen, D. L. Allara, J. D. Andrade, E. A. Chandross, S. Garoff, J. Israelachvili, T. J. McCarthy, R. Murray, R. F. Pease, J. F. Rabolt, K. J. Wynne, and H. Yu, *Langmuir* **3**, 932 (1987).
13. J. A. Armstrong and G. M. Whitesides, *Chem. Eng. News* **64**, 22 (1986). *Research Briefings 1986;* National Academy: Washington DC (1986); *Proc. Natl. Acad. Sci.* **84**, 4665–4748 (1987).
14. G. M. Whitesides and G. S. Ferguson, *Chemtracts Org. Chem.* **1**, 171 (1988).
15. J. R. Rasmussen, E. R. Stedronsky, and G. M. Whitesides, *J. Am. Chem. Soc.* **99**, 4736 (1977).
16. J. R. Rasmussen, D. E. Bergbreiter, and G. M. Whitesides, *J. Am. Chem. Soc.* **99**, 4746 (1977).
17. S. R. Holmes-Farley and G. M. Whitesides, *Langmuir* **2**, 266 (1986).
18. S. R. Holmes-Farley, R. H. Reamey, T. J. McCarthy, J. Deutch, and G. M. Whitesides, *Langmuir* **1**, 725 (1985).
19. S. R. Holmes-Farley, C. D. Bain, and G. M. Whitesides, *Langmuir* **4**, 921 (1988).
20. S. R. Holmes-Farley and G. M. Whitesides, *Langmuir* **3**, 62 (1987).
21. R. G. Nuzzo and G. Smolinsky, *Macromolecules* **17**, 1013 (1984).
22. K.-W. Lee and T. J. McCarthy, *Macromolecules* **21**, 309 (1988).
23. K.-W. Lee and T. J. McCarthy, *Macromolecules* **20**, 1437 (1987); K.-W. Lee and T. J. McCarthy, *Macromolecules* **21**, 2318 (1988).
24. A. J. Dias and T. J. McCarthy, *Macromolecules* (1985), **18**, 1826; A. J. Dias and T. J. McCarthy, *Macromolecules* **20**, 2068 (1987).
25. D. E. Bergbreiter, H.-P. Hu, M. D. Hein, *Macromolecules* **22**, 654 (1989).
26. D. E. Gregonis, R. Hsu, D. E. Buerger, L. M. Smith, and J. D. Andrade, in *Macromolecular Solutions* (R. B. Seymour and G. A. Stahl, eds.), p. 120, Pergamon, New York (1978).

27. H. Schonhorn, *Macromolecules* **1**, 145 (1968).
28. A. Baszkin and L. Ter-Minassian-Saraga, *J. Polym. Sci. C* **34**, 243 (1970).
29. T. Tsunoda, T. Seimiya, and T. Sasaki, *Bull. Chem. Soc. Japan* **35**, 1570 (1962).
30. S. R. Holmes-Farley, R. H. Reamey, R. Nuzzo, T. J. McCarthy and G. M. Whitesides, *Langmuir* **3**, 799 (1987).
31. A. Baszkin and L. Ter-Minassian-Saraga, *J. Colloid Interface Sci.* **43**, 190 (1973).
32. A. Baszkin, M. Nishino, and L. Ter-Minassian-Saraga, *J. Colloid Interface Sci.* **54**, 317 (1976).
33. A. Baszkin and L. Ter-Minassian-Saraga, *Polymer* **15**, 759 (1974).
34. I. Langmuir, *Science* **87**, 493 (1938).
35. E. M. Cross and T. J. McCarthy, *Polym. Prepr., Am. Chem. Soc. Div. Polym. Chem.* **29**, 285 (1988); T. J. McCarthy, *Org. Coatings Appl. Polym. Sci. Prepr.* **48**, 520 (1983); D. R. Gagnon and T. J. McCarthy, *J. Appl. Polym. Sci.* **29**, 4335 (1984).
36. H. Yasuda, A. K. Sharma, and T. Yasuda, *J. Polym. Sci. Phys.* **19**, 1285 (1981).
37. J. D. Andrade, S. M. Ma, R. N. King, and D. E. Gregonis, *J. Colloid Interface Sci.* **72**, 488 (1979).
38. R. J. Good and E. D. Kotsidas, *J. Colloid Interface Sci.* **66**, 360 (1978).
39. J. F. M. Pennings and B. Bosman, *Colloid Polym. Sci.* **257**, 720 (1979).
40. F. J. Holly, M. F. Refojo, *J. Biomed. Mater. Res.* **9**, 315 (1975).
41. E. Ruckenstein and S. H. Lee, *J. Colloid Interface Sci.* **120**, 153, 529 (1987).
42. S. H. Lee and E. Ruckenstein, *J. Colloid Interface Sci.* **117**, 172 (1987).
43. E. Ruckenstein and S. V. Gourisankar, *J. Colloid Interface Sci.* **109**, 557 (1986).
44. E. Ruckenstein and S. V. Gourisankar, *J. Colloid Interface Sci.* **107**, 488 (1985).
45. L. Lavielle and J. Schultz, *J. Colloid Interface Sci.* **106**, 438 (1985).
46. R. J. Good, in *Surface and Colloid Science* (R. J. Good and R. R. Stromberg, eds.), Vol. 11, p. 1, Plenum, New York (1979).
47. A. W. Neumann and R. J. Good, in *Surface and Colloid Science* (R. J. Good and R. R. Stromberg, eds.), Vol. 11, p. 31, Plenum, New York (1979).
48. P. G. de Gennes, *Rev. Mod. Phys.* **57**, 827 (1985).
49. R. D. Vold and M. J. Vold, *Colloid and Interface Chemistry*, Addison-Wesley, Reading, MA (1983).
50. A. W. Adamson, *Physical Chemistry of Surfaces*, 4th ed., Wiley, New York (1982).
51. J. D. Andrade, in *Surface and Interfacial Aspects of Biomedical Polymers* (J. D. Andrade, ed.), Vol. 1, Chap. 5, Plenum, New York (1985).
52. K. Knutson and D. J. Lyman, in *Surface and Interfacial Aspects of Biomedical Polymers* (J. D. Andrade, ed.) Vol. 1, Chap. 6, Plenum, New York (1985).
53. D. J. Angier, in *Chemical Reactions of Polymers* (E. M. Fettes, ed.), Chap. 12, Interscience, New York (1964).
54. J. D. Andrade, D. E. Gregonis, and L. M. Smith, in *Surface and Interfacial Aspects of Biomedical Polymers* (J. D. Andrade, ed.) Vol. 1, Chap. 2, Plenum, New York (1985).
55. C. D. Bain, E. B. Troughton, Y.-T. Tao, J. Evall, G. M. Whitesides, and R. G. Nuzzo, *J. Am. Chem. Soc.* **111**, 321 (1989).
56. M. D. Wilson and G. M. Whitesides, unpublished results.
57. R. E. Johnson, Jr. R. H. Dettre, *Adv. Chem. Ser.* **43**, 112 (1964); R. H. Dettre and R. E. Johnson, Jr., *Adv. Chem. Ser.* **43**, 136 (1964).
58. M. E. R. Shanahan, A. Carre, S. Moll, and J. Schultz, *J. Chim. Phys.* **83**, 351 (1986); A. Carre, S. Moll, J. Schultz, and M. E. R. Shanahan, in *Adhesion* (K. W. Allen, ed.), Vol. II, Chap. 6, Elsevier, New York (1987).
59. J. Rault, A. Aouinti, M. Goldman, and A. Goldman, *C.R. Acad. Sci. Paris* **309**, 333 (1989).
60. Y. C. Ko, B. D. Ratner, and A. S. Hoffman, *J. Colloid Interface Sci.* **82**, 25 (1981).
61. T. Young, *Philos. Trans. R. Soc.* **95**, 65 (1805).
62. R. E. Johnson, Jr., R. H. Dettre, *Surf. Colloid Sci.* **2**, 85 (1969).
63. J. D. Andrade, L. M. Smith, and D. E. Gregonis, in *Surface and Interfacial Aspects of Biomedical Polymers* (J. D. Andrade, ed.), Vol. 1, Chap. 7, Plenum, New York (1985).
64. J. F. Oliver, C. Huh, and S. G. Mason, *Colloids Surf.* **1**, 79 (1980).
65. R. N. Wenzel, *Ind. Eng. Chem.* **28**, 988 (1936); *J. Phys. Colloid Chem.* **53**, 1466 (1949).
66. R. Shuttleworth and G. L. J. Bailey, *Discuss. Faraday Soc.* **3**, 16 (1948).
67. A. B. D. Cassie, *Discuss. Faraday Soc.* **3**, 11 (1948); A. B. D. Cassie and S. Baxter, *Trans. Faraday Soc.* **40**, 546 (1944).

68. R. J. Good, *J. Am. Chem. Soc.* **74**, 5041 (1952); J. D. Eick, R. J. Good, and A. W. Neumann, *J. Colloid Interface Sci.* **53**, 235 (1975).
69. H. Schonhorn and F. W. Ryan, *J. Phys. Chem.* **70**, 3811 (1966).
70. W. A. Zisman, *Adv. Chem. Ser.* **43**, 1 (1964), and references therein.
71. W. C. Bigelow, D. L. Pickett, and W. A. Zisman, *J. Colloid Sci.* **1**, 513 (1946).
72. R. G. Nuzzo, F. A. Fusco, and D. L. Allara, *J. Am. Chem. Soc.* **109**, 2358 (1987).
73. M. D. Porter, T. B. Bright, D. L. Allara, and C. E. D. Chidsey, *J. Am. Chem. Soc.* **109**, 3559 (1987).
74. R. G. Nuzzo and D. L. Allara, *J. Am. Chem. Soc.* **105**, 4481 (1983).
75. J.-E. Sundgren, P. Bodoe, B. Ivarsson, and I. J. Lundstroem, *J. Colloid Interface Sci.* **113**, 530 (1986).
76. J. M. Stouffer and T. J. McCarthy, *Macromolecules* **21**, 1204 (1988).
77. H. O. Finklea, S. Avery, M. Lynch, and T. Furtsch, *Langmuir* **3**, 409 (1987).
78. T. Diem, B. Czajka, B. Weber, and S. L. Regen, *J. Am. Chem. Soc.* **108**, 6094 (1986).
79. D. L. Allara, A. F. Hebard, F. J. Padden, R. G. Nuzzo, and D. R. Falcone, *J. Vac. Sci. Technol. A* **1**, 376 (1983).
80. L. Strong and G. M. Whitesides, *Langmuir* **4**, 546 (1988).
81. C. E. D. Chidsey, G.-Y. Liu, P. Rowntree, and G. Scoles, *J. Chem. Phys.* **91**, 4421 (1989).
82. C. D. Bain and G. M. Whitesides, *J. Am. Chem. Soc.* **110**, 5897 (1988).
83. L. E. Janes, P. E. Laibinis, K. L. Prime, and G. M. Whitesides, unpublished results.
84. J. Smith, *J. Colloid Interface Sci.* **88**, 51 (1980); M. E. Schrader, *J. Phys. Chem.* **74**, 2313 (1970).
85. M. D. Wilson, G. S. Ferguson, and G. M. Whitesides, *J. Am. Chem. Soc.* **112**, 1244 (1990).
86. C. D. Bain and G. M. Whitesides, *Science* **240**, 62 (1988); C. D. Bain and G. M. Whitesides, *J. Am. Chem. Soc.* **111**, 7164 (1989).
87. P. E. Laibinis, R. G. Nuzzo, and G. M. Whitesides, unpublished results.
88. M. D. Wilson and G. M. Whitesides, *J. Am. Chem. Soc.* **110**, 8718 (1988).
89. P. G. de Gennes, *Scaling Concepts in Polymer Physics*, Cornell University, Ithaca, NY (1979).
90. M. Tirrell, *Rubber Chem. Technol.* **57**, 523 (1984).
91. A. E. Chalykh and V. B. Zlobin, *Russ. Chem. Rev.* **57**, 504 (1988).
92. P. F. Green and E. J. Kramer, *MRS Bull.* Nov. 16/Dec. 31, 42 (1987).
93. J. Klein and B. J. Briscoe, *Proc. R. Soc. London A* **365**, 53 (1979).
94. C. D. Bain and G. M. Whitesides, *J. Am. Chem. Soc.* **110**, 6560 (1988).
95. D. Szabo Sarkadi and J. H. De Boer, *Rec. Trav. Chim.* **76**, 628 (1957).
96. H. W. Fox and W. A. Zisman, *J. Colloid Sci.* **7**, 428 (1952).
97. *Polymer Handbook* (J. Brandrup and E. H. Immergut, eds.), Wiley, New York (1975).
98. R. A. L. Jones, E. J. Kramer, M. H. Rafailovich, J. Sokolov, and S. A. Schwarz, *Phys. Rev. Lett.* **62**, 280 (1989).
99. I. Schmidt and K. Binder, *J. Phys. (Paris)* **46**, 1631 (1985).
100. H. Nakanishi and P. Pincus, *J. Chem. Phys.* **79**, 997 (1983).
101. D. H. Pan and W. M. Prest, Jr., *J. Appl. Phys.* **58**, 2861 (1985).
102. Q. S. Bhatia, D. H. Pan, and J. T. Koberstein, *Macromolecules* **21**, 2166 (1988).
103. K. B. Wiberg, *Oxidation in Organic Chemistry*, Part A, p. 119, Academic Press, New York (1965).

7

Block Copolymers and Hydrophilicity

Atsushi Takahara

1. INTRODUCTION

Block copolymers are composed of more than two homopolymer components within the same molecular chain that are usually incompatible; that is, they tend to separate into microscopic domains. This phenomenon is called *microphase separation.*[1] The state of microphase separation depends on the molecular weight of each component, its molecular weight distribution, and the relative molecular weight ratios of the polymer components. These polymeric blocks can also be arranged in various structures as shown schematically in Fig. 1. A block copolymer that consists of two homopolymer components is called a *diblock copolymer*. A typical example is polystyrene (PS)–polybutadiene (PBD) (SB). A *triblock copolymer* has two possible configurations, such as A-B-A or A-B-C. A PS–PBD–PS (SBS) triblock copolymer is commercially available. Charge mosaic membranes can be obtained from A-B diblock, A-B-C triblock, or A-B-C-A-B pentablock copolymers. A *multiblock copolymer* consists of alternating oligomeric segments, segmented poly(urethanes) being well-known examples. In addition, recent macromonomer developments enable block copolymers to be prepared with graft side chains.[2]

Applications of block copolymers are becoming increasingly important. New applications are taking maximum advantage of the inherent capabilities of block copolymers. Table 1 summarizes these applications. The surface structure and surface hydrophilicity are especially important for biomedical applications,[3] surface modification,[4] and adhesion.[5] For example, in surface coating, block copolymers offer a combination of substrate adhesion through one of the segments and overall coating property characteristics through a second segment.

For adhesive applications, the block copolymer structure should be designed to achieve good adhesion between two different substrata that may have different

Atsushi Takahara • Department of Chemical Science and Technology, Faculty of Engineering, Kyushu University, Higashi-ku, Fukuoka 812, Japan.

Modern Approaches to Wettability: Theory and Applications, edited by Malcolm E. Schrader and George Loeb. Plenum Press, New York, 1992.

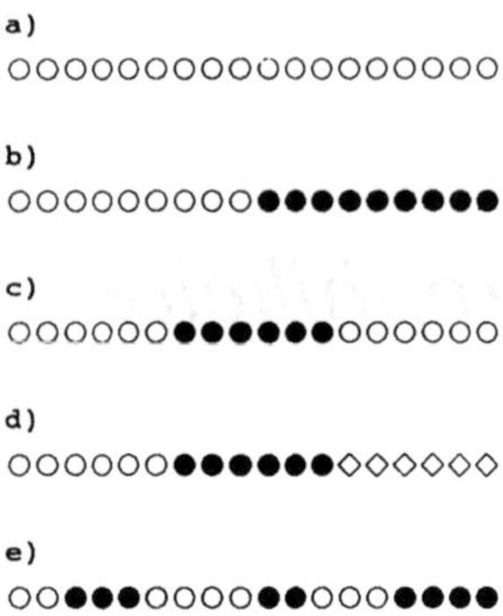

FIGURE 1. Schematic representation of various block copolymer architectures: (a) homopolymer; (b) A-B diblock copolymer; (c) A-B-A triblock copolymer; (d) A-B-C triblock copolymer; (e) multiblock copolymer.

surface characteristics. Since a block copolymer is composed of more than two segments that have different cohesive energy, these polymers can show surface activity such that the surface chemical composition becomes different from that in bulk. Thus, for the application and molecular design of block copolymers, it is necessary to investigate their surface hydrophilic character.

This chapter reviews the methods of characterizing the surface hydrophilicity of block copolymers and summarizes the relationships between their surface chemical composition and hydrophilicity.

2. CHARACTERIZATION METHODS OF POLYMER HYDROPHILICITY

2.1. Spreading Phenomena

The phenomenon of microphase separation has been well established for block copolymer systems. The theory of bulk microphase separation of diblock and triblock copolymers with narrow molecular weight distribution has been reviewed by several researchers.[1, 6] The driving force for phase separation is entropic or energetic. Figure 2 illustrates the various phase structures of diblock copolymers.[6] The domain structure changes from A spheres to A cylinders to A-B lamellae with an increase in the fraction of the A component. These structures represent the ideal block copolymer morphology.

The surface chemical composition of block copolymers is different from the

TABLE 1. Typical Applications of Block Copolymers

High-impact-strength polymer
Elastic fiber (Spandex)
Elastomer (Hytrel)
Surfactant (Pluronic)
Permselective membrane
Biomaterials (Biomer)
Adhesive
Sealants
Surface modifier

FIGURE 2. Schematic illustration of various phase structures of diblock copolymers composed of components A and B. (Reprinted with permission from Ref. 6.)

bulk due to the difference in cohesive energy of each homopolymer component. The polymer surface can also be highly mobile. Such surface dynamics permit the interface to restructure or reorient in response to different environments. This is particularly important in block copolymer systems. In a well-phase-separated block copolymer, the molecular mobility of a particular segment of the copolymer may be relatively independent of another component. The "environmental" effect is particularly pronounced in an aqueous environment, where the polarity of the aqueous phase provides a large driving force for the restructuring of the surface in order to minimize the interfacial free energy. In vacuum and air, the polymer orients with its hydrophobic components toward the interface.[7, 8] These polymer surface dynamics are closely related to the hydrophilicity of the block copolymer surface. The spreading phenomenon is well characterized in liquids.[9] The theory of spreading can be applied to predict the surface chemical composition of a multiphase polymer system.

The interfacial free energy between A and B segments is given by the equation[9]

$$\begin{aligned}\gamma_{AB} &= \gamma_A + \gamma_B - 2\sqrt{\gamma_A^d \gamma_B^d} - 2\sqrt{\gamma_A^p \gamma_B^p} \\ &= (\sqrt{\gamma_A^d} - \sqrt{\gamma_B^d})^2 - (\sqrt{\gamma_A^p} - \sqrt{\gamma_B^p})^2 \qquad (1)\end{aligned}$$

where γ_A and γ_B are the surface free energies of the A and B components, respectively. The superscripts d and p represent the dispersion and polar components. Figure 3 shows the schematic representation of spreading of liquid B over liquid A. The free-energy change for the spreading of a film of liquid B over liquid A is given by

$$S_{B/A} = \gamma_A - \gamma_B - \gamma_{AB} \qquad (2)$$

where $S_{B/A}$, the spreading coefficient of B on A, is positive if spreading is accompanied by a decrease in surface free energy; that is, spreading is spontaneous. From the definition of the work of adhesion and cohesion, the spreading coefficient can be given by the difference between the work of adhesion of A to B and the work of cohesion of B:

$$S_{B/A} = W_{AB} - W_{BB} \qquad (3)$$

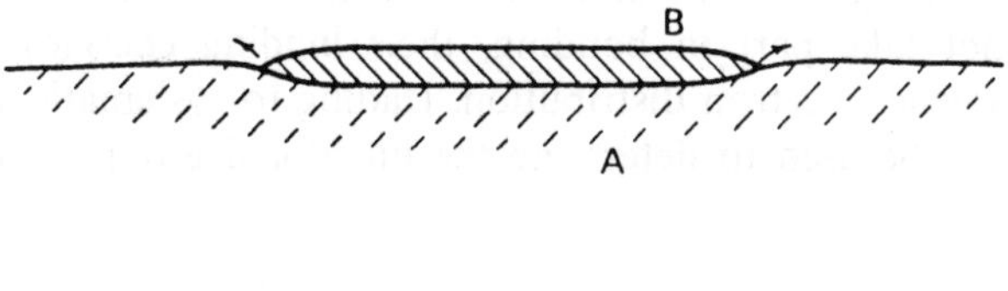

FIGURE 3. Schematic representation of spreading of component B over component A.

If $W_{AB} > W_{BB}$, the A-B interaction is sufficiently strong to promote the wetting of A by B. This is the significance of a positive spreading coefficient. Conversely, no wetting occurs if $W_{BB} > W_{AB}$. If the calculated spreading coefficients $S_{B/A}$ and $S_{A/B}$ are large negative values, then neither component can spread over the other component. The concept of a spreading coefficient can be used to predict the surface-spreading behavior of one segment (component) of a multiphase polymeric system spreading over another segment (component).

Matsuda and co-workers applied this theory to estimate the surface chemical composition of bisphenol A polycarbonate (PC)–poly(ethylene oxide) (PEO) multiblock copolymer system.[10] In air, the PEO segment has a slightly higher surface free energy than the PC segment in air, but has a much lower surface energy in water. The spreading coefficients in air give $S_{PC/PEO} = -11.2\,\text{mN m}^{-1}$ and $S_{PEO/PC} = -8.2\,\text{mN m}^{-1}$. The large negative values of the spreading coefficients indicate that neither component can spread over the other component, resulting in the formation of a microphase-separated structure on the surface. The driving force for the marked reduction in interfacial free energy with water is due to the large wetting pressure ($\Delta F = \gamma_S - \gamma_{SL}$), where γ_S and γ_{SL} indicate the surface free energy of solid and interfacial free energy between solid and liquid (water), respectively. The spreading coefficient under water is positive for $S_{PC/PEO}$ ($= 15.8\,\text{mN m}^{-1}$) but negative for $S_{PEO/PC}$ ($= -35.3\,\text{mN m}^{-1}$), indicating that the PEO segment can spread over the PC phase but the PC phase cannot spread over the PEO phase in water. The interfacial free energy between the water–PEO interface is smaller than that of the PEO–PC interface. These values suggest that the PEO segment will be enriched and fully hydrated at the water–multiphase polymer interfaces. Also, this approach has been utilized for the surface characterization of segmented poly(urethaneureas).[11] The calculation of the free-energy change during spreading does not include the entropic change related to the junction of two segments. Both the enthalpic and entropic changes must be taken into account to explicitly predict the surface chemical composition of block copolymers.

2.2. Surface Characterization Techniques

Table 2 summarizes the techniques employed to characterize the surface hydrophilicity of block copolymers. Various analytical methods have been developed for the surface characterization of polymeric materials, and many papers have reviewed them.[12–14]

2.2.1. X-ray Photoelectron Spectroscopy

X-ray photoelectron spectroscopy (XPS) is based on the photoelectric effect and gives information used to describe the chemical composition of the surface.[12–14] The kinetic energy of the emitted photoelectrons depend on their binding energy and the photon source used. This energy dependence allows the element emitting the photoelectron to be identified. In addition, since the core electrons do not take part in bonding, their binding energies are slightly dependent on the valence electron distribution, leading to systematic shifts in the peak position. This can be used to determine the functional group on the surface. Angular-dependent

TABLE 2. Characterization Method of Surface Hydrophilicity

Spectroscopic techniques	
	X-ray photoelectron spectroscopy (XPS)
	Auger electron spectroscopy (AES)
	Secondary ion mass spectroscopy (SIMS)
	Ion scattering spectroscopy (ISS)
	ATR Fourier transform IR (ATR FT IR)
Microscopy	
	Scanning electron microscopy (SEM)
	Scanning tunneling microscopy (STM)
	Atomic force microscopy (AFM)
	Transmission electron microscopy (TEM)
Contact angle measurements	
	Contact angle (droplet method)
	Captive-bubble method
	Wilhelmy plate method

XPS is a useful technique to characterize the depth profile of the surface layer. Figure 4 shows the relationship between the analytical depth (d) and the emission angle θ of a photoelectron, where θ is defined as the angle between the direction of the analyzer and the surface. The analytical depth (d) can be given by the following equation:

$$d = 3\lambda \sin\theta \tag{4}$$

where λ is the photoelectron mean free path in solids. A decrease in emission angle decreases analytical depth, thereby increasing the surface sensitivity.

Since XPS is measured under high-vacuum conditions, the surface structure may be different from that in water. To characterize the surface chemical composition in water, researchers examined hydrated specimens in a frozen state. Ratner and co-workers measured the surface structure of radiation-grafted hydrogels on silicone rubber in a dry state and in a hydrated condition, using XPS.[15] In the hydrated state, signals from the hydrogel were observed, whereas in the dry state only signals from the base silicone rubber were observed. This technique is useful for characterization at a water–solid interface.

2.2.2. Secondary Ion Mass Spectroscopy

Secondary ion mass spectroscopy (SIMS) is one of the most recently developed characterization methods and has been applied to the surface characterization of

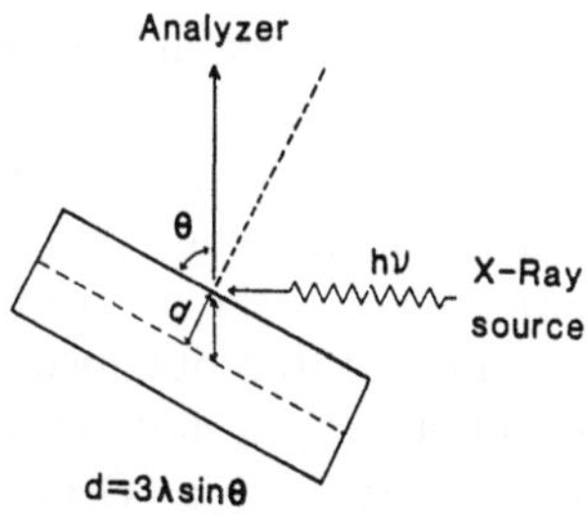

FIGURE 4. Relationship between incident x-ray beam and direction of analyzer: θ, λ, and d represent emission angle of photoelectron, electron mean free path in the solid, and analytical depth, respectively.

polymeric materials. A surface under vacuum is bombarded with an accelerated ion beam. Fragments of surface molecules are generated due to the energy transfer from the ions to the atoms in the surface region. Similar to conventional mass spectroscopy, the mass-to-charge ratio (*m/c*) of these fragments can be measured. The plot of intensity as a function of *m/c* gives information about the surface chemical composition. The advantages of SIMS are its high surface sensitivity (<1 nm) and the large amount of information obtained. The SIMS characterization technique is described in several reviews.[16, 17]

2.2.3. Infrared Spectroscopy

The recent development of Fourier transform infrared (FTIR) spectroscopy enables the measurement of surface chemical structure. By using the attenuated total reflection (ATR) crystal, one can obtain the surface IR spectra.[18] Due to the relatively large wavelength of the infrared ray, the surface depth probed ranges from 0.5 to 3 μm, depending on the wave number, the crystal, and incident angle. The advantage of IR spectroscopy is the large amount of information obtained, such as types of molecular bonds, state of hydrogen bonding, and orientation of dipole moments.[19]

2.2.4. Microscopy

Light microscopy, scanning electron microscopy (SEM), and transmission electron microscopy (TEM) are all useful techniques in the surface analysis of block copolymer systems. The resolution range required generally determines which method is used. SEM, which utilizes secondary electron imaging, is useful for characterizing surface topography. Characterization of the block copolymer morphology has been done by selective staining, thin sectioning, and TEM.[20] However, especially for the characterization of the outermost surface, thin sectioning perpendicular to the film surface is necessary. This technique, however, is very difficult for routine work.

Recent developments in SEM instruments enable the observation of backscattered electron (BSE) images[21] and low-voltage secondary electron images. Since backscattering electrons come from the surface layer due to the limited electron mean free path, the surface morphology of block copolymers can be observed. Backscatter efficiency depends on the atomic number of the surface atoms; thus, selective staining of one phase can enable us to observe the surface morphology of multiphase polymer systems. Andrade and co-workers applied this technique to examine the surface morphology of polystyrene (PS)–polybutadiene (PBD) blends.[22] This method also can be used to study the surface morphology of block copolymers containing hydrophilic and hydrophobic segments.

Low-voltage scanning electron microscopy (LVSEM) is another technique for observing surface topography. Its major advantages are that the secondary electrons that form the image are generated by the electron beam interactions within a few nanometers of the probing electron beam. Therefore, surface sensitivity of LVSEM is similar to that obtained with XPS. This technique is also used to observe surface morphology of a multiphase system of which one phase was selectively stained with· heavy metal such as OsO_4.

2.2.5. Contact Angle Measurements

Contact angle measurements give a direct measure of surface wettability. The simplest and most useful wettability measurement involves placing a probe liquid on a surface and measuring the angle that the liquid makes with the surface of interest. The wettability of a surface is often characterized by a critical surface tension, γ_c, which is obtained by measuring the contact angle with a series of homologous liquids, plotting $\cos\theta$ against γ_{LV}, the liquid–vapor surface tension, and extrapolating to the value of γ_{LV} corresponding to $\cos\theta = 1$.[23] When measured using organic liquids on organic polymers, γ_c should closely approximate the surface free energy of the solid, γ_S. A method for measuring the surface free energy of solids and for resolving the surface energy into contributions from dispersion and dipole–hydrogen bonding forces has been developed by Owens and Wendt.[24] It is based on the measurement of contact angles with water and methylene iodide (see Good & Van Oss, Chapter 1).

The surface characteristics of a water-equilibrated surface have been obtained through a technique developed by Hamilton,[25] which introduces an octane droplet at the polymer–water interface. Andrade and co-workers[26] and Hoffman and co-workers[27] applied this method to deduce interfacial energetics at a gel–water interface.

The dynamic contact angle observed upon advancement of the liquid front is called the advancing contact angle, θ_a, and the angle of the receding liquid front is the receding contact angle, θ_r. Contact angle hysteresis is expressed as $\theta_a - \theta_r$. Both θ_a and θ_r can be measured by direct observation of the droplet on the surface[28] or by the Wilhelmy plate technique.[29] Hysteresis with movement of the liquid front is often a result of surface heterogeneity or roughness.[30] Heterogeneity usually causes much larger hysteresis effects than roughness does. The relationship between surface heterogeneity and contact angle was reviewed by Johnson and Dettre.[31] Since block copolymers show surface heterogeneity and roughness due to the difference in surface free energy of each component, the dynamic contact angle hysteresis measurement is a useful method for characterizing block copolymer hydrophilicity.

A block copolymer that contains a hydrophilic segment may show variations in its contact angle upon immersion in water. This is attributed to surface changes, such as swelling of the solid by absorption of liquids and deformation or reorientation of the solid surface to lower the solid–liquid interfacial tension. The measurement of this time-dependent process is used to characterize the dynamic nature of the polymer surface.

3. *DIBLOCK AND TRIBLOCK COPOLYMERS*

3.1. *Polystyrene–Polydiene Diblock and Triblock Copolymer and Their Derivatives*

Anionically prepared block copolymers of PBD and PS exhibit a well-developed microphase separated structure. Vanzo studied the surface morphology of solution-cast PS–PBD diblock copolymer films. He found that the regularly

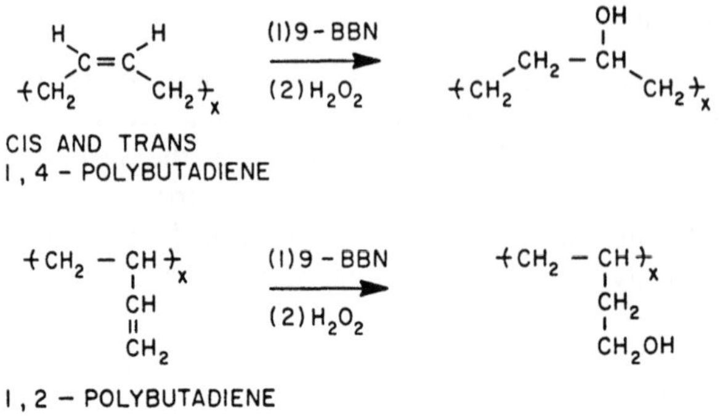

FIGURE 5. Hydroxylation of butadiene segment by hydroboration procedure: 9-BBN = 9-borobicyclo[3,3,1]nonane. (Reprinted with permission from Ref. 35.)

spaced structure correlated with the molecular weight. This reflects the domain structure of elastomers in bulk.[32] Hashimoto and Hasegawa studied the surface morphology of a PS–polyisoprene (PI) diblock copolymer at a free surface by transmission electron microscopy.[33] They observed a thin PI layer at the polymer–air interface as expected, since the critical surface tension for PI (30–32 mN m^{-1}) is smaller than that for PS (33–36 mN m^{-1}). The enrichment of PI on the surface might be due to the difference in surface free energy between those two segments.

The PS–PBD–PS triblock copolymer (SBS elastomer) can be used to prepare a copolymer with hydrophilic and hydrophobic segments by oxidation of the butadiene block either by peracid followed by hydrolysis[34] or by a hydroboration procedure.[35] The butadiene block of SBS elastomers of different PS–PBD ratios were hydroborated using 9-borobicyclo[3,3,1]nonane (9-BBN) in tetrahydrofuran, as shown in Fig. 5. The equilibrium water content and the contact angle results of these polymer systems as a function of butadiene content are shown in Fig. 6. The captive-air-bubble underwater contact angle closely corresponds to the receding

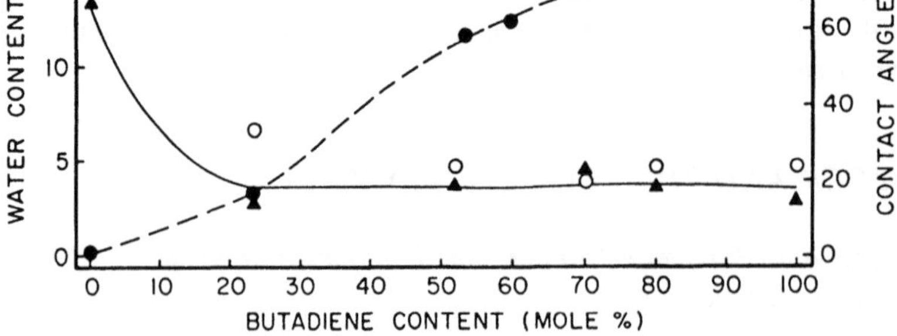

FIGURE 6. Contact angle measurements at 24 h hydration and equilibrium water content of polystyrene–poly(hydroxylated butadiene)–polystyrene triblock copolymers. (Reprinted with permission from Ref. 35.)

contact angle. The receding angle does not correlate to the bulk water content of the polymers. The bulk water content increases almost linearly with PBD content; however, the receding contact angle decreases sharply with a small amount of hydroxylated PBD and then remains constant as the PBD content is increased. This may suggest that the underwater surface of all the block copolymers is dominated by the hydroxylated PBD phase in order to minimize interfacial free energy at the water–polymer surface. However, an advancing angle did not show any dependence on the PBD content, since the surface was enriched with hydrophobic PS even if the (hydroxylated PBD) content was high.

3.2. Polystyrene–Polytetrahydrofuran Diblock and Triblock Copolymers

PS–polytetrahydrofuran (PTHF) block copolymers show unique characteristics due to the crystallization and surface activity of the PTHF segment.[36] PS–PTHF diblock and PS–PTHF–PS triblock copolymers were synthesized by an ion-coupling reaction between the living ends of PS anions and PTHF cations. The morphology observed by TEM showed the formation of PTHF fibrils covered with PS for PTHF-rich block copolymers and the accumulation of PTHF blocks for PS-rich block copolymers. The observed morphologies can be explained by considering that, at high PTHF contents, crystallization of the PTHF excludes PS. At high PS concentration, the surface activity of the PTHF units provides the driving force to coat the surface. The PTHF block was found to have a surface activity 3 mN m^{-1} lower than the PS block. The contact angle of water droplets was a linear function of surface composition, and water wettability of the PS-rich block copolymers due to enrichment of the surface by PTHF blocks was observed.

3.3. Polystyrene–Polydimethylsiloxane Diblock and Triblock Copolymers

PS–polydimethylsiloxane (PDMS) block copolymers were synthesized by sequential anionic polymerization of styrene and hexamethylcyclotrisiloxane. PDMS has an extremely low surface tension and surface free energy due to its very large molar volume, very low cohesive energy density, and high flexibility. Gaines and Bender[37] studied the surface tensions of melts of PS blends containing 0.05–5 wt% of PS–PDMS diblock copolymers at 458–473 K. Their results indicated that the surface of the blends containing 0.2 wt% or more of copolymer were almost completely covered with PDMS. Similar behavior has been observed for PS–PDMS–PS triblock copolymer systems.

Surface properties and chemical compositions of PS–PDMS copolymers were studied by Clark and co-workers[38] and Thomas and co-workers,[39] using contact angle measurements and XPS. Clark and co-workers observed the surface enrichment of the PDMS components. By comparing the XPS intensities for PDMS and PS, they determined the thickness of the PDMS outer layer. This was found to vary between 1.3–4.0 nm, depending on the copolymer film fabrication method. Thomas *et al.* studied the effect of the PDMS block length (DP = 3–37) on the extent of surface modification. PDMS contents of the copolymers were less than 20 wt%. The surface tension for the PS control was determined to be 30 mN m^{-1}. In copolymers with short PDMS blocks (DP = 3–9), surface tension gradually decreased from 32 to $25\,mN\,m^{-1}$. For large segment lengths (DP > 17), the surface tensions were 22–23 $mN\,m^{-1}$, which indicated complete surface coverage with PDMS.

3.4. Polystyrene–Poly(Ethylene Oxide) Diblock Copolymer

A PS–PEO copolymer readily undergoes microphase separation and domain formation due to the large difference in cohesive energy of PS and PEO. The surface tensions of PS and PEO in the solid state are 33–36 $mN\,m^{-1}$ and 44 $mN\,m^{-1}$, respectively. PS–PEO diblock copolymers were prepared by a two-stage anionic polymerization reaction. Thomas and O'Malley studied the surface chemical composition of PS–PEO diblock copolymers by angular-dependent XPS.[40] The XPS results clearly indicate that the copolymer surface compositions are significantly different from the overall bulk compositions. An excess of PS at the surface was observed for all copolymers cast from chloroform, ethylbenzene, and nitromethane. Angular-dependent XPS indicates that the copolymer surfaces are laterally non-homogeneous in PS and PEO and that the PS component is often raised above the PEO domain. This result was confirmed by replication electron micrographs of the copolymer films.

3.5. Poly(2-Hydroxyethyl Methacrylate)–Polystyrene Triblock Copolymer

Poly(2-hydroxyethyl methacrylate) (PHEMA) exhibits amphiphilic properties, because of the difference in hydrophilicity between the backbone chains and the hydroxyethyl group. PHEMA–PS–PHEMA triblock copolymers have unique properties due to the difference in cohesive energy between PHEMA and PS. Figure 7 shows the chemical structure of the PHEMA–PS–PHEMA triblock copolymer studied by Okano and co-workers.[41, 42] This material was prepared from an amino-terminated PHEMA and bis-isocyanate-terminated styrene oligomer. Figure 8 shows the wetting characteristics determined by using a water droplet on the dry polymer surface of the PHEMA–PS–PHEMA triblock copolymer and corresponding HEMA–styrene (St) co-oligomers. The wettability was found to increase for a PHEMA mole fraction of approximately 0.8 in the co-oligomer system and of approximately 0.9 for a triblock copolymer system. This increase corresponds to the morphological change in the domain structure from lamellae to cylinder (sphere) as observed by transmission electron microscope.

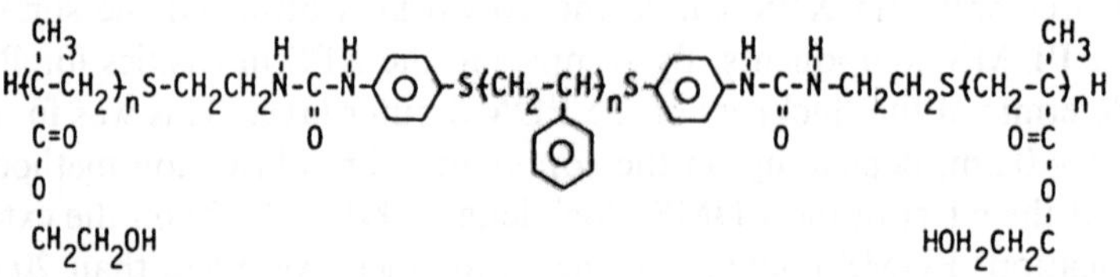

FIGURE 7. Chemical structure of PHEMA–PS–PHEMA A-B-A block copolymer. (Reprinted with permission from Ref. 35.)

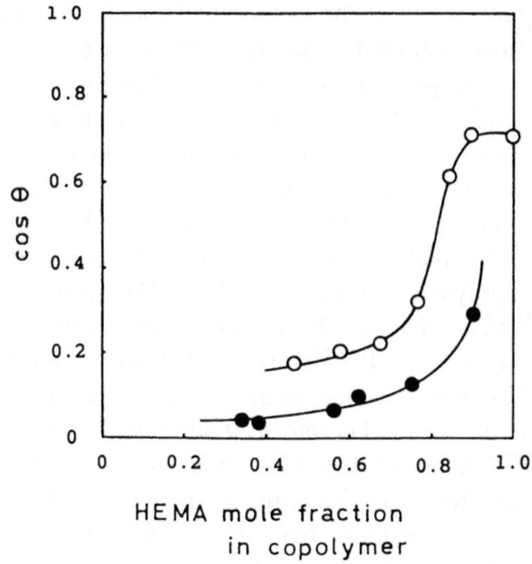

FIGURE 8. Relationships between wettability and copolymer composition for (●) PHEMA–PS–PHEMA triblock copolymer and (○) HEMA-styrene co-oligomer systems. (Reprinted with permission from Ref. 41.)

3.6. Block Copolypeptides

The unique properties of block copolypeptides are attributed to the aggregation state of the polypeptide domain. A-B-A triblock copolymers with a polypeptide as the A component and various oligomers as the B components have been synthesized.[43–46] Amine-terminated oligomers, such as aliphatic secondary amine-terminated PBD and amine-terminated PDMS, have been employed as the B component. The *N*-carboxy anhydride (NCA) of the amino acid composing the polypeptide chain was polymerized at the amine group of the B component. Table 3 summarizes chemical constituents of block copolypeptides. A detailed analysis of the physicochemical properties of the poly(ε-*N*-benzyloxycarbonyl-L-lysine) (PBCL)–PBD–PBCL block copolypeptide (LBL) has been done. Circular dichroism measurements and the infrared spectra showed that the polypeptide chain in the copolymer is in its α-helical conformation, as observed for the homopolymer chain. The microphase-separated structure of the block copolypeptide was examined by TEM. The micelle dimension of copolypeptides was evaluated from thermodynamic considerations, taking into account the conformational

TABLE 3. Chemical Composition of A-B-A Triblock Copolypeptides

A component	B component	Ref.
Poly(ε-*N*-benzyloxycarbonyl-L-lysine) (PBCL) $[R = -(CH_2)_4NHCOOCH_2Ph]$	PBD	43
Poly(γ-methyl-L-glutamate) (PMLG) $[R = -(CH_2)_2COOCH_3]$	PBD	44
Poly(γ-benzyl-L-glutamate) (PBLG) $[R = -(CH_2)_2COOCH_2Ph]$	PDMS	45, 46

[a] Peptide unit $-(NH-CHR-CO)_n-$

parameters of the component block, and were found to be in good agreement with those obtained from the electron micrograph.

Kugo and co-workers studied the surface structure and properties of A-B-A tri-block copolymers (LBL).[43] Replication electron micrographs revealed information about the surface morphology of membranes. Some of the block copolymers had convex domains raised above a relatively flat matrix phase. Thermodynamic considerations suggest that the convex domain and matrix phase are PBD and PBCL, respectively. Figure 9 shows Zisman's plots for PBCL and the LBL-5 block copolymer (11.9 wt% PBD). The wettability of the LBL-5 is better than that of PBCL; that is, the contact angles for LBL-5 are smaller than for PBCL. From Fig. 9, both PBCL and LBL-5 were found to have critical surface tensions (γ_c) of 41 mN m^{-1}. The value of γ_c for poly(*trans*-1,4-butadiene) has been reported as 31 mN m^{-1}. Therefore, it may be concluded that PBD is more hydrophobic than PBCL. Since the LBL block copolymer is composed of PBCL and PBD, it is expected that the contact angles for LBL should be between those of PBCL and PBD. As evident from Fig. 9, however, the wettability of LBL is better than that of PBCL. Two possible factors contributing to this enhanced wettability may be surface roughness and the presence of many residues capable of forming hydrogen bonds at the LBL surface. The XPS spectra of this block copolymer surface showed PBD enrichment at the outermost surface. A surface structural model shown in Fig. 10 was proposed based on the TEM, XPS, and contact angle measurements.

Surface roughness is expressed in terms of Wenzel's "roughness factor," r, which can be evaluated as

$$\cos \theta_w = r \cos \theta \tag{5}$$

where θ_w and θ are the apparent and true contact angles, respectively, and r is defined as

$$r = \frac{\Omega}{A} \tag{6}$$

where Ω is the area of the actual surface and A is the projection of the actual

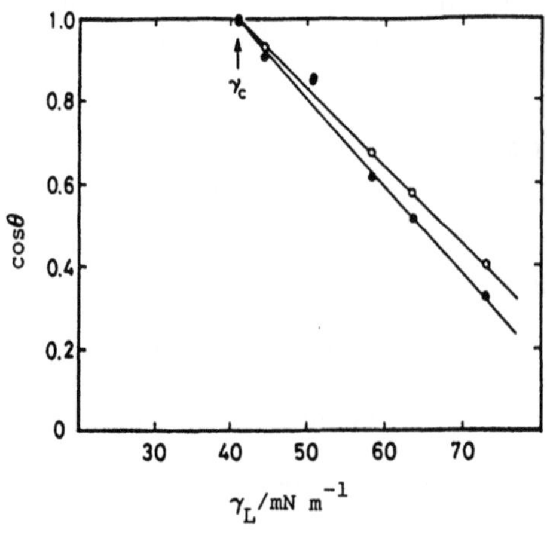

FIGURE 9. Zisman plots for block copolypeptide (LBL-5) and polypeptide (PBCL, filled circles). (Reprinted with permission from Ref. 43.)

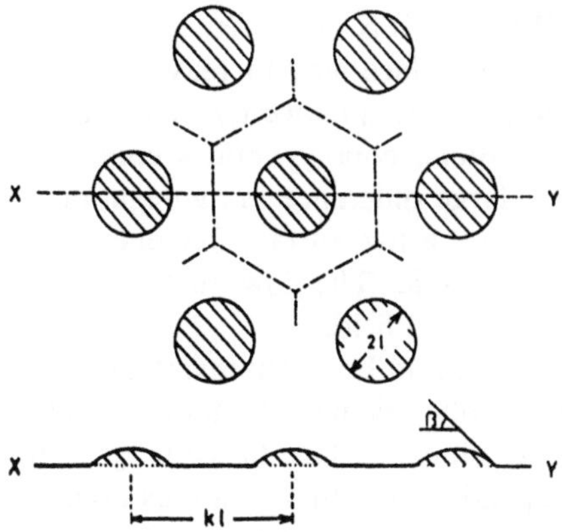

FIGURE 10. Model for surface topography of the copolypeptide surface. (Reprinted with permission from Ref. 43.)

surface on the plane.[47] The surface roughness was estimated from the replication electron micrographs to be $r = 1.063$. Thus, the true contact angle of water was calculated to be 67.9. This magnitude of the contact angle of LBL is smaller than that of PBCL. Thus, the contribution of surface roughness is not sufficient to explain the good wettability of the LBL copolymer.

The good wettability of LBL must be attributed to the high concentration of residues capable of forming hydrogen bonds at the surface. This agrees with XPS results. These block copolymers are considered to be negatively adsorbed onto the solution surface due to the small critical surface tension of cast solvents (2,2,2-trifluoroethanol $\gamma_{(TFE)} = 20.6 \text{ mN m}^{-1}$). Therefore, it seems reasonable to assume that the formation of micelles takes priority over the molecular adsorption on the solution surface. The shape of an LBL micelle is spherical, and its outer surface is PBCL. The α-helical PBCL rods should be stabilized when arranged parallel to the solution surface. Due to the difference in surface tension between PBCL and PBD, PBD domains may deform, thereby forming a lenslike structure at the polymer surface as shown in Fig. 11. This surface structure contributes to the surface excess of

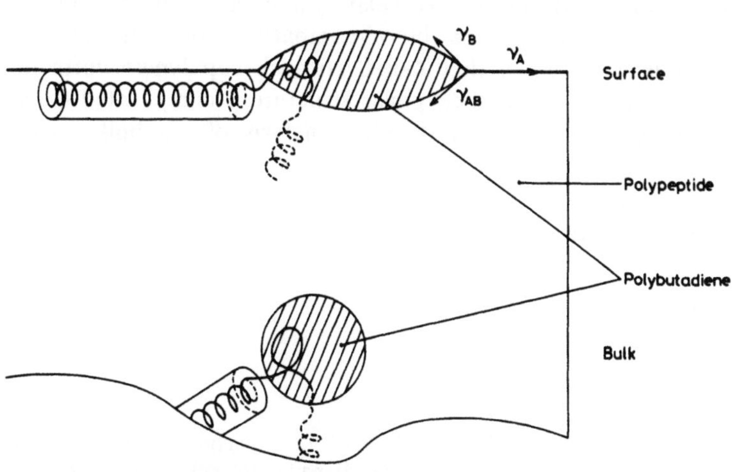

FIGURE 11. Cross-sectional model for the copolypeptide surface. (Reprinted with permission from Ref. 43.)

PBD on XPS measurement. In addition, the orientation of PBCL helices parallel to the surface may increase the concentration of the amide carbonyl groups at the surface and enhance the wettability of the block copolymer.

Block copolypeptides containing PDMS also have unique properties, due to the large difference in aggregation state of each component. PDMS is amorphous and has a low surface free energy, whereas the polypeptide block is crystalline and hydrophilic. The low surface tension of PDMS leads to its application in the biomedical field.

Ito and Imanishi investigated the surface properties and blood compatibility of the poly(γ-benzyl-L-glutamate) (PBLG)–PDMS–PBLG block copolypeptide.[46] The critical surface tension, γ_c, of these block copolymers was found to be slightly dependent on their composition. The γ_c values ranged from 20 to 21 $mN\,m^{-1}$, which indicates a nonwettable surface enriched with PDMS. Surface properties of the PBLG–PDMS–PBLG block copolypeptides were also studied by Kugo *et al.*[45] They reported that the surface free energy value of the block copolypeptide ranged between 31 and 32 $mN\,m^{-1}$ and did not show a dependence on composition.

The spreading coefficient of PDMS over PBLG in air is a large positive value ($S_{PDMS/PBLG} = 15.4\ mN\,m^{-1}$), while $S_{PBLG/PDMS}$ is a large negative value, $-53.4\ mN\,m^{-1}$. This indicates that PDMS will spread over the air-facing surface of a block copolypeptide. These results are confirmed by the increase in absorbance of Si–O stretching and the decrease in analytical depth of ATR-IR. In water, however, $S_{PDMS/PBLG}$ and $S_{PBLG/PDMS}$ are large and negative, indicating that both components cannot spread on the solid–water interface. Underwater contact angles of the block copolypeptides indicate that the surface enrichment of PDMS is preserved in water. This might be related to the entropic factor for the surface reorganization or lack of change in structure on immersion for this copolymer of water insoluble components.

3.7. A-B-C Triblock Copolymers

A-B-C triblock copolymers have been synthesized primarily for their surface modification applications, such as blood-compatible polymers and permselective membranes. It is well known that A-B-C triblock copolymers form microphase-separated structures; however, their phase structure is complicated compared with an A-B-A triblock copolymer. A detailed analysis of the bulk phase structure

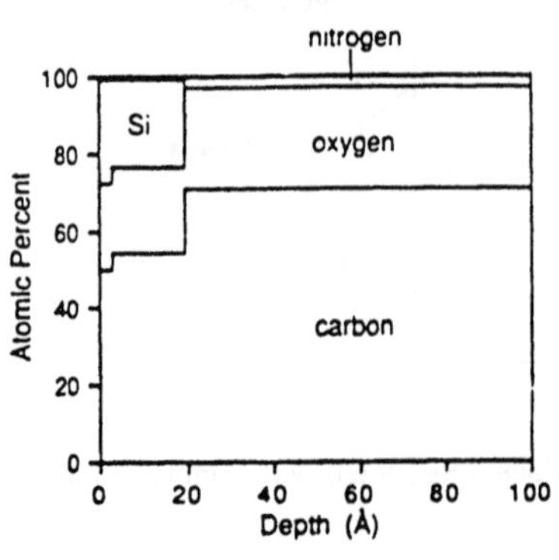

FIGURE 12. Composition depth profile of PDMS–PEO–Heparin triblock copolymer obtained from angular-dependent XPS. (Reprinted with permission from Ref. 49.)

TABLE 4. Dynamic Contact Angles of PDMS–PEO Diblock and PDMS–PEO–Heparin Triblock Copolymers

Specimen	Hydration (h)	θ_a	θ_r	$\Delta\theta$
PDMS–PEO	0	100.0 ± 4.5	53.6 ± 1.1	46.0
	1	88.7 ± 2.1	49.4 ± 4.4	40.0
	12	58.7 ± 1.9	53.5 ± 1.4	35.0
PDMS–PEO–Hep	0	109.8 ± 2.4	45.8 ± 1.1	64.0
	1	58.1 ± 11.8	48.5 ± 1.5	10.0
	12	58.7 ± 1.9	48.9 ± 2.1	10.0

From Ref. 50.

of PS–PBD–poly(4-vinylpyridine) (PVPy) film was done by Kotaka and co-workers.[(48)] Only a few studies have been reported on surface properties of A-B-C triblock copolymers.

Kim and co-workers studied the surface structure and blood compatibility of PDMS–PEO–heparin (Hep) triblock copolymers[(49, 50)] prepared via a series of coupling reactions that used functionalized prepolymers, diisocyanates, and derivatized heparin. Angular-dependent XPS measurements were performed for spin-cast films of PDMS–PEO–Hep triblock copolymers. A compositional profile of PDMS–PEO–Hep film under high vacuum is shown in Fig. 12. The surface composition of the PDMS–PEO–Hep triblock copolymers is very different from their bulk composition. PDMS, with its low surface free energy, dominated the surface at the lowest emission angle θ, with an increase in PEO and heparin signals at greater sampling depths.

Table 4 summarizes the result of dynamic Wilhelmy plate measurements. The advancing angles from the first immersion curve indicate that the surface is very hydrophobic due to the enrichment of PDMS at the air–solid interface. However, during immersion, the surface composition of PDMS–PEO–Hep copolymers reorganized so that subsequent advancing contact angles are significantly decreased. The small hysteresis after hydration is ascribed to the stable hydrated surface layer of the heparin block. The stability of the heparin block in water might be closely related to the antithrombogenicity of PDMS–PEO–Hep triblock copolymers.

4. CHARGE MOSAIC MEMBRANE

Charge mosaic membranes, which are composed of cation and anion domains, have received much attention for their transport properties and biocompatibility. Earlier methods for preparing charge mosaic membranes were reviewed by Leitz.[(51)] The formation of polyion complexes destroys the charge mosaic structure. To avoid this, Fujimoto *et al.*[(52)] and Kotaka *et al.*[(53)] reported preparing a charge mosaic membrane from tri- or pentablock copolymers. Fujimoto and co-workers prepared a pentablock copolymer of PI–PS–PI–poly(4-vinylbenzyldimethylamine) (PVBDMA)–PI by anionic polymerization. These polymers form well-defined microphase-separated structures. A positive charge was introduced into the nitrogen atom of PVBDMA by using methyl iodide. The PI block was cross-linked

TABLE 5. Interfacial Properties and XPS Results of Charge Mosaic Membranes Prepared from Polystyrene–Polybutadiene–Poly(4-vinylpyridine) Triblock Copolymer[a]

	Underwater contact angle			ESCA			
Sample[b]	θ_{Air}	θ_{Octane}	γ_{SW} (mN m^{-1})	N/C	S/C	Br/C	Cl/C
SBP/C	33	30	1.6	0.020	—	—	—
SBP/B	33	43	2.4	0.010	—	—	—
SBP(Q)/C	18	Spread	—	0.084	—	0.089	—
SB(X)P(Q)/C	13	Spread	—	0.0033	0.010	0.0048	0.0074
S(S)B(X)P(Q)/C	15	Spread	—	0.0034	0.039	0.0053	0.0083

[a] $\overline{Mn}$(PS) = 13,300, $\overline{Mn}$(PBD) = 15,600, $\overline{Mn}$(PVPy) = 24,200.
[b] C = cast from chloroform; B = cast from 9:1 mixture of butylaldehyde and chloroform; Q = quaternized; X = cross-linked; S = sulfonated.

by sulfur chloride to maintain the microphase structure of the film during the violent sulfonation reaction. Cross-linking of the PI domain increased the membrane rigidity. Sulfonation of the PS block was carried out by using chlorosulfonic acid. These charge mosaic membranes were highly permeable to sodium chloride in mixed aqueous solutions of sodium chloride and organic species.

Kotaka and co-workers prepared charge mosaic membranes from PS–PBD–PVPy (SBP) triblock copolymers.[53, 54] A positive charge was introduced to the PVPy domain by vapor phase quaternization using methyl bromide. The PBD phases were cross-linked using sulfur chloride to avoid polyion complex formation. The final step involved sulfonation of the PS blocks using a chlorosulfonic acid vapor. The surface properties of this membrane were investigated by XPS and the underwater captive-bubble technique. These results are summarized in Table 5. The solid–water interfacial energies calculated for the unmodified polymers are low, indicating that they are rather hydrophilic surfaces. The chloroform-cast materials (SBP/C) were slightly more hydrophilic than the same polymers cast from a mixture of butylaldehyde and chloroform. XPS results indicate that the SBP/B contain less surface nitrogen and less PVPy phase on the surface than do the same materials cast from chloroform. The PVPy phase is more hydrophilic than PS and PBD, and thus a lower interfacial free energy is observed for SBP/C. The modified

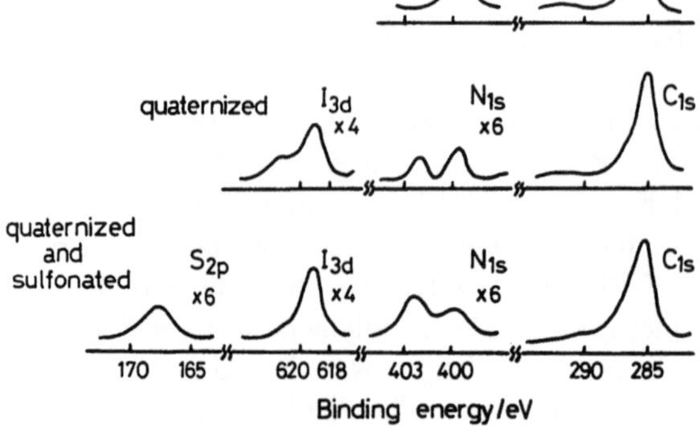

FIGURE 13. Variation of XPS spectra with introduction of positive and negative charges for PS–PVPy(30/70) diblock copolymers (cast from TCE). (Reprinted with permission from Ref. 55.)

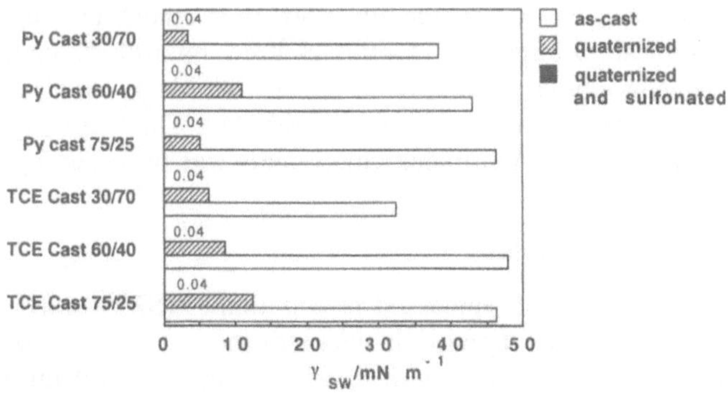

FIGURE 14. Effect of introduction of ionic groups on interfacial free energy of PS–PVPy diblock copolymers cast from trichloroethane and pyridine. Magnitudes of γ_{SW} for quarternized and sulfonated are indicated in this figure.

surface showed spreading of octane bubbles; however, the magnitudes of the water–air contact angles were low. These results indicate that modified surfaces were more hydrophobic than unmodified surfaces. XPS results showed a corresponding decrease in surface nitrogen content.

Takahara and co-workers prepared charge mosaic membranes from PS–PVPy diblock copolymers.[55] To retain the microphase-separated structure after introduction of a mosaic charge, they performed quaternization and fixing of the PVPy blocks by reacting PVPy with tetramethylene diiodide. The existence of surface microphase separation after introducing positive and negative charges was confirmed by transmission electron microscopy. The state of the surface charges was investigated by XPS. Figure 13 shows the XPS spectra for C_{1s}, N_{1s}, I_{3d}, and S_{2p} of PS–PVPy(30/70). During quaternization, the intensity of the N_{1s} peak from the neutral nitrogen decreased, while that from the quaternized nitrogen and I_{3d} peak increased. The intensity of the S_{2p} peak increased with sulfonation. The existence of I^- after introduction of SO_3^- indicates that the mosaic charge is introduced without destroying the microphase-separated structure on the film surface.

Underwater contact angles were measured for the PS–PVPy diblock copolymers and their charge mosaic membrane. Figure 14 shows the water–solid interfacial energy, γ_{SW}, for neutral, cationic, and charge mosaic membrane surfaces. This energy was estimated by the procedure of Andrade and co-workers,[26] and decreased with quaternization. The introduction of negative and positive charges reduced the water–solid interfacial free energy almost to zero, indicating that the surface of the charge mosaic membrane is completely hydrophilic. These results are quite different from those of Kotaka and co-workers.[53] This may be attributed to the difference in surface chemical composition as revealed by XPS.

5. MULTIBLOCK COPOLYMERS

Multiblock copolymers (Fig. 1e) consist of two types segments that are incompatible due to different cohesive energies. Various types of multiblock copolymers have been prepared from bifunctional oligomers. The polymer chain therefore consists of many short blocks. Segmented poly(ester ether) and segmented polyurethanes are typical examples of commercialized multiblock copolymers. The driving force of microphase separation of some multiblock copolymer systems is attributed to the crystallization of one type of segment. The thermodynamic theory of microphase separation of multiblock copolymers was proposed by Krause.[56]

5.1. Segmented Polyurethanes

The surface properties of segmented polyurethanes (SPUs) have been studied extensively, because they are applied in biomedicine, as surface coatings, and as adhesives. They can be prepared with a wide range of mechanical properties, ranging from flexible elastomers to rigid plastics.

The preparation of polyurethanes is based primarily on the reaction of an isocyanate and an active hydrogen of a diol or diamine. SPUs are prepared by the reaction of three components: diisocyanates, polyols, and chain extenders. Figure 15 shows the typical chemical components used for the preparation of

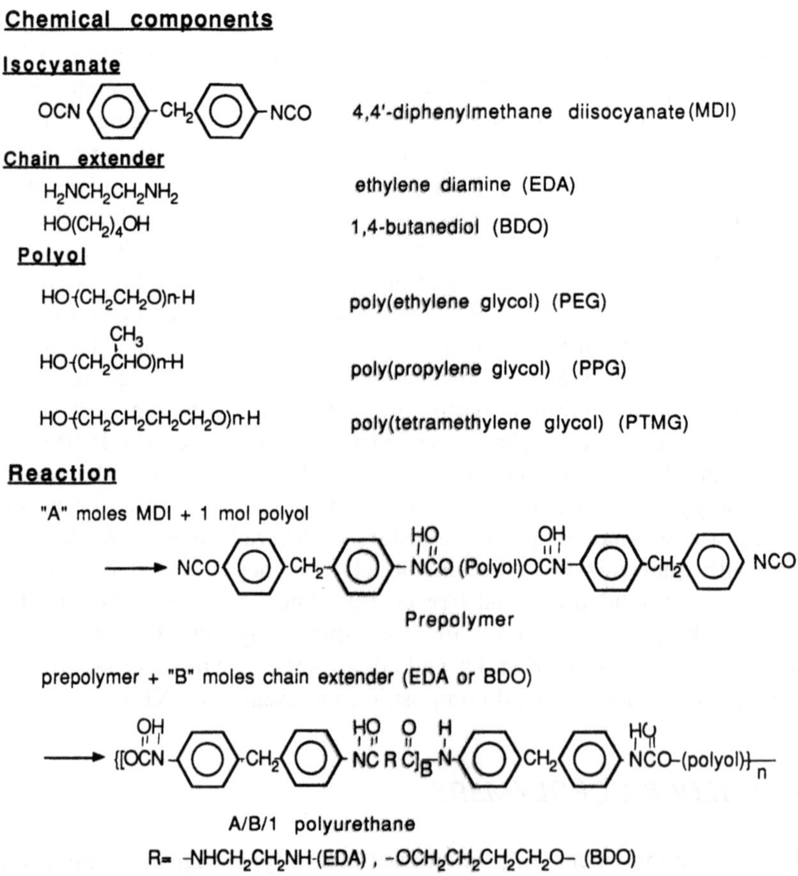

FIGURE 15. Chemical structure of components of segmented poly(urethanes) and segmented poly(urethaneureas).

SPUs. The diisocyanates often used are 4,4′-diphenylmethanediisocyanate (MDI) and toluenediisocyanate (TDI). The polyols are linear polyesters, polyethers, and hydroxy-terminated polybutadiene with molecular weights from 500 to 5000. The polyols commonly used are PEO, poly(propylene oxide) (PPO), and poly(tetramethylene oxide) (PTMO). The chemical structures and physical properties of the polyols influence the surface hydrophilicity of SPUs. Typical chain extenders are either an aliphatic diol or an aliphatic diamine. An aliphatic diol chain extender such as 1,4-butanediol (BDO) results in a poly(etherurethane), whereas an aliphatic diamine, such as ethylenediamine (EDA), produces poly(urethaneureas). For large-volume production of SPUs, the diisocyanate, polyols, and chain extender are mixed in the absence of a solvent. For a small-batch reaction, however, SPUs are generally prepared by a two-step process.[57] One mole of polyol is end-capped with 2 moles of diisocyanate in the first stage. Then 1 mole of chain extender is added in the second stage. The segment with the polyol and diisocyanate is called soft. The higher the molecular weight of the polyol, the softer and more elastic is the resulting SPU. The hard segment is composed of the diisocyanate and chain extender. An increase in hard-segment content results in an increase in modulus. The microphase-separated structure of an SPU results from the incompatibility of the segments and aggregation of the hard segments into a paracrystalline domain. Figure 16 is a schematic representation of the microphase-separated structure of an SPU.[58]

The study of the surface properties of SPUs has been of primary importance because of their application in biomedicine. Surface chemical analysis of a variety of SPUs has been performed by using XPS,[59–61] SIMS,[62] and FTIR.[19] Since most of the measurements have been carried out in a high-vacuum or air environment, the surface under these conditions is enriched with the lower-surface-energy component. The depth profile of an SPU containing a hydrophobic soft segment has been measured by angular-dependent XPS. The angular-dependent XPS data was analyzed with the Paynter algorithm[59] and inverse theory, developed by Ratner and co-workers.[61] The depth profiles obtained indicate that within the top 2 nm of the air–solid interface there is a significant excess of polyether component

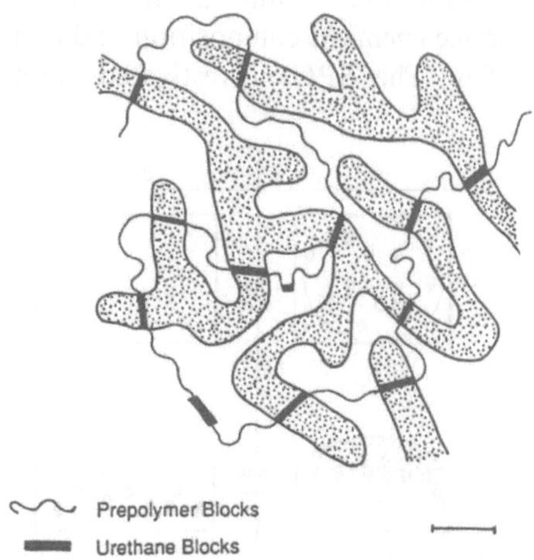

FIGURE 16. Schematic representation of microphase-separated structure of segmented poly(urethanes). The scale bar is approximately 5 nm. (Reprinted with permission from Ref. 58.)

compared with the average bulk composition, in conjunction with the relative exclusion of the nitrogen-containing hard-segment domains. In addition, the depth profile revealed that an impurity such as PDMS, which is commonly used as a lubricant, is enriched at the outermost layer.[59] These findings support the contention that when cast, the SPU surface consists mainly of the hydrophobic polyether component in accordance with a minimization of interfacial free energy.

Ratner and co-workers investigated the relationship between the bulk microphase-separated structure and surface chemical composition of an SPU composed of a fluorocarbon-containing diol chain extender.[62] The surface composition and fluorine depth profile data are closely correlated to the differential scanning calorimetric (DSC) and IR results and showed a gradual morphological transition with the hard-segment content. No lateral inhomogeneity of the surface composition was detected with XPS, due to the small domain size compared with the XPS analyzing area. Annealing induced the migration of soft segments to the surface, thus improving the phase separation of SPU. This result indicated that the surface free energy of a soft segment was smaller than that of a fluorine-containing hard segment. The surface free energy in air was estimated from the contact angle data measured by Owens and Wendt's method.[24] However, even with the enrichment of PTMO at the surface, the surface free energy of the SPU was larger than that of PTMO. The lack of correlation between the XPS data and the surface free energy values was attributed to the nonequilibrium nature of the wetting experiment for this particular polymeric system. (See, however, Ferguson and Whitesides, Chapter 6, on sub-θ interphase, where it is shown that polar groups may be stable a short distance below the surface and can be detected by XPS.)

Figure 17 schematically represents the surface structural change that occurs during a characteristic wetting measurement.[63] The surface free energy under hydrophobic air conditions is minimized by allowing the more hydrophobic soft-segment matrix to dominate at the surface. The presence of water, a polar solvent, on the SPU's surface induces the polar segment to migrate to the surface to minimize the interfacial free energy. Dynamic contact angle measurements are required to understand this behavior.

Dynamic contact angle measurements of SPUs have been performed by Andrade *et al.*[63] and Takahara *et al.*[64, 65] Andrade and co-workers investigated the surface chemical composition and dynamic contact angles of commercially available SPUs. The SPUs investigated were Vialon 510X (Deseret Polymer Res. Co.)

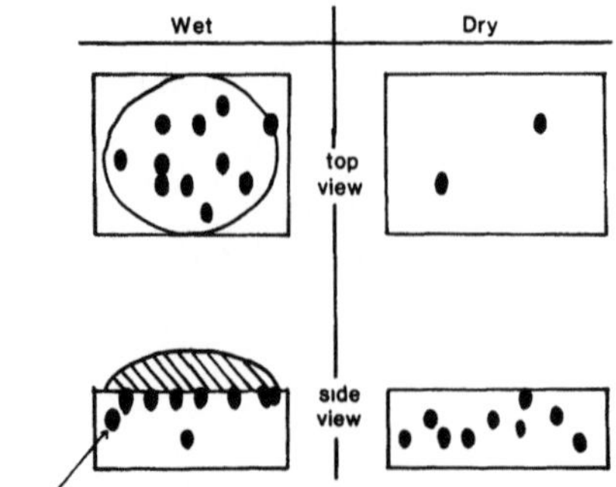

FIGURE 17. Schematic representation of the surface structural change induced by environmental change. (Reprinted with permission from Ref. 63.)

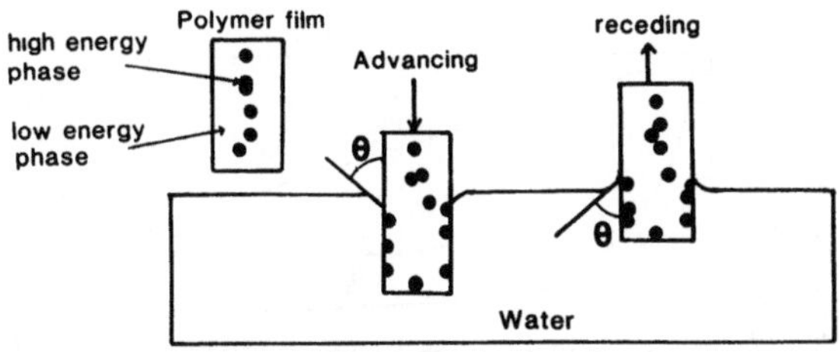

FIGURE 18. Dependence of surface chemical composition on environment during dynamic contact angle measurement. (Reprinted with permission from Ref. 63.)

containing a PTMO soft segment with different hard-segment contents and Vialon H/H (Deseret Polymer Res.), which contained blocks of PTMO and PEO varying from 0 to 100%. XPS results indicated that the surface enhancement of the soft segment increased as the hard-segment content increased. This is due to the improvement of the bulk phase separation, which occurs with an increase in hard-segment content. For the H/H series, increase in the PEO content of the soft segment decreased the enhancement of soft segments at the air–solid interface. The degree of phase separation between hard and soft segments decreased as PEO content increased. Since phase mixing reduced the molecular mobility of soft segments, their enrichment at the surface hardly occurs for SPUs with high PEO content. Also, the higher surface free energy of PEO may be another factor influencing the surface enhancement of the polyether phase.

Dynamic contact angle measurements have been carried out for these specimens. These copolymers showed a large hysteresis in their contact angles. Figure 18 indicates the dynamic nature of the surface being measured using dynamic contact angle measurement, based on the hypothesis that the mobile phase reorients to minimize the interfacial free energy.[63] Figure 19 shows the decrease in advancing angle for commercially available SPUs with hydration time. This indicates that long time-scale reorientation may take place to present a more hydrophilic surface.

The existence of surface heterogeneity of SPU has been discussed by several researchers. XPS analysis of an SPU containing a large $\overline{Mn}$, hydrophobic soft segment indicated complete domination of the surface with a polyether soft-segment overlayer. Takahara and co-workers measured angular-dependent XPS for SPUs with polyether segments of various $\overline{Mn}$.[60] XPS results indicated that SPUs with polyether soft segments with $\overline{Mn} < 2000$ were not completely covered with a

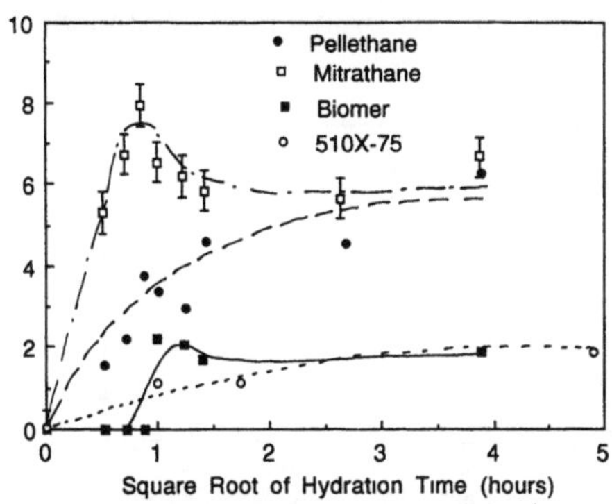

FIGURE 19. Variation of decrease in advancing contact angle of various SPUs with hydration time. (Reprinted with permission from Ref. 63.)

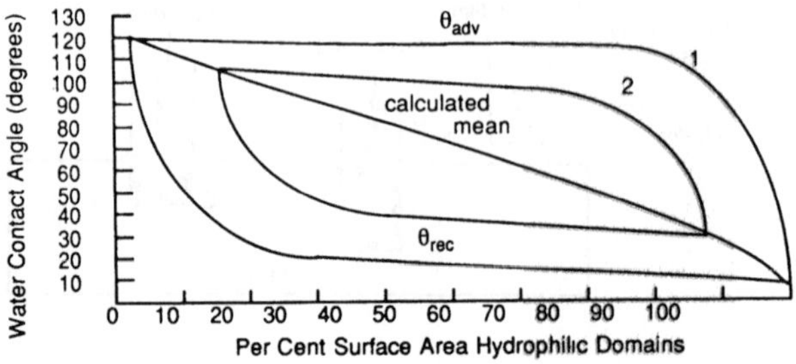

FIGURE 20. Dynamic contact angle of heterogeneous surfaces based on the model of Johnson and Dettre. Heterogeneous domain size decreases in the order from 1 to 2. (Reprinted with permission from Ref. 63.)

polyether overlayer. Knutson and Lyman measured FT ATR-IR spectra of SPUs.[19] The state of hydrogen bonding dependend on the IR sampling depth, indicating that there was greater microphase separation on the surface. These results indicate that surface heterogeneity exists for SPUs with certain $\overline{Mn}$ of the polyether soft segment. Direct evidence of the surface microphase separation has been obtained by low-voltage high-resolution SEM. Li and Cooper observed surface microphase separation of PBD-based SPU after staining PBD blocks with osmium tetroxide.[66]

The relationship between surface heterogeneity and contact angle hysteresis has been discussed by Johnson and Dettre,[31] who modeled a heterogeneous surface and observed that the contact angle hysteresis varied with the percentage of low-contact-angle component surface coverage. Figure 20 shows the relationships between the water contact angle and the percent of surface area composed of hydrophilic domains.[63] The contribution of the hydrophobic component is large for the advancing angle, whereas the contribution of the hydrophilic component is large for the receding angle. A decrease in the size of the heterogeneity decreases the hysteresis.

Andrade and co-workers studied the relationship between advancing and receding angles with hydrophilic group content in Vialon H/H (PEO–PTMO containing SPU).[63] Figure 21 shows the relationships between contact angle and weight percentage of PEO in a soft segment of Vialon H/H SPU. The variation of the contact angles with percent PEO in the soft segment showed similar trends to those determined by the mopel of Johnson and Dettre. However, at very high and very low PEO content, the decrease in the advancing angle and increase in the receding angle with PEO content is not as profound as the values predicted by the model.[31] Since the contact angle hysteresis can be attributed to many factors, including environmentally dependent molecular motion, deformation of the surface,

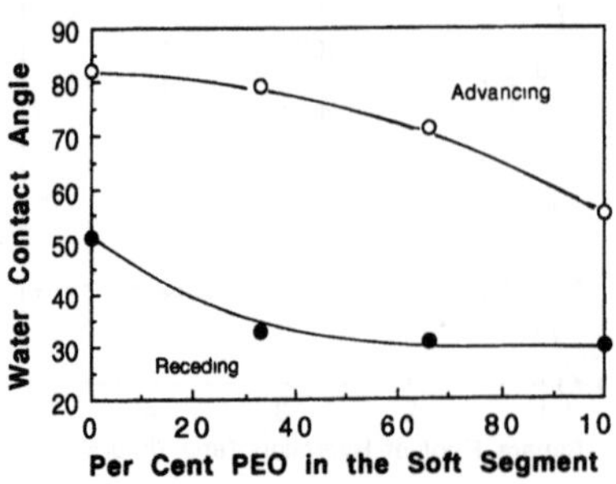

FIGURE 21. Water contact angle of PEO–PTMO-based SPU of similar composition or relative size of heterogeneous domains. (Reprinted with permission from Ref. 63.)

and water absorption, these results indicate that hysteresis does not occur from a single origin.

Takahara and co-workers observed the time-dependent change of the relative surface tension for SPUs containing various soft segments after exposure to a water environment.[64, 65] The changes of the surface tension minimize the interfacial free energy between surface and environment. A large relaxation of the surface tension is observed for SPUs containing large $\overline{Mn}$ polyether soft segments (PPO $\overline{Mn} = 3000$) or hydrophilic polyether soft segments such as PEO. Since the SPU containing a large $\overline{Mn}$ polyether soft segment showed a lower T_g compared with an SPU with small $\overline{Mn}$ polyether soft segment, these results indicate that the surface molecular mobility reflects the bulk molecular mobility. Water acts as a plasticizer for the hydrophilic SPU. The water absorbed by the hydrophilic SPU enhanced the molecular mobility at the surface.

Direct evidence of surface reorganization upon environmental change has been observed by means of freeze-dried XPS. Takahara and co-workers observed variations in the XPS spectra with hydration and dehydration processes for several SPUs. Figure 22 shows XPS spectra for an SPU composed of a fluorine-containing diol chain extender in both the hydrated and dehydrated states.[65] Since the F_{1s} signal comes from hard-segment regions, this can be used as a marker of hard segments. In the hydrated state, the intensity of the hard-segment signal increased, since PEO can be hydrated and become mobile on the surface. However, after dehydration the surface concentration of the hard segment is decreased, since the surface free energy of the hard segment is lower than PEO. This change in surface chemical composition indicates molecular reorganization at the surface of SPU. Figure 23 shows a schematic model of the surface structure of an SPU with hydrophobic and hydrophilic polyether soft segments. In air, the surface layer is organized to minimize the interfacial free energy between the air and the solid. Thus, the lower-surface-free-energy component is enriched on the air-facing surface. In water, the surface is enriched with a higher-surface-free-energy component. For SPUs containing a PEO soft segment, the PEO chain is hydrated and expanded on the surface. The XPS data indicate that the surface heterogeneity may exist for SPU containing an intermediate $\overline{Mn}$ polyether soft segment.

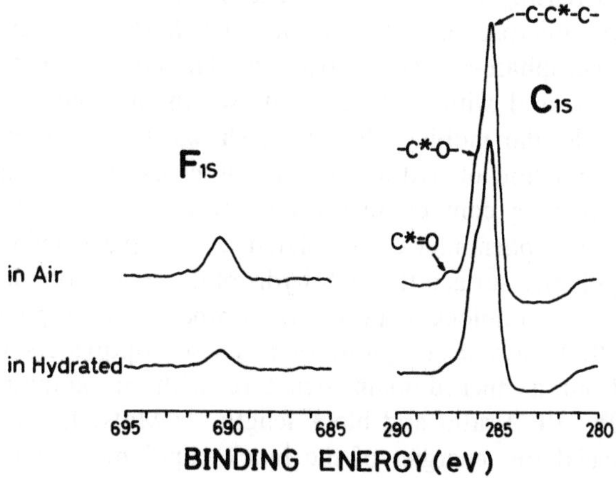

FIGURE 22. XPS spectra of SPU chain extended with fluorine containing diol in the hydrated and dehydrated states. $\overline{Mn}$ of PEO is 2000, $\theta = 90°$. (Reprinted with permission from Ref. 65.)

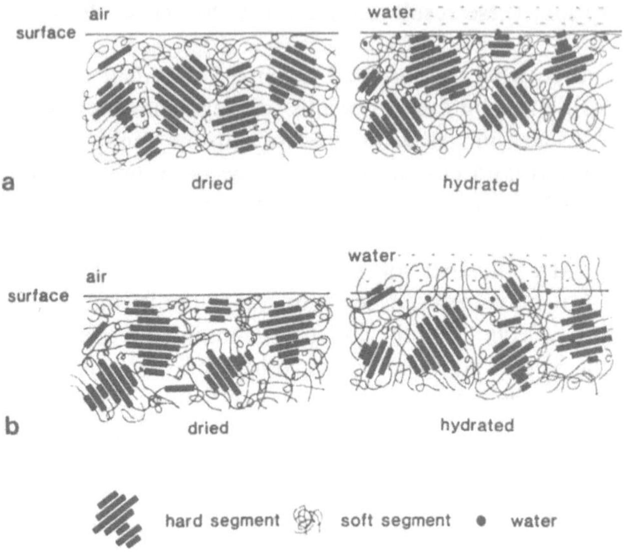

FIGURE 23. Schematic representation of the surface structure of SPU in the hydrated and dehydrated states. (a) SPU with hydrophobic polyether; (b) SPU with hydrophilic polyether. (Reprinted with permission from Ref. 64.)

5.2. Other Multiblock Copolymers

Figure 24 shows the chemical structures of multiblock copolymers used for surface-related applications. Kim *et al.* studied the surface properties of multiblock copolymers of hydrophilic PEO and hydrophobic PS.(67) This copolymer was synthesized by the polyaddition reaction of isocyanate–telechelic oligostyrene and dihydroxy-terminated PEO. The surface of this polymer appeared to have a microphase-separated structure. The dynamic contact angles were measured by the dynamic Wilhelmy plate technique. As the fraction of hydrophobic PS in the bulk composition decreased, receding angles decreased dramatically, while advancing angles changed slightly. The contact angle hysteresis increased substantially with increasing PEO content. Since the surface of this copolymer is smooth, as verified by microscopic observation, the hysteresis may be attributed to the surface microphase-separated structure. The contact angle hysteresis of the hydrated state increased with PEO content, which may be ascribed to the increase in surface molecular mobility. Figure 25 shows the variation of the receding contact angle as a function of hydration time for block copolymers with various PS composition. Receding contact angles steadily decreased with an increase in hydration time. These phenomena are related to the plasticizing effect of water penetrating the polymer surface through hydrophilic PEO domains.

Multiblock copolymers composed of bisphenol A polycarbonate (PC) and PEO have been synthesized. These copolymers have unique properties resulting from a microdomain structure with an amphiphilic nature depending on the PC–PEO ratio and block length.(68) Matsuda and co-workers measured the interfacial free energies of the block copolymers at the water–block copolymer inter-

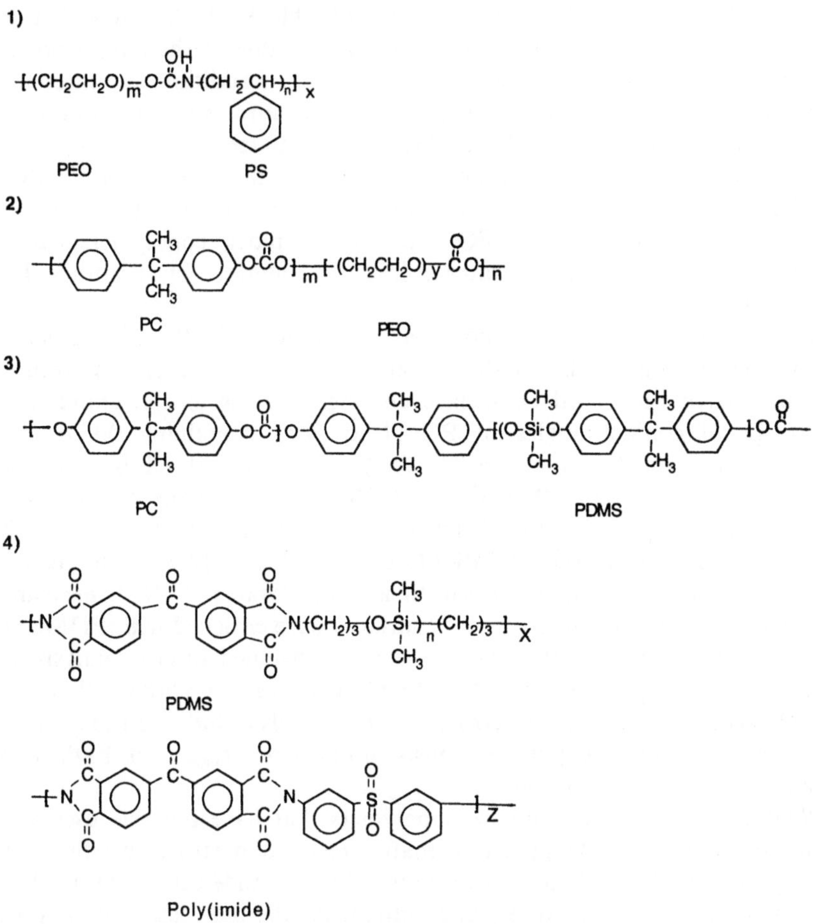

FIGURE 24. Chemical structures of multiblock copolymers for surface related application. (1) PEO–PS, (2) PC–PEO, (3) PC–PDMS, (4) Poly(imide)–PDMS.

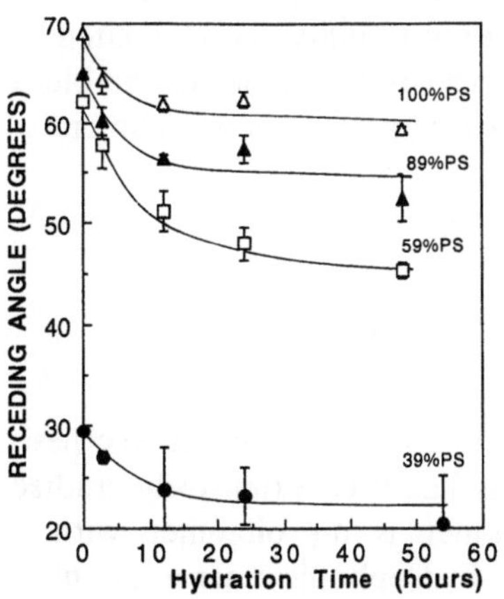

FIGURE 25. Variation of receding contact angle of PS–PEO multiblock copolymer as a function of hydration time. (Reprinted with permission from Ref. 67.)

face.[10] Higher PEO content and longer PEO block length showed a marked reduction in γ_{SW}, which resulted from stabilization due to the strong polar–polar interaction between the PEO segments and water. In addition, the interfacial free energies at the interface between the water–chloroform solution containing 5% polymer were measured by the Teflon Wilhelmy plate technique. Block copolymers with lower PC content had a markedly reduced interfacial free energy with water. The effect of PEO block length on the interfacial free energy is found to be much smaller than that of the PC–PEO ratio. These block copolymers can act as surfactants, which stabilize the water–chloroform interface by orienting the PEO blocks to the water phase.

Various segmented block copolymers containing PDMS soft segments have been synthesized. Since PDMS shows very low surface free energy, those block copolymers have been used as surface coatings. Gaines and Legrand measured ethylene glycol contact angles of PC blend films containing 0.1 to 4.0 wt% PDMS–PC segmented copolymers.[69] They observed that all blends displayed contact angles very close to that of pure PDMS. This behavior was also found to be insensitive to the PDMS block lengths if molecular weights were 1500 or higher. The surface properties of the PDMS–PC multiblock copolymers and their blends with PC were also studied by McGrath and co-workers.[70] XPS measurements clearly demonstrated that all copolymer surfaces were enriched with PDMS regardless of their bulk PDMS content (10–75 wt%). A detailed surface analysis of these PDMS–PC block copolymers was done by Gardella and co-workers.[71] The results of the XPS and ion scattering spectroscopy (ISS) analysis indicated that the surface morphology of the block copolymer consisted of discrete regions of PDMS and PC oriented perpendicular to the surface.

Poly(imide)–PDMS multiblock copolymers that have possible uses as tough and environmentally stable structural matrix resins and structure adhesives have been prepared by McGrath and co-workers.[72] Poly(amide acid)–PDMS thin films were cast from reaction mixtures, and subsequent thermal dehydration produced poly(imide)–PDMS block copolymers. These copolymers showed a microphase-separated structure due to the incompatibility of PDMS and poly(imide). Their excellent thermal properties reflect the thermal stability of the PDMS and poly(imide) blocks. The XPS results demonstrate that the PDMS components dominate the surface of the copolymer independent of the weight percent of the siloxane incorporated into the copolymer. The surface wettability was estimated by means of water contact angle measurement. The water contact angles of the block copolymers were comparable to that of PDMS, which indicated the enrichment of PDMS on the surface.

6. BLOCK COPOLYMERS WITH GRAFT SIDE CHAINS

The macromonomer (macromolecular monomer) technique for the synthesis of graft copolymers with well-defined graft segments has been extensively studied in the last decade. By definition, a macromonomer is an oligomer with a polymerizable end group. A variety of hydrophilic or hydrophobic macromonomers with radically copolymerizable end groups, such as styrene, methacryate, and vinylsilane derivatives, has been prepared.[73–75] The copolymers of those macromers show surface activity due to their amphiphilic nature. Also, macromonomers designed to give polyurethane copolymers via a polyaddition reaction have been reported by Yamashita[73, 74] and Tezuka.[75–78] Tezuka and co-workers reported that uniform size PDMS macromonomers containing a diol end group were synthesized by hydrosilylation of PDMS containing a silane end group with 3-allyloxy-1,2-propanediol.[76] Figure 26a shows the chemical structure of a polyurethane–PDMS graft copolymer prepared from this uniform-size PDMS macromer. XPS measurement of these surfaces showed an enrichment of PDMS on the surface. Angular-dependent XPS showed a change in thickness of the surface PDMS layer from 2 to 10 nm, as the $\overline{Mn}$ of PDMS and PDMS blocks increased. Figure 27 shows the change of the underwater contact angle of air as a function of

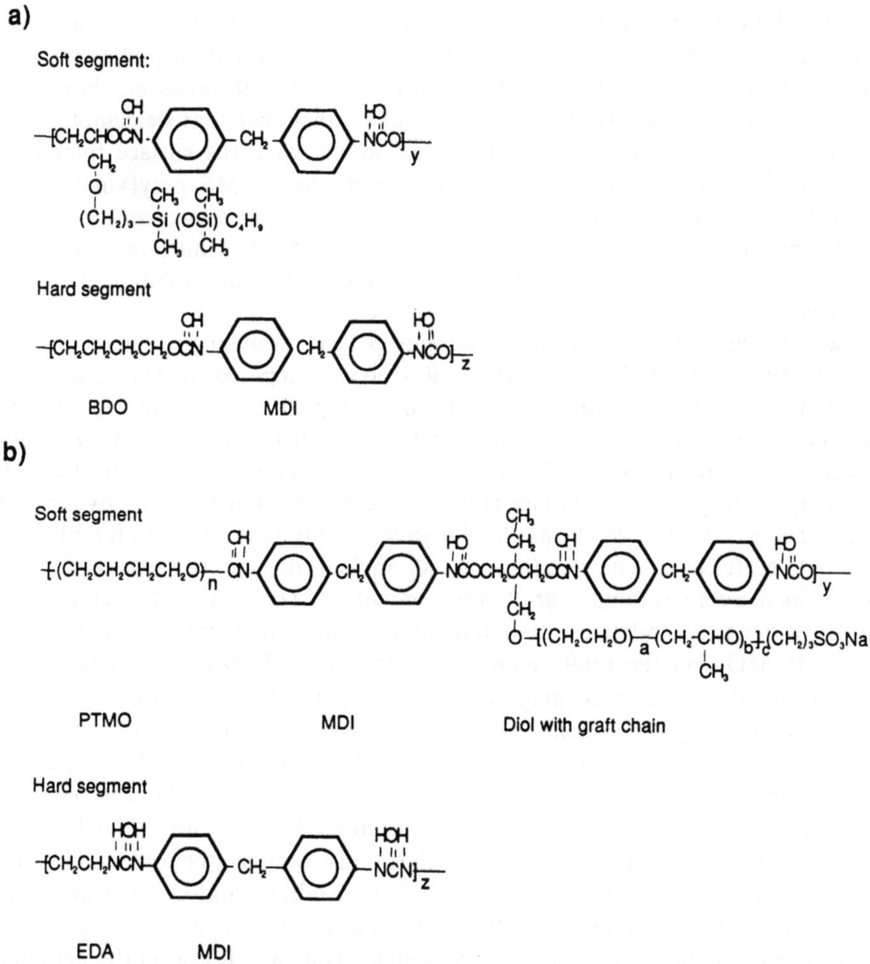

FIGURE 26. Chemical structure of segmented poly(urethanes) with graft chain: (a) PDMS graft, (b) hydrophilic graft.

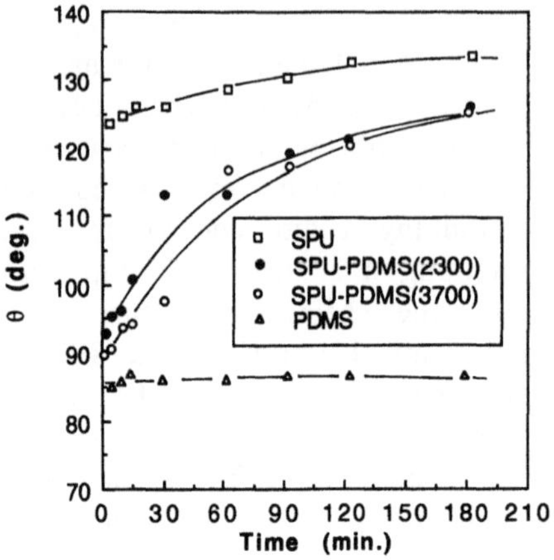

FIGURE 27. Variation of underwater contact angle of air with hydration time for SPU, PDMS, and PDMS grafted SPU. (Reprinted with permission from Ref. 77.)

hydration time. The SPU containing a short-graft PDMS chain and low PDMS content had a contact angle similar to that of PDMS at the initial time. The contact angle increased with an increase in hydration time. SPUs with large PDMS chains ($\overline{Mn} = 6900$, not shown in Fig. 27) and high PDMS content, however, showed only a slight dependence on hydration time. Tezuka concluded that detectable surface reorganization due to environmental change occurs when the surface thickness of the PDMS layer is less than 10 nm.(77) However, the PDMS–poly(vinyl alcohol) (PVA) graft copolymer showed a fast surface structural reorganization even for PDMS thicknesses greater than 10 nm.(78) These results indicate that surface structural reorganization is closely related to the molecular mobility of the bulk components.

Commercially available diols containing a long graft chain can be used for the preparation of SPUs. Takahara and Cooper prepared SPUs containing a hydrophilic graft chain having a sulfonate end group(79) as shown in Fig. 26b. XPS measurement indicated that the surface concentration of sulfur was very small at the solid–air interface, due to the high surface free energy of the graft chain. The free energy at the air–solid interface is minimized by enrichment of the hydrophobic PTMO segment. Figure 28 shows the dynamic wetting tension loops for SPU with a sulfonate graft chain. The receding contact angle decreased; i.e., tension increased with an increase in sulfonate graft diol content.(79) The large advancing contact angle (low tension) indicates that, due to the large difference in hydrophilicity between PTMO and the graft chain, the hydrophobic PTMO is enriched on the air–solid interface. The small magnitude of the receding contact angle reflects the hydrophilic property of the graft chain. Large-contact-angle hysteresis may indicate high mobility of the graft chain, which responds to environmental change. Since this polymer has an ionic end group on the graft chain, the "buoyancy" slope during immersion is greater than during retraction. However, after NaCl is added to the water phase, the slope ratio approaches unity. Similar phenomena have been observed for poly(sulfoalkyl methacrylates)(80) by Andrade and Chen. The addition of salt suppresses the tendency for the counterion in the film surface to diffuse into the water, quenching the thickness of the double layer, which lowers the net charge in the polymer backbone. These effects result in the depression of chain expansion; that is, the contraction of the polymer chain in the salt solution makes the chain

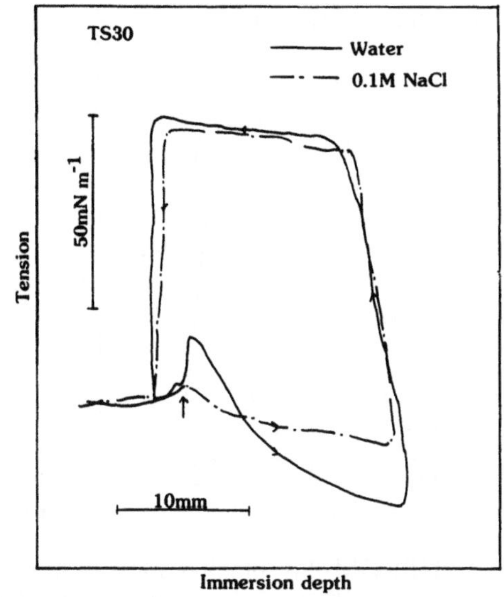

FIGURE 28. Dynamic contact angle loops for SPU with ionic graft chains. (Reprinted with permission from Ref. 79.)

behave as "normally." Also, this behavior indicates that the graft chain is soluble in water.

Methoxy poly(ethylene glycol) methacrylate (MPEGM) is a commercially available macromer that can be used for preparation of graft copolymers with various degrees of hydrophilicity. The hydrophilicity of this monomer can be changed by altering the length of the grafted PEG chain. Nagaoka and co-workers prepared graft copolymers based on MPEGM and investigated the molecular motion of the graft PEG chains in an aqueous environment.[81, 82] The molecular motions of the graft chains were investigated by C^{14} NMR and revealed that with an increase in PEG chain length the mobility of the PEG chain increased.[81, 82]

Takahara and co-workers prepared diblock copolymers of PMPEGM–PS and poly[hydroxy poly(ethylene glycol) methacrylate] (PHPEGM)–PS[83–85] by radical living copolymerization.[86] Figure 29 shows the chemical structures of PMPEGM–PS and PHPEGM–PS diblock copolymers. A styrene telechelic oligomer was prepared, and MPEGM and HPEGM were polymerized by a PS iniferter. Also, the corresponding random copolymer was prepared by solution radical polymerization.

Transmission electron micrographs of these diblock copolymer films cast from toluene showed a microphase-separated structure, whereas the corresponding random copolymer showed no evidence of microphase separation. Figure 30 shows C_{1s} spectra of PMPEGM–PS and PHPEGM–PS diblock copolymers with 50% PS content. The satellite peak corresponding to the π-π^* of PS is clearly observed for

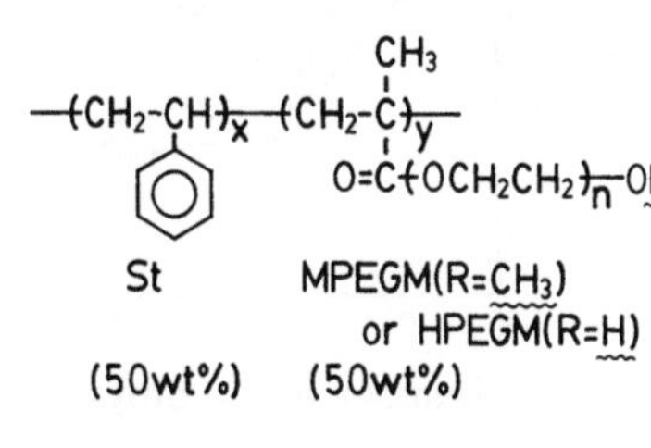

FIGURE 29. Chemical structures of PS–PMPEGM and PS–PHPEGM diblock copolymers. (Reprinted with permission from Ref. 84.)

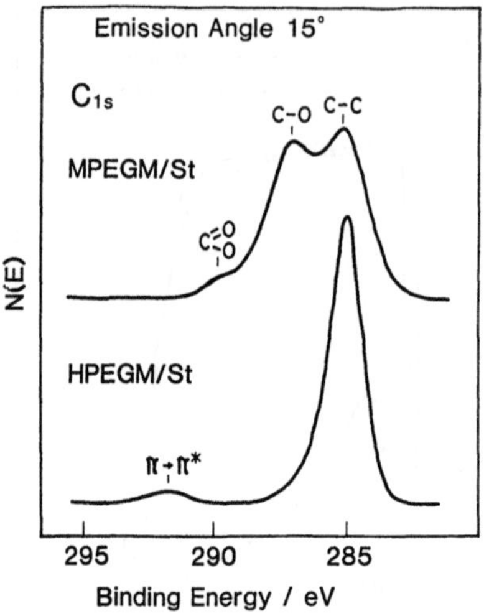

FIGURE 30. XPS spectra of PS–PMPEGM and PS–PHPEGM diblock copolymers. (Reprinted with permission from Ref. 84.)

the PHPEGM–PS block copolymer, whereas the intensity of the ether carbon signal is strongest for the PMPEGM–PS block copolymer. This indicates the surface enrichment of PS for the PHPEGM–PS block copolymer. On the other hand, the PMPEGM–PS block copolymer showed an enrichment of the methoxy group at the air–solid interface. The concentration of the methoxy group at the air–solid interface decreased with an increase in $\overline{Mn}$ of the graft PEG chain. The difference in the orientation of the end groups of HPEGM- and MPEGM-based diblock copolymers is ascribed to the difference in hydrophilicity of the end group. The corresponding random copolymers, however, did not show any difference in surface and bulk compositions.

A solvent effect was observed for the surface composition of PMPEGM–PS

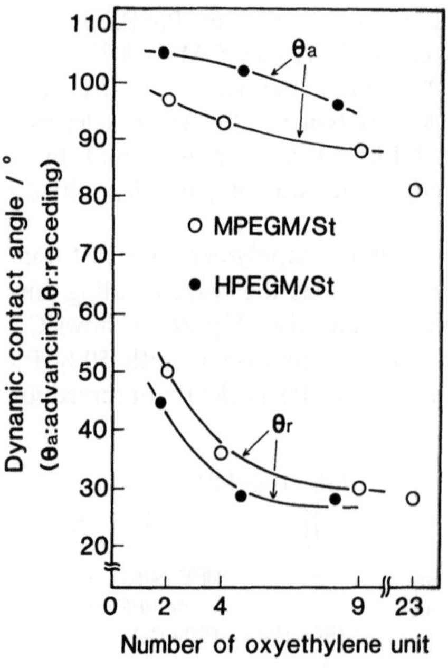

FIGURE 31. Variation of advancing and receding angles for PS–PMPEGM and PS–PHPEGM diblock copolymers with PEG graft chain length. (Reprinted with permission from Ref. 84.)

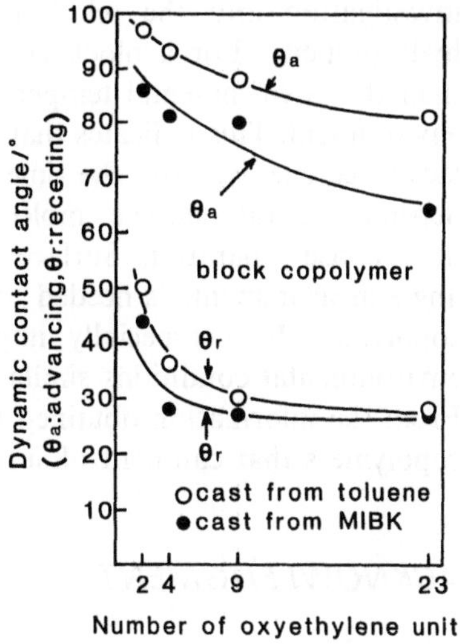

FIGURE 32. Effect of casting solvent on variation of advancing and receding angles for PS–PMPEGM diblock copolymer with PEG graft chain length. (Reprinted with permission from Ref. 84.)

diblock copolymers. The PMPEGM–PS block copolymer film cast from toluene showed a higher concentration of PS, since toluene is a good solvent for PS.

The surface molecular mobility and hydrophilicity of the block and random copolymers have been investigated by the dynamic Wilhelmy plate technique. Figure 31 shows the variation of the advancing and receding contact angles as a function of PEG length for PMPEGM–PS and PHPEGM–PS block copolymers with 50 wt % PS. The advancing and receding angles decreased with an increase in $\overline{Mn}$ of the PEG graft chain. The advancing angle for the HPEGM-based block copolymer showed a larger magnitude than that for the MPEGM-based block copolymer. This is ascribed to the higher surface concentration of PS for the PHPEGM–PS block copolymer. However, the receding contact angle for the HPEGM block copolymer had a lower magnitude due to the difference in its end group. The advancing contact angle did not change with an increase in PEG chain length, which corresponds to the XPS result. Since the advancing contact angle reflects the concentration of the hydrophobic components, these results indicate that the surface concentration of PS is not affected by an increase in the $\overline{Mn}$ of the PEG graft chains.

The effect of casting solvent has been observed for the PMPEGM–PS block copolymer. Figure 32 shows the variation of the advancing and receding angles for the PHPEGM–PS block copolymer cast from toluene and methylisobutylketone (MIBK). The advancing and receding contact angles for block copolymer films cast from toluene were larger than the angles of the films cast from MIBK. Since toluene is a good solvent for PS, the PS segment will be enriched on the surface of the block copolymer cast from it.

7. CONCLUSION

The surface hydrophilicity of block copolymers is an important property that is closely related to practical applications. However, there has been very little investigation into the relationships between surface chemical composition and hydrophilicity. For a block copolymer with a glass transition temperature lower than the environmental temperature, the surface may respond to a change in the environment. This indicates that the surface structure reorganizes to minimize interfacial free energy. Also, the time required for the rearrangement of the surface may depend on the surface molecular aggregation state, such as the degree of microphase separation. Surface analysis using spectroscopic methods and contact angle measurements is needed to characterize the surface hydrophilicity of a block copolymer. It is especially important that the surface be characterized under environmental conditions similar to the conditions under which it will be applied. From the information obtained from those analyses, the molecular design of block copolymers that can control surface hydrophilicity is possible.

ACKNOWLEDGMENT

The author acknowledges Mr. T. Teraya and Professor T. Kajiyama (Kyushu University, Fukuoka, Japan) for some of the experimental information in this chapter and for their helpful discussions. Also, discussions with Professor S. L. Cooper and Dr. A. Z. Okkema (University of Wisconsin, Madison, WI) are greatly acknowledged.

REFERENCES

1. G. E. Molau, ed., *Colloidal and Morphological Behavior of Block and Graft Copolymers*, Plenum Press, New York (1971).
2. Y. Yamashita and Y. Tsukahara, *J. Macromol. Sci.-Chem. A* **21**, 974 (1984).
3. B. D. Ratner, in *Surface and Interfacial Aspects of Biomedical Polymers*, Vol. 1, *Surface Chemistry and Physics* (J. D. Andrade, ed.), pp. 373–394, Plenum Press, New York (1985).
4. D. E. Leyden and W. T. Collins, eds., *Chemically Modified Surfaces in Science and Technology*, Gordon and Breach, New York (1987).
5. A. Noshay and J. E. McGrath, *Block Copolymers, Overview and Critical Survey*, Academic Press, New York (1977).
6. S. L. Aggarwal, *Block Copolymers*, Plenum Press, New York (1970).
7. J. D. Andrade, *Med. Dev. Diagn. Ind.*, June, p. 22 (1980).
8. J. D. Andrade and W.-Y. Chen, *Surf. Interface Anal.* **8**, 253 (1986).
9. A. W. Adamson, *Physical Chemistry of Surfaces*, 5th ed., Wiley, New York (1990).
10. T. Matsuda, T. Akutsu, T. Suzuki, and T. Kotaka, *Rep. Prog. Polym. Phys. Jpn* **25**, 857 (1982).
11. T. Matsuda and T. Akutsu, *ACS Org. Coat. Appl. Polym. Sci.* **48**, 647 (1983).
12. J. D. Andrade, in *Surface and Interfacial Aspects of Biomedical Polymers* (J. D. Andrade, ed.), p. 105, Plenum Press, New York (1985).
13. B. D. Ratner, *Ann. Biomed. Eng.* **11**, 313 (1983).
14. D. Briggs and M. P. Seah, eds., *Practical Surface Analysis*, Wiley, New York (1983).
15. B. D. Ratner, P. K. Weathersby, A. S. Hoffman, M. A. Kelly, and L. H. Scharpen, *J. Appl. Polym. Sci.* **22**, 643 (1978).
16. A. Benninghoven, *Surf. Sci.* **35**, 427 (1973).
17. D. Briggs, *Surf. Interface Anal.* **4**, 151 (1982).
18. N. J. Harrick, *Ann. NY Acad. Sci.* **101**, 928 (1963).
19. K. Knutson and D. J. Lyman, in *Biomaterials: Interfacial Phenomena* (S. L. Cooper and N. A. Peppas, eds.), *Adv. Chem. Ser.* No. 199, p. 109 Am. Chem. Soc., Washington, DC (1982).

20. L. C. Sawyer and D. T. Grubb, *Polymer Microscopy*, Chapman and Hall, New York (1987).
21. V. N. E. Robinson, *J. Phys. E.* **7**, 650 (1974).
22. J. D. Andrade, D. L. Coleman, and D. E. Gregonis, *Makromol. Chem. Rapid Commun.* **1**, 101 (1980).
23. W. A. Zisman, in *Contact Angle, Wettability, and Adhesion* (F. M. Fokes, ed.), *Adv. Chem. Ser.* No. 43, p. 1 Am. Chem. Soc., Washington, DC (1964).
24. D. K. Owens and R. C. Wendt, *J. Appl. Polym. Sci.* **13**, 1741 (1969).
25. W. C. Hamilton, *J. Colloid Interface Sci.* **40**, 219 (1972).
26. R. N. King, J. D. Andrade, S. M. Ma, D. E. Gregonis, and L. R. Brostrom, *J. Colloid Interface Sci.* **103**, 62 (1985).
27. Y. C. Ko, B. D. Ratner, and A. S. Hoffman, *J. Colloid Interface Sci.* **82**, 25 (1981).
28. G. Zografi and B. A. Johnson, *Int. J. Pharm.* **22**, 159 (1984).
29. L. Smith, C. Doyle, D. E. Gregonis, and J. D. Andrade, *J. Appl. Polym. Sci.* **26**, 1269 (1982).
30. J. D. Andrade, L. M. Smith, and D. E. Gregonis, in *Surface and Interfacial Aspects of Biomedical Polymers*, Vol. 1, *Surface Chemistry and Physics* (J. D. Andrade, ed.), pp. 249–292, Plenum Press, New York (1985).
31. R. E. Johnson and R. H. Dettre, *Surf. Colloid Sci.* **2**, 85 (1969).
32. E. Vanzo, *J. Polym. Sci. A-1* **4**, 1727 (1966).
33. H. Hasegawa and T. Hahimoto, *Macromolecules* **18**, 589 (1985).
34. M. V. Sefton and E. W. Merril, *J. Biomed. Mater. Res.* **10**, 33 (1976).
35. D. E. Gregonis, R. Hsu, D. E. Buerger, L. M. Smith, and J. D. Andrade, in *Macromolecular Solution* (R. B. Seymour and G. A. Stahl, eds.), pp. 120–130, Pergamon Press, New York (1982).
36. Y. Yamashita, *J. Macromol. Sci.-Chem. A* **13**, 401 (1979).
37. G. L. Gaines, Jr. and G. W. Bender, *Macromolecules* **5**, 82 (1972).
38. D. T. Clark, J. Peeling, and J. M. O'Malley, *J. Polym. Sci. Polym. Chem. Ed.* **14**, 543 (1976).
39. D. Shuttleworth, J. G. Vab Dusen, J. J. O'Malley, and H. R. Thomas, *ACS Polym. Prepr.* **20**(2), 499 (1979).
40. H. R. Thomas and J. J. O'Malley, *Macromolecules* **12**, 323 (1979).
41. T. Okano, M. Katayama, and I. Shinohara, *J. Appl. Polym. Sci.* **22**, 369 (1978).
42. T. Okano, T. Aoyagi, K. Kataoka, K. Abe, Y. Sakurai, M. Shimada, and I. Shinohara, *J. Biomed. Mater. Res.* **20**, 919 (1986).
43. K. Kugo, Y. Hata, T. Hayashi, and A. Nakajima, *Polym. J.* **14**, 401 (1982).
44. K. Kugo, T. Hayashi, and A. Nakajima, *Polym. J.* **14**, 391 (1982).
45. K. Tamaki, N. Kodera, K. Kugo, and J. Nishino, *Mem. Konan Univ. Sci. Ser.* **35**(2), 199 (1988).
46. I.-K. Kang, Y. Ito, M. Shishido, and Y. Imanishi, *Biomaterials* **9**, 138 (1988).
47. R. N. Wenzel, *Ind. Eng. Chem.* **28**, 988 (1936).
48. K. Arai, T. Kotaka, Y. Kitano, and K. Yoshimura, *Macromolecules* **13**, 1670 (1980).
49. D. W. Grainger, S. W. Kim, and J. Feijen, *J. Biomed. Mater. Res.* **22**, 231 (1988).
50. D. Grainger, J. Feijen, B. D. Ratner, D. Castner, Y. K. Sung, T. Okano, and S. W. Kim, *Trans. Soc. Biomater.* **11**, 285 (1988).
51. F. B. Leitz, in *Membrane Separation Process* (P. Meares, ed.), p. 261, Elsevier, Amsterdam (1974).
52. T. Fujimoto, K. Ohkoshi, Y. Miyakaki, and M. Nagasawa, *Science* **224**, 74 (1984).
53. M. D. Lelah, S. L. Cooper, H. Ohnuma, and T. Kotaka, *Polym. J.* **17**, 841 (1985).
54. H. Ohnuma, I. Kudose, T. Shinohara, and T. Kotaka, *Rep. Prog. Polym. Phys. Jpn* **25**, 237 (1982).
55. M. Katayose, A. Takahara, and T. Kajiyama, *Rep. Prog. Polym. Phys. Jpn* **30**, 851 (1987).
56. S. Krause, *Macromolecules* **3**, 84 (1970).
57. D. J. Lyman, C. Kwan-Gett, H. H. J. Zwart, A. Bland, N. Eastwood, J. Kawai, and W. J. Kolff, *Trans. Amer. Soc. Artif. Intern. Organs* **17**, 406 (1971).
58. G. M. Estes, R. W. Seymour, and S. L. Cooper, *Macromolecules* **4**, 452 (1971).
59. R. W. Paynter, B. D. Ratner, and H. R. Thomas, in *Polymer as Biomaterials* (S. W. Shalaby, A. S. Hoffman, B. D. Ratner, and T. A. Horbett, eds.), p. 121, Plenum Press, New York (1984).
60. A. Takahara and T. Kajiyama, *Nippon Kagakukaishi* **1985**, 1293 (1985).
61. R. S. Yih and B. D. Ratner, *J. Electron Spectrosc. Relat. Phenom.* **43**, 61 (1987).
62. M. J. Hearn, B. D. Ratner, and D. Briggs, *Macromolecules* **21**, 2950 (1988).
63. K. G. Tingey, J. D. Andrade, C. W. McGary, Jr., and R. J. Zdrahala, in *Polymer Surface Dynamics* (J. D. Andrade, ed.), pp. 153–170, Plenum Press, New York (1988).
64. A. Takahara, N. J. Jo, and T. Kajiyama, *J. Biomater. Sci. Polym. Ed.* **1**, 17 (1989).

65. A. Takahara, N. J. Jo, K. Takamori, and T. Kajiyama, in *Progress in Biomedical Polymers* (C. G. Gebelein and R. Dunn, eds.), pp. 217–227, Plenum Press, New York (1990).
66. C. Li, S. L. Goodman, R. M. Albrecht, and S. L. Cooper, *Macromolecules* **21**, 2367 (1988).
67. D. Grainger, T. Okano, and S. W. Kim, in *Advances in Biomedical Polymers* (C. G. Gebelein, ed.), p. 229, Plenum Press, New York (1987).
68. T. Suzuki, H. Chihara, and T. Kotaka, *Polym. J.* **16** (1984).
69. D. C. Legrand and G. L. Gaines, Jr., *ACS Polym. Prepr.* **11**, 442 (1968).
70. J. E. McGrath, D. W. Dwight, J. S. Riffle, T. F. Davidson, D. C. Webster, and R. Viswanathan, *ACS Polym. Prepr.* **20**(2), 528 (1978).
71. R. L. Schmitt, J. A. Gardella, Jr., J. H. Magill, L. Salvati, Jr., and R. L. Chin, *Macromolecules* **18**, 2675 (1985).
72. R. H. Bott, J. D. Summers, C. A. Arnold, L. T. Taylor, T. C. Ward, and J. E. McGrath, *J. Adhesion* **23**, 67 (1987).
73. Y. Kawakami, Y. Miki, T. Tsuda, R. A. N. Murpphy, and Y. Yamashita, *Polym. J.* **14**, 913 (1982).
74. Y. Chujo, T. Tatsuda, and Y. Yamashita, *Polym. Bull.* **8**, 239 (1982).
75. Y. Tezuka, A. Fukushima, S. Matsui, and K. Imai, *J. Colloid Interface Sci.* **114**, 16 (1986).
76. H. Kazama, T. Ono, Y. Tezuka, and K. Imai, *Polymer* **30**, 553 (1989).
77. Y. Tezuka, T. Ono, and K. Imai, *Polym. Prepr. Jpn* **37**, 1466 (1988).
78. Y. Tezuka, S. Matsui, A. Fukushima, M. Miya, K. Imai, K. Kataoka, and Y. Sakurai, *Kobunshi Ronbunshu* **42**, 629 (1985).
79. A. Takahara, H. D. Wabers, A. Z. Okkema, and S. L. Cooper, *J. Biomed. Mater. Res.* **25**, 1095 (1991).
80. E.-Y. Chen and J. D. Andrade, *J. Colloid Interface Sci.* **110**, 468 (1987).
81. S. Nagaoka, Y. Mori, H. Takiuchi, K. Yokota, H. Tanzawa, and S. Nichiumi, *ACS Polym. Prepr.* **24**, 67 (1983).
82. S. Nagaoka, Y. Mori, H. Takiuchi, K. Yokota, H. Tanzawa, and S. Nishiumi, in *Polymer as Biomaterials* (S. W. Shalaby, A. S. Hoffman, B. D. Ratner, and T. A. Horbett, eds.), p. 361, Plenum Press, New York (1984).
83. T. Teraya, A. Takahara, and T. Kajiyama, *Preprint of IUPAC Macro88*, p. 370 (1988).
84. T. Teraya, A. Takahara, and T. Kajiyama, *Polymer* **31**, 1149 (1990); A. Takahara, T. Teraya, and T. Kajiyama, *Kobunshi Ronbunshu* **47**, 395 (1990).
85. T. Kajiyama, A. Takahara, and T. Teraya, *Polym. Bull.* **24**, 333 (1990).
86. T. Otsu, A. Kuriyama, and M. Yoshida, *Kobunshi Ronbunshum* **40**, 583 (1983).

8

Wettability and Bioadhesion in Ophthalmology

Frank J. Holly

1. INTRODUCTION

The wetting of a solid or liquid by another liquid involves a process by which an intimate molecular contact is established between the two phases. To achieve such a close contact, at least one of the phases has to be fluid. In living systems, the omnipresent liquid is water, which is by far the major and most vital component of organisms. Hence, the water wettability of cellular structures and tissue interfaces is of the utmost importance in biology.

Wettability of a given phase by the liquid is usually important when the wetting has not yet taken place. In a functional living organisms, often the molecular contact between water and other structures is already established. Still, the wetting characteristics of the nonaqueous phases are important as they directly affect the stability of the aqueous film sandwiched between two adjacent structures. The interaction of these structures will depend on the state and stability of the water layer between them.

Cellular adhesion, contact inhibition, the functionality of tissue membranes, even the functioning of intracellular structures will be greatly influenced by the properties of the hydration shell surrounding them, which in turn will depend on the interfacial properties of the wetted boundary.

It had been long suspected, and now is widely accepted, that the properties of water wetting a solid surface are different in the close proximity of the solid–water interface.[1] The properties of this "vicinal" water[2] are determined by the superficial molecules of the solid and can be manipulated by subtle changes in the solid boundary. At surface boundaries a thin (1 nm) layer of water may even behave as a "soft" ice; its viscosity may be a millionfold higher than that of bulk water. Its dielectric constant and other physical properties can also be greatly altered through structural changes in the water itself.[3]

Frank J. Holly • Dry Eye Institute, Lubbock, Texas 79499.

Modern Approaches to Wettability: Theory and Applications, edited by Malcolm E. Schrader and George Loeb. Plenum Press, New York, 1992.

In such water, the dissociation coefficients for a given functional group or electrolyte would be vastly different. It has been estimated that near a solid–water interface, the hydrogen ion concentration would be a hundredfold higher than in the bulk of the solution; i.e., the interfacial pH would be two units lower than that of the bulk.[4]

The thickness of such an altered water layer at the interface is not known, but the alteration is certain to gradually decrease away from the phase boundary. Depending on the structure and composition of the solid surface, such changes could still be significant at 10 to 100 nm from the solid. The volume of the living cells is about 1000 μm^3 (10^{-6} cm^3), and most of it consists of water. If the total area of interfaces within the cell is at least 100,000 μm^2 (10^{-1} cm^2) (and it has been estimated to be much greater than that figure), then most of the water in the cell is altered, vicinal water, to which bulk chemistry is not directly applicable.

It follows from the aforesaid that our knowledge of physiological processes would increase tremendously, if biochemists, molecular biologists, and other scientists studying biological systems would constantly keep in mind the inapplicability of bulk parameters obtained with components removed from the original structure and then dissolved to form aqueous solution.

In the last 20 years, such basic, surface chemical considerations have been applied while studying human and mammalian animal eyes.[5] This approach has contributed significantly to our understanding of ocular physiology. In this chapter, this basic progress in ophthalmology as related to wetting and bioadhesion will be briefly discussed.

2. BASIC CONSIDERATIONS

2.1. Thin Aqueous Film Stability

The thermodynamics of equilibrium wetting has been discussed in many textbooks and reference sources. The reader unfamiliar with this topic should consult one of them. Here we shall only discuss those aspects that are important in ocular surface chemistry and have not been extensively studied elsewhere.

When a liquid is placed on a solid, it may spontaneously spread and coat the solid. Then the liquid is said to have a positive spreading coefficient on the solid; $S>0$. The spreading coefficient[6] is defined as

$$S=F_{sv}-(F_{lv}+F_{sl}) \tag{1}$$

where F is the specific surface free energy for the solid–vapor, liquid–vapor, and solid–liquid interfaces, respectively, It can also be expressed as the difference between the adhesive energy of the liquid to the solid and the cohesive energy of the liquid. Spontaneous spreading occurs when the attraction between the solid and liquid molecules are stronger than among the liquid molecules.

However, more often that not, the liquid will not spread over the solid, but will form a droplet with a certain contact angle between the solid and the liquid surface at the three-phase boundary line. For such a partial wetting, the Young–Duprè holds,[7] which expresses the contact angle, θ, in terms of the specific surface free energies:

$$F_{\mathrm{lv}} \cos \theta = F_{\mathrm{sv}} - F_{\mathrm{sl}} \tag{2}$$

Note that the left side of the equation, the product between the liquid surface tension and the cosine of the contact angle, is an experimentally measurable quantity, often referred to as adhesion tension. Neither quantity on the right side, on the other hand, can be directly measured.

In such a case the spreading coefficient can be written as

$$S = F_{\mathrm{lv}}(\cos \theta - 1) \tag{3}$$

which is clearly negative except for zero contact angle, when $S = 0$, in which case there is complete wetting.

When more liquid is added to a sessile drop, its height will only increase to a limiting thickness, h. Once this thickness has been reached, additional liquid will only increase the wetted area, i.e., the diameter of the liquid lens. The thickness of such a "free" lens resting on the solid depends on the magnitude of the (negative) spreading coefficient and the density of the liquid, ρ:[8]

$$h = \left[\frac{2(-S)}{\rho g}\right]^{1/2} \tag{4}$$

where g is the gravitational acceleration.

The limiting thickness of such free water lenses can be quite high on hydrophobic solids. For values calculated for Teflon, paraffin, poly(ethylene), and poly(methyl methacrylate) (PMMA), see Table 1.

An interesting situation arises, when the liquid lens is formed in a circular trough and becomes large enough so that it comes in contact with the solid wall. Basically the same situation arises when the lens is surrounded by bulk water if the supporting solid is in a larger container filled with water (Fig. 1). When this "confined" liquid film is thinned, i.e., by evaporation, its thickness becomes considerably smaller than the limiting thickness of a "free" lens. Eventually, at a critical

TABLE 1. Critical Thickness of Rupture of Aqueous Films Confined by a Solid or by Bulk Water and the Limiting Thickness of a Free-Water Lens Resting on the Same Type of Solid Surface

	Critical thickness of rupture (μm)		
Solid surface	Confined by solid	Confined by liquid	Limiting thickness of free lens (μm)
Teflon	790	510	4,440
Paraffin	—	310	4,380
Polyethylene	420	—	3,960
PMMA	390	—	3,110

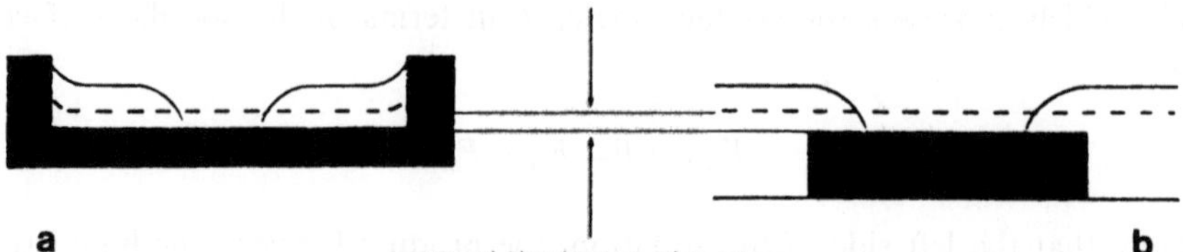

FIGURE 1. Confined films over hydrophobic surfaces and their rupture: (a) confined by solid, (b) confined by bulk water.

thickness, the aqueous film will suddenly rupture and recede toward the boundaries until the limiting thickness is reached.[9] This critical thickness should depend on the cohesive energy density of the liquid, and appears to depend on the ions in solution.[10] So far the magnitude of this critical thickness has not been expressed in terms of surface or bulk parameters.

It is interesting that the critical thickness of rupture observed in solid troughs and over submerged solid "islands" appear to be comparable in magnitude,[10,11] indicating that it may not depend directly on the spreading coefficient. Table 1 contains the values of critical thickness obtained in the two systems. It is puzzling, to say the least, that such relatively thick films can be ruptured by an effect initiated at the solid–water interface several hundred microns distant from the surface of the confined film. The stability of such framed aqueous films is directly relevant to tear film stability in the eye that affects visual acuity and the well-being of the underlying tissue.

2.2. *Meniscus-Induced Local Thinning*

Another interesting phenomenon occurring with a confined liquid "lens"[11] can be observed using the same circular trough containing an aqueous film. Since the solid is only partially wettable, the water layer forms a finite contact angle with the solid frame around it. The amount of water in this meniscus surrounding the film depends on the magnitude of the contact angle, the surface tension, and the density of the liquid, as well as the length of the meniscus and can be expressed as:[12]

$$W = 2\pi R F_{\mathrm{lv}} \cos\theta \tag{5}$$

where R is the radius of the trough in centimeters and W is the weight of the liquid comprising the meniscus in dynes.

If the bottom of the trough is made hydrophilic, e.g., by coating the plastic with a monolayer of hydrophilic macromolecules, then the thickness of the film will decrease without compromising its continuity. When the total amount of water in the trough decreases to the value, W, as given in Eq. (5), an unexpected thing happens. Instead of having no visible film in the center and a liquid ring around the periphery forming an isolated meniscus, at a film thickness of about 10–20 μm, a circular local thinning occurs where the meniscus and the film adjoin.[13] At this location the film thins to approximately 1% of its thickness. This local thinning can be made visible by staining the water with nonsurface active (pure) sodium fluorescein and viewing it under blue light. Due to its appearance in comparison to the brightly lit film and meniscus, the local thinning has been referred to as a "black line" (Fig. 2).

There is no obvious explanation for this phenomenon. The film profile at the thinning is contrary to what is expected, the convex curvature of the fluid surface with corresponding increased pressure in the fluid is adjoined by a lower concavely curved surface under which the pressure is negative (cf. Fig. 2). Such an arrangement is clearly not at equilibrium, and liquid flow should occur to eliminate the local depression in the fluid surface. However, such a black line appears to be stable for extended periods of time. One qualitative explanation of this occurrence may be that at this point the available fluid is insufficient to form an equilibrium meniscus at the edge of the trough, where the negative pressure due to the curvature and tension of the fluid surface is compensated by increased hydrostatic pressure due to elevation of fluid level. Such a "thirsty" meniscus would induce fluid flow from the film,[11,13] but would become self-limiting due to the increased hydraulic resistance. The incipient flow would constrict the film at the point adjacent to the edge of the meniscus, thereby increasing hydraulic resistance to flow (if the thickness decreases 100-fold, the resistance would increase 10^4-fold). Hence, further draining of the film to the meniscus would occur at such a negligible rate that the system would appear to be at an equilibrium. Such a black line was first observed in the open eye adjacent to the tear meniscus formed against the lid.[13,14]

2.3. Formation of Lipid Layers on Water

The spreading of an immiscible liquid such as lipid on water is basically the same as the wetting of a solid surface. The initial spreading coefficient is again given by Eq. (1). The lipid will spread spontaneously if the initial spreading coefficient is positive. After spreading, however, both the lipid and the water phases will become mutually saturated, changing their surface tensions. As a result, the value of the spreading coefficient will also change. This final spreading coefficient[15] may be written as

$$S = F_{\text{bv}}(a) - F_{\text{av}}(b) - F_{\text{ab}} \tag{6}$$

where phase b is spreading on phase a. The letters in parentheses indicate that the particular phase is saturated with respect to the phase indicated by the letter.

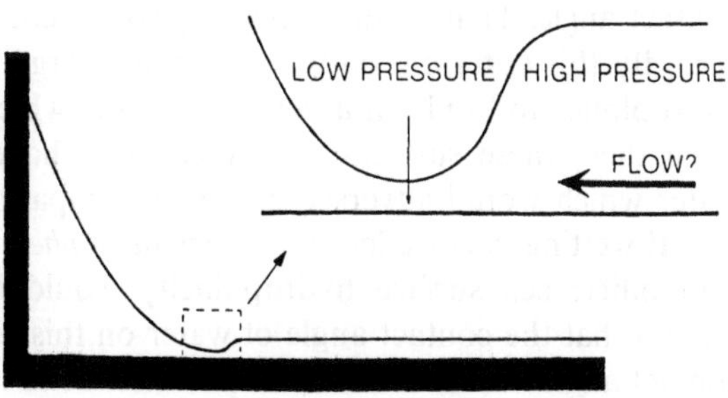

FIGURE 2. Meniscus-induced local thinning in a confined film over a hydrophilic surface and its cross-sectional profile (pressure indicated is due to interfacial curvature and surface tension).

For most pure phases, the final spreading coefficient usually becomes negative. The superficial phase will initially spread, then the excess will recede forming lenses. The rest of the surface will be covered by a monolayer lowering the surface tension of liquid a, F_{av}, to $F_{av}(b)$. This means that the sum of the surface and interfacial tensions of a multimolecular lipid layer will be higher than the monolayer-covered water; thus, such a thick film would not be stable. Harkins attempted to prove that for all systems consisting of two single-component phases, the stable arrangement is a lens in equilibrium with a monolayer.[(16)] His proof, however, has been challenged, as apparently there are exceptions to this rule.[(17)]

Stable multimolecular films can easily be produced over water if the lipid phase contains at least two components of different polarities. The more polar component will preferentially adsorb at the lipid–water interface, lowering its tension. Langmuir observed[(18)] that certain lipids with long hydrocarbon chains formed monomolecular layers over water that were similar in properties to multimolecular lipid films; i.e., the hydrocarbon part of the monolayer behaved as a distinct liquid phase. Such films are classified as "lipid-expanded" type. Langmuir named these monomolecular layers duplex films, because of their strong similarity to thin hydrocarbon layers lying over water, which have a gaseous film of highly polar components present at the lipid–water interface. We shall use the term *duplex film* exclusively for multimolecular layers consisting of a large variety of lipids, when discussing the superficial lipid layer of the tear film later in this chapter.

There are two lipid films of practical usefulness in studying the properties of other surface films. One is a monolayer-lens type called the *piston oil*, the other is a duplex film named *indicator oil*. Piston oil[(19)] consists of one component that maintains a given film pressure as long as excess oil in the form of a lens is present. Such materials are used to maintain a surface film under investigation at a constant film pressure even when the amount of spread substance changes during the experiment. Indicator oils,[(20)] on the other hand, form multimolecular layers, and their film pressure increases with increasing thickness. Thus, indicator oils can be used to estimate the film pressure of adjacent invisible films over the water surface on the basis of their interference colors.

2.4. Water Wettability of Solid Surfaces

The hydrophilic nature of solid surfaces is usually described in terms of wetting by water.[(21)] Wetting can be considered as a *spreading phenomenon*, where the advancing contact angle of sessile water droplets is determined. The surface is considered hydrophilic if water spontaneously spreads on it or at least has a zero contact angle. That is, the spreading coefficient of water is equal to or greater than zero. By this criterion, only a few, high-energy solid surfaces could be classified as hydrophilic, and only in a very pure state, which is extremely difficult to maintain. Even then, these surfaces are expected to have a high interfacial tension against water which would adversely affect biocompatibility.

If wetting is considered as a *capillary phenomenon*, then the condition of water wettability; i.e., surface hydrophilicity would be a positive capillary rise, which implies that the contact angle of water on this solid is less than 90°. At this limiting contact angle value, the solid–vapor interfacial tension is exactly equal to the solid–

water interfacial tension, making the spreading coefficient equal to the negative value of the water surface tension [cf. Eq. (1)]. This criterion implies that if the surface exhibits a slightly higher degree of attraction to condensed water than vapor, the surface is hydrophilic. Since the density ratio between water and its vapor is about 1000:1, only completely apolar and inert surfaces would be classified as hydrophobic by this criterion.

Preferential wetting may be the most realistic test of surface hydrophilicity. In this test, the solid surface is exposed to two liquid phases, one of them water, the other a hydrophobic liquid immiscible with water. The criterion of hydrophilicity is the requirement that the solid should exhibit more affinity toward water than toward the hydrophobic liquid. If the nonpolar liquid chosen has the same surface tension as the dispersion force component of water surface tension (21.8 mN/m at 25°C), e.g., *n*-octane, then the cosine of the two-phase contact angle will be proportional to the magnitude of polar and hydrogen-bond-forming interaction across the interface between the solid and water.[22] Since the molecular interaction across the interface inversely affects the magnitude of the interfacial tension, this criterion of water wettability may be the most useful one for assessing biocompatibility.

A solid surface is often characterized by its critical surface tension, which was defined by Zisman.[23] Contrary to expectation and occasional misconceptions, this is not the surface free energy of the solid; it is not even a sole property of the solid surface. For a low-energy solid, it can be determined by measuring the advancing contact angle of a series of liquids, usually pure hydrocarbons, which were termed *diagnostic liquids* by Zisman, who empirically found a rectilinear relation between the cosine of the contract angles and the surface tension of the liquids. If the straight line is extrapolated to zero contact angle ($\cos\theta = 1$), the abscissa of the intersection will yield the critical surface tension. It follows from the definition that liquids having surface tensions at or below this value will completely wet the solid.

Since the Zisman method uses hydrophobic liquids to determine the contact angles, the critical surface tension obtained is characteristic only of the dispersion force field component of the specific surface free energy of the solid. Thus, only for a completely apolar surface will it be relevant to water wettability or surface hydrophilicity. For other solids, the magnitude of the critical surface tension will be misleading as to the actual water wettability. This is especially true where the solid has polar binding sites or contains water. For example, the critical surface tension of polyethylene consisting mostly of methylene groups is about the same as that of a polyacrylamide gel containing 78% water at equilibrium hydration. On the other hand, the water contact angle on polyethylene is 94°, while it is only 16° on poly(acrylamide).[24]

The sessile drop contact angles are also poor indicators of surface hydrophilicity, even if one uses the criterion of zero capillary rise. Poly(methyl methacrylate), a solid polymer that can absorb only 1.5% water at equilibrium hydration, exhibits an advancing water contact angle of 73°. Poly(2-hydroxyethyl methacrylate) (PHEMA), a hydrogel with an equilibrium water content of 38–42%, can also exhibit a water advancing contact angle of similar magnitude even when its surface is fully hydrated.[25]

Contrary to expectation, hydrogel surfaces appear to be slightly less wettable when the water contact angles are measured by the captive-bubble technique than when measured by the sessile drop technique.[25] This phenomenon is even more pronounced with copolymers of siloxane and methyl methacrylate.[26] No explanations have been offered yet of this intriguing phenomenon. The criterion of preferential wetting by water when compared to an apolar liquid appears to be the most realistic for deciding the degree of hydrophilicity of a solid surface, especially when hydrogels are involved.[27] Table 2 contains the water-in-octane advancing contact angles for polyethylene, PMMA, PHEMA, poly(2-hydroxyethyl acrylate) (PHEA), poly(glyceryl methacrylate) (PGMA), and poly(acrylamide) (PAA). Using 90° as the dividing value, we can see from the data that by this criterion both polyethylene and PMMA have hydrophobic surfaces, while all hydrogels have hydrophilic surfaces of varying degrees.

2.5. Hydrophilic–Hydrophobic Surface Transition

When describing a solid surface, one often presumes or implies that its surface characteristics are well-defined, inherent, and invariant properties of the solid that will only change if the composition of the solid changes or if adsorption takes place. Unfortunately, this assumption is usually untrue. Often the measurement itself employed to characterize the surface will effect some surface change. For example, in contact angle goniometry, the liquid used may affect the surface. Water may hydrate the surface layer, may hydrolyze surface molecules, or cause stereochemical changes. All of these processes will change the original properties of the surface.

One indication of such changes is the contact angle hysteresis often observed in real systems. There can be a considerable difference between the advancing contact angle formed by gradually increasing the size of the droplet in contact with the solid and the receding contact angle observed when the liquid is gradually withdrawn form the sessile droplet. Surface roughness is a well-known cause of contact angle hysteresis. Physical or chemical interaction between the liquid and the solid will also cause contact angle hysteresis, since the solid surface in contact with the droplet is different than the surface outside the droplet.[28]

There is an additional possible cause of contact angle hysteresis, which was

TABLE 2. Advancing Contact Angle Value of Water in Air and in Octane on Solid Polymers and Hydrogels

		Advancing contact angle of water	
Solid or gel	Water content (%)	In air (deg)	In octane (deg)
Polyethylene	0	94	153
PMMA	1.5	73	120
PHEMA	38.5	60	88
PHEA	71.9	44	79
PGMA	73.3	41	63
PAA	77.7	10	12

PMMA = poly(methyl methacrylate), PHEMA = poly(hydroxyethyl methacrylate), PHEA = poly(hydroxyethyl acrylate), PGMA, poly(glyceryl methacrylate), PAA = poly(acrylamide).

first recognized in the middle 1970s.[21,27] In certain solids, the surface molecules have sufficient mobility to change their conformation when the liquid comes in contact with it or when the polarity of the liquid phase, in contact with the solid, changes. Polymeric surfaces,[29] hydrogels,[21] and biosurfaces[24] exhibit large contact angle hysteresis due to such an effect. The same has been found to be true for plasma-treated solid surfaces[30] and for the surface of hydrophilic polymers grafted onto hydrophobic substrate.[31]

Cellular surfaces and tissue interfaces usually exhibit pronounced hydrophobic–hydrophilic transition. Almost any solid surface can acquire the same property via the adsorption of an amphipathic polymer. On the other hand, if the polymer has negligible amphipathic character, by adsorbing it onto a given surface one can render the surface properties more invariant.

An interesting empirical relationship was found[21] between the relative contact angle hysteresis defined as

$$H_R = \frac{\theta_A - \theta_R}{\theta_A} \tag{7}$$

and the advancing contact angle, when the "diagnostic" liquid remained the same, i.e., water, while some property of the solid was changed, i.e., equilibrium water content of a given hydrogel, or the component ratio in a copolymer. (Even if the surface change resulting in different contact angles is due to unknown factors and is not introduced willfully.)

When H_R is plotted against θ_A for a given type material, often a rectilinear relationship is found either with a positive or a negative slope. In the first case, the relative contact angle hysteresis increases with decreasing wettability; in the second case, the hysteresis increases with increasing wettability. Most simple hydrogels exhibit the former wettability dependence. This is explained[21] by increasing segmental mobility of amphipathic surface polymer chains. Such a mobility increase would increase hysteresis as well as the magnitude of the advancing contact angle, since hydrophobic groups can be more completely exposed to the air at the surface at higher segmental mobility. It has been observed that the slope of the straight line in the H_R versus θ_A diagram is usually greater if the polymer matrix does not contain the highly hydrophobic methyl side groups (e.g., PHEA and PAA), because the lack of methyl group imposes a lower ceiling on the degree of hydrophobicity than can be achieved in air or a nonpolar medium.

When a copolymer comprised of a hydrophobic and an amphipathic polymer is examined, the relative contact angle hysteresis will decrease with decreasing wettability. The reason for this negative hysteresis slope is that only one polymer contributes to hysteresis and to wettability. Thus, increasing the ratio of the amphipathic component to the hydrophobic component will increase both wettability, as evidenced by smaller advancing contact angles, and contact angle hysteresis. The copolymer siloxane with poly(vinyl pyrrolidone) is a typical example of such a surface behavior.[21] This technique can be quite useful in evaluating polymer surfaces that have been exposed to ionic bombardment and where the surface composition and stability are usually unknown.[32]

3. APPLICATIONS TO THE OCULAR SYSTEM

All the basic concepts considered in the preceding part of this chapter were a by-product of a basic investigation of tear film physiology and contact lens biocompatibility that commenced in our laboratory in 1968 and is still going on. It has been said that the preocular tear film is a biological system that can be used for the demonstration of most, if not all, surface chemical concepts such as wetting, spreading, dewetting, adsorption, thin film stability, etc. Thus, it is proper to start this applied section with a discussion on the preocular tear film emphasizing the surface chemical aspects related to wetting and adhesion.

3.1. Preocular Tear Film

The exposed part of the eye consists of the conjunctiva overlying the exposed part of sclera (the white of the eye) and the cornea, which is the clear window through which the colored iris and the dark pupil can be seen. The cornea, due to its curvature and the large difference in refractive index across the external surface, is the most powerful refractive surface (lens) in the eye. Its net power of refraction ranges between 44–47 diopters. The outermost layer of the cornea consists of epithelial cells approximately four to five cells deep. This cellular layer is covered by a tear layer 5–10 μm thick, which in turn is coated with a duplex film of lipids from the meibomian glands of the eyelids.

This tear film provides the cornea with an optically smooth surface and also ensures the well-being of the corneal epithelium. The aqueous layer of the tear film lubricates the lids during blinking. Since tears contain bacteriocidal components, the tear film also serves as a first line of defense microbial invasion.

3.1.1. Wettability of the Corneal Epithelium

The corneal epithelium is not a well-defined surface. Basically it consists of epithelial cell membranes that form numerous microvilli and microridges. These surface formations quickly disappear when the epithelium is exposed to a hydrophobic liquid, to a surface-active anesthetic agent, or a cationic surfactant, the latter is used as a preservative in eyedrops. The cellular surface is believed to be coated with a thin layer of mucous glycoproteins, most of which are secreted by the conjunctival goblet cells located around the cornea. The epithelial cells are also known to secrete a small amount of such glycoproteins. In addition, both the main and the accessory lacrimal glands secrete glycoproteins, which are present in a dissolved state in the aqueous tears.

While most animal cells have low interfacial tension when immersed in body fluids, the cell and tissue boundaries exhibit a low-energy surface, only partially wettable by water, when exposed to air, even to air saturated with water vapor. This is so because the amphipathic macromolecules and lipids at the surface are capable of significant hydrophilic–hydrophobic transition (cf. Section 1.5), which lowers the surface energy to a minimum: hydrophobic in air or in nonpolar liquids, and hydrophilic in water. The critical surface tension of the corneal surface when free of excess mucus was found to range from 28 to 32 mN/m, depending on how successfully the mucin layer had been removed from the surface.[33,34] This critical surface tension compares well with the value of 28 mN/m obtained for the blood vessel intima.[35] Due to the considerable hydrogen bonding capability of the epithelial surface, the advancing contact angle of physiological saline is much lower than is expected from the critical surface tension ranging from 30° to 50°.[33,36]

Lipids secreted by the meibomian glands that are located at the lid edges have an average molecular weight of 720 daltons as opposed to the mean molecular weight of sebum, which is 490 daltons.[37] These consist mainly of long-chain waxy and cholesteryl esters. The polar fraction is small and consists of free fatty acids and alcohols and minute amounts of phospholipids and tryglycerides. These lipids also have a critical surface tension of 28 mN/m when adsorbed on a solid substrate, indicating that the cornea without its mucus layer has a lipidlike surface when exposed to air or nonpolar liquids.[38]

Mucin-coated corneas or other mucin-coated substrates such as polyethylene or glass, on the other hand, have a considerably higher critical surface tension, 38 mN/m, as determined by the Zisman method.[33] The advancing contact angle of saline on mucin-coated surfaces is low[36] but still finite (Table 3). However, the surface tension of mucin dissolved in saline at physiological concentration is much lower than that of saline, 37–38 mN/m,[33] so aqueous tears should spontaneously spread over a mucin-coated surface except when it is dehydrated.

3.1.2. The Superficial Lipid Layer

The meibomian lipids form a duplex film over the aqueous tear layer in the open eye having an approximate thickness of 100 nm. When its thickness increases, first-order interference colors can be observed at the tear film surface. While there may not be interference colors in the wide-open eye, when the eyelids begin to close, they compress the lipid layer until the interference colors can be seen.

When the eyelids are closing, they compress the superficial lipid layer similarly to the moving barrier of a Langmuir trough, which only compresses the insoluble surface film. In the closed eye, the thickness of the compressed lipid layer, that can still be maintained between the adjacent lids, is about 0.1 mm.[39]

When meibomian lipids are placed on physiological saline at 34–37°C, they spontaneously spread, forming a duplex film similar to an indicator oil. The equilibrium final spreading pressure is approximately 12–13 mN/m, so that the surface

TABLE 3. Water Wettability and the Critical Surface Tension of Corneal Epithelium and Adsorbed Meibomian Lipids

Substrate	Advancing contact angle of saline (deg)	Zisman's critical surface tension (mN/m)
Mucin-coated cornea	15–20	38
Clean[a] cornea	20–50	28–31
Cultured corneal epithelium	50	27
Adsorbed meibomian lipids	50	28

[a] Excess mucus removed by mechanical means.

tension of the lipid-covered saline is about 58–60 mN/m.[40] When such a film is compressed, the film pressure increases only moderately in a reversible manner.

The aqueous tear fluid is known to contain over 50, and possibly as many as 80, different proteins, including glycoproteins and enzymes. They have various levels of surface activity: the least surface active appears to be lysozyme, while, surprisingly, the most surface active components are the mucous glycoproteins.[41] However, even the lysozyme is more surface active than the meibomian lipids. Thus, the meibomian lipids are not able to spread on tears as they do on saline.[40] Still a continuous superficial lipid layer exists as the external layer of the tear film.

This apparent contradiction is easy to explain. When the eyelids open at a speed of 10–16 cm/s, they create a new aqueous tear–air interface.[38] Since apparently all the surface-active components of the aqueous tears are macromolecular, initially the surface has the surface tension of pure saline. It takes several seconds before sufficient dissolved glycoprotein molecules can migrate to the surface and lower its surface tension. The meibomian lipids have an initial spreading velocity of approximately 25–30 cm/s, faster than the speed of the opening eyelid.[42,43] Thus, the aqueous tear surface is never exposed to air.

The spreading of the first lipid monolayer is soon (within 1 s) followed by the spreading of excess lipid establishing a multilayer with an initial film pressure of 12 mN/m. After several seconds, the film pressure starts to increase due to lipid–protein interaction at the lipid–tear interface. When the interacting macromolecules are mucin, the film pressure may triple to become 36 mN/m. With whole tears, the final lipid film pressure will become about 32 mN/m.[38,40] When the lipid layer is compressed during eyelid closure, due to the fast lid motion, most of the protein interacting with the lipid will also be compressed, since the protein desorption from the lipid–tear interface is relatively slow. Thus, upon the opening of the eyelid, the spreading of the lipid will be assisted by the spreading tendency of the compressed protein (glycoprotein) layer at the lipid–tear interface.

It has been a matter of controversy whether the superficial lipid layer retards the evaporation of the aqueous tear layer underneath.[44,45] Briefly, the retarding effect is always there, but whether it can be observed will depend the conditions in the adjacent air layer. If this is stagnant, its resistance will be high because the diffusion of water molecules away from the surface will be the limiting factor of evaporation. However, in turbulent dry air, evaporation will be rapid due to a high water concentration gradient at the surface, so the resistance of the lipid layer to evaporation will be the determining factor.

3.1.3. Tear Film Stability

The surface chemical condition of stability of a thin (less than 100 μm thick) fluid film over a solid is that the sum of the surface and interfacial tensions of the fluid film, the so-called free energy, has to be lower than the surface free energy of the solid–vapor interface. The same conditions have to be fulfilled for the tear film if it is to remain continuous:

$$F_{sv} > [\mathbf{F}_{lv} + \mathbf{F}_{sl}] \tag{8}$$

As we have seen, the epithelial surface will increase its surface free energy via mucin adsorption. Due to the strongly hydrophilic characteristics of the mucin, the interfacial tension against water will decrease de to intense hydrogen bonding across the interface while the surface tension of the tear film will decrease due to the high lipid film pressure resulting from protein–lipid interaction. Thus, mucin increases the value of the left side of the equation and lowers both terms on the right side in the Eq. (8), thereby ensuring that the inequality is achieved. This is why mucous glycoproteins are suspected to act as the lacrimal surfactant, the existence of which was also predicted from hydrodynamic analysis of tear film formation.[42]

3.1.4. Description of the Lacrimal Surfactant

The surface characterization of the various tear proteins has not been completed. There are various problems associated with tear research. Tear composition appears to be related to tear secretion rate.[46,47] Only a very small sample of human tears can usually be collected without irritating the eye. Tears obtained from irritated eyes contain a large amount of plasma proteins as a result of leakage through the walls of the dilated conjunctival blood vessels.[48,49] Furthermore, the protein separation has to be done with methods that do not denature the proteins. Gel filtration chromatography is one of the least denaturing methods, but with this method only four to five peaks can be obtained.[50] On the other hand, the surface activity of the various fractions show a more intense fluctuation (Fig. 3).

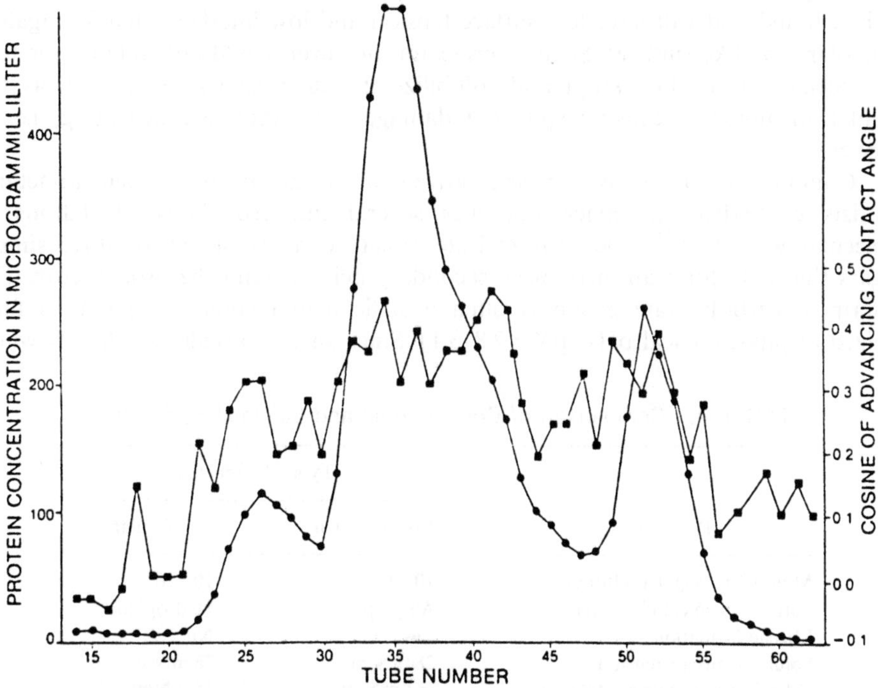

FIGURE 3. Surface activity of the various fractions of human tears obtained by gel filtration chromatography measured as the cosine of advancing contact angle on polyethylene. Circles show protein concentration measurements; squares show cosine of contact angle.

It is known with a fair amount of certainty that the most effective surfactant in the tears is contained in the large molecular weight fraction and that it most likely consists of mucous glycoproteins. The surface properties of conjunctival mucin were found to be similar to those of salivary glycoproteins,[33] in particular to the commercially available bovine submaxillary mucin (BSM).

The surface activity of BSM has been characterized in detail.[41,51] It appears to be more surface active than cellular fibronectin, serum albumin, gamma globulin, or lysozyme. Lactoferrin, an intrinsic tear protein exhibits quite high surface activity, but only at a 100-fold higher concentration than expected.[50] This may indicate that the surface active agent is not the lactoferrin but a minor component strongly associated with lactoferrin.

Some surface activity has been observed in the low molecular weight fraction of the tear proteins (M.W. < 5000 daltons), and this fraction appears to increase during incubation. It is most likely the result of some proteolysis, and thus this fraction may not be present in an appreciable quantity in the eye.[50,52] Since at present, the most likely candidate for the lacrimal surfactant is mucin, in Table 4 the characteristic properties of salivary mucin and those of conventional ionic surfactants are compared. They both function the same way by lowering the surface and interfacial tension of water in order to achieve wetting and, thus, film stability. The mechanism, by which complete wetting is achieved, is different, however.

The conventional surfactants consist of small amphipathic molecules that are closely packed at the water surface and at the solid–water interface, presenting a low-energy, hydrocarbon surface. Then the water surface will behave as a hydrocarbon liquid and will have low surface tension and low interfacial tension against nonpolar, lipidlike surfaces. Such a surfactant, however, could not stabilize the tear film because it would disrupt and solubilize the superficial layer as well as the cellular membranes, causing epithelial damage and eventually dewetting, i.e., a "dry eye."[40]

Conjunctival mucin, as a representative component for the lacrimal surfactant, consists of hydrophilic macromolecules several hundred thousand daltons in molecular weight.[53,54] The carbohydrate moiety consists of two or three simple sugars that may form an incomplete secondary helix around the protein core. The outermost carbohydrate group is often a sialic acid (sialomucins) that is fully ionized at physiological pHs ($pK = 2.8$–3.1). Since such molecules are highly water

TABLE 4. Comparison of Conventional and Lacrimal Surfactants

Property	Type of surfactant: Conventional	Type of surfactant: Ocular
Molecular weight (daltons)	10^2–10^3	10^5–10^6
Surface chemical character	Amphipathic	Hydrophilic
Micelle formation	Yes	No
Solution surface tension	28 mN/m	38 mN/m
Oil-solution interfacial tension	<1 mN/m	15 mN/m
Effect on superficial lipid layer	Disruptive	Stabilizing
Effect on cell membranes	Disruptive	Not detrimental

soluble and resist denaturation, it is not clear why they exhibit such a high degree of surface activity.

Due to their high molecular weight, the sialomucins are capable of adsorbing at almost any interface, even if the adsorption process is enthalpically unfavorable. The adsorption of large molecular weights polymers is usually entropy driven, since each adsorbed mucin molecule will displace thousands of water molecules from the interface.[55] Once at the solid–liquid interface, these molecules would lower the solid–water interfacial tension by providing numerous sites for hydrogen bonding without affecting the cell membrane integrity or the continuity of the superficial lipid layer. Actually, it is through secreting glycoproteins that the cells are capable of maintaining a low interfacial tension in an aqueous environment.

The mucous glycoproteins thus lower the interfacial tension at the epithelium–tear interface and increase the film pressure through interaction with the lipid molecules at the superficial lipid–tear interface i.e., lowering film free energy, thereby stabilizing the tear film.

3.1.5. Spontaneous Rupture of the Tear Film

The preocular tear film in a healthy eye is sufficiently stable to remain continuous between two consecutive blinks. The average blink rate for humans is approximately 12 to 15 blinks per minute, which corresponds to an average blinking interval of 4–5 s. Occasionally the blinking interval can be much longer during periods of stare, even as long as 20 to 30 s.

When blinking is prevented or voluntarily restrained, the tear film will rupture and "dry spots" of increasing area are formed.[56] This spotwise drying of the corneal surface was first believed to be due to local evaporation of the tears from the tear film.[57] Such rupture, however, takes in a relatively short time interval measured in seconds, while the drying time for the tear film is measured in minutes.[45] These dry spots now are viewed as local nonwetting as the tear film appears to form a finite contact angle at the periphery of dry spots.[33,38,58] It is known that surfactants, including polar lipids, are able to dewet solids by migrating to and adsorbing at the solid–water interface, making it hydrophobic even when their solubility is limited. Lipids possess slight water solubility, which is enhanced by lipid–protein interaction. The flux of the lipid molecules arriving at the interface increases inversely with the square of the distance. Local thinning, due to either local surface tension gradients or the presence of minute epithelial irregularities, would considerable enhance lipid contamination at certain locations, producing hydrophobic spots over which the tear film would recede.

This tentative mechanism of tear film rupture was proposed in the early 1970s[33] and is supported by the formation of lipid-laden mucous strands in the eye, which are removed by blinking, by *in vitro* experiments on the effect of polar lipids on thin aqueous film stability,[58] and by challenging the tear film *in vivo* by highly polar lipids such as the sebum.[59] Some controversy still exists on the nature of the discontinuity in the tear film. It is not known whether the base of the dry spot is "dry" or whether it is covered by an extremely thin autophobic fluid film. Such a film would be unwettable by the tear film proper.

Theoretical work has been done on the tear film rupture, but its relevance to the actual occurance has not been shown. Lin and Brenner presumed that the tear film rupture is due to instability caused by van der Waals forces between the whole tear film and the epithelium.[60] Sharma and Ruckenstein[61] correctly point out that if that were the case, the tear film breakup time should be measured in months. The authors also suggest a mechanism based on the rupture of the mucus layer coating the epithelium.[61] Since its thickness is several hundred times smaller than that of the whole tear film, the calculated breakup time is correspondingly much shorter. Unfortunately, the uncertainty in the numerical value of the Hamaker constant in their equation is too large to make a meaningful comparison between observed and calculated tear film breakup times.

It is perhaps of interest to note that the mucus layer appears to be able to mask the hydrophobic character of lipid contamination to a certain degree.[58] However, when the lipid content of mucus becomes the major component, the interfacial free energy increases so that the contaminated mucus layer will have a tendency to decrease its surface area. The shear forces created by the rapidly moving eyelids thus tend to roll the layer up into long strands and threads that will be carried by the tear flow toward the puncta, the opening of the tear drainage system on the nasal side of the lid edges.[62] If the thread is too large so that it cannot pass through the puncta, then it will accumulate in the canthus, the corner of the eye, as "sleep," where it is usually removed manually.

3.1.6. Conclusions Concerning the Preocular Tear Film

The formation of the tear film over the ocular surface involves lipid spreading over aqueous tears that have momentarily waterlike surface tension. The resulting film is framed by the tear meniscus formed against the lids. In the wide-open eye, local thinning adjacent to the meniscus can be observed, indicating that an uncompensated negative pressure may develop in the tears comprising the meniscus. The resulting tear film is so thin that gravity has a negligible effect on it, so that no flow is produced when the film is vertical (standing or sitting positions). The film pressure of the superficial lipid layer, a duplex film, is high despite the low spreading pressure of meibum because of the lipid–protein interaction at the lipid layer–tear layer interface. The epithelium–aqueous tear layer interface also has low interfacial tension due to adsorbed mucus that provides abundant sites for hydrogen bond formation across the interface.

Thus, the conjunctival mucins appear to be instrumental in stabilizing the tear film over a hydrophobic epithelium by lowering both the surface and interfacial tensions of the tear film, thus acting as a surfactant but without the detrimental effect on the superficial lipid layer or on cell membranes that conventional surfactants would have. The lipids contained in the superficial lipid layer can also be detrimental to tear film stability by overwhelming the mucus layer and increasing interfacial tension. This does not usually interfere with vision, since the tear film has to remain stable only until the next blink. An unstable tear film, which ruptures before the next blink occurs, can cause the painful and debilitating dry eye syndrome, which can result in blindness. Thus, complete wetting of the ocular surface between two consecutive blinks is indispensable for good visual acuity and the well-being of the whole eye, the organ of vision.

3.2. Contact Lens Wear

During the past 30 years, the contact lens has achieved an ever-growing popularity. Without doubt, the invention and the gradual perfection of contact represents a major breakthrough in the methods of refractive error correction for the eye. The contact lens is small and thin, approximately the size of the cornea. Such a lens having the appropriate refractive power is placed directly over the cornea, thereby retaining full peripheral vision as well as minimizing the change in image size. The latter is especially important in unilaterally aphakic eyes, where the positive correction needed is so high that the image obtained cannot be fused with that of the fellow eye and thus the binocular vision is lost.

We have seen that the surface of the cornea, the most powerful refractive medium of the eye, is coated with the preocular tear film to provide an optically smooth refractive surface and to ensure the well-being of the corneal epithelium. In addition, the tear film (*sans* the lipid layer) also lubricates the eyelids so that their rapid motion does not result in cellular damage. When a contact lens is placed in the eye, a tear film also has to form over the lens. This so-called prelens tear film is practically identical to the precorneal tear film. In a well-fitted lens, there also has to be an aqueous tear layer underneath the lens to prevent damage to the underlying epithelial surface.[63]

3.2.1. Contact Lens Materials

The first contact lens was made of high-quality optical glass from Jena (now in Germany). Soon thereafter, the choice of material became poly(methyl methacrylate), the transparent polymeric material that became popular under the trade name Plexiglass at the time of the Second World War. Such lenses are called hard lenses and, even today, are still used by many thousands of people.

Two other types of lenses were developed during the 1960s. Both were soft lenses. One was made of the hydrogel poly(2-hydroxyethyl methacrylate) that was cross-linked and could absorb 38% to 42% water at equilibrium hydration. The other was made of a silicone elastomer, a derivative of polymeric dimethyl siloxane. The PHEMA lenses provided a high degree of comfort, while the silicone lenses had high oxygen permeability.

Of these two, the PHEMA lenses soon became commercially available as polymacon lenses. Manufactured by Bausch and Lomb, they monopolized the soft lens market for years. The development of the silicone lenses was much slower, and only in the 1970s did they become commercially available, first in Europe and then in the United States.

During the 1970s, several other types of hydrogel lenses were developed. They were mostly acrylic polymers or copolymers of acrylic monomers with vinyl pyrrolidone. Silicone was also grafted with poly(vinyl pyrrolidone) to improve its water wettability.[64] One hydrogel consisting of the copolymer of methyl methacrylate and glyceryl methacrylate was found to have especially favorable mechanical properties.[65]

With the newer hydrogel lenses the equilibrium water content was increased in order to increase wettability and oxygen permeability. Some of the lenses have an equilibrium water content as high as 80% with a corresponding loss in mechanical strength. Contact lenses have also been made out of cellulose acetate butyrate, even though the popularity of such lenses has waned. Then a whole family of hard, gas-permeable lenses was created that consisted basically of silicone and methacrylic acid and related compounds. Such "silicone acrylate" contact lens materials are quite difficult to protect by patents because of their ill-defined chemical composition. Minor changes in composition can easily be achieved, resulting in "new" materials that may not be protected by previous patents.

There are two basic properties of contact lens materials that are considered to be of overriding importance for biocompatibility.[63] One is good water wettability; the other is high oxygen permeability. Good wettability is thought to ensure the stability of the tear film over and under the lens, thus increasing wearing comfort. The sufficiently high oxygen permeability is needed to supply the corneal epithelium with oxygen. Since, the cornea is an avascular tissue, no oxygen through blood supply is available to these cells.

3.2.2. Water Wettability of Contact Lenses

The classical hard contact lens material, PMMA, is only partially wetted by water. The advancing contact angle of water is 70–80°, depending on the purity of the polymer. The receding water contact angle is about 50°.[21,25] The material imbibes only 1.5% water at equilibrium hydration, so it is not considered to be a hydrogel. Over the years various attempts have been made to increase water wettability of PMMA. Decades ago a thin, transparent titanium dioxide layer was deposited on the lens surface. This high-energy surface layer was initially water wettable; however, it attracted and strongly bound lipids and denatured proteins due to its *high* water–solid interfacial tension. Ionic bombardment to hydrophilize the surface has also been tried with limited success.

For several years, the hydrogel lenses were made exclusively of PHEMA having an equilibrium water content of 38–40% water. PHEMA hydrogels that contain less or more water than this narrow range lose their transparency. Despite this considerable water content, the advancing contact angle of water on PHEMA gel is comparable to that on PMMA. The receding contact angle, on the hand, is much lower on PHEMA, about 25–28°.[25] The greater hysteresis is thought to be due hydrophilic–hydrophobic surface conversion. Contact angle values obtained with the captive-bubble technique as opposed to the sessile drop method are slightly higher for PHEMA hydrogels.[25] The reason for this reproducible phenomenon is not known.

In a detailed study of hydrogel wettability, it was found that the equilibrium gel water content does not indicate the degree of wettability.[21,66] When the equilibrium water content of PHEA was varied by changing the cross-link density in the gel matrix, it was found that the magnitude of the advancing water contact angle actually *increased* with increasing hydration. This is in agreement with the previously discussed observations that the relative contact angle hysteresis increases with increasing advancing contact angle for a given hydrogel.[21]

When several types of hydrogels of approximately the same water content were studied,[21] the magnitude of the advancing contact angle of water depended on the hydrophilicity of the polymer matrix. This demonstrates the fallacy of subconsciously equating bulk properties with surface properties. A hydrogel containing 40% water at equilibrium hydration does not necessarily have a composite surface consisting of 40% water.

Silicone elastomer is an extremely hydrophobic material.[32] The advancing contact angle of water on this material can be as high as 105–108°, almost as high as on Teflon. Hence, poly(dimethyl siloxane) is seldom used. One of the methyl groups on the siloxane chain could be substituted by a more hydrophilic group. Alternatively, a hydrophilic polymer such as poly(vinyl pyrrolidone) can be grafted onto the silicone.[64] Surface treatment consisting of ionic bombardment in the presence of water vapor is often used. This process implants hydrophilic groups on the surface, which can be maintained if the material is stored under water. Dry storage accelerates the decomposition of such surface groups, and then the surface of the treated material reverts to the highly hydrophobic state.

The water wettability of hard, gas-permeable materials depends on the content, or rather on the surface concentration, of methacrylic acid. Since the chemical composition of the copolymer, especially on the surface, is poorly defined and can vary from sample to sample, a definitive study has yet to be published on their water wettabilities. The advancing water contact angle is similar to that obtained on PMMA and is usually 65–85°. The receding contact angle values show an even wider range, from 20° to 50°. Some preliminary data indicate[26] that the difference between contact angles obtained by the sessile drop method and those obtained by the captive-bubble method are quite different for these materials, the captive-bubble method yielding the higher values. No explanation for this puzzling phenomenon has been given.

The "wetting angles" publicized by the manufacturers are obtained by ill-defined procedures under uncontrolled conditions where deliberate attempts to contaminate the surfaces are also made. These values are much lower than even the receding water contact angle values and have no theoretical, and only limited practical, significance.

3.2.3. Oxygen Permeability of Contact Lenses

The oxygen permeability of PMMA is negligibly small. The oxygen permeability of the hydrogel lenses depends on their water content in an exponential manner.[67] Therefore, the hydrogel lenses designed for extended wear contain 57–78% water at equilibrium hydration. With increased water content, however, the thickness of the lens has to be increased to maintain mechanical integrity, which in turn decreases permeability.

The thickness of a contact lens usually ranges between 100 and 300 μm. The so-called ultrathin lenses have thicknesses that fall between 40 a 70 μm. Even these are several times thicker than the tear film and are hard to handle. Silicone-type elastomers are known for high oxygen permeability, and this is the primary reason they are considered as contact lens materials. It was initially believed that cellulose acetate butyrate lenses also had a high oxygen permeability. However, controlled studies[86] demonstrated that their oxygen permeability was comparable to that of PHEMA hydrogels, the least oxygen-permeable hydrogel.

Silicone-containing contact lens materials have sufficiently high oxygen permeability to supply adequate oxygen to the epithelial cells. There are clinical signs, such as regression of neovascularization, that indicate the superiority of such lenses in this respect. In the overall clinical performance, however, the silicone lenses do not fare better than other type lenses and in certain respects they may even fare worse.

3.2.4. Lens–Tear Film Interaction

As expected, when a contact lens is placed over the cornea, it will interfere with the functioning of the lacrimal system.[69] Due to its considerable thickness when compared with that of the tear film, the lens will adversely affect the lid–globe congruity, especially if the lens edge is not designed to taper gradually. In a poorly designed lens, the lid will bridge over a portion of the conjunctiva where the ocular mucus layer will not be regularly resurfaced and cleansed from lipid components. Due to the vertical motion of the lid, these areas are at the three o'clock and nine o'clock positions and will show increased vital staining, indicating slightly damaged surface epithelium.

Even if the edges are well designed, there will be a tear meniscus around the periphery of the contact lens. If the tear volume in the eye is limited, at the edge of this meniscus there may also appear a "black line" due to meniscus-induced local thinning. Due to this unsaturated meniscus, tear film formed over the lens often drains, thereby rapidly accelerating its breakup. Once a contact lens has been worn successfully for several minutes, it can be demonstrated that a mucus layer, similar to that on the corneal epithelium, forms on the anterior surface of the lens, thus enhancing its wettability and biocompatibility. In a well-fitted and worm lens, the tear film breakup time of the prelens tear film may be similar to the tear film breakup time measured over the cornea.

The stability of the prelens tear film is quite important for visual acuity, for comfort, and for the prevention of deposit formation on the lens surface. The most common problem in contact lens wear is the formation of a lipid-containing denatured protein deposit on the lens surface. Occasionally the deposit may contain calcium salts.[70] Such deposits interfere with vision, decrease comfort, may induce allergic reaction, and can be harmful to the lens material as well. If the prelens tear film prematurely ruptures over the contact lens, deposit formation will eventually take place.

As mentioned, there is an aqueous tear layer under the lens. Its thickness could vary from 0.2 to 7 μm. When such a postlens tear layer exists, the lens moves a few millimeters with every blink, then quickly centers over the cornea, due to the corneal curvature, which is higher than that of the surrounding conjunctiva. If the disjoining pressure of the postlens tear layer is positive, this lens motion is ensured. Most of the lenses also flex some, resulting in the exchange of the tear trapped under the lens. This pumping action varies for different lenses.[71] It is quite pronounced for hard PMMA lenses and thus contributes to the tolerance of such an impermeable lens by the eye. The pumping is the least for thin hydrogel lenses, which is due to the thinness of the postlens tear layer and the lack of elasticity of the material. The pumping action is quite important. It contributes to the oxygen supply and, even more importantly, removes metabolic products and cellular and other debris from under the lens.

Occasionally the disjoining pressure in the postlens tear layer becomes negative. Then the posterior surface of the contact lens will come into intimate contact with the underlying tissue. The motion of the lens ceases and contact adhesion will take place.

Initially the immobilized lens feels quite comfortable, but this situation is dangerous for the eye. Since there is no more tear pumping, the metabolic waste and dying cells will accumulate under the lens and create a toxic environment for the tissue. This will accelerate cellular breakdown, and the resulting enzymes will degrade the underlying stroma, creating an ulcer. Fortunately, the accompanying severe discomfort usually forces the wearer to have the lens removed. When removed gently, possibly under topical anesthesia, the weakest link will be broken, resulting in the cohesive failure of the epithelium.

The reason for the strong adhesive bond between the lens and the epithelium is the fact that at least one of the adherends is pliable, and close molecular contact an thus be established. The adhesive force at such an interface is known to be stronger than the cohesiveness of at least one of the adherends.[72]

3.2.5. Lens Surface Properties and Biocompatibility

We have seen that the stability of the prelens and postlens tear layers is important for ensuring good biocompatibility. One would conclude that the surface properties of the contact lens will be decisive in affecting biocompatibility, and that the water wettability of the anterior and posterior surfaces of the lens will directly affect the stability of these tear layers. Unfortunately, due to the complex nature of the bioenvironment, the problem of designing a biocompatible lens surface is not so simple.[69]

The minimum interfacial tension hypothesis of biocompatibility states that if the water–solid interfacial tension can be minimized, preferably as close to zero as possible, then the material (prosthesis) will be biocompatible, and even blood compatible.[73] In a study,[74] protein and lipid adsorption by acrylic hydrogels was determined together with an estimate of the water–gel interfacial tension values. The results appear to support the minimum interfacial free energy hypothesis of biocompatibility.

As discussed, water wettability does not yield direct information on the magnitude of water–solid interfacial tension. High-energy solids (such as the aforementioned TiO_2-coated PMMA lens) will be completely wettable by water. However, due to high interfacial tension, protein adsorption and denaturation and lipid adsorption will take place at the interface, initiating deposit formation.

The surface properties of lens materials are usually determined in a clean system by using pure liquids. When the lens is placed in the eye, it is surrounded by tears and may come in contact with the superficial lipid layer or become contaminated with sebum from the fingers. Once the lens is surrounded by tear fluid, an aqueous solution of complex biopolymeric composition, adsorption takes place, which may or may not lead to an equilibrium.[75] When equilibrium is reached, i.e., the adsorption is reversible and self-limiting, the resulting interfacial properties will depend on the composition and molecular configuration of the adsorbate and will likely be quite different from the original surface or even from the original water–solid interface.

In an examination of the adhesion tension versus solution surface tension plots of biopolymeric solutions on various hydrogels, an interesting study demonstrated that the Gibbs surface excess concentration can be negative at the water–gel interfaces of the more hydrophilic hydrogels such as PHEA and PAA.[76]

The adsorption process, at least initially, will depend on the original interfacial properties as well as on the composition of the surrounding medium and the dynamic conditions of the process. Unfortunately, we do not know enough about adsorption processes from such a complex medium as tears to be able to predict the properties of the solid–liquid interface from the original properties of the solid surface after the adsorption of solutes had taken place and equilibrium has been reached.

There are, hever, approximate guidelines gleaned mainly from the study of tear film physiology and that of the ocular surface. It appears that, if given a chance, mucous glycoproteins will preferentially adsorb at the lens–tear interface, thereby lowering the interfacial tension to such a degree that further adsorption of other components is discouraged. The mucous adsorbed layer has the additional advantage of masking lipid contaminants that bombard the surface. The severely contaminated mucus will be removed by the shear forces created by blinking.[58]

Such mucous layers should form both on the anterior and the posterior surface of the lens, thereby stabilizing both the prelens and postlens tear layers. The removal of the contaminated mucus is admittedly more difficult from the posterior surface of the lens. It is fortunate that this posterior surface also receives much less lipid contamination, since the full thickness of the lens separates it from the superficial lipid layer.[63] Such a serendipitous arrangement is probably the reason why so many different types of contact lens material will function fairly well in the eye. Mucous glycoproteins also appear to have negligibly small amphipathic character; thus, the wettability of an adsorbed mucus layer will not depend much on the nature of the surface of the solid adsorbent or that of the surrounding medium.

The aforesaid refers to a well-designed, well-fitted, and well-acclimatized lens that is already in equilibrium with the surrounding tears. The first minutes after the insertion of a contact lens is therefore critical. This is why various wetting and cushioning solutions are used to coat the lens prior to insertion, and this is why the formulation of those solutions and the handling of the lenses may decide the tolerance for the contact lens.

3.3. Adhesion and Erosion of Corneal Epithelium

The cornea is a connective tissue of about 0.5 mm thick. Its external surface is covered with an epithelial layer four to five cell layers thick. Underneath the epithelium, there is a thin basement membrane less than 50 nm in thickness that consists of collagen fils and glycosaminoglycans. The thickest layer of the cornea is called the stroma, which—with the exception of a few keratocytes—is an acellular structure consisting of parallel collagen fibers arranged in layers surrounded by a ground substance of proteoglycans and glycosaminoglycans such as chondroitin, chondroitin sulfate, and keratan sulfate. Descemet's membrane forms the posterior boundary of the stroma. This structure is similar to the basement membrane, *albeit* somewhat thicker. Relative to its thickness it is mechanically the strongest structure of the cornea. The innermost layer of the cornea is a monolayer of endothelial cells that plays an important role in maintaining corneal hydration.

It is of importance that the stroma has an imbibition pressure of about −40 torr, corresponding to a colloidal osmolality of 2 mOsm/kg, including the pressure difference due to the Gibbs–Donnan effect. If the stroma is allowed to imbibe water, then it loses its transparency. When the cornea is edematous, the epithelial layer will also contain intercellular water that interferes with its cohesiveness and possibly with its adhesion to the underlying tissue.[(77)]

Normally the corneal epithelium strongly adheres to the underlying basement membrane. Mechanical scraping will usually damage the basement membrane. In 1969, the author studied the effect of typical abhesive (antonym of adhesive) compounds such as long-chain alkyl alcohols on the adhesion of epithelium. The most appropriate member of the homologue series was *n*-heptyl alcohol, when applied to the corneal surface, decreased epithelial adhesiveness to such a degree that the cells could be wiped off by a dry Q-tip, leaving an intact basement membrane behind. Lower homologues penetrated the stroma due to their higher water solubility, while higher members of the homologues penetrated the epithelium slower. This method of epithelium removal has become popular among researchers for use in experimental animals.[(78)] Occasionally *n*-heptanol is also used in humans by corneal surgeons to remove epithelium, instead of the more destructive cocaine–iodine method that damages the basement membrane, although such use has not yet been reported in the literature.

Recurrent epithelial erosion is a painful condition, which can be quite incapacitating for the patient. Usually as a result of a blink, a few square millimeters of the corneal surface becomes denuded of epithelium. Since the nerve endings of the highly sensitive cornea reside in the epithelium, the creation of such an epithelial defect is quite painful and thus is accompanied by tearing, photophobia, and general demise. The defect heals within 24 to 48 h, when the eye is patched, but the adhesiveness of the new epithelial layer remains impaired. A new defect can occur practically any time and without warning. It is understandable then that this condition can be quite demoralizing to the patient suffering from the disorder.

Some *in vivo* work on epithelial adhesion has been reported.[(79)] The results indicate that the strength of the adhesive joint between the epithelium and the stroma depends on the condition or presence of the basement membrane.

3.3.1. Water as a Bioabhesive Substance

The primary cause of adhesive failure in industrial or everyday practice is the presence of a weak boundary layer at the interface of the adherends or the adhesive and the adherend.[(72)] Such an abhesive layer could consist of trapped gas, adsorbed water, or lipid contamination. It is not surprising that in biological systems, where water and lipids are omnipresent, and where the surface free energy of the boundaries can be changed by subtle changes in surface molecular configurations or composition, a weak boundary layer can readily develop under a certain set of conditions.[77,80]

The stability of a thin layer of aqueous fluid between two solid surfaces will depend on the way the film free energy (sum of the interfacial free energies) varies with the decreasing distance between the solids. This negative energy gradient has been named the disjoining pressure by Derjaguin.[81] When the two solids have hydrophilic surfaces, the disjoining pressure is positive, and considerable amount of external pressure is needed to thin the aqueous layer and bring the two solids together. Under such conditions, the aqueous layer acts as an effective lubricant to prevent adhesion.

On the other hand, when the solid surfaces are hydrophobic, the disjoining pressure will be negative. The aqueous film in this case will spontaneously thin and recede from the gap between the solids. If the solids are pliable, close molecular contact will be established, resulting in "contact adhesion" even in the absence of a liquid adhesive. The immobilization of a contact lens on the corneal surface when the postlens tear layer is unstable is a good example of such a contact adhesion. In biological systems, the conditions for the formation and retention of a weak boundary layer of water are usually favorable even if only one of the adherend surfaces is hydrophilic. In such cases, poor adhesion is observed.

Biosurfaces are usually capable of hydrophilic–hydrophobic transition that can be initiated by the proximity of a stable solid surface or a fluid having a character opposite to that of the solid surface. This complicates the problem and can cause unwanted adhesion or adhesive failure *(vide infra)*.

3.3.2. Wettability of the Corneal Interfaces

The water wettability of the various corneal layers has been measured by using an aqueous dextran solution.[77] The concentration and the molecular weight of dextran were such that the colloidal osmolality of the diagnostic solution was equal to that of the deturgescent stroma in order to avoid swelling of the tissues examined.

The following surfaces were examined in rabbit cornea: (i) surface of the blotted mucus layer over the epithelium, (ii) the demucinized spithelial surface, (iii) the basement membrane (deepithelialized cornea), and (iv) the bare stroma (keratectomized cornea). Table 5 contains the advancing contact angle of the

TABLE 5. The Advancing Contact Angle of Water[a] and the Critical Surface Tension of Various Corneal Interfaces

Surface	Advancing contact angle (deg)	Critical surface tension (mN/m)
Mucus layer	20	38
Epithelial surface[b]	81	32.5
Basement membrane	70	38
Bare stroma	44	38

[a] Five percent dextran solution having a surface tension of 73 mN/m was used to avoid tissue swelling.
[b] One or two epithelial cell layers were removed prior to measurements to ensure the absence of excess mucus.

dextran solution in physiologic saline and the critical surface tension of these interfaces obtained by the Zisman method.[77]

As expected the most wettable surface is the mucus surface. The demucinized epithelial surface is the most hydrophobic. The surface of the basement membrane is almost as hydrophobic as that of the mucin-free epithelia. Even the bare stroma, basically a hydrogel, has a surface that was less hydrophilic than mucin, but much more so than either of the hydrophobic surfaces. The critical surface tension was found to be about the same for all the surfaces except the demucinized epithelium, indicating that the hydrophobic structures of these tissue boundaries are surface chemically similar as far as the dispersion force field is concerned. Demucinized epithelium has a low critical surface tension characteristic of a lipid-rich surface.[33]

3.3.3. Adhesion and Abhesion of the Corneal Tissue Layers

The presence of mechanical interlinking in the form of hemidesmosomes at the adhesive joint of two adjacent biological tissues such as the epithelium and the basement membrane is well established,[82] and the strength of the adhesive biological joint is usually attributed to these anatomical structures. Actually, the small cohesive strength of these structures and their relatively small area of contact may render these cell-coupling structures secondary in importance in determining the stress and shear resistance of the cellular adhesive joint.

In our view, the presence of such tight junctions signifies the absence of a continuous weak boundary layer (i.e., it indicates close molecular contact between structures); therefore, the adhesion should be unimpaired. Thus, desmosomes and hemidesmosomes are more indicators of close molecular contact, and hence good adhesion, than the actual cause of that adhesive strength. Once, in applied engineering, it was also thought necessary to have mechanical interlinking between the two solids (e.g., by roughening and compressing the surfaces) in order to have sufficient adhesive strength. Progress in interface science and rheology of adhesion has all but eliminated this erroneous view from the physical sciences, and hopefully will do so in biological sciences in the foreseeable future.

It has been proposed[77] that the adhesive joint of two adjacent tissue layers such as the corneal epithelium and stroma and, as we shall see in the next section, the retina and the choroid can be represented by a simple model consisting of the two adherent tissues joined by a thin layer of viscous adhesive, which could consist of the unusually thick glycocalyx of the boundary cell layer.[83] In such a system, the resistance to separation, i.e., the adhesiveness, depends only on the rheological properties of the adhesive layer.[72]

The integrity of the adhesive bond in such systems is determined by its *tackiness*, which is the product of stress and the length of time the stress had to be applied to achieve separation.[72] When the stress is created by a pulling force, the tackiness varies inversely with the second power of the adhesive layer thickness and linearly with its viscosity.

When a *flexible* adherend is peeled off, a different relation is obtained.[84] When a tangential shear is applied, such as by the lid motion through the tear film, the integrity of the adhesive joint is determined by the shear resistance of the intertissue adhesive. When the intertissue adhesive is diluted by a thin aqueous layer so that the thickness of the layer will increase while its viscosity decreases to the same order of magnitude as water, the shear resistance will become negligibly small and the adhesive bond will be broken.

If the cause of the recurrent epithelial erosion is indeed the presence of a weak boundary aqueous layer between the epithelium and the basement membrane, then the topical application of a colloidal solution having sufficient osmolality to remove such water, with large enough molecular size for the solutes to prevent penetration of the damaged and therefore "leaky" epithelium, should be of therapeutic value.

This is apparently the case, as has been shown[85] in the testing of high molecular weight dextran solutions with colloidal osmotic pressures of 40–50 torr in patients with severe cases of recurrent epithelial erosion. Such preparations have also been effective when applied in patients suffering from superficial trauma of dendritic keratitis that were accompanied by stromal and epithelial edema. In addition, colloidal osmotic solutions topically applied every 2 h to rabbit corneas denuded of their epithelium effectively diminished corneal edema end accelerated healing.[86]

3.3.4. Epithelial Trauma during Tonography

The aforementioned considerations can be useful when considering other problems that may occur during the application of invasive diagnostic methods, such as tonography, used for testing the outflow facilities of the ocular aqueous chamber in patients suspected of glaucoma. This method is designed to measure the excess pressure inside of the eye by pressing a plastic footplate against the anesthetized cornea and monitoring the decay of the increased intraocular pressure with time.[87]

It is clear that the stability of the tear layer between the footplate and the cornea will be highly important for the protection of the corneal epithelium. If contact adhesion were to occur, epithelial damage would be unavoidable. Actually, after undergoing the procedure, the patients experience a decrease in visual acuity, some discomfort, and corneal edema that may last for days.

Since the mucus layer coating the epithelium is not prone to hydrophilic–hydrophobic conversion, when the hydrophobic footplate is pressed against the cornea, the corneal surface should remain hydrophilic, provided the mucus layer is continuous and the lipid contamination is minimal.

However, this is often not the case. Then a hydrophobic conversion will take place at the corneal surface induced by the proximity of the hydrophobic footplate. The disjoining pressure in the tear layer will become negative, and the footplate will adhere to the cornea. Since minute eye movements are impossible to prevent during the procedure, which lasts several minutes, epithelial cells will be sheared off during the procedure.

It has been shown in human patients that if a drop of aqueous solution of hydrophilic polymers is topically applied to the eye prior to the procedure, the damage to the epithelium and the accompanying unpleasant aftereffects are minimized.[88] Such formulations are expected to be especially effective in mucin-deficient "dry" eyes, where the potential for epithelial damage during tonography is the greatest.

3.4. Retinal Adhesion and Its Failure

Retinal detachment is a fairly common cause of partial or complete blindness, since the retina separated from the choroid loses its blood supply and therefore deteriorates and becomes nonfunctional.

The posterior part of the ocular globe is filled with the vitreous gel that primarily consists of large molecular weight hyaluronic acid and some accompanying proteins. The sensory retina covers the posterior part of the wall of the globe that consists of the choroid and the sclera and is connected to the choroid through a layer of pigmented epithelial cells.

There are no tight junctions between the pigment epithelium and the neural retina. Contact is maintained across the interphotoreceptor matrix (IPM), which is of irregular geometry. Thin (20–50 nm) villous apical projections of the pigment epithelial cells ensheathe the distal half of photoreceptor outer segments. The macromolecular composition of the IPM is over 95%, mostly uncharacterized proteins. The remaining 5% includes at least three types of glycosaminoglycans, namely chondroitin sulfate, sialoglycans, and hyaluronic acid, all of less than 10,000 daltons in molecular weight. It has been hypothesized that these materials can act as a "viscoelastic adhesive," which would oppose retinal detachment.[89]

The adhesion of the retina, and especially the loss of adhesiveness of the retina is of utmost importance to vision. Much, mostly unsuccessful, work has been done to elucidate the mechanism of adhesion and that of adhesive failure that occurs in retinal detachment. It is a generally held belief that the mechanical pressure difference across the retina due to the positive intraocular pressure (IOP) promotes or is even responsible for the adhesion of the retina to the choroid. It can readily be shown by theoretical considerations that the stress required to separate the two membranes is larger if the pressure difference across the retina is negligible. It turns out that the stress of adhesive failure is directly related to the pressure difference created while separating the membranes, and that this pressure difference depends on (a) the rate of detachment, (b) the viscosity and thickness of the intermembrane fluid, and (c) the hydraulic conductivity of the membranes.[84]

In this section, we will briefly review a simple model of retinal adhesion that considers only the retina and the choroid (sclera) as two leaky, elastic, adjacent curved membranes with some intermembrane space between them that contains IPM as an adhesive. This model is useful to estimate the relative importance of the pressure distributions due to mechanical equilibrium and to steady-state flow of fluid, but it does not take into account osmotic forces or possible active transport. This model was first proposed during an ophthalmology seminar at Yale in 1972 and has not yet been submitted for publication.[90]

The IOP is defined as the pressure difference between the interior of the eye and the atmospheric pressure. In the normal eye, it ranges from 12 to 18 torr. Thus, there is a pressure drop of this magnitude the retina and the choroid (sclera) of the eye. The model of retinal adhesion thus may consist of two concentric, elastic, spherical membranes that have finite hydraulic conductivities. The inner membrane is the retina with an elasticity modulus E_1, hydraulic conductivity K_1, and thickness d_1. The outer membrane is the choroid with corresponding parameters

E_2, K_2, and d_2. Let the stresses, due to the pressure distribution in these two membranes, be S_1 and S_2. (Fig. 4).

The way the hydrostatic pressure difference, IOP, will be distributed between the two membranes will depend on their respective elasticities. As a first approximation, we may assume the retina and the choroid to be thin membranes so that linear stress distribution can be assumed. Let the pressure in the intermembrane region be P'. We may then write

$$\text{for the retina} \qquad S_1 = \frac{(\mathrm{IOP} - P')\,R}{2d_1} \tag{9}$$

$$\text{for the choroid} \qquad S_2 = \frac{P'R}{2d_2} \tag{10}$$

where R is the radius of the curved membranes assumed to be spherical. If we assume that Hooke's law holds and the strains (y) are approximately the same, then

$$S_1 = E_1 y \qquad \text{and} \qquad S_2 = E_2 y \tag{11}$$

Then by combining these equations, we obtain the intermembrane pressure

$$P' = \frac{E_2 d_2 \cdot \mathrm{IOP}}{E_1 d_1 + E_2 d_2} \tag{12}$$

For the special case where the thicknesses of the membranes are comparable ($d_1 = d_2$):

$$P' = \frac{E_2 \cdot \mathrm{IOP}}{E_1 + E_2} \tag{13}$$

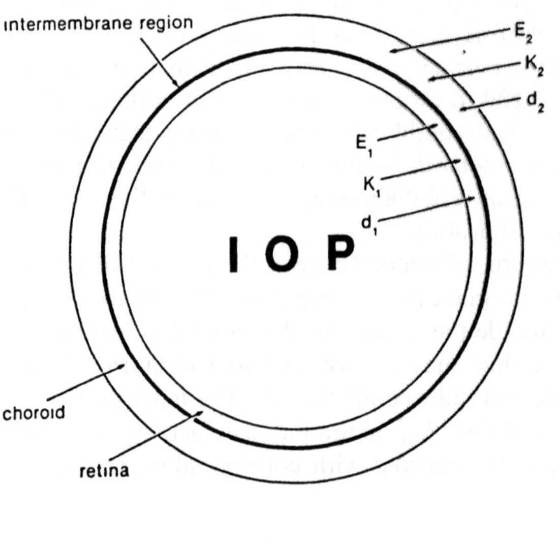

FIGURE 4. Double-membrane model of retinal adhesion to choroid. The membranes are both elastic and have finite hydraulic conductivities: E = elastic modulus; K = hydraulic conductivity; d = thickness.

If the retina and the choroid have hydraulic conductivities, K_1 and K_2, respectively, the water flux at steady state through each membrane can be written as

$$J_1 = \frac{K_1(\mathrm{IOP} - P'')}{vd_1} \tag{14}$$

$$J_2 = \frac{K_2 P''}{vd_2} \tag{15}$$

where v is the viscosity of the fluid and P'' is the steady-state pressure between the membranes. In a steady state

$$J_1 = J_2 \tag{16}$$

and, hence,

$$P'' = \frac{K_1 d_2 \cdot \mathrm{IOP}}{K_1 d_2 + K_2 d_1} \tag{17}$$

When the thicknesses of the two membranes are comparable,

$$P'' = \frac{K_1 \cdot \mathrm{IOP}}{K_1 + K_2} \tag{18}$$

Now we have two pressure values for the intermembrane region. Pressure P' results from the stress distribution at mechanical equilibrium and is given by the ratio of elasticities according to Eq. (13). Pressure P'' results from the steady-state hydraulic flow and is given by the ratio of hydraulic conductivities according to Eq. (18). It is clear that to prevent fluid accumulation or depletion in the intermembrane region, the following has to hold:

$$\frac{K_1}{K_1 + K_2} = \frac{E_2}{E_1 + E_2} \tag{19}$$

Since elasticities and hydraulic conductivities are two independent properties of membranes, it is clear that the probability that Eq. (19) holds is exceedingly small. Instead, there is probably an inequality between those ratios. If

$$\frac{K_1}{K_1 + K_2} < \frac{E_2}{E_1 + E_2} \tag{20}$$

then $P'' < P'$. Thus, fluid will be depleted from the intermembrane region, and this process favors adhesion. If the inequality of Eq. (20) is reversed, then $P'' > P'$; fluid accumulates in the intermembrane region, and adhesion is impaired.

If the elasticity of the retina is smaller than that of the choroid, $E_1 < E_2$, and this is indeed the case, then most of the stress will be borne by the choroid. For this reason the retina will be pressed against the choroid by the IOP, and thus its adhesion to the underlying tissue will be further strengthened, especially if the thickness of the intermembrane region is small.

The hydraulic conductivity and the thicknesses of the retina and choroid are known and can be used to calculate the magnitude of P'', the subretinal pressure. Since the hydraulic conductivity of the retina is 20,000 times greater than that of the choroid, and since the choroid is also six times thicker than the retina, the subretinal pressure is practically identical to that of the intraocular pressure. If, e.g., the IOP = 18.0000 torr, then $P'' = 17.9994$ torr. Indeed, attempts to measure the subretinal pressure *in vivo* by ingenious techniques could not detect any difference between that and IOP.[91]

While the elasticity of the retina is not known accurately, an estimate of the relative elasticities of the retina and choroid suggests that actually the reverse of inequality (20) is true; that is,

$$\frac{K_1 d_2}{K_1 + K_2} > \frac{E_2 \cdot d_1}{E_1 d_2 + E_2 d_1} \tag{21}$$

so it is expected that fluid would accumulate in the intermembrane region from hydraulic flow alone. The observed colloidal osmolality of this region would also likely be higher than that of the choroidal side or the retinal side, thus further contributing to the dilution of the intermembrane adhesive. Hence, an "active" pump is probably needed to maintain the membrane region at a relatively low degree of hydration, to enhance adhesiveness by the higher viscosity and smaller intermembrane thickness.

The characteristics of the adhesive failure of the retina and choroid junction have been studied at various rates of pull in rhesus monkeys.[84] Interestingly, it was observed that when the animal was sacrificed during the traction experiment, within minutes after death the traction force drastically diminished. This finding appears to imply that an energy-dependent process plays a role in adhesion, e.g., active transport of water out of the intermembrane region as discussed above.

Such a simple double-membrane model, which has been presented here, does not include all the factors that appear to be operative in the living system. However, it clearly defines the role of membrane elasticity and hydraulic conductivity, and predicts the need for an active process of water removal from the intermembrane region. The model also predicts the way changes in hydraulic conductivity or elasticity could promote or impair adhesion in the absence of major deteriorative pathological changes.

3.5. *Wetting in Intraocular Surgery*

An interesting side effect during the implantation of intraocular lenses is directly related to hydrophilic–hydrophobic transition, instability of aqueous films, and contact adhesion. Similar phenomena have been shown to contribute to tissue damage in abdominal surgery when elastic sterile surgical gloves were used, which are hydrophobic.

The implantation of a PMMMA intraocular lens in the eye following cataract extraction has been a major advance in ophthalmic techniques to improve vision. These plastic lenses have fixed focal lengths and are thus unable to accommodate. The same is true for thick contact lenses designed for aphakic eyes, or the thick spectacles that traditionally had to be worn after the cataract removal. Actually, the crystalline lens of the eye also loses its ability to accommodate after the age of 60 years.[92]

As we have already mentioned, if an aphakic spectacle of around +12 diopters is worn, it considerably increases image size to such a degree that, if only one eye is aphakic, the images cannot be fused by the brain and double vision will occur. The image size difference will be considerably smaller with contact lenses. However, elderly people often have trouble in manipulating, inserting, or removing contact lenses. With an intraocular lens (IOL), there is no difference in image size. Even if additional visual correction is needed for either near of far vision, it is only minor and will have no major effect on the image size.

Once an intraocular lens is implanted, it is imperative that the lens be biocompatible with the surrounding fluid and tissue. An inflammatory reaction should not be elicited, and the lens–fluid interfacial tension should be low enough so that protein adsorption, denaturation, and eventual accumulation does not take place, and thus the clarity of the lens is not jeopardized.

3.5.1. Contact Adhesion between IOL and Corneal Endothelium

In addition to the aforesaid, another, originally unforseen, complication may arise during the implantation of an IOL. The endothelial cell layer constituting the posterior surface of the cornea may be damaged during implantation to such a degree that the cornea becomes edematous, and a corneal transplant (penetrating keratoplasty) could eventually be required.

The corneal endothelium consists of a monocellular layer with negligible mitotic activity. Occasional cell loss is compensated by the spreading of the neighboring endothelial cells to preserve the continuity of the cell layer. Once the continuity of the endothelial layer is impaired, the cornea loses its deturgescent state and becomes permanently edematous, resulting in a dramatic decrease of visual acuity.

The surgical trauma to the corneal endothelium could be directly traced to intimate contact with the IOL. Endothelial cell fragments could be seen attached to the PMMA lens surface by electron microscopy. Lubricating the IOL with various aqueous solutions occasionally proved to be helpful, especially if they were able to form a stable continuous aqueous layer over the lens.[93] Unfortunately, such solutions made the prostheses quite slippery and therefore were difficult for the surgeon to handle.

In view of the foregoing discussion, it is quite clear what must transpire during the implantation procedure of an IOL. The PMMA surface is hydrophobic, even when it is presoaked in water (receding contact angle of water is 50°). The aqueous humor is relatively poor in proteins; e.g., its oncotic pressure is less than 0.2 torr.[94] The endothelial surface of the cornea is not protected by a mucus layer; thus is expected to exhibit considerable hydrophilic–hydrophobic transition.

When the hydrophobic surface of the IOL comes into close proximity with the endothelium, surface transformation will take place, converting the endothelial surface into a hydrophobic one. The fluid layer between the two solids then becomes unstable due to the negative disjoining pressure thus created and will spontaneously recede. The lens and the endothelium will then establish an intimate molecular contact assisted by the pliability of the tissue. Once the contact adhesion is established, attempts to break it when the lens is moved into place will result in the rupture of the weakest link, and cohesive failure in the endothelium will take place, thus destroying the cells.

Solutions capable of wetting the IOL completely will drastically lower the lens–water interfacial tension. Such an interface then will not elicit a hydrophobic conversion of the endothelium. The intersolid fluid film remains stable, effectively lubricating the sliding surfaces and protecting the endothelium provided the shear forces are moderate and the viscosity of the fluid film is low.

These considerations are valid not only for the IOL implantation but for practically any intraocular surgical procedure that involves the anterior segment of the eye, since contact adhesion can occur between surgical instruments and vulnerable tissue. The harmful effects of irrigating solutions, air, and other media also cannot be ruled out. The average endothelial cell loss after a routine cataract extraction has been reported to range between 8% and 13%, compared with the average endothelial cell loss range of 14% to 64% following IOL implantation.[95–97]

3.5.2. Viscoelastic Surgical Aids

Since the protection of the endothelium is of the utmost importance during intraocular surgery involving the anterior segment of the eye, considerable applied research has been conducted concerning the role of natural biopolymers, more specifically certain glycosaminoglycans. These substances would act as lubricants and abhesives, and also as spacers separating tissues and maintaining the anterior chamber once the globe is penetrated and the intraocular pressure is lost.

The first, and most often used, substance is sodium hyaluronate (Healon), the major constituent of the vitreous humor. This material is also found in rooster combs and umbilical cords of the newly born.[98] Healon is used in a 1% solution, which is clear and highly viscous. Since its viscosity is far from Newtonian, surgery involving the use of such materials has been termed *viscoelastic* surgery. Other materials have started to appear, such as Viscoat, which contains sodium hyaluronate and sodium chondroitin sulfate.[99] Even such a mundane preparation as the aqueous solution of methyl cellulose has been tried[100] and now is available commercially.

While these preparations undoubtedly make the surgery easier by decreasing manipulations within the anterior chamber, and thereby lessening endothelial damage, there are still problems associated with their use. Temporary elevation of intraocular pressure following surgery has been commonly observed. Some damage to the endothelium still occurs, the extent of which depends on the skill of the surgeon.

A scanning of the numerous articles and books on this subject[101] reveals that the research in this area is highly empirical and that the surface chemistry and rheology of the problem are not clearly understood. Increased viscosity is usually viewed as a desired property, while the well-known adverse effect of increased viscosity on lubrication is ignored. It should be possible to design a solution–gel system that would be biocompatible and protect the endothelium and other delicate intraocular structures by increasing lubrication during manipulation. It would also possess sufficient rigidity in the absence of shear to act as a spacer in the anterior chamber. As awareness of such concepts increases among applied researchers, further rapid progress resulting in more efficacious products can be expected.

3.5.3. Role of Contact Adhesion in Other Areas of Surgery

It is of interest that despite clear evidence indicating the importance of wetting and the damaging effect of contact adhesion in surgery, such considerations are ignored by surgeons and by manufacturers of surgical instruments and aids. A comparative study of canine abdominal surgery between surgeons using highly hydrophobic rubber gloves and bare hands clearly indicated that due to contact adhesion between the hydrophobic gloves and the intestinal surfaces, the tissue damage was considerable compared to the lack of such damage when the surgeons did not use gloves.[102] One may suspect that unwanted tissue adhesion following surgery could also be avoided if the irrigation solutions, sponges, and other surgical aids would be designed with such considerations in mind.

4. CONCLUSIONS

In this chapter, the importance of wetting and its role in adhesion was reviewed in ophthalmology. Even though the topic was restricted to a relatively small area of medicine and biology, it appears that the significance of such considerations would be high in practically any area of medicine, dentistry, and cellular biology, where intracellular processes, as well as cellular interactions will greatly depend on wetting. In these processes, the properties of the vicinal water and their influence on microwetting, including the formation and rupture of the hydration shell, will become the dominating factor.

As progress in biology resulted in the shift in emphasis from cellular biology to molecular biology, interestingly enough such biophysical and surface chemical considerations have been neglected rather than gaining acceptance. Perhaps it is human nature, or the limitation of the human mind that approaches problems only in a certain way, using specific techniques and concepts, to obtain and interpret results. The unquestioning loyalty to a scientific discipline or a school of thought appears to exclude a multidisciplinary approach and the consideration of another scientific discipline, especially if its techniques and concepts tend to point out potential conflicts and weakness in the accepted schools of thought.

As federal support becomes more and more important in basic and applied research, and as the granting agencies heavily rely on the opinion of study committees consisting almost exclusively of faithful disciples of the present school of thought in vogue, the expansion of our knowledge does not increase in proportion to expenditure. Daring, pioneering, innovative proposals are rarely funded, and related papers are difficult or impossible to publish.

As Szent-Györgyi stated:[103] "There can be no doubt about the wonderful achievements of the molecular approach (to life sciences). However, the molecular is but one of the three dimensions involved in the mechanism of life: the molecular, the infra- and the supra-molecular. Life can be understood only by their combination." It is perhaps puzzling why water and wetting, despite their enormous significance in living systems, have received relatively little attention from biologists. According to Szent-Györgyi, possibly the correct answer was given by Sir Oliver Lodge, who stated that the very last thing a deep-sea fish would discover is water. Possibly the omnipresence of the substance in the living system has restrained us from giving it sufficient attention.

In this chapter, the progress in ophthalmology based on the surface chemical and rheological considerations of certain problems has been reviewed. It is hoped that these modest, but successful, attempts will inspire other scientists well versed in these principles to initiate biologically related research fully exploiting their expertise, thereby contributing to progress and to our understanding of these complex but highly important and fascinating systems.

REFERENCES

1. C. F. Hazelwood, in *Cell-Associated Water* (W. Drost-Hansen and J. Clegg, eds.), p. 165, Academic Press, New York (1979).
2. W. Drost-Hansen, *Ind. Eng. Chem.* **61**, 10 (1969).
3. J. T. Davies and E. K. Rideal, *Interfacial Phenomena*, 2nd ed., p. 369, Academic Press, London (1963).
4. R. A. Peters, *Proc. Soc. A1* **33**, 140 (1931).
5. F. J. Holly, D. W. Lamberts, and J. A. Buesseler, *Plastic Reconstr. Surg.* **74**(3), 438 (1984).
6. J. T. Davies and E. K. Rideal, *Interfacial Phenomena*, 2nd ed., p. 39, Academic Press, London (1963).
7. J. T. Davies and E. K. Rideal, *Interfacial Phenomena*, 2nd ed., p. 35, Academic Press, London (1963).
8. J. T. Davies and E. K. Rideal, *Interfacial Phenomena*, 2nd ed., p. 50, Academic Press, London (1963).
9. D. J. Doughman, F. J. Holly, and C. H. Dohlman, *Preliminary Studies on the Stability of Aqueous Films* on Solids, paper presented at the Assoc. Res. Vis. Ophthalmol. Annual Mtg., Sarasota, FL. (1971).
10. J. Padday, *Spec. Discuss Faraday Soc.* **1**, 64 (1970).
11. F. J. Holly and T. Rázgha, Meniscus-induced thinning in confined films (unpublished study).
12. A. W. Adamson, *Physical Chemistry of Surfaces*, 4th ed., p. 12, Wiley, New York (1982).
13. J. E. McDonald and S. Brubaker, *Am. J. Ophthalmol.* **72**, 139 (1971).
14. D. Maurice, Stanford University (personal communication).
15. J. T. Davies and E. K. Rideal, *Interfacial Phenomena*, 2nd ed., p. 24, Academic Press, London (1963).
16. W. D. Harkins, *The Physical Chemistry of Surfaces*. Reinhold, New York, 1952.
17. A. W. Adamson, *Physical Chemistry of Surfaces*, 4th ed., p. 103, Wiley, New York (1982).
18. I. Langmuir, *J. Chem. Phys.* **1**, 756 (1933).
19. I. Langmuir and V. J. Schaeffer, *Chem. Rev.* **24**, 181 (1939).
20. J. Blodgett, *J. Opt. Soc. Am.* **24**, 313 (1934).
21. F. J. Holly and M. F. Refojo, in *Hydrogels for Medical and Related Applications* (J. D. Andrade, ed.), ACS Symp. Ser. No. 31, p. 252 Am. Chem. Soc., Washington, DC (1976).
22. W. C. Hamilton, *J. Colloid Interface Sci.* **40**, 219 (1972).
23. H. W. Fox and W. A. Zisman, *J. Colloid Sci.* **5**, 514 (1950).

24. F. J. Holly in *Physicochemical Aspects of Polymer Surfaces* (K.L. Mittal, ed.), Vol. 1, p. 141, Plenum, New York (1983).
25. F. J. Holly and M. F. Refojo, *J. Biomed. Mater. Res.* **9**, 315 (1975).
26. F. J. Holly, unpublished data.
27. F. J. Holly and M. F. Refojo, in *Colloid and Interface Science* (M. Kerker, ed.), Vol. 3, p. 321, Academic Press, New York (1976).
28. N. K. Adam, in *The Physics and Chemistry of Surfaces*, p. 180, Dover, New York (1968).
29. H. Yasuda and A. K. Sharma, *J. Polym. Sci. Polym. Phys. Ed.* **19**:1285 (1981).
30. H. Yasuda, H. C. March, S. Brandt, and C. N. Reilly, *J. Polym. Sci. Polym. Chem. Ed.* **15**, 991 (1977).
31. B. D. Ratner, P. K. Weatherby, A. S. Hofman, M. A. Kelly, and L. H. Sharpen, *J. Appl. Polym. Sci.* **22**, 643 (1978).
32. F. J. Holly and M. J. Owen in *Physicochemical Aspects of Polymer Surfaces* (K.L. Mittal, ed.), Vol. 2, p. 625, Am. Chem. Soc., Washington, DC (1982).
33. F. J. Holly and M. A. Lemp, *Exp. Eye Res.* **11**, 39 (1971).
34. F. J. Holly, in *Wetting, Spreading, and Adhesion*, (J. Padday, ed.), p. 439, Academic Press, London (1978).
35. R. E. Baier, R. C. Dutton, and V. L. Gott, *Adv. Expl. Med. Biol.* **7**, 235 (1970).
36. M. A. Lemp, F. J. Holly, S. Iwata, and C. H. Dohlman, *Arch. Ophthalmol.* **83**, 89 (1970).
37. J. S. Andrews, in *The Preocular Tear Film and Dry Eye Syndromes* (F. J. Holly and M. A. Lemp, ed.), *Int. Opthalmol. Clin.* **13**(1), p. 23, Little, Brown, & Co., Boston (1973).
38. F. J. Holly, *Exp. Eye Res.* **15**, 515 (1973).
39. F. J. Holly, *Trans. Ophthalmol. Soc. UK* **104**, 374 (1985).
40. F. J. Holly, *Contact Intraocl. Lens Med. J.* **4**(3), 52 (1978).
41. F. J. Holly, *J. Colloid Interface Sci.* **49**, 221 (1974).
42. R. E. Berger and S. Corrsin, *J. Biomech.* **7**, 225 (1974).
43. M. G. Doane, *Am. J. Opthalmol.* **89**, 507 (1980).
44. S. I. Brown and D. G. Derjaguin, *Arch. Opthalmol.* **82**, 537 (1969).
45. S. Iwata, M. A. Lemp, F. J. Holly, and C. H. Dohlman, *Invest. Ophthalmol.* **8**, 613 (1969).
46. D. Dartt, I. Knox, A. Palau, and S. Y. Botelho, *Invest. Ophthalmol Vis, Sci.* **19**(11), 1342 (1980).
47. A. Berta, *Am. J. Ophthalmol.* **96**, 115 (1983).
48. P. T. Janssen and O. P. van Bijsterveld, *Invest. Ophthalmol. Vis, Sci.* **24**, 623 (1983).
49. R. N. Stuchell, J. J. Feldman, R. L. Farris, and I. D. Mandel, *Invest. Opthalmol. Vis, Sci.* **25**(3), 374 (1984).
50. F. J. Holly and B. S. Hong, *J. Optom. Physiol. Opt.* **59**(1), 43 (1982).
51. F. J. Holly, K. Dołowy, and K. W. Yamada, *J. Colloid Interface Sci.* **100**(1), 210 (1984).
52. P.-K. Tsung, B.-S. Hong, and F. J. Holly, *Suppl. Invest. Ophthalmol. Vis, Sci* (ARVO Abstracts), p. 212 (1981).
53. S. Iwata and I. Kabasawa, *Exp. Eye Res.* **12**, 360 (1971).
54. C. W. Chao, J. P. Vergnes, and S. I. Brown, *Exp. Eye Res.* **36**, 139 (1983).
55. I. Koral, R. Ullman, and F. Enrich, *J. Phys. Chem.* **62**, 541 (1958); *Clincs*, **13**(1), 97, Little, Brown, Boston (1973).
56. F. J. Holly, in *The Preocular Tearfilm in Wealth, Disease, and Contact Lens Wear* (F. J. Holly, D. W. Lamberts and D. L. MacKeen, eds.) p. 634, Dry Eye Institute, Lubbock (1986).
57. S. Mishima and D. M. Maurice, *Exp. Eye Rres.* **1**, 39 (1961).
58. F. J. Holly, in *The Preocular Tear Film and Dry Eye Syndromes* (F. J. Holly and M. A. Lemp, ed.), *Int. Opthalmol. Clinics* **13**(1), 73, Little, Brown, Boston (1973).
59. J. E. McDonald, *Am. J. Ophthalmol.* **76**, 56 (1969).
60. S. P. Lin and H. Brenner, *J. Colloid Interface Sci.* **89**, 226 (1982).
61. A. Sharma and E. Ruckenstein, *J. Colloid Interface Sci.* **106**, 12 (1985).
62. M. S. Norn, *Acta Ophthalmol. (Kbh)* **43**, 557 (1965).
63. F. J. Holly, *Am. J. Optom. Physiol. Opt.* **58**, 324 (1981).
64. A. Baszkin, M. M. Boissonnade, J. E. Proust, S. D. Tschaliovska, L. Ter-Minassian-Saraga, and G. Wajs, *J. Bioeng.* **2**, 527 (1978).
65. M. F. Refojo, in *Modern Plastic Encyclopedia*, Vol. 54, No. 10A, p. 653, McGraw-Hill, New York (1975).

66. F. J. Holly and M. F. Refojo, in *Proc. Int. Conf. Coll. Surf. Sci.* (E. Wolfram and T. Szekrényessy, eds.), Vol. 2, Akadémiai Kiadó, Budapest, Hungary (1976).
67. F. J. Holly and M. F. Refojo, *J. Am. Optom. Assoc.* **43**(11), 1173-1180 (1972).
68. M. J. Refojo, F. J. Holly, and F. L. Leong, *Cont. Intraocul. Med. J.* **3**, 27 (1977).
69. F. J. Holly, *Colloids Surf.* **10**, 343 (1984).
70. G. E. Lowther and J. A. Hilbert, *Am. J. Optom. Physiol. Opt.* **52**, 687 (1975).
71. K. A. Polse, *Invest. Opthalmol. Vis, Sci.* **18**, 209 (1979).
72. J. J. Bikerman, *Trans. Soc. Theol.* **1**, 3 (1957).
73. J. D. Andrade, *Med. Instrum. (Arlington)* **7**(2), 110 (1973).
74. F. J. Holly, *J. Polym. Sci. Polym. Symp.* **66**, 409 (1979).
75. J. E. Proust, A. Baszkin, and M. M. Boisonnade, *J. Colloid Interface Sci.* **94**, 421 (1983).
76. F. J. Holly, and M. F. Refojo, in *Hydrogels for Medical and Related Applications* (J. D. Andrade, ed.), ACS Symp. Ser., No. 31, p. 267, Am.Chem. Soc., Washington, DC (1976).
77. F. J. Holly, *Invest. Opthalmol. Vision Sci.* **17**, 552 (1978).
78. C. Cintron, L. Hassinger, C. L. Kublin, and J. Friend, *Ophthalmic Res.* **11**, 90 (1979).
79. A. A. Khodadoust, A. M. Silverstein, K. M. Kenyon, and J. E. Dowling, *Am. J. Opthalmol.* **65**, 339 (1968).
80. R. E. Baier, E. G. Shafrin, and W. A. Zisman, NRL Report #6691, Naval Research Laboratory, Washington, DC (1968).
81. B. V. Derjaguin, *Trans. Faraday Soc.* **36**, 203 (1940).
82. J. N. Goldman and T. Kuwabara, *Int. Opthalmol. Clin.* **8**, 561 (1968).
83. K. Dołowy and F. J. Holly, *J. Theoret. Biophys.* **75**, 373 (1978).
84. H. deGuillebon, and F. J. Holly, *Invest. Opthalmol.* **11**, 46 (1972).
85. G. N. Foulks, *Ophthalmology* **88**, 801 (1981).
86. D. W. Lamberts, G. N. Foulks, and F. J. Holly, Effect of colloidal osmotic solution on the hydration of de-epithelialized cornea in the rabbit (as manuscript).
87. Alcon Applanation Pneumatonograph®, Alcon Laboratories, Inc., Fort Worth, TX 76101.
88. D. W. Lamberts, Y Do, and F. J. Holly, Prevention of epithelial trauma during tonography (as manuscript).
89. A. J. Adler and K. M. Severin, *Exp. Eye Res.* **32**, 755 (1981).
90. F. J. Holly, Double membrane model of retinal adhesion (unpublished work).
91. J. Salmon, H. Zauberman, and D. M. Maurice, *Exp. Eye Res.* **12**, 212 (1971).
92. R. A. Moses, *Adler's Physiology of the Eye: Clinical Applications*, 15th ed., p. 368, C. V. Mosby, St. Louis (1970).
93. G. A. Peyman and K. Zweig, *Am. J. Ophthalmol.* **87**(4), 561 (1979).
94. C. Kirkland, E. D. Esquivel, and F. J. Holly, Oncotic pressure and protein composition of aqueous humor (as manuscript).
95. W. M. Bourne and H. E. Kaufman, *Am. J. Ophthalmol.* **82**, 44 (1976).
96. L. W. Hirst, R. C. Snip, W. J. Start, and A. E. Maumenee, *Am. J. Opthalmol.* **84**, 775 (1977).
97. W. M. Bourne and H. E. Kaufman, *Am. J. Opthalmol.* **81**, 482 (1976).
98. E. A. Balázs, U.S. Patent 4,141,973 (1979).
99. CILCO, Inc. Viscoat®®, *Product Monograph*, CILCO, Huntington, WV (1984).
100. P. U. Fechner and M. U. Fechner, *Br. J. Opthalmol.* **67**, 259 (1983).
101. D. Miller and R. Stegmann, eds, Healon, A Guide to Its Use in Ophthalmic Surgery, Wiley, New York (1983).
102. E. Goldberg, University of Florida, personal communication.
103. A. Szent-Györgyi, in *Cell-Associated Water* (W. Drost-Hansen and J. Clegg, eds.), p. 1, Academic Press, New York (1979).

9

Wettability of Surfaces in the Oral Cavity

H. J. Busscher

1. THE IMPORTANCE OF SURFACES IN THE ORAL CAVITY

The importance of surfaces in the oral cavity is enormous. Both the tooth surface and the soft mucosal tissues form a substratum for a large variety of microorganisms.[1] Bone and periodontal tissue adhere to root surfaces in order to ensure good and permanent anchorage of teeth.[2] If cavities occur, restorative materials have to adhere to the surfaces of the cavity walls[3] to create a good sealing.

The adhesion of microorganisms to surfaces in the oral cavity is generally preceded by adsorption of salivary constituents, forming a proteinaceous organic film called the acquired pellicle.[4] Although by definition, the pellicle is essentially free of microorganisms, it forms a major element in the adhesion, as well as growth and colonization, of microorganisms to teeth and, thus, in the formation of dental plaque. Plaque is recognized as the main etiological cause for caries and periodontal diseases. In some people plaque calcifies extremely rapidly, leading to the adhesion of dental calculus[5] to enamel and root surfaces, which creates another stimulus for periodontal diseases.

Periodontal problems arise as a result of bacterial infections of the pockets between root surfaces and periodontium, originating from loss of periodontal tissue attachment. Here the importance of root surfaces as a substratum for cellular adhesion and spreading is obvious.

If irreversible damage has been done to enamel and/or dentine surfaces due to decalcification by acids produced by adhering microorganisms (caries), the resulting cavities are filled with restorative materials such as amalgams or composites. Microleakage along the cavity walls can lead very easily to bacterial adhesion under a restoration, causing so-called secondary caries, which is practically impossible to detect by the dentist until pain develops. Good adhesion between the restorative

H. J. Busscher • Laboratory for Materia Technica, University of Groningen, 9713 AV Groningen, The Netherlands

Modern Approaches to Wettability: Theory and Applications, edited by Malcolm E. Schrader and George Loeb. Plenum Press, New York, 1992.

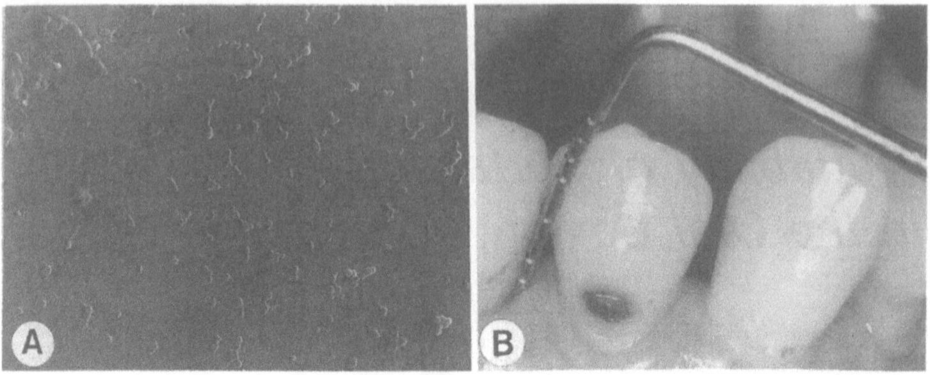

FIGURE 1. Illustrations of various adhesion phenomena in the oral cavity: (A) Initial adhesion of oral bacteria (diameter $\sim 10^{-3}$ mm) to an enamel surface. (B) Needle probing the attachment of gingival tissue to a root surface, as common in periodontal practice and adhesion of a restorative material to a cavity wall (amalgam filling).

material and cavity wall is required, whereas bacterial adhesion to the restorative materials should be small.

In summary, we have distinguished four distinct adhesion phenomena in the oral cavity:

1. Bacterial adhesion to pellicle-coated enamel and dentine surfaces, restorative materials, and soft tissues
2. Adhesion of calcified plaque (calculus) to tooth surfaces
3. Adhesion of periodontal tissue cells to root surfaces
4. Adhesion of restorative materials to enamel and dentine

Whereas the former two phenomena stimulate oral diseases and are therefore unwanted, the latter two are clearly required. The above phenomena are illustrated in Fig. 1.

The importance of surface science in the oral cavity becomes clear from the above discussion and was first really recognized in the 1960s. Pioneering work on the wettability of dental and oral surfaces has been done by Glantz[6, 7] and Baier,[8] leading eventually to the active interest in the application of surface science in the oral cavity existing nowadays.

2. WETTABILITY OF ORAL SURFACES

In this section, contact angle data on oral surfaces from the literature are summarized and discussed. As far as the wetting liquids are concerned, the data presented are restricted to water as a polar probing liquid and α-bromonaphthalene as an almost completely apolar liquid.[9] The oral surfaces to be discussed can be subdivided into these classes:

1. Enamel and dentine surfaces (*in vitro* and *in vivo*)
2. Oral microbial and mucosal surfaces
3. Surfaces of restorative materials

2.1. Enamel and Dentine Surfaces

A major constraint for the application of colloid and surface science in dental research is the limited size of available enamel and dentine specimens. Most investigators use material of whale, bovine, or, sometimes, human origin. The limited size of available samples, typically 3 mm × 3 mm, explains the popularity of the sessile drop method[10] among dental researchers.

Table 1 presents a summary of literature data on the wettability of enamel and dentine surfaces by water and α-bromonaphthalene.

Water contact angles reported for polished enamel appear independent of the origin of the enamel and somewhat higher than reported for hydroxyapatite (25°[11–20]). This is most likely due to the influence of small spots of the organic matrix exposed at the enamel surface, in which the hydroxyapatite crystals are embedded.[21]

The wettability of the enamel surface depends greatly on the amount of material removed by grinding and polishing prior to contact angle measurements.[14] In most studies the enamel surface with a fully matured pellicle present is slightly more hydrophobic than of polished enamel. Exposure to salivary proteins decreases the water contact angles within minutes to approximately 30°, after which time interval the pellicle slowly matures by selective adsorption and conformational changes of adsorbed proteins to an essentially more hydrophobic protein layer with a water contact angle around 60° (see also Fig. 2). Only one study presents low contact

TABLE 1. Summary of Literature Data on the Wettability of Enamel and Dentine Surfaces by Water and α-Bromonaphthalene (α-br)

Material	Contact angle (deg)		Reference
	Water	α-br	
BE[a] polished	51	25	11
BE polished	50	—	12
HE[b] polished	40	16	7
HE polished	54	—	12
HE polished	48	26	13
HE −150 μm from anatomical surface	47	—	14
HE −6 μm from anatomical surface	53	—	14
HE intact	60	—	14
HE *in vivo* clinically registered	21–28	—	15
HE *in vivo* clinically registered	64	30	16
HE with 5-min pellicle	28	20	17
HE with 2-h pellicle	31	21	17
with 7-d pellicle	65	28	17
WD[c] polished	46	18	7
HD[d] polished	45	17	7
HD polished	60	23	18
HD with 3-h pellicle	43	31	19

[a] bovine enamel
[b] human enamel
[c] whale dentine
[d] human dentine

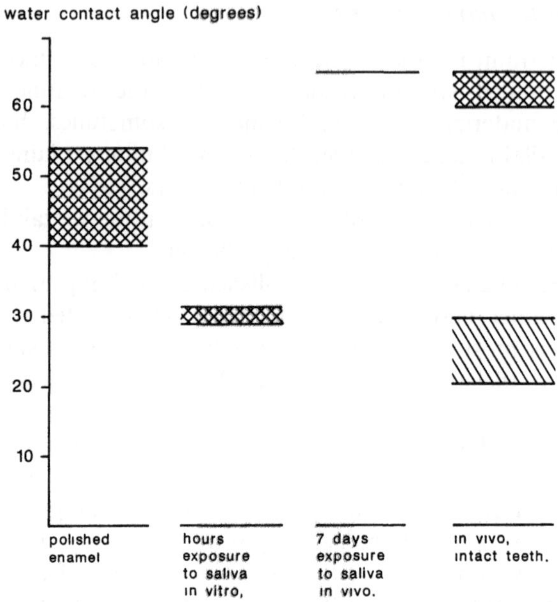

FIGURE 2. Water wettability of enamel as a function of the exposure time to saliva. Data taken from Table 1.

angles between 20° and 30° for clinically registered water contact angles on intact teeth *in vivo*, probably because contact angles were measured immediately after wetting the teeth with saliva by means of the tongue and no special precautions were taken to remove free water and/or unbound proteinaceous substances.

The changes in water contact angles on enamel with exposure time to saliva as shown in Fig. 2 are in accordance with the general notion in dentistry that, although a pellicle forms within hours,[22] its complete maturation takes several days.[23]

At this stage it is also important to note that drying of specimens to remove unbound water is required in order to measure physiologically relevant contact angles on biological substances. This point was stressed several years ago by Van Oss[24, 25] for lawns of bacteria, which need approximately $1\frac{1}{2}$ h before stable contact angles can be recorded dependent on the experimental conditions.[26] For protein layers on solid substrata, similar drying times have been reported.[27] As such prolonged drying times are impossible to apply *in vivo* for the clinical registration of contact angles in voluntary test persons, Perdok *et al.*[16, 28] have suggested the use of negative pressure by means of a vacuum suction tip, held just above the tooth surface. Using this procedure, the time required to obtain stable, so-called plateau contact angles in test persons is reduced to approximately 1 min.

Water contact angles on dentine surfaces are slightly higher than on enamel, which could be a reflection of the higher content of organic material in dentine.

2.2. Oral Microbial and Mucosal Surfaces

For many of the oral adhesion phenomena previously mentioned, oral microbial cell surfaces are as important as the enamel and dentine surfaces. Although van

Oss[24, 25] pioneered the field of contact angle measurements on nonoral microorganisms along time ago, it is only recently that oral microorganisms have become the subject of extensive contact angle measurements.[26, 29–31] As a brief outline of the method, bacteria are stacked up to 50–100 layers on membrane filters using negative pressure. Subsequently, contact angles are measured as a function of time to monitor the drying process. Stable, plateau contact angles are usually found after 1–3 h drying. These angles are considered throughout as the physiologically relevant ones.

Table 2 lists some of the available contact angles on oral microorganisms[30–32] and oral mucosa.[33] The collection of oral streptococci shown is predominantly hydrophilic, with *Streptococcus mitis* being a clear exception. The water contact angle on *S. mitis* strains is astonishingly high and approaches values measured on waxy materials. Also within the collection of candida strains reported by Minagi *et al.*[32] some strains appear extremely hydrophobic. For a collection of 6–10 oral streptococcal strains, x-ray photoelectron spectroscopy on freeze-dried cells has demonstrated that the origin of the high water contact angles is essentially its high protein content at the surface.[34, 35] For the candida strains, the origin of the extreme hydrophobicities is unknown.

Data on the wettability of oral mucosa are scarce, and the only data observed in the literature were from Vissink *et al.*,[33] studying the wettability of mucosal surfaces by artificial saliva substitutes.

2.3. Surfaces of Restorative Materials

Amalgams, gold alloys, and poly(methyl methacrylate) are the traditional restorative materials used in dentistry. Nowadays amalgams are frequently replaced by composites, mainly for cosmetic reasons. Table 3 lists wettability data for several of the above-mentioned materials. All restorative materials without any exception exhibit a moderate wettability rather than extreme hydrophobicity or hydrophilicity, which might present a compromise between low plaque formation and a good adhesion to enamel and dentine surfaces (see Section 4).

TABLE 2. Summary of Literature Data on the Wettability of Oral Microbial and Mucosal Surfaces

Type of surface	Contact angle (deg)		Reference
	Water	α-br	
S. mutans	34–54	35–57	30
S. mutans	16–32	—	31
S. sanguis	25–47	30–56	30
S. sanguis	37–44	—	31
S. mitis	66–111	33–47	30
Candida stellatoidea	46	—	32
C. albicans	51	—	32
C. krusei	94	—	32
C. tropicalis	119	—	32
Human mucosa *in vivo*	70–75	—	33

TABLE 3. Summary of Literature Data on the Wettability of Dental Restorative Materials

Materials	Contact angle (deg) Water	α-br	Reference
Amalgam	40		7
Amalgam	69–73	—	31
Au–Ag–Pd alloy	83	—	31
Gold			7
Poly(methyl methacrylate)	76	15	36
Silar	66	—	31
P-10	64	—	31
Clearfil F3	65	—	31
Microrest	69	—	31
Concise	53	15	37
Concise White	65	10	37
Delton Sealant	69	19	37
Prisma Shield	77	12	37
Helioseal	60	13	37

3. SURFACE FREE ENERGIES OF ORAL TISSUES

The pathway from measured contact angles to estimated surface free energies is slippery but necessary if one wants to make progress in a thermodynamic approach towards various adhesion phenomena. No decisive breakthrough has been made on this point ever since the introduction of the empirical concept of critical surface tensions by Zisman,[38] although various useful methods are reported nowadays, among which are

- Several equation of state approaches[39, 40]
- The concept of dispersion and polar components[41] based on either a geometric, harmonic, or geometric–harmonic mean,[42] with or without spreading pressures taken into account[43–46]
- A recent recognition of van Oss *et al.*,[47–49] in which a so-called Lifshitz–van der Waals surface free energy component, representing London, Debye, and Keesom interactions is distinguished next to a so-called acid–base component, representing hydrogen-donating and hydrogen-accepting interactions

The issue of spreading pressures is especially crucial for the estimation of enamel surface free energies. Slow cleavage experiments[50] and analysis of dislocation cores in hydroxyapatite[51] have indicated independently from wettability data that the surface free energy of hydroxyapatite is approximately 100 mJ m^{-2}. Hence, we decided in the past that for the estimation of enamel surface free energies it was necessary to account for the spreading pressure term. Following Dann,[43, 44] we employed the geometric mean equation

$$\gamma_l(\cos\theta + 1) = 2(\gamma_s^d \gamma_l^d)^{1/2} + 2(\gamma_s^p \gamma_l^p)^{1/2} - \pi_e \quad (1)$$

in which θ is the contact angle, γ_l and γ_s are the liquid and solid surface free energies, respectively, with d and p denoting dispersion and polar, respectively, and π_e is the equilibrium spreading pressure, defined as

$$\pi_e = \gamma_s - \gamma_{sv} \tag{2}$$

where γ_{sv} is the solid surface free energy against a saturated vapor of the liquid employed. Generally, π_e is neglected without much proof, but its numerical value may become considerable for high-surface-free-energy solids, and some authors[52, 53] indicated long ago that spreading pressures of water may assume appreciable values even on Teflon. If π_e is neglected, Eq. (1) can, in principle, be solved when the contact angles of two liquids are known, as only two unknowns are involved. If, however, π_e is taken into account, the number of unknowns involved is always more than the number of available contact angle values, as, in principle, π_e depends on the solid–liquid combination. Because for a large variety of materials the spreading pressures of water and *n*-propanol are about equal as demonstrated ellipsometrically,[46] we have suggested that this problem could be solved in a first-order approximation by using water, water–*n*-propanol mixtures, and α-bromonaphthalene as wetting agents. Fig. 3 shows a comparison between accepted literature values for the surface free energies of a series of solids obtained by least-square fitting of contact angles with the above-mentioned wetting agents to Eq. (1), assuming π_e constant. Clearly, deviations only occur for surface free energies higher than 45–50 mJ m^{-2}, which makes the method extremely suitable for the characterization of both low as well as of high-surface-free-energy materials.

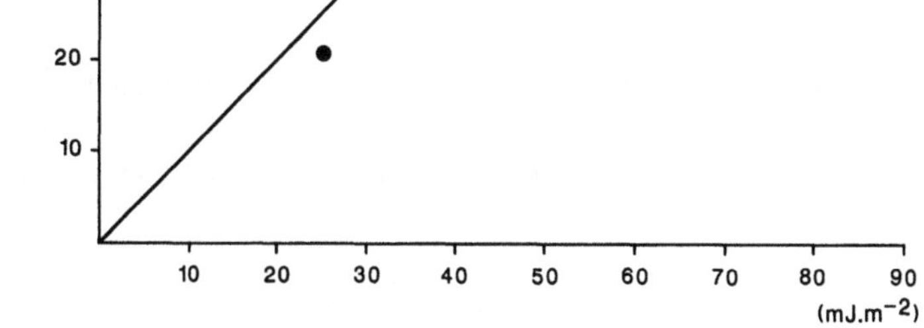

FIGURE 3. A comparison of solid surface free energies estimated from contact angle data employing the approach of polar and nonpolar components with and without spreading pressures taken into account: *y* axis: Values calculated from Eq. (1) using pure liquid contact angles while neglecting spreading pressures. These values correspond to commonly accepted literature values. *x* axis: Values calculated from Eq. (1) using contact angles measured with water, water–*n*-propanol mixtures, and *α*-bromonaphthalene while accounting for spreading pressure. Note the much larger *x* axis compared with the *y* axis.

TABLE 4. Surface Free Energies (in mJ m^{-2}) of Surfaces with Dental Relevance Estimated from the Concept of Dispersion and Polar Components with Spreading Pressures Taken into Account

Material	γ_s	Reference
HE, polished	88	13
HE, *in vivo* clinically registered	77	28
HE, with 2-h pellicle	110	17
HD, polished	92	19
S. mutans	104	30
S. sanguis	102	30
S. mitis	45	30
Poly(methyl methacrylate)	56	45
Concise	89	37
Concise White	71	37
Delton Sealant	67	37
Prisma Shield	56	37
Helioseal	73	37

See Eq. (1) employing contact angles of water, water–*n*-propanol mixtures, and α-bromonaphthalene from Refs. 45 and 46.

Table 4 summarizes surface free energies of materials with dental relevance, estimated in this way. The wide range of surface free energies that can be treated in this way is due to the fact that spreading pressures were not neglected.

4. ORAL ADHESION PHENOMENA AND SURFACE FREE ENERGIES

As pointed out in Section 1, the major relevance of surface free energies in dentistry is for adhesion phenomena. Thermodynamically, adhesion between any two materials 1 and 2 in a medium 3 is energetically favorable if the free energy of adhesion

$$\Delta F_{\text{adh}} = \gamma_{12} - \gamma_{13} - \gamma_{23} \tag{3}$$

is most negative, presuming all other interactions can be neglected.[54]

Using the approach based on separation of dispersion and polar components and on the geometric mean equation, the interfacial free energies in Eq. (3) can be expressed as[26]

$$\gamma_{ij} = \gamma_i + \gamma_j - 2(\gamma_i^{\text{d}}\gamma_j^{\text{d}})^{1/2} - 2(\gamma_i^{\text{p}}\gamma_j^{\text{p}})^{1/2} \tag{4}$$

Fig. 4 demonstrates linear relations between the interfacial free energy of adhesion and the adhesion of three strains of oral streptococci to bare solid substrata,[55] including dental enamel.[56] The cause of the different slopes for the various strains lies within their different microbial cell surface properties and has been discussed in detail previously.[57, 58] Although Fig. 4 convincingly shows the role of interfacial thermodynamics in bacterial adhesion *in vitro*, it also immediately raises the

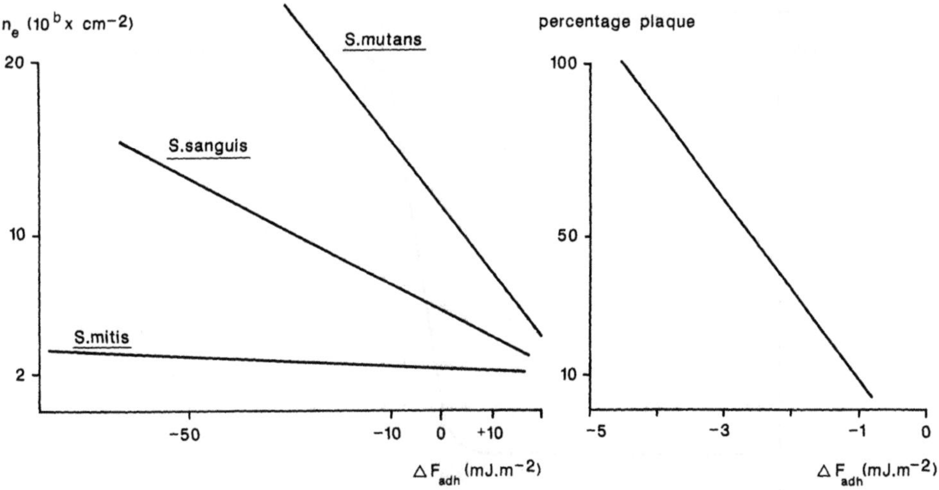

FIGURE 4. The role of interfacial thermodynamics in oral bacterial adhesion *in vitro*[55, 56] and plaque formation *in vivo*.[59] Numbers of adhering bacteria as well as the percentage plaque coverage of teeth *in vivo* are given as a function of the interfacial free energy of adhesion. For bacterial adhesion, ΔF^{adh} has been calculated on basis of the bare substratum surface free energies and those of the streptococcal strains,[30] whereas for plaque the mean surface free energies of the strains isolated from the plaques were used.[60].

question whether these relations hold under *in vivo* conditions in the oral cavity, with its continuous presence of saliva, fluctuating shear forces, and bacterial multiplication. In studies by Glantz[6] and Quirijnen *et al.*,[59] low-surface-free-energy materials *in vivo* accumulated significantly smaller amounts of plaque than the high-surface-free-energy tooth surface. The composition of the plaque accumulated on various materials was the same at the species level, albeit that low-surface-free-energy materials attracted microorganisms with a relatively low surface free energy.[60] By consequence, plaque formation *in vivo* obeys interfacial thermodynamics in a way similar to bacterial adhesion *in vitro* (see also Fig. 4). From this persistent influence of bare substratum surface properties in the presence of an adsorbed protein film, we concluded in the past that substratum properties are transferred by the adsorbed protein layer to the interface with adhering cells and bacteria.[61]

In many people, plaque calcifies extremely rapidly, especially on the front incisors, and regular scaling or ultrasonic treatment by a dentist is required to remove calculus. Despite the fact that calculus formed in only slightly smaller quantities on polytetrafluoroethylene (PTFE)–Teflon than on teeth, its removal was much easier from low- than from high-surface-free-energy substrata.[62]

Fig. 5 illustrates the relation observed between cell spreading and substratum surface free energies.[63] Such relations hold for various cell types both in the absence and in the presence of preadsorbed protein films,[64] again pointing to a transferral of substratum properties by selective adsorption or conformational changes of adsorbed proteins induced by the surface free energy of the substrata.

Finally, as the last example of oral adhesion phenomena, we present the tensile

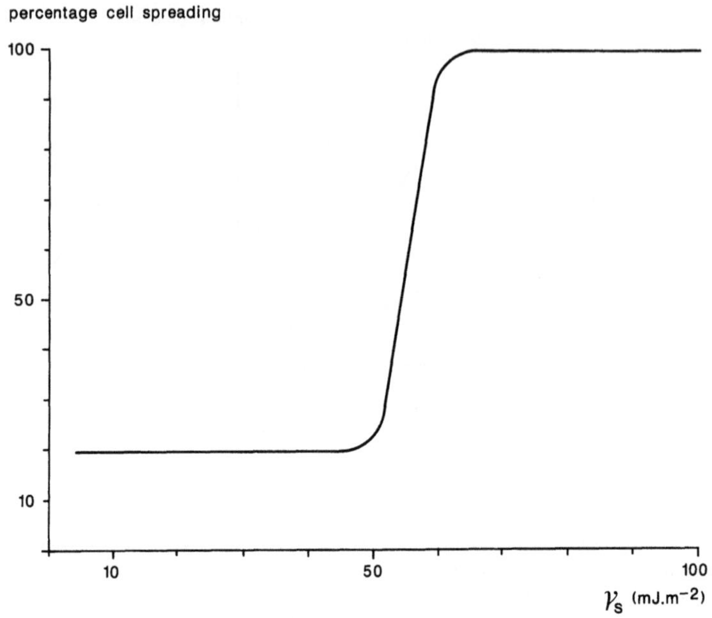

FIGURE 5. Spreading of human skin fibroblasts on various solid surfaces as a function of substratum surface free energies.[63, 64]

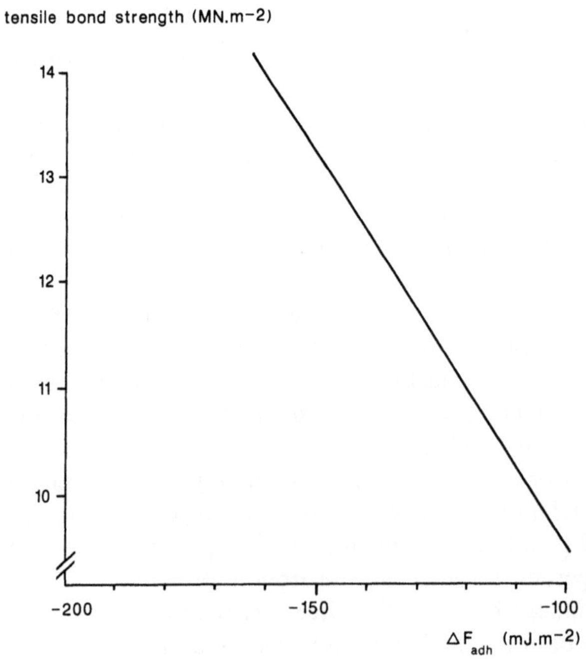

FIGURE 6. Tensile bond strengths to etched enamel of various composite restorative materials as a function of the interfacial free energy of adhesion.[37]

bond strength of composite restorative materials to enamel as a function of the interfacial free energy of adhesion, ΔF_{adh}[37] (see Fig. 6). This relation, too, is in complete accordance with interfacial thermodynamics.

5. *MODIFICATION OF THE PHYSICOCHEMICAL PROPERTIES OF TOOTH SURFACES*

In previous sections, it has become obvious that changing the wettability of teeth by appropriate surfactants can yield great progress in dentistry. Presently, however, there is not much use of surface science in clinical dentistry. Yet the possibilities are numerous:

- Application of protective, low-surface-free-energy coatings on caries-susceptible sites of teeth and molars could locally reduce the occurrence of bacterial adhesion and plaque formation.
- Front incisors could be coated to create conditions for easier removal of calculus.
- Spreading and adhesion of periodontal tissues after periodontal surgery on root surfaces could be enhanced by increasing the surface free energy of dentine.
- Firmer adhesion of restorative materials to the enamel, but also a reduction of plaque formation on these restorations, could be established by the application of surface science.

Most dental problems would be avoided, if not solved, if plaque formation could be prevented. As a preventive measure, additional to tooth brushing, many dentifrices and mouth rinses contain surfactants with an antimicrobial effect,[65] such as, e.g., chlorhexidine,[66] cetyl pyridiniumchloride,[67] or aminefluorides.[68] The main working mechanism of these products is killing the bacteria rather than decreasing their adhesion to teeth, and although most products adsorb onto the enamel surface and more or less profoundly change the enamel surface properties,[16, 69] aminefluoride is presently the only surfactant yielding a sizable enamel surface free energy reduction from 88 mJ m^{-2} to values between 35 and 52 mJ m^{-2}.[70]

6. *CONCLUDING REMARKS*

Summarizing, the importance of wettability in the oral cavity has been demonstrated in relation to different adhesion phenomena, and a literature survey has been given. The possibilities of changing the wettability of oral surfaces to improve oral health conditions and as an additional preventive measure have been outlined.

ACKNOWLEDGMENTS

H. J. Busscher gratefully acknowledges Mrs. Marjon Schakenraad-Dolfing for the preparation of this manuscript.

REFERENCES

1. R. J. Gibbons, in *Microbial Adhesion to Surfaces* (R. C. W. Berkeley, J. M. Lynch, J. Melling, P. R. Rutter, and B. Vincent, eds.), pp. 351–388, Ellis Horwood, England (1980).
2. B. F. Lowenberg, J. E. Aubin, D. A. Deporter, J. Sodek, and A. H. Melcher, *J. Dent. Res.* **64**, 1106 (1985).
3. D. H. Retief, *J. Oral Rehabil.* **1**, 265 (1974).
4. K. H. Eggen and G. Rölla, in *Cariology Today* (B. Guggenheim, ed.), pp. 109–110, Karger, Basel (1984).
5. H. E. Schroeder, in *Formation and Inhibition of Dental Calculus*, Hans Huber, Vienna (1969).
6. P. O. Glantz, *J. Colloid Interface Sci.* **37**, 281 (1971).
7. P. O. Glantz, *Odontol. Revy* **20**, 5 (1969).
8. R. E. Baier, *Swed. Dent. J.* **1**, 261 (1977).
9. F. M. Fowkes, *Ind. Eng. Chem.* **56**, 41 (1964).
10. A. W. Neumann and R. J. Good, in *Surface and Colloid Science* (R. J. Good and R. R. Stromberg, eds.), vol. 2, pp. 31–91, Plenum Press, New York (1979).
11. H. P. de Jong, *Surface Free Energies of Enamel–In Vivo and In Vitro*, Thesis, Groningen, The Netherlands (1984).
12. T. Bartels, A. W. J. van Pelt, H. P. de Jong, and J. Arends, *Caries Res.* **16**, 51 (1982).
13. A. W. J. van Pelt, H. P. de Jong, H. J. Busscher, and J. Arends, *J. Biomed. Mater. Res.* **17**, 637 (1983).
14. H. P. de Jong, A. W. J. van Pelt, and J. Arends, *J. Dent. Res.* **61**, 11 (1982).
15. M. D. Jendresen, R. E. Baier, and P. O. Glantz, *J. Colloid Interface Sci.* **100**, 233 (1984).
16. J. F. Perdok, H. C. van der Mei, M. J. Genet, P. G. Rouxhet, and H. J. Busscher, *Caries Res.* **23**, 297 (1989).
17. H. P. de Jong, P. de Boer, H. J. Busscher, A. W. J. van Pelt, and J. Arends, *Caries Res.* **18**, 408 (1984).
18. H. P. de Jong, P. de Boer, A. W. J. van Pelt, H. J. Busscher, and J. Arends, *J. Periodontal Res.* **19**, 540 (1984).
19. H. J. Busscher, H. M. Uyen, and J. Arends, in *Dentine and Dentine Reactions in the Oral Cavity* (A. Thylstrup, S. A. Leach, and V. Qvist, eds.), pp. 225–234, IRL Press, England (1987).
20. H. J. Busscher, H. P. de Jong, and J. Arends, *Mater. Chem. Phys.* **17**, 553 (1987).
21. J. Arends and W. L. Jongebloed, *Swed. Dent. J.* **1**, 215 (1977).
22. T. Sönju and G. Rölla, *Caries Res.* **7**, 30 (1973).
23. M. J. Levine, L. A. Tabak, M. Reddy, and I. D. Mandel, in *Molecular Basis of Oral Microbial Adhesion* (S. E. Mergenhagen and B. Rosan, eds.), pp. 125–130, Am. Soc. Microbiology, Washington, DC (1985).
24. C. J. van Oss, C. F. Gillman, and A. W. Neumann, *Phagocytic Engulfment and Cell Adhesiveness*, Marcel Dekker, New York (1975).
25. C. J. van Oss and C. F. Gillman, *J. Reticuloendothel. Soc.* **12**, 23 (1972).
26. H. J. Busscher, A. H. Weerkamp, H. C. van der Mei, A. W. J. van Pelt, H. P. de Jong, and J. Arends, *Appl. Environ. Microbiol.* **48**, 980 (1984).
27. J. van Dijk, F. Herkströter, H. Busscher, A. Weerkamp, H. Jansen, and J. Arends, *J. Clin. Periodontol.* **14**, 300 (1987).
28. J. F. Perdok, A. H. Weerkamp, L. J. van Dijk, J. Arends, and H. J. Busscher, *J. Dent. Res.* **67**, 701 (1988).
29. H. C. van der Mei, A. H. Weerkamp, and H. J. Busscher, *J. Microbiol. Methods* **6**, 277 (1987).
30. A. W. J. van Pelt, H. C. van der Mei, H. J. Busscher, J. Arends, and A. H. Weerkamp, *FEMS Microbiol. Lett.* **25**, 279 (1984).
31. J. Satou, A. Fukunaga, N. Satou, H. Shintani, and K. Okuda, *J. Dent. Res.* **67**, 588 (1988).
32. S. Minagi, Y. Miyake, Y. Fujioka, H. Tsuru, and H. Suginaka, *J. Gen. Microbiol.* **132**, 1111 (1986).
33. A. Vissink, H. P. de Jong, H. J. Busscher, J. Arends, and E. J. 's-Gravenmade, *J. Dent. Res.* **65**, 1121 (1986).
34. H. C. van der Mei, A. J. Leonard, A. H. Weerkamp, P. G. Rouxhet, and H. J. Busscher, *Colloids Surf.* **32**, 297 (1988).
35. H. C. van der Mei, A. J. Leonard, A. H. Weerkamp, P. G. Rouxhet, and H. J. Busscher, *J. Bact.* **170**, 2462 (1988).

36. H. J. Busscher, A. W. J. van Pelt, P. de Boer, H. P. de Jong, and J. Arends, *Colloids Surf.* **9**, 319 (1984).
37. H. J. Busscher, D. H. Retief, and J. Arends, *Dent. Mater.* **3**, 60 (1987).
38. W. A. Zisman, in *Contact Angle, Wettability and Adhesion* (R. F. Gould, ed.), pp. 1–51, Am. Chem. Soc., Washington, DC (1964).
39. A. W. Neumann, R. J. Good, C. J. Hope, and M. Sejpal, *J. Colloid Interface Sci.* **49**, 291 (1974).
40. D. F. Gerson, *Colloid Polym. Sci.* **260**, 539 (1982).
41. D. K. Owens and R. C. Wendt, *J. Appl. Polym. Sci.* **13**, 1741 (1969).
42. S. Wu, *J. Adhesion* **5**, 39 (1973).
43. J. R. Dann, *J. Colloid Interface Sci.* **32**, 302 (1970).
44. J. R. Dann, *J. Colloid Interface Sci.* **32**, 321 (1970).
45. H. J. Busscher, A. W. J. van Pelt, H. P. de Jong, and J. Arends, *J. Colloid Interface Sci.* **95**, 23 (1983).
46. H. J. Busscher, G. A. M. Kip, A. van Silfhout, and J. Arends, *J. Colloid Interface Sci.* **114**, 307 (1987).
47. C. J. van Oss, R. J. Good, and M. K. Chaudhury, *J. Colloid Interface Sci.* **111**, 378 (1986).
48. C. J. van Oss, R. J. Good, and M. K. Chaudhury, *Adv. Colloid Interface Sci.* **28**, 35 (1987).
49. C. J. van Oss, M. K. Chaudhury, and R. J. Good, *Chem. Rev.* **88**, 927 (1988).
50. M. Aning, D. D. Welch, and B. S. H. Royce, *Phys. Lett.* **37A**, 253 (1971).
51. J. Arends and W. L. Jongebloed, *Caries Res.* **11**, 186 (1977).
52. M. E. Tadros, P. Hu, and A. W. Adamson, *J. Colloid Interface Sci.* **49**, 184 (1974).
53. P. Hu and A. W. Adamson, *J. Colloid Interface Sci.* **59**, 605 (1977).
54. D. R. Absolom, F. V. Lamberti, Z. Policova, W. Zingg, C. J. van Oss, and A. W. Neumann, *Appl. Environ. Microbiol.* **46**, 90 (1983).
55. I. H. Pratt-Terpstra, A. H. Weerkamp, and H. J. Busscher, *J. Gen. Microbiol.* **133**, 3199 (1987).
56. I. H. Pratt-Terpstra, A. H. Weerkamp, and H. J. Busscher, *J. Dent. Res.* **68**, 463 (1989).
57. I. H. Pratt-Terpstra, A. H. Weerkamp, and H. J. Busscher, *Curr. Microbiol.* **16**, 311 (1988).
58. I. H. Pratt-Terpstra, A. H. Weerkamp, and H. J. Busscher, *J. Colloid Interface Sci.* **129**, 568 (1989).
59. M. Quirynen, M. Marechal, H. J. Busscher, J. Arends, P. L. Darius, and D. van Steenberghe, *J. Dent. Res.* **68**, 796 (1989).
60. A. H. Weerkamp, M. Quirijnen, M. Marechal, H. C. van der Mei, D. van Steenberghe, and H. J. Busscher, *Microb. Ecol. Health Dis.* **2**, 191 (1989).
61. H. J. Busscher, A. H. Weerkamp, H. C. van der Mei, D. van Steenberghe, M. Quirijnen, I. H. Pratt-Terpstra, M. Marechal, and P. G. Rouxhet, *Colloids Surf.* **42**, 331 (1989).
62. H. M. Uyen, L. J. van Dijk, and H. J. Busscher, *J. Clin. Periodontol.* **16**, 9 (1989).
63. P. van der Valk, A. W. J. van Pelt, H. J. Busscher, H. P. de Jong, Ch. R. H. Wildevuur, and J. Arends, *J. Biomed. Mater. Res.* **15**, 807 (1983).
64. J. M. Schakenraad, H. J. Busscher, Ch. R. H. Wildevuur, and J. Arends, *Cell Biophys.* **13**, 75 (1988).
65. M. Pader, in *Surfactants in Cosmetics* (M. M. Rieger, ed.), pp. 293–348, Marcel Dekker, New York (1985).
66. P. Gjermo, *J. Clin. Periodontol.* **1**, 143 (1974).
67. J. Llewelyn, *Br. Dent. J.* **148**, 103 (1980).
68. H. Schmid, *Dtsch. Zahnaerztl. Z.* **1/83**, 45 (1983).
69. H. C. van der Mei, J. F. Perdok, M. Genet, P. G. Rouxhet, and H. J. Busscher, *Clinical and Preventive Dentistry* **12**, 25 (1990).
70. H. J. Busscher, H. M. Uyen, H. P. de Jong, G. A. M. Kip, and J. Arends, *J. Dent.* **16**, 166 (1988).

10

Wettability as a Surface Signal for Sessile Aquatic Organisms

James W. Mihm and George Loeb

1. INTRODUCTION

The consequences of biological colonization affect shipping and other industrial activities, such as offshore oil recovery, cooling loops for power stations, and mariculture. Difficulties arise from increased resistance to flow past the colonized surfaces and through the apertures of pens that confine cultured organisms, reduced heat conductivity, and biologically influenced corrosion. Colonization in such contexts is known as *biofouling*. On the other hand, mariculture of oysters and mussels requires that the larvae of these organisms attach at the desired locations. Although discussions of these phenomena are beyond the scope of this chapter, they are sufficiently important commercially to motivate serious efforts to understand and control such colonization.

Populations of both microbial forms and of larger organisms exist in aquatic environments. The primary defense against undesired aquatic organisms in the past has been the use of toxins and disinfectants, but environmental problems are now receiving the attention of the public and of regulatory authorities, which in turn motivates increased interest in the understanding of nontoxic influences on bioattachment. One influence on the behavior of organisms is their physicochemical interaction with the attachment substratum.

2. AQUATIC ORGANISMS

Aquatic organisms may be free-living for the major portion of their life cycle, or they may belong to the group whose members attach and spend the subsequent part of their life cycle fixed to their chosen site. For this permanently attached type

---

James W. Mihm and George Loeb • David Taylor Research Center, Ship Materials Engineering Department, Annapolis, Maryland 21402.

Modern Approaches to Wettability: Theory and Applications, edited by Malcolm E. Schrader and George Loeb. Plenum Press, New York, 1992.

of organism, one important aspect of survival is location and occupancy of a suitable home site. Because immense numbers of larvae are produced and released into the surrounding waters, they seemed sufficient to occupy suitable habitats on a purely chance basis. Therefore, it was thought that random settlement of larvae on all available surfaces occurred, with colonization of suitable surfaces (and thus survival) ensured by force of overwhelming numbers. However, as investigations of larval behavior and colonization progressed, it became apparent that preferences for different materials were exhibited by the settling larvae, and the random settlement postulate was not sufficient to explain the observed phenomena.[1]

2.1. Reversible Attachment

There are two basic patterns of attached life. One may be described as optional, and is to some degree reversible. In this mode an organism may attach to a surface and remain for a considerable time, but no drastic change in its anatomical form occurs. If the attachment is broken, either by an outside force or by the organism itself if conditions become unsuitable, the detached organism is capable of resuming a mobile existence, and perhaps eventually finding another site for colonization. Many marine microbes fit into this category.

2.2. Permanent Attachment

The second mode involves a much more drastic process of metamorphosis, in which the organism undergoes an irreversible change to a form that the uninitiated observer would not recognize as related to the larva. Part of this irreversible change is the secretion of a permanent attachment cement or structure, which must be broken by an external force before the organism can be removed from the substratum to which it has attached. (Although it may be possible for the organism to survive under very favorable conditions after this trauma, it is incapable of swimming or exploring the environment as it could before its initial attachment.) Therefore, finding a suitable site for its permanent home is a vital function for the larva of such a creature.

3. WETTABILITY

The criteria for site selection include a number of environmental variables, such as current velocity, orientation of the substratum, surface roughness, concavity or convexity, color, light intensity, and water chemistry. However, this chapter is concerned with the surface physical chemistry of the substratum.

The wettability of surfaces is determined in terms of the contact angles with liquids, commonly designated θ, as described in Chapters 1 and 5. The work of adhesion may also be used and is given by the formula

$$W^{A} = \gamma_{L}(1 + \cos\theta) \tag{1}$$

Here γ_L is the surface tension of the liquid used and may be easily measured by any of several well-known techniques.[2] A commonly used liquid is water, or the aqueous medium pertinent to the biological situation.

The quantity W^A represents the reversible work required to separate unit area of the liquid–solid interface so as to form unit areas of solid and liquid surfaces. As shown in Chapters 1 and 5, the contact angle formed on a solid by a particular liquid may be predicted if the surface energies of the liquid and the solid surface are sufficiently well known. The γ_c parameter of Zisman was an empirical indication of the surface tension of a liquid that just wets and spreads on a solid surface. A low value of this parameter indicates a relatively hydrophobic nonwettable material with low surface energy, while a high value indicates a hydrophilic easily wetted high-surface-energy solid. Zisman and co-workers developed correlations between the chemical composition of solids and the value of γ_c.[3]

Zisman's group was also active in development of the concept that the γ_c parameter was a valuable parameter characterizing surface effects in biology and medicine. Baier, while working in Zisman's group at the Naval Research Laboratory in Washington, DC (see the preface), and later in Buffalo, NY, developed correlations between the value of γ_c and biological interactions with surfaces.[4, 5]

Initially working with tissue cells, he found a relationship of cell spreading to γ_c, with a minimum in the region of $20 < \gamma_c < 30$ mN/m, called the *biocompatible range*. Although the empirical nature of the γ_c parameter hindered rigorous interpretation of the correlation, coincidence of this value with the γ_c of water implied that this was a significant relationship. The concept of resolution of the surface energy into van der Waals and polar components allowed identification of the γ_c parameter with the van der Waals component of solid surface energy,[6] because the low γ_c liquids used to define the low-energy region of the Zisman plot are van der Waals liquids. Schrader[7] considered this relationship and postulated that a minimum of adhesive strength with low-energy materials in water should occur at values of γ_c of 20–30 mJ/m^2 because this minimizes van der Waals interactions in aqueous media. The γ_c of silicone rubber is in this biocompatible range, which is consistent with a surface layer of close-packed methyl groups, which in turn arises from the minimum free-energy configuration of the dimethyl silicone polymer system. (Many medical implants are based on silicone materials.) Other classes of materials whose surface energies lie in this range are close-packed $-CH_3$ terminal fatty acids and amines and certain partially fluorinated materials.

Several investigations have examined the relationship between colonization of immersed surfaces and wettability, or solid surface energy, in a number of contexts. Microorganisms and the larger invertebrates such as barnacles, mussels, and bryozoa have been studied, as well as some algae.

4. *MICROORGANISMS*

The marine and freshwater bacteria have been extensively studied.[8] These studies have established definite preferences of certain types of organisms for surfaces of particular surface energy, or wettability, ranges.

The demonstrated utility of Zisman's γ_c in the context of adhesion and cell spreading led to a natural extension to the aquatic environment. A study by Dexter *et al.* of attachment of bacteria to surfaces of various γ_c values immersed in the sea[9] indicated a dependence of bacterial attachment on surface energy. Although not apparent at short times after immersion, a minimum at this range of γ_c (or van der Waals surface energy component values) was noticed after a day of immersion.

Other studies of bacterial attachment also showed a strong dependence on the substratum wettability. Thus, Fletcher and Loeb[10] found that a marine pseudomonad showed great preference for hydrophobic rather than hydrophilic substrata. Loeb,[11] in a study performed during an oceanographic cruise in the Atlantic and Caribbean, found that hydrophobic surfaces (polystyrene) were somewhat more subject to colonization by marine bacteria than surfaces prepared by treating the polystyrene material with an electric glow discharge or by sulfonation in the manner used to encourage tissue cell adhesion. Such treatments result in a considerable decrease of contact angles, especially receding angles. However, no overwhelming quantitative difference in colonization was observed; there was less than an order of magnitude difference between the populations attached to treated (hydrophilic) versus untreated (hydrophobic) surfaces.

Further exposure of treated and untreated polystyrene in Atlantic coastal waters revealed differences in the populations attached to these samples. Strains were isolated from each material, which, when cultured in the laboratory, maintained their preferences for the substratum type from which they were originally isolated.[12] Fletcher and Marshall[13] and Fletcher and Pringle[14] isolated strains of attaching organisms from both marine and freshwater environments and described their preferences in terms of the work of adhesion of the substratum materials with water. Each organism exhibited an attachment maximum at a particular value of W_a, but except for the observation that hydrophobic substrata seemed to be preferred, no particular value within this range was significantly preferred.

The data available, taken as a whole, may indicate that a diverse population exists in natural waters and that the population attached to an immersed material reflects the population present at the time of exposure.

5. THERMODYNAMIC INTERPRETATION

A more specific accounting for the variety of results obtained with bacterial attachment studies requires an analysis that takes the surface properties of the liquid medium, the substratum, and the adhering entity into account. The adsorption of soluble material to the substratum can have a strong influence upon γ_{SL}, and the possibility of competitive adsorption processes among the variety of soluble materials found in natural waters greatly complicates the issue. Dexter[15] made an attempt to account for the variety of such data in terms of the effect of substratum surface energy and adsorption of dissolved organic matter in the sea[16] upon bacterial adhesion. As he pointed out, however, insufficient data were available to allow a clear analysis of the situation at that time. Detailed studies of adsorption from natural waters are needed to illuminate this area.

The third surface involved in the process is the microorganism itself. The surface energies of bacteria have been estimated in several ways. Direct measurement of contact angles of aqueous media with the surface of bacterial "lawns" on agar or membrane filters has been described[17] and the surface energy estimated using the postulated equation of state approach.[18] Values of θ from 66° to 70° were obtained. Measurements of γ_c on bacterial lawns using nonaqueous liquids have also been made.[19]

Alternatively, the microbial surface energy may be estimated by the following approach, which is based on the postulates that (a) the number of organisms attached to a particular substratum is closely related to the free energy of adhesion, ΔF^a, of the organism–liquid–substratum combination used, and (b) that the equation of state approach to determining surface energies from contact angles is valid.[18]

Defining

$$\Delta F^A = \gamma_{BS} - \gamma_{BL} - \gamma_{SL}$$

where B, S, and L represent bacterium, substratum, and liquid medium, respectively, we can express the terms in this equation in terms of γ_{SV}, γ_{LV}, and γ_{BV} by using the formalism

$$\gamma_{Bi} = f(\gamma_{BV}, \gamma_{iV})$$

where i denotes a third phase. (The determinations of γ_{SV} via contact angle measurements and of γ_{LV} do not involve the microorganisms.)

Choosing a medium of a particular γ_{LV}, we expose a series of solid substrata of varying γ_{SV} to the microorganism culture and then assayed for adhering cells. The slope of the (number attached) versus (γ_{SV}) curve is then obtained, and the experiment is repeated for liquids of different γ_{LV} (Fig. 1a). The slopes so obtained are then plotted versus γ_{LV} (Fig. 1b). The value of γ_{LV} at which the slope is zero corresponds to the value of γ_{BV}.

When this procedure was followed for several strains of bacteria, the values of γ_{BV} obtained agreed well with the value obtained by direct contact angle measurement with drops of aqueous medium on lawns of these bacteria. This agreement was taken as strong evidence for the validity of the hypotheses.

The method's assumptions are subject to evaluation. The surface equation of state approach used to evaluate γ_{BV} allows determinations of the organism surface energy to be estimated using the same medium as is used for the attachment studies. This avoids problems involving possible medium effects on the cells and allows measurement of γ_{SV} in the same environment also. However, the validity of the determination of γ_{SL} by contact angle with only a single liquid has been questioned (Chapter 1). If this point can be validated, or if an alternative method for obtaining γ_{LB} that does not change the cell surface becomes available, this question will be resolved.

The second assumption is that ΔF^a is closely correlated to the number of cells attached after a certain time of exposure. The free-energy change for a process can define the equilibrium and is associated with the reversible work. Therefore, a relationship to strength of adhesion may be expected. This parameter can be measured

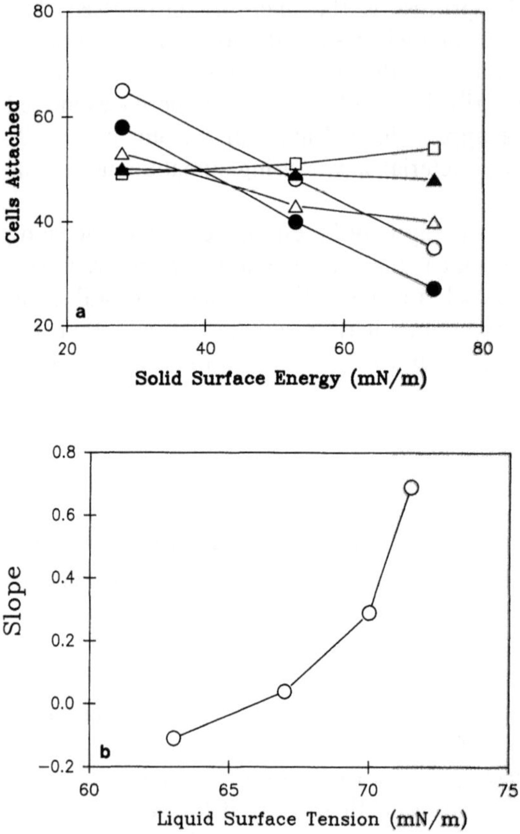

FIGURE 1. Determination of bacterial surface energy. (a) Plot of counts on substratum vs. substratum surface energy at several values of liquid surface tension: ○, 73 mN/m; ●, 71.5 mN/m; △, 70 mN/m; ▲, 67 mN/m; □, 63 mN/m. (b) Plot of slopes of lines of Fig. 1a vs. liquid surface tension. [Replotted from data of Fletcher and Pringle.[14]]

by the work required to prevent attachment or the work required for detachment. Detachment is the more usual measurement, but it is difficult to ensure that separation is at the level of the desired interface—i.e., that a cohesive failure within one of the two adherends has not occurred. Therefore a measurement that prevents attachment seems more relevant. Attachment can be prevented by application of a known shear stress at the interface, and several methods for shear generation have been used for this purpose.

Fowler and McKay[20] have described a device, the radial flow growth chamber, and cited earlier work with other devices. The radial flow growth chamber has been used to compare such dynamic adhesion assessments for several materials and cell cultures. Their results indicated less adhesion tendency for siliconized glass (a hydrophobic surface) than for the hydrophilic uncoated glass. Studies using parallel plate shear cells have also been performed, and surfaces of different wettability were compared with respect to adhesion.

Other flow studies[21] showed that three species of bacteria had different preference patterns for clean vs siliconized glass. One showed no preference, a second showed preference for clean glass by a factor of 3, and the third showed a similar preference by a factor of 10. To the author's knowledge, experiments of this type have not yet shown cases where low-energy surfaces have been preferred, although hydrophobic surfaces are frequently preferred in static assays. This inconsistency is an obstacle to clear understanding of the strength of adhesion at this time. It has been suggested that the hydrophobic low-energy region and the biocompatible zone could indicate a minimum in resistance of the bioadhesive joint to the detachment stresses that exist in natural environments, rather than a minimum in the tendency to attach in static conditions.[11, 19] Unfortunately, many attachment experiments designed to evaluate attachment to a variety of substrata have not been performed in defined flow regimes so that data of this type are insufficient to evaluate these inferences on a broad scale.

6. MACROSCOPIC ATTACHMENT

6.1. Life Cycles and Metamorphosis

The microscopic forms are perceptible to the casual observer as a substance that is well characterized by its common description: slime. This is not as noticeable as the macroscopic sessile organisms, which have long been recognized as major members of the "fouling community." These include barnacles, tubeworms, hydrozoa, bryozoa, and macroscopic algal growths, which are found on immersed surfaces. They are usually deleterious; however, mussels and oysters are also surface-dwelling creatures that are the products of a growing maricultural industry. The controlled settlement and attachment of these organisms on convenient substrata is an important aspect of mariculture.

The semimicroscopic swimming larvae of these macroscopic organisms settle on and attach to a solid substratum and undergo a drastic metamorphosis, after which they remain attached as sessile adults. (Mussels are somewhat different in that they may break away parts of their attachment apparatus and regenerate replacement parts.) The larval forms of the barnacle are best known because of their strong effect on the shipping and electric power generation industry.

6.2. Barnacles and Mussels

The settlement of barnacles and mussels to several materials has been described by Crisp *et al.*,[22] who give references to earlier work.

6.2.1. Larval Behavior

The larval barnacle stage just prior to settlement is called a cyprid. Its behavior includes exploration of surfaces. The two antennules, which are prominent appendages at the forward end of the cyprid, are used in a "walking" manner. This seems to be a testing of the adhesion of the antennules to the prospective homesite surface on which it finds itself. A cyprid will leave a surface not passing the test. However, if it finds the situation satisfactory, a permanent bond is formed. Its cement glands discharge a quantity of cement that bonds the antennules and nearby body to the surface of the substratum, and the cyprid undergoes metamorphosis to become an adult.

Young larva are quite discriminating in their choices and do not settle on

TABLE 1. Force for Detachment of Cyprid Antennules from Materials

(N/m^2)	
0.66×10^5	Beeswax
0.68×10^5	Glass
1.11×10^5	Poly(methyl methacrylate) (PMMA)
1.26×10^5	Slate
1.45×10^5	Reinforced phenolformaldehyde (PF)
1.70×10^5	Poly(vinyl chloride) (PVC)

unfavorable surfaces, although discrimination decreases with age. The exploring larvae must make use of a "temporary" adhesive, which is "weak" enough to allow removal of the antennules by the barnacle, but retentive enough to allow the walking and surface testing to be performed despite the flow of seawater along the surface. Crisp and co-workers have determined the temporary adhesion of cyprid antennules to several prospective substrata by directly measuring the force required to detach cyprids from them. They found that greater force was required for those surfaces that the cyprids preferred.

The ranking of detachment force is shown in Table 1.

The method by which the walking cyprid removes its antennule from the surface against these forces was not fully understood, although a "peeling" action was postulated.

A study of barnacle settlement on a set of siliconized glass surfaces has been reported.[(23)] The wettability was measured by the spreading behavior of alcohol solutions on the siliconized surfaces. This wettability scale was compared with surface energies determined by contact angle measurements on well-characterized surfaces and gave similar values. The surfaces used in this work differed only in the very thin silane layer, which minimized differences in surface roughness across the wettability spectrum. The silicone materials used are listed in Table 2.

The barnacle settlement pattern was determined both in a laboratory setting, with cultured barnacle larvae, and in the estuary at Beaufort, NC. Settlement in the field experiments was similar to the results reported by Crisp *et al.*[(22)] in that a positive correlation of settlement with wettability or surface energy was found for the range of surfaces used. (Fig. 2, circles) The laboratory experiments with cultured animals revealed a less pronounced variation with wettability.

TABLE 2. Wettability of Silane-Coated Glass

Wettability	Silicone coating on glass
10.9	(Heptafluoroisopropoxy)propylmethyldichlorosilane
20.1	Trimethylchlorosilane
29.1	Diphenyldichlorosilane
31.2	3-Chloropropyltrimethoxysilane
34.0	Aminopropyltriethoxysilane

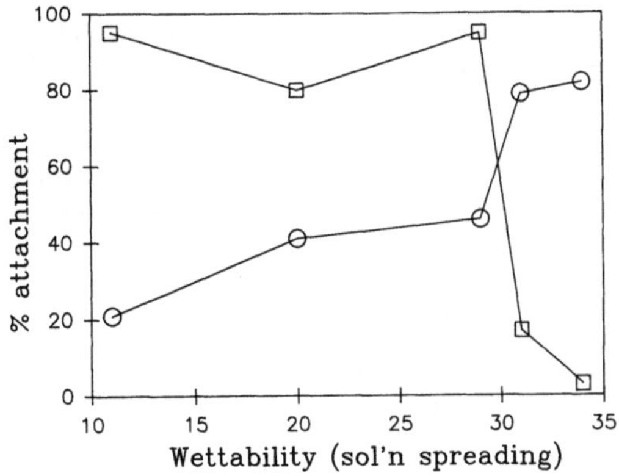

FIGURE 2. Attachment of barnacles and bryozoans to substrata of different wettability: □, bryozoans; ○, barnacles. [Data of Rittschoff and Costlow.[23]]

6.2.2. Adult Barnacle Detachment

Measurements were also made[22] of the detachment force for barnacles after metamorphosis from the same materials as used in the study of cyprid detachment (Section 6.2.1). The measurements made on slate indicated detachment forces of the same order of magnitude as for cyprids (5×10^5), but cohesive failure within the cement was found. On glass, lower forces (2×10^5) and some adhesive failure occurred.

A more extensive correlation of detachment force with wettability was carried out for the purpose of evaluating the possibility of developing self-cleaning surfaces.[24] A number of commercial plastic panels were exposed in the sea off Key West, FL, and a barnacle population was permitted to attach. The force per unit base area required to detach the adult barnacles was measured. The contact angles with water, methylene iodide, and glycerol were measured on duplicate unexposed plastic panels. From this data, the van der Waals and polar components of the solid surface energies were determined. As shown in Table 3 and Fig. 3, there is a clear trend of detachment force with surface energy for the homogeneous polymers. The detachment stress was much more closely related to the van der Waals component of γ_S than to the polar component as found by this treatment.

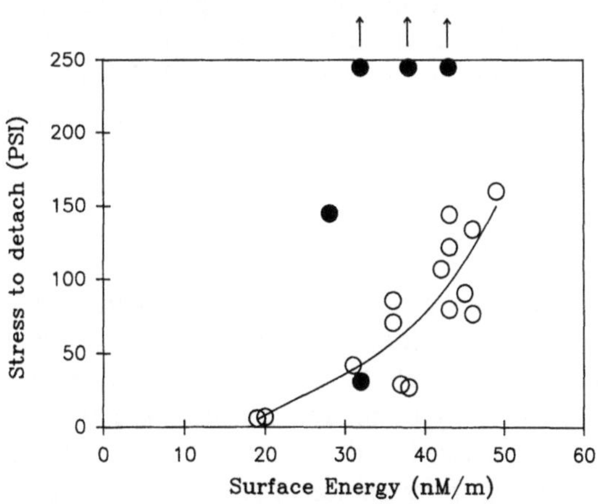

FIGURE 3. Detachment force for barnacles from polymer surfaces of different surface energies: ○, homogeneous polymers; ●, blends or alloys. [Data of Becka and Loeb[24] replotted.]

TABLE 3. Detachment Force for Adult Barnacles from Commercial Plastics

Surface energy			Force	
γ_S^W	γ_S^P	γ_{ST}	(mN/m^2)	Material
18	1	19	6	TFE Teflon
15	5	20	7	FEP Teflon
30	19	49	160	Nylon 66
24	14	38	27	Cellulose acetete butyrate
25	12	37	29	Polysulfone
25	6	31	42	Polypropylene
25	7	32	31	*MoS_2-filled nylon
33	3	36	71	High-density polyethylene
41	2	42	107	Poly(fluorotrichloroethylene)
17	11	28	145	*High-impact polystyrene
31	13	43	80	Polyacetal
33	3	36	86	High molecular weight polyethylene
22	24	46	77	Abrasion-resistant lucite
35	11	45	91	Polyvinylidine fluoride
36	8	43	144	Poly(vinyl chloride)
15	17	32	544	*Polyacrylamide–butadiene–styrene
29	9	38	280	*Acrylic styrene PVC alloy
31	12	43	533	*TFE–polyacetal
33	10	43	420	*Polyester–glass laminate
37	10	46	134	Polycarbonate
35	8	43	122	Phenylene oxide

Several materials are polymer alloys or blends. They are marked with an asterisk in the table and by filled symbols in Fig. 3 and, except for the MoS_2–nylon, showed significant positive deviations from the curve. The increased adhesiveness developed by these materials is not well understood, but may be a reflection of microscopic regions of particularly favorable surface composition or conformation that are formed at microgradients in these materials. MoS_2-filled nylon appears to have its surface coated with a layer of MoS_2, and so has a homogeneous surface and fits the general correlation.

The data presented for cyprid adhesion and for adult detachment are consistent with a correlation between force of adhesion and choice of substratum for attachment and metamorphosis. Although more data are clearly needed, the data available do not show differences between cyprid and adult cements in their relative adhesion to different materials.

6.2.3. Mussels

Mussels attach in a different manner than most other organisms. The adult form is not as intimately attached because it uses long threadlike structures (byssi) to attach, which have an adhesive pad at their extremity. Any thread of this structure may be broken and the animal essentially unharmed, while a new thread can be secreted and attached with its new adhesive pad. This animal also shows preference in its choice of byssus attachment sites and, like the barnacle, chooses higher-surface-energy areas for attachment. Thus, when presented with a checkerboard pattern of materials for attachment, more threads are fastened to higher-energy spots (with the exception of glass surfaces). Measurements of the force to detach the adhesive pads showed higher detachment force from the more favored substrata, as was true for the barnacle cyprids. The low detachment forces on glass (which is a high-surface-energy material) obtained with both barnacles and mussels were attributed to its exceptionally smooth surface in comparison with the other materials tested.[22]

6.2.4. Bryozoa

Bryozoa are another group of fouling organisms, some of which have a different preference pattern of substratum colonization. Studies of Bowerbankia[25] and Bugula,[11, 26, 27] which form bushy colonies, and Watersipora,[11] which is an encrusting organism, indicated that they prefer more hydrophobic materials for colonization, in contrast to the barnacle and mussel patterns discussed in Sections 6.2.2 and 6.2.3.

Attachment Bugula neritina larvae were given the opportunity to choose among different substrata for settlement in what were called "choice" experiments, or, alternatively, larvae were placed in a glass dish that contained only one type of material for what were called "no-choice" trials.[26] The results of such trials showed clearly that clean polystyrene, which had a hydrophobic surface characterized by water contact angle greater than 90°, was overwhelmingly preferred, and the number of organisms settling on clean glass, or on polystyrene treated with an electric glow discharge, was smaller by an order of magnitude or more. In addition, it was observed that there was a clear difference in the spatial distribution of attached larvae on the two different surfaces. The animals settled on hydrophobic polystyrene were well spread out. On hydrophilic glass, however, most of the larvae were attached in groups. (Such behavior has been termed *gregarious.*) Close observation indicated that their attachment areas were touching. This was interpreted as indicating that the adhesive of the first organism to settle made the local surface much more attractive than the glass surface, and that the apparently gregarious behavior was an attempt to locate an attachment site with favorable surface properties rather than a response to some other attractive signal associated with other settled larvae.

Attachment of Bugula stolonifera to a series of polyethylene derivatives fluorinated to different extents was also determined.[27] The materials and their critical surface tensions are shown in Table 4.

On clean surfaces, preference was for the polymers, on which 80% of the larvae attached, rather than for the high-surface-energy glass, with the exception of the poly(vinylidene fluoride) PVF_2. PVF_2 was not preferred more than the glass samples (Fig. 4a). After a preliminary exposure to filtered seawater, with its dissolved load of organic matter, the contact angles on the Teflons were greatly reduced and the surfaces were no longer attractive, while high settlement was observed on the other polymers (Fig. 4b). This change is consistent with adsorption of marine dissolved organic matter.[16]

TABLE 4. Materials Used in Bryozoan Trials

γ_c	Material
22	Poly(perfluoroethylene-propylene) (FEP Teflon)
18	Poly(tetrafluoroethylene) (TFE Teflon)
26	Poly(vinylidene fluoride) (PVF_2)
28	Poly(vinyl fluoride) (PVF)
31	Poly(ethylene) (PE)
33	Polystyrene
46	Glass

Low molecular weight (< 5 kdaltons) fractions of the dissolved organic matter did not cause a different attachment pattern than clean samples. Exposure to the high molecular weight fraction of dissolved organic matter, however, resulted in a settlement pattern similar to that of the seawater exposure. Because high molecular weight polymers are more strongly adsorbed than lower molecular weight homologous polymers, we may speculate that the absence of an effect of low molecular weight fractions is a result of their easy displacement by the outer layers of attaching or exploring organisms.

Studies of bryozoan attachment using the same series of silicone-modified glass surfaces as was used for the barnacle studies described in Section 6.2.1 were also performed by Rittschoff and Costlow.[23] The settlement patterns observed in this work for the bryozoa agreed with the observations of Mihm *et al.*[26] and Mihm and Dexter[27] in that the lower-energy surfaces were preferred. On the other hand, this is in contrast to the barnacle settlement studies in that the higher-surface-energy surfaces were more attractive for barnacle larvae (Fig. 2, squares).

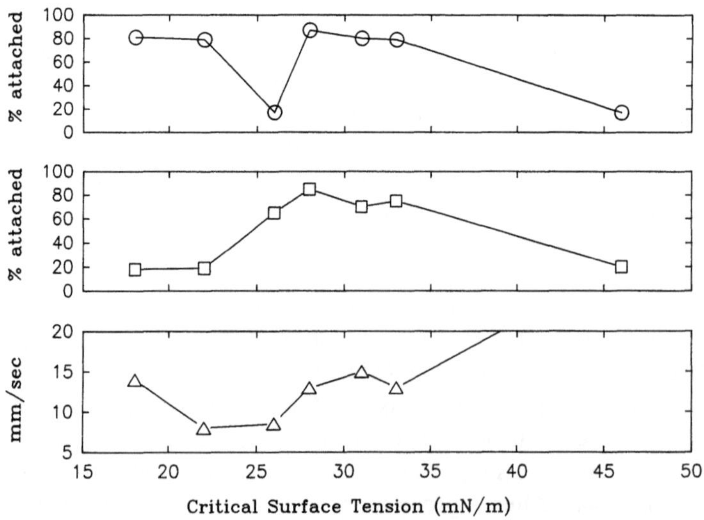

FIGURE 4. Bryozoan attachment and detachment—substrata of different critical surface tension: (top) attachment to clean surfaces; (center) attachment to surfaces filmed with dissolved organic matter; (bottom) flow required for detaching settled organisms. [Data of Mihm and Dexter.[27] (1990).]

Detachment Mihm and Dexter made use of a water jet to apply increasing shear forces to attached Bugula on their set of surfaces (Fig. 4c). In contrast to the situation with barnacles, where settlement preferences followed the force of detachment of both the larval antennules and the permanently attached adults,[22, 24] the force required for detachment of the bryozoans increased with critical surface tension or wettability, although hydrophobic surfaces were preferred for settlement. These data emphasize the need to examine the relationship between the thermodynamic predictions and adhesion with great care.

6.2.5. Macroalgae

Fletcher and co-workers[28] examined the effects of a range of surface energies upon the form of algae developing from spores that had been settled on glass plates, some of which had been treated with silanes to produce a range of critical surface tensions. The values of γ_c were <20, 20–30, 30, 30–40, and >70. Six types of common fouling algae were used. The results of this study indicate that the morphology of the basal structures, which form the contact with the substratum, are different on the different types of surfaces. It was found that a filamentous type that spread extensively and rapidly was not usually as adherent as a type that formed a disk of tightly branched filaments. However, there was no "model response" found for all the species with respect to the surfaces, so that it was not possible to make a strong association between morphology, adherance, and γ_c for all the algae investigated.

7. *PHYSICAL FORCES VERSUS BEHAVIOR*

If thermodynamic parameters such as work of adhesion were to account for bioattachment, behavioral responses of the organisms in the absence of other signals should not cause significant deviations from predictions based on measurements of surface and interfacial tension, as discussed by Neumann[18] and Good,[6] or adhesion measurements. The attachment patterns reported by Neumann and co-workers and by Fletcher and Pringle seem to agree with interfacial tension arguments for the microorganisms with which they worked. The work described by Crisp *et al.*[22] with barnacle and mussel organs of attachment (Section 6.2) indicate that the resistance of the attachment organ to detachment was consistent with its use as an important signal for substratum organism preference. Not many studies of this type have been carried out, however, and other organisms are more inconvenient for such manipulation.

An alternative measure of adhesive tendency is the hydrodynamic force required to separate the interface into two phases, but this approach has its difficulties, as discussed by Fowler and McKay.[20] The plane of separation is not, in general, the same as the plane at which the adhesive contact is made—at the outer surface of the animal. The separation will occur at the plane of the weakest layer of the structure. Frequently, this is within the organism. If that is so, it is often possible to observe "footprints," which are cellular or extracellular material left behind at the substratum. (Flow cells and the radial flow growth chamber of

Fowler and McKay avoid this problem for microfouling by allowing the settlement to take place in the regions of a flow field where the adhesive interaction is enough to overcome the flow.)

Although aware of these problems, Mihm and Dexter[27] used seawater flows to estimate the relative detachment stress for bryozoa attached to their series of materials (Fig. 4c). Although relative rather than absolute values of shear stress were obtained, it was clear that higher-energy material (glass), which was not preferred by the bryozoan larvae, retained the settled organism against the strongest flows used. The materials of intermediate surface energy, poly(vinyl fluoride) (PVF), polyethylene (PE), and polystyrene (PS) remained attractive and their populations had fairly strong resistance to flow. The changes occurring with poly(tetrafluoroethylene) (TFE), fluorinated ethylene-propylene (FEP), and PVF_2 are of great interest. TFE and FEP Teflons showed high attraction for the larvae when clean, but low when filmed with the dissolved organic matter from seawater, while PVF_2 changed its unattractive signal in the clean state to an attractive signal when filmed. However, the resistance of the larvae to detachment by flow remained the same on the filmed surfaces as on the clean materials. Thus, the initial attraction of the surface for the organism is not reflected in the bond strength. It is unlikely that a weak layer in the organism is responsible for the joint failure for the following reasons: the strength of the joint with glass withstands much greater stresses, there is a variation in detachment stress with material, surfaces with microbial films supported larger stresses than clean surfaces (not shown), and microscopic examination of the settlement site with histological stain did not reveal perceptible residues. Therefore, at least for the low-surface-energy materials, the correlation between detachment strength and settlement appears to break down.

However, it should be borne in mind that the selection of a site is accomplished by the larva before metamorphosis occurs. If a process of testing joint quality is used by the larva, some property of the larval form would be pertinent, rather than the cement of the adult. According to Ryland:[29] "The larva selects a suitable spot for attachment with the aid of the plume cilia ... adhering to it (the substratum) by a secretion coming in part from the pyriform organ..."

Perhaps the function of the antennular secretion of barnacle cyprids[22, 30] is served by the pyriform organ secretion, as was also indicated by observations of M. J. Loeb and Walker.[31] The question of interest would then be whether the pyriform organ secretion has adhesive interactions with this range of low-energy surfaces similar to the adult cement, as with barnacles.

8. SUMMARY

The thermodynamic treatments based on the free energy of adhesion have shown promise for explaining some aspects of microbial adhesion, and have shown correlation with more direct measurements of surface energies of several types of bacteria. However, measurements of adhesive strength have so far not indicated strong enough interaction with hydrophobic substrata to be consistent with observed preferences of a number of microorganisms. An expansion of the very limited data base using controlled shear or other known force regimes during biofilm growth to explore this relationship is needed to resolve this inconsistency. Studies that deal with aspects of the problem separately may be misleading because living organism surface properties may change with age or respond to variations in culture conditions.

The larger sessile aquatic organisms whose adhesion has been studied carefully also show preferences for substratum wettability. With barnacles and mussels, the apparent strength of adhesion of the attachment organs to the substratum is strongly correlated with the choice of a permanent home. A third class of organism, some bryozoans, shows the inverse preference pattern. It is not yet known whether this negative correlation with attachment strength reflects a difference in the characteristics of larval and adult adhesives or some other sensory signal. As with microbes, it will be important to investigate the dynamic forces of adhesion during the part of their life cycle when the decision to attach permanently is made. If in fact the work of adhesion cannot account for the substratum preferences, metabolic and behavioral responses of other types will have to be explored. Finally, the role of dissolved material in natural waters, which is always present, but as of yet ill-defined and of a variable nature, will have to be explored. Although several studies have indicated its importance, characterization of the molecules involved in the process, and their interactions, is just beginning.

ACKNOWLEDGMENT

Support from the Office of Naval Research and the Energy Program Office of the David Taylor Research Center is gratefully acknowledged.

REFERENCES

1. D. J. Crisp, in *Marine Biodeterioration: An Interdisciplinary Study* (J. D. Costlow and R. C. Tipper, eds.), pp. 103–126, Naval Institute Press, Annapolis, MD (1984).
2. A. W. Neumann and R. J. Good, in *Surface and Colloid Science* (R. J. Goode and R. R. Stromberg, eds.), vol. 11, pp. 31–91, Plenum Press, New York (1979).
3. W. A. Zisman, in *Contact Angle, Wettability and Adhesion* (R. F. Gould, ed.), pp. 1–51, Am. Chem. Soc., Washington, DC (1964).
4. R. E. Baier, in *Adhesion in Biological Systems* (R. S. Manly, ed.), pp. 15–48, Academic Press, New York (1970).
5. R. E. Baier, *J. Biomech. Eng.* **104**, 257–271 (1982).
6. R. Good, this volume, chap. 1.
7. M. E. Schrader, On adhesion of biological substances to low-energy surfaces; *J. Colloid Interface Sci.* **88**, 296–297 (1982).
8. K. C. Marshall, in *Bacterial Adhesion* (D. C. Savage and M. Fletcher, eds.), Chap. 6, pp. 133–162, Plenum Press, New York (1985).
9. S. C. Dexter, J. D. Sullivan, Jr., J. Williams III, and S. W. Watson, *Appl. Microbiol.* **30**, 298–308 (1973).
10. M. Fletcher and G. Loeb, *Appl. Environ. Microbiol.* **37**, 67–72 (1979).
11. G. I. Loeb, U.S. Naval Res. Lab Rept. M-3665, pp. 1–9 (1977).
12. G. Loeb, C. Bailey, and S. Wajsgras, *Int. Biodeterior. Bull.* **19**, 79–82 (1983).
13. M. Fletcher and K. C. Marshall, *Appl. Environ. Microbiol.* **44**, 184–192 (1982).
14. M. Fletcher and J. H. Pringle, *J. Colloid Interface Sci.* **104**, 5–14 (1985).

15. S. C. Dexter, *J. Colloid Interface Sci.* **70**, 346–354 (1979).
16. G. Loeb and R. Neihof, in *Applied Chemistry at Protein Interfaces* (R. E. Baier, ed.), pp. 319–335, Am. Chem. Soc., Washington, DC (1975).
17. D. R. Absolom, F. V. Lamberti, Z. Polikova, W. Zingg, C. J. van Oss, and A. W. Neumann, *Appl. Environ. Microbiol.* **46**, 90–97 (1983).
18. A. W. Neumann, this volume, chap. 5.
19. R. E. Baier, in *Adsorption of Microorganisms to Surfaces* (G. Bitton and K. C. Marshall, eds.), pp. 59–104, Wiley, New York (1980).
20. H. W. Fowler and A. J. McKay, in *Microbial Adhesion to Surfaces* (R. C. W. Berkely, J. M. Lynch, J. Melling, P. R. Rutter, and B. Vincent, eds.), pp. 143–162, Soc. Chem. Ind., London (1980).
21. K. Pedersen, C. Holmstrom, A. K. Olssen, and A. Pedersen, *Arch. Microbiol.* **145**, 1–8 (1986).
22. D. J. Crisp, G. Walker, G. A. Young, and A. B. Yule, *J. Colloid Interface Sci.* **104**, 40–50 (1985).
23. D. Rittschoff and J. Costlow, *Eur. Mat. Biol. Soc. J.* **22**, 411–416 (1989).
24. A. Becka and G. Loeb, *Biotechnol. Bioeng.* **24**, 1245–1251 (1984).
25. R. Eiben, *Mar. Biol.* **37**, 249–254 (1976).
26. J. W. Mihm, W. C. Banta, and G. I. Loeb, *J. Exp. Mar. Biol. Ecol.* **54**, 167–179 (1981).
27. J. Mihm and S. Dexter, Ph.D. Thesis, University of Deleware, Lewes (1985).
28. R. L. Fletcher, R. E. Baier, and M. S. Fornalik, in *Proc. 6th Int. Congr. Marine Corrosion and Fouling Athens* (1984).
29. J. S. Ryland, *Bryozoans*, Hutchinson Univ. Library, London (1970).
30. A. B. Yule and G. Walker, *J. Mar. Biol. Assoc. U.K.* **65**, 707–712 (1985).
31. M. J. Loeb and G. Walker, *Mar. Biol.* **42**, 37–46 (1977).

11

Wettability of Clay Minerals

Shmuel Yariv

1. INTRODUCTION

Clay minerals, hydrated oxides, and hydroxides, mainly of Si, Al, Fe, and Mn, and some organic macromolecules are responsible for most wettability properties of geologic systems that constitute the sedimentary cycle. They are the soil components that exert the dominant influence on chemical and physical properties of a soil, and a detailed examination of the nature of their wettability by water is essential for a complete understanding of a soil's potential with respect to agricultural and engineering applications.[1–3] The clay minerals, which generally constitute the greatest part of the colloid fraction in these systems, are named after the term used by sedimentologists and soil scientists for the fraction of particles having a very small size, with an equivalent diameter smaller than 2 μm, the "clay fraction." Although most clay minerals occur as particles too small to be resolved by the ordinary microscope, x-ray diffraction analysis shows that most of them, even in their finest size fraction, are composed of crystalline particles and that the number of crystalline minerals likely to be found is limited. Furthermore, a wide distribution of particle size is frequently present and certain clay deposits contain well-defined crystalline particles with diameter greater than 2 μm.

In this chapter a distinction is made between clays and clay minerals. The former term is used for the small-size particles found in soils and sediments, including crystalline and amorphous oxides and hydroxides of various metals, whereas the latter is used for a certain group of layer crystalline silicate minerals.

Wettability of clay minerals is a crucial problem, faced in various industrial technologies, such as metallurgy (molding sands), drilling fluids, painting, oil refining and decolorization, water clarification, catalysis, and production of ceramics, papers, adhesives, portland cements, plastics, rubber, medicines, cosmetics, etc.[4]

Shmuel Yariv • Department of Inorganic and Analytical Chemistry, The Hebrew University of Jerusalem, Jerusalem 91904, Israel.

Modern Approaches to Wettability: Theory and Applications, edited by Malcolm E. Schrader and George Loeb. Plenum Press, New York, 1992.

In any treatment on clay–water interactions and on the wettability of clay minerals, one should take into consideration the existence of three different states of a clay–water system. The first state is the dry clay. A ground dry clay on being touched gives a greasy feeling. In this state water molecules are located on the surface of the solid clay in the adsorbed state, interacting with active functional groups. The second state is the plastic clay. Plasticity may be defined as the property of a material that permits it to be deformed under stress without rupturing and to retain the shape produced after the stress is removed. This is a colloid system in which microdrops of free water (liquid phase) are dispersed between the solid clay particles. The liquid water that fills the voids between the clay particles, serves as a lubricant that permits some movement between the particles under the application of a deforming force, thus giving the system its plastic properties. The third state is an aqueous colloid solution or suspension of clay. Here liquid water serves as a continuous dispersing phase in which the solid clay particles are dispersed. In the first state the water and the silicate together form a single solid phase, whereas in the second and third states liquid water forms a separate phase in addition to water molecules that are adsorbed on the clay surface and thereby belong to the silicate solid phase.

In this chapter the current state of the surface chemistry of hydration of clay minerals is reviewed. For colloidal properties and plasticity of clay aqueous systems and for water in clay micropores the reader is referred to Refs. 1–3.

2. STRUCTURES OF CLAY MINERALS

The clay minerals are essentially hydrous layer aluminosilicates (phyllosilicates), with magnesium and iron acting as proxy wholly or in part for the aluminum in some of the minerals and with alkali metals and alkaline earth metals present as essential constituents in some of them.[5–7] The magnesium- or aluminosilicate layer is essentially composed from two types of sheets, a tetrahedral and an octahedral, designated as T and O, respectively. Each sheet is composed of planes of atoms, arranged one above the other, a plane of oxygens above a plane of silicons or aluminums, and the latter above another plane of oxygens, and so on.

2.1. The Tetrahedral Sheet

A continuous linkage of SiO_4 tetrahedra by the sharing of three O atoms with three adjacent tetrahedra produces a sheet with a planar network (Fig. 1). In such a sheet the silica tetrahedral groups are arranged to form a hexagonal network, which is repeated indefinitely to form a layer of composition $[Si_4O_{10}]^{4-}$. The tetrahedra are arranged so that all their apices point in the same direction and all their bases are in the same plane. In this plane the oxygens form an open hexagonal network, which is often referred to as the hexagonal plane or the perforated plane of oxygens. In reality the silicon tetrahedra are slightly distorted; consequently, the cavities that are bordered by six oxygens are ditrigonal rather than hexagonal. This perforated plane of oxygens is an important contributor to the surface properties of

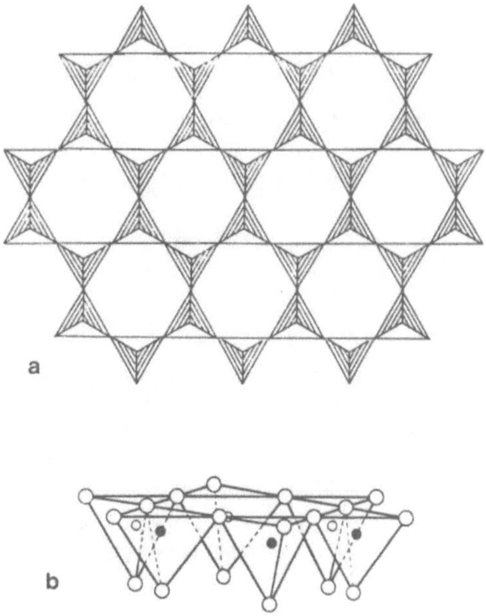

FIGURE 1. The tetrahedral sheet, $[Si_4O_{10}]^{4-}$ (a) top view; (b) side view.

the clay minerals. Each oxygen of this plane is covalently bound to two silicons, thus playing the active component of a siloxane group.

Silicon atoms in the tetrahedral sheet can be replaced by aluminum atoms. Substitution of Al for Si contributes a negative charge to the tetrahedral sheet. In many minerals this substitution is very small, but in micas 25% of the possible silicon sites are substituted by Al. As we will show, the substitution of Al for Si changes the surface properties of the perforated oxygen plane, since Si–O–Al groups are better donors of electron pairs than are Si–O–Si groups.

2.2. The Octahedral Sheet

Such a sheet is obtained by the condensation of single $Mg(OH)_6^{4-}$ or $Al(OH)_6^{3-}$ octahedra (Fig. 2). Each O atom is shared by three octahedra, but two octahedra can share only two neighboring O atoms. In this sheet the octahedra are arranged to form a hexagonal network, which is repeated indefinitely to form a layer of composition $[Mg_6O_{12}]^{12-}$ or $[Al_4O_{12}]^{12-}$. The minerals brucite $Mg(OH)_2$ and gibbsite $Al(OH)_3$ have such a sheet structure. This sheet structure is densely packed, being composed of a dense hexagonal plane of Mg or Al atoms sandwiched between two dense hexagonal planes of hydroxyls. The latter is named *the hydroxyl plane.* All the octahedra are filled with Mg atoms in brucite or in its clay derivatives, but only two thirds of the octahedra are filled with Al atoms in gibbsite and in its clay derivatives. Derivatives of brucite and gibbsite are called tri- and dioctahedral clay minerals, respectively. In some clay minerals the hydroxyl plane is an important contributor to the surface properties of the mineral.

Aluminum atoms in the octahedral sheet of a dioctahedral mineral can be replaced by iron or magnesium atoms. Similarly, magnesium atoms in a

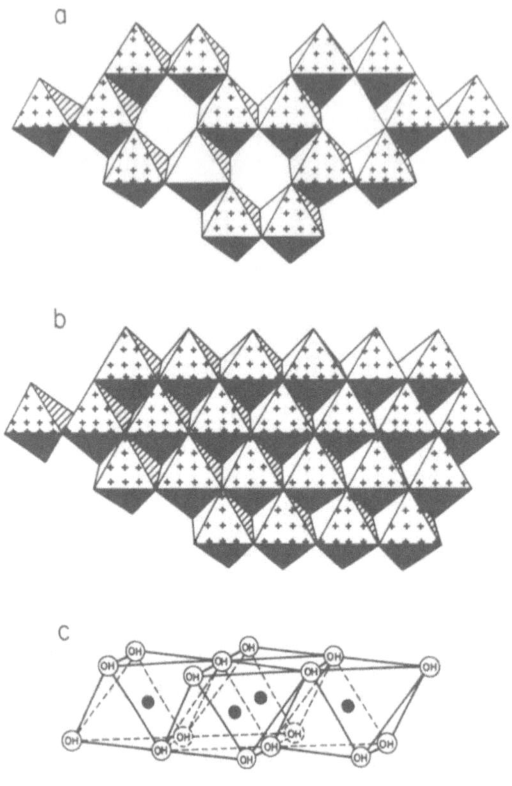

FIGURE 2. The octahedral sheet: (a) gibbsite $Al(OH)_3$ (top view); (b) brucite, $Mg(OH)_2$ (top view); (c) side view.

trioctahedral mineral can be replaced by trivalent aluminum, di- or trivalent iron, and monovalent lithium atoms. Substitution of a trivalent cation by a divalent cation or a divalent one by a monovalent cation contributes a negative charge to the octahedral sheet.

2.3. The TO-Type Layer Silicates

Minerals of the serpentine–kaolin group are composed of a single tetrahedral sheet condensed with a single octahedral sheet into one unit layer designated by 1:1 or by TO. In this unit layer the apices of the silica tetrahedra and one of the OH planes together form a single plane that becomes common to both tetrahedral and octahedral sheets. Two thirds of the O atoms in the common plane are shared by Si and Mg or Al atoms, and the remainder are shared by protons and Mg or Al atoms. In serpentine the octahedral sheet is brucitic, whereas in kaolinite it is a gibbsite-like layer. The ideal structural formulas of layers of serpentine and kaolinite are $[Mg_6Si_4O_{10}](OH)_8$ and $[Al_4Si_4O_{10}](OH)_8$, respectively.

The minerals of the kaolin subgroup consist of dioctahedral layers continuous in the *a* and *b* directions and stacked one above the other in the *c* direction (Fig. 3). The variation between members of this subgroup consists mainly in the way in which the TO unit layers are stacked one above the other. These minerals expose surfaces of oxygens and hydroxyls. For most kaolinites a specific surface area of

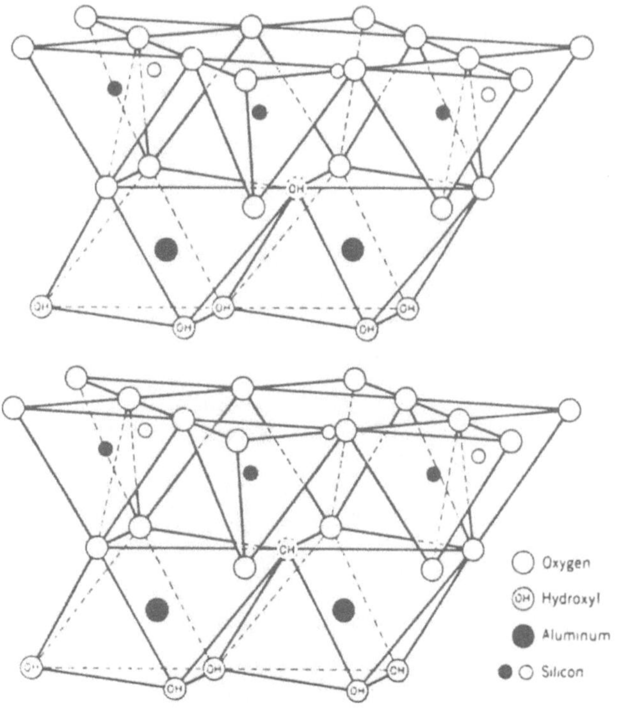

FIGURE 3. A structural scheme of a TO clay mineral (kaolinite or serpentine).

about 10 $m^2 g^{-1}$ was determined. The edges of kaolinite usually consitute 15–20% of its total area, and thus the two faces should each make about 40% of the external surface.[8] In halloysite parallel TO unit layers are separated by a water monolayer. Some of the minerals of the serpentine subgroup, e.g., lizardite, show a stacking of continuous trioctahedral layers similar to that of kaolinite. In the case of chrysotile the trioctahedral extending TO unit layer is rolled up, and a tubular fiber is obtained from this spiral rolling. This mineral exposes surfaces of hydroxyls.

2.4. The TOT-Type Layer Silicates

2.4.1. Talc and Pyrophyllite

Minerals of the talc–pyrophillite group are composed of two tetrahedral sheets condensed to a central octahedral sheet forming one unit layer designated by 2:1 or by TOT. The sheets are combined so that the apices of the tetrahedra of each Si sheet and one of the hydroxyl planes of the octahedral sheet form a plane common to both sheets. Two common planes are obtained on both sides of the octahedral sheet. These common planes are the same as the common plane obtained for the serpentine–kaolin group. The ideal structural formula of layers of talc and pyrophyllite are $[Mg_6Si_8O_{20}](OH)_4$ and $[Al_4Si_8O_{20}](OH)_4$, respectively. A crystal of the mineral consists of TOT layers continuous in the *a* and *b* directions and stacked one above the other in the *c* direction (Fig. 4). The layers are held together by van der Waals attractions. These minerals expose perforated oxygen planes that are highly hydrophobic.

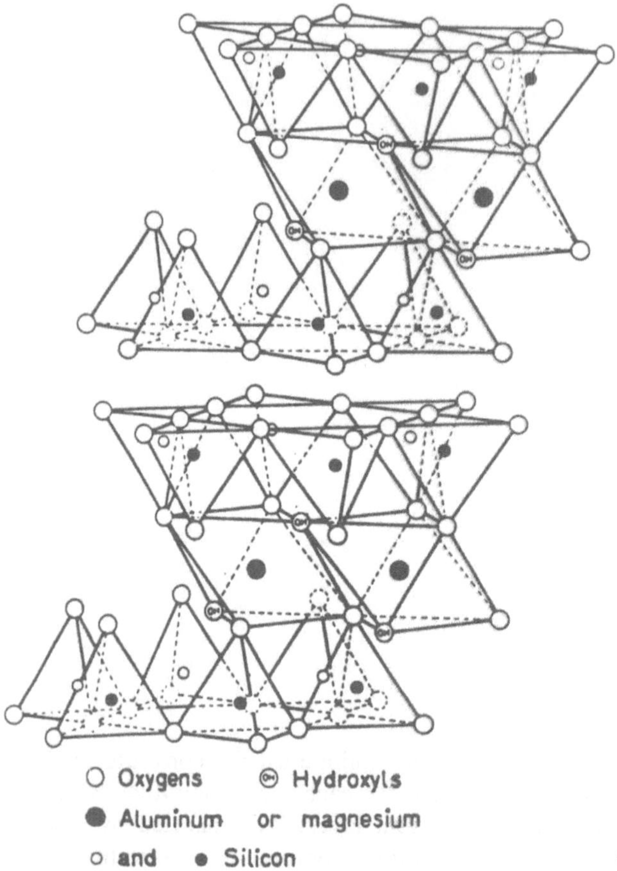

FIGURE 4. A structural scheme of a nonexpanding TOT clay mineral (talc or pyrophyllite).

2.4.2. Smectites

Minerals of the smectite group (known sometimes as the montmorillonite group) also consist of TOT layers. They differ from talc and pyrophyllite in that a small fraction of the tetrahedral Si atoms is substituted by Al atoms, and/or octahedral atoms (Al or Mg) are substituted by atoms of a lower oxidation number. The resulting charge deficiency is balanced by exchangeable cations, mainly Na, Ca, and Mg, which are located between parallel layers. The negative charge per unit cell from isomorphous substitution ranges between 0.5 and 1.3 electron charges. Water and other polar molecules may penetrate between the layers, causing the expansion of the structure in a direction perpendicular to the layers (Fig. 5).

Beidellite and saponite are di- and trioctahedral smectites, respectively, with mainly tetrahedral substitution, whereas montmorillonite and hectorite are di- and trioctahedral smectites, respectively, with mainly octahedral substitution. Nontronite is an iron-rich smectite. Montmorillonite is the most common mineral of this group. For most smectites an interior specific surface area of about 750–800 $m^2\ g^{-1}$ on the oxygen cleavage planes was determined. The exterior surface area for most natural samples is less than 20% of the interior surface area.

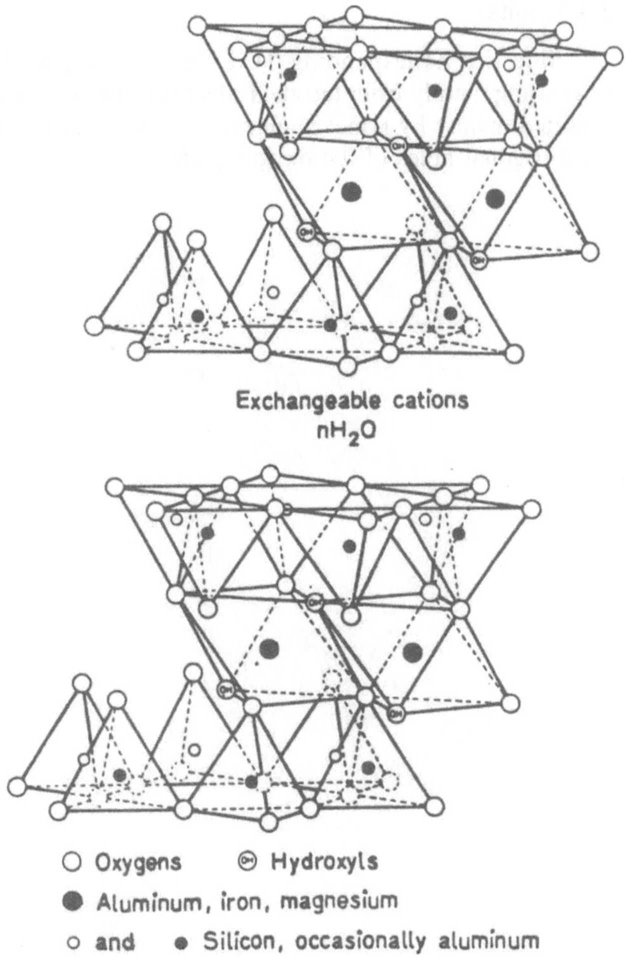

FIGURE 5. A structural scheme of an expanding TOT clay mineral (smectite or vermiculite).

2.4.3. Vermiculites

Vermiculites also consist of TOT unit layers in which some of the structural Si, Mg, or Al is isomorphically substituted by atoms of a lower oxidation number. Vermiculites usually have greater layer charge densities than smectites, and the charge originates mainly from tetrahedral substitution. The negative charge per unit cell from isomorphous substitutions ranges between 1.1 and 2.0 electron charges. In the natural mineral the balancing exchangeable cation is Mg, sometimes with a small contribution from Ca or Na. Similar to the smectites, water and other polar molecules may penetrate between the layers, causing the expansion of the structure in a direction perpendicular to the layers. However, due to the higher negative charge of the vermiculite layer and the higher density of cations in the interlayer space, the expansion of vermiculites is smaller compared with that of the smectites. The total interior surface area determined for most vermiculites was about $750\ m^2\ g^{-1}$, whereas the total exterior surface area for most natural samples is not more than a few square meters per gram.

2.4.4. Illites

Illite is a nonexpanding TOT clay mineral, mostly dioctahedral, with charge originating mainly from tetrahedral substitutions. The negative layer charge is compensated mainly by nonexchangeable K ions located in the interlayer, embedded in the ditrigonal holes of the oxygen plane.

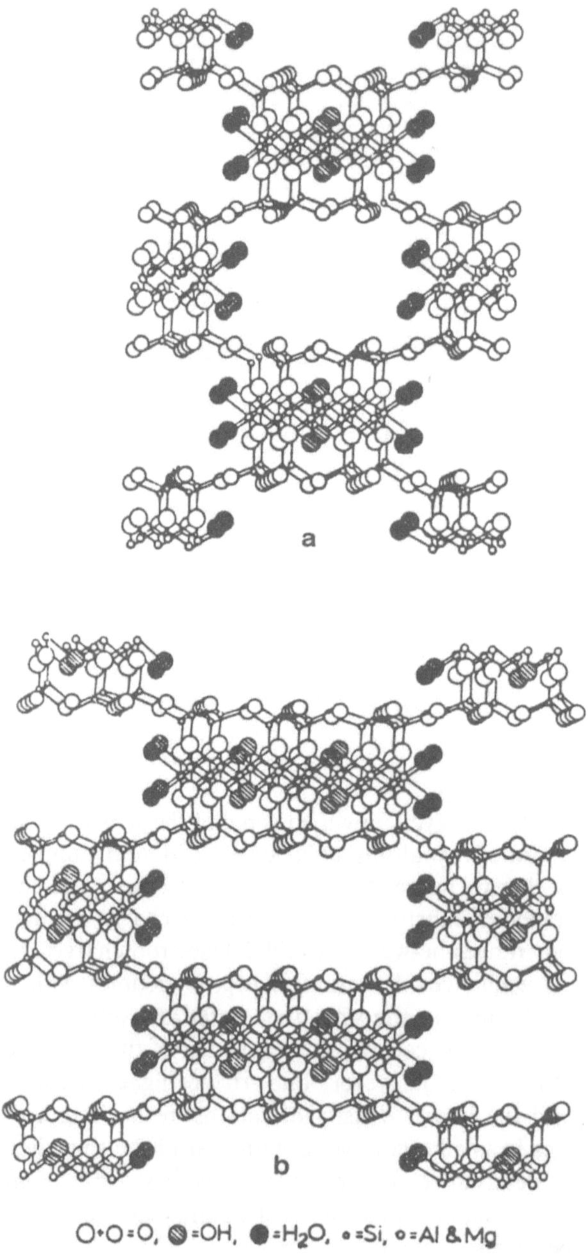

FIGURE 6. Structural schemes of channel clay minerals: (a) palygorskite and (b) sepiolite.

2.4.5. Sepiolite and Palygorskite

Sepiolite and palygorskite are unique among the TOT clay minerals in having channel structure. This structure is obtained from the repeated inversion of the silicate layer. The two minerals differ in the frequency of inversion, sepiolite having wider channels (Fig. 6).

3. WETTABILITY OF THE OXYGEN PLANE

Hydrophilicity of the oxygen plane of the silicate layer is determined by possible interactions between the O atoms of the siloxane groups and water molecules that tend to donate protons and to form hydrogen bonds. For this to happen the siloxane oxygen atom must donate a lone pair of electrons to the water proton. Only nonbonding hybridized orbitals are able to serve as electron pair donors, because in the hybridized orbital the electron density is high on one side of the O nucleus, and this orbital may essentially overlap with the *s* orbital of the water hydrogen, whereas in a nonhybridized orbital the lone-pair electrons are equally distributed on both sides of the nucleus.

In covalent compounds of oxygen one would expect that the oxygen atom would undergo an sp^3 hybridization, giving rise to four equal hybridized orbitals with a minimum repulsion between the four electron pairs that fill the valence shell of this oxygen. This hybridization permits two nonbonding hybridized orbitals with lone-pair electrons to serve as sites for hydrogen bonds. Theoretically, an sp^3 hybridization should result in an Si–O–Si angle slightly smaller than 109°.

Determination of Si–O–Si angles in different silicates revealed that this angle ranges between 120° and 180°. In most silicates the value of 139–140° was determined.[7] An angle of 180° or 120° is an indication of an *sp* or sp^2 hybridization on the oxygen atom, respectively. If the angle ranges between these values, it should be due to a resonance of two canonic structures, one with *sp* and one with sp^2 hybridization on the oxygen atom (Fig. 7). Hydrophilicity of the oxygen plane decreases with decreasing electron density in the nonbonding hybridized orbitals, and, consequently, sp^2 hybridization on the oxygen decreases hydrophilicity compared with sp^3, whereas *sp* hybridization leads to hydrophobicity. To conclude, the Si–O–Si angle can give information on the hydrophilicity of a siloxane group, the hydrophilicity increasing with decreasing Si–O–Si angle. In the following section we will show that hybridizations *sp* and sp^2 on the O atom occur in order to enable this element to be involved in a d_π–p_π bond.

3.1. The d_π–p_π Bond

The elements of the third row of the periodic table are characterized by the involvement of the d_π–p_π bond in the overall bonding system.[9, 10] This bond has been found to occur between Si, P, S, or Cl and N, O, or F from the second row of the periodic table. Because the *d* orbitals have a considerable sideways extension, they can overlap with *p* orbitals on the second atom to form π bonds. However, because the *d* orbitals have higher energy than the *p* orbitals, they are used in π

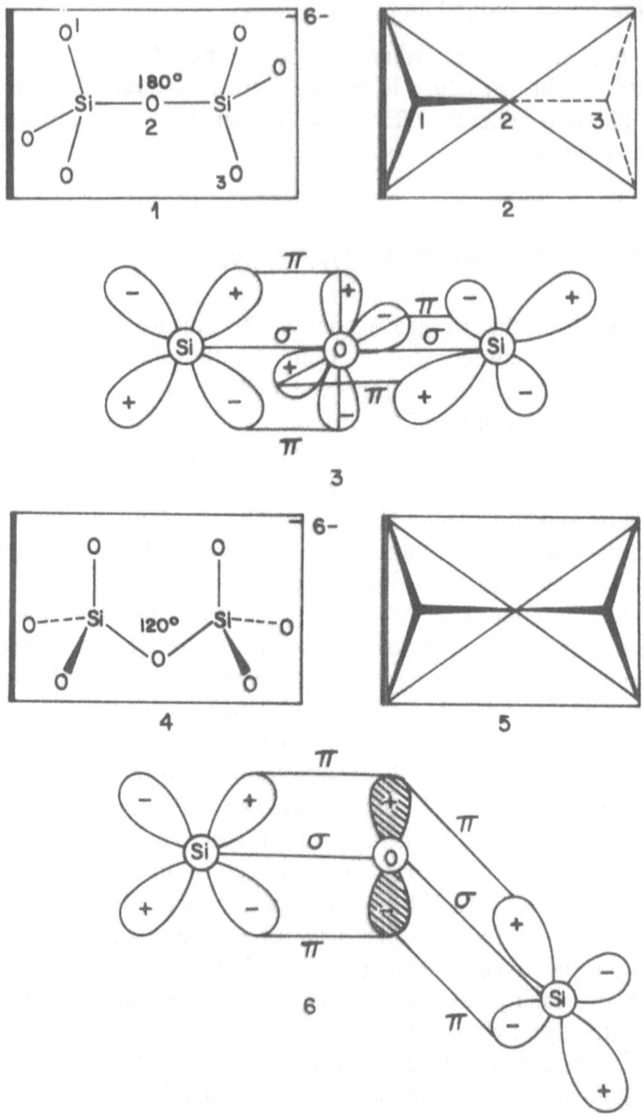

FIGURE 7. Two canonic structures of a siloxane group giving rise to an angle ranging between 120° and 180°. (1–3) and (4–6) *sp* and sp^2 hybridizations, respectively, on the O atom of an Si–O–Si group. (From Ref. 15. Reproduced by permission of *Clay Minerals*.)

bonding mainly to active ligands such as N, O, or F; π bonds involving *d* orbitals probably add less on average to overall stability of the bond than does the d_π–p_π bonding of the second-row elements. In the formation of this bond, silicon, which has empty *d* orbitals, serves as an acceptor of electron pairs (Lewis acid) and the element of the second row, which has a relatively higher electron density in its valence shell, serves as a donor of electron pair (Lewis base). Since this bonding is an addition to the σ bonding, the Si–O bond is considered to have a partial-double-bond character.

The first three elements in the third row of the periodic table (Na, Mg, and Al) do not take part in the d_π–p_π bondings. These elements also have empty 3d orbitals. Low charges of their nuclei, compared with those of Si, P, S, or Cl, lead to their behavior as weak acids, so that their empty 3d orbitals never accept electron pairs from N, O, or F atoms. As well, the first four elements of the second row (Li, Be, B, and C) do not take part in d_π–p_π bondings. Atoms of these elements have relatively low electron densities in their valence shells and, consequently, are weak Lewis bases, not donating electron pairs to form this kind of bonding.

3.2. The Double-Bond Character of the Si–O Bond of Siloxanes in Clay Minerals

The coordination number of Si in clay minerals is 4, involving sp^3 hybridization. Three of the oxygens that coordinate the silicon atom belong to the perforated oxygens plane, whereas the fourth oxygen atom belongs to the O, OH plane, which is common to the tetrahedral and octahedral sheets. The Si atom uses vacant d orbitals to form π bonds with O atoms. The O atom that bridges between the tetrahedral and octahedral sheets is involved in an sp^3 hybridization, and, consequently, its contribution to a d_π–p_π bond is negligible. On the other hand, oxygens that belong to the perforated O plane, are the principal contributors to the d_π–p_π bonds. Each O atom needs at least one or two nonhybridized p orbitals in order to form π bonds with Si. Consequently, the bridging O atoms in a siloxane group would be expected to show sp or sp^2 hybridization, giving rise to Si–O–Si angles of 180° or 120°, respectively. An sp hybridization enables two p orbitals of an O atom to overlap with d orbitals of two Si atoms. Two Si–O double bonds are obtained, each consisting of a σ and a localized π orbital. An sp^2 hybridization enables only one p orbital of an O atom to overlap with d orbitals of two Si atoms. These three atomic orbitals overlap to form a three-centered π bond orbital. The π bond order in the three-centered bond is less than in the localized π bond.

Very little information is available from the literature on x-ray determinations of the Si–O–Si angle in clay minerals, due to the absence of single crystals, which are essential for this determination. Lee and Guggenheim[11] observed that the Si–O–Si angle in pyrophillite was equal to $\approx 136°$. From this data it can be concluded that a substantial structure results from the two canonic structures with sp and sp^2 hybridizations on the O atom, and that the latter makes the principal contribution to the resonance in the double-bond system. Lin and Guggenheim[12] studied the structure of a brittle mica where almost 50% of the tetrahedral Si was substituted by Al or Be, which do not participate in a d_π–p_π bond. They found that the Si–O–Al (or Be) angle was equal to 120.9°. From their observation it can be concluded that in this mineral the O is involved in an sp^2 hybridization only. According to our assumption on the double-bond character of the Si–O bond in the siloxanes, the difference between the two samples can be attributed to the substitution of Si by Al or Be. In the brittle mica each O atom in the perforated O plane is simultaneously bound to one Si and one Al (or Be) atoms. The Si–O bond consists of a σ and a localized π orbital, whereas the Al–O bond is a pure σ bond and does not use a nonhybridized p orbital. In this case O atoms undergo sp^2 hybridization in order to minimize repulsion between bonding and nonbonding electrons in the valence shell. According to the present model hydrophilicity of brittle mica should be higher compared with that of pyrophyllite. This is demonstrated in Fig. 8, showing that the electron density in the nonbonding orbital of an oxygen belonging to the perforated oxygens plane is higher in the Si—O—Al group as compared with an Si—O—Si group.

Antigorite (a serpentine subgroup mineral), sepiolite, and palygorskite show an occasional inversion of the silicate layer with Si—O—Si groups, which serve as bridges between the alternating ribbons of alumino–Mg–silicates. An Si—O—Si angle of $\approx 180°$ is supposed to be found in these bridging groups.[13, 14] The double-bond character of both Si—O bonds in this siloxane group is revealed from an IR absorption band at $\approx 1200\ cm^{-1}$.[15] According to the present model, water molecules cannot form hydrogen bonds with the O atoms that are located at the inversion sites of these minerals.

3.3. Wettability of Talc, Pyrophyllite, and Vermiculite

Schrader and Yariv[16] determined the surface characteristics of the perforated oxygen plane from hydration properties of pyrophyllite and talc. Advancing contact angles of water were compared with those of tetrabromoethane on freshly prepared surfaces of these minerals. The advancing contact angle of water on the relatively

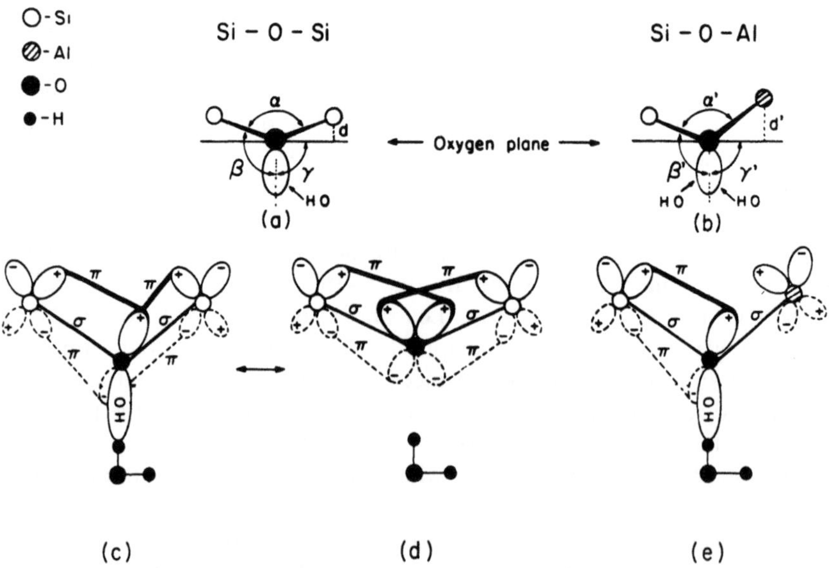

FIGURE 8. Bonds and atomic orbitals in Si—O—Si and Si—O—Al groups in exposed surfaces of tetrahedral sheets of layer silicates and possible hydrogen bonding interactions between these groups and a water molecule. (a) and (b) are spacial representations of Si—O—Si and Si—O—Al groups, respectively. H.O. = hybridized orbital; (c), (d) and (e) give nonhybridized (+ −) and hybridized (H.O.) atomic orbitals of oxygen (2*p* orbitals) and of silicon or aluminum (3*d* orbitals) and bonding orbitals between oxygen and silicon or aluminum (σ and π). (c) and (d) demonstrate possible hydrogen bonding interactions with a hybridized orbital. (From Ref. 16. Reproduced by permission of the *Journal of Colloid and Interface Science*.)

randomly oriented minerals is about 61°, and on the crystals on tape backing that have a preponderance of cleavage plane orientation, the advancing angle is about 83°. The latter value seems to be the most representative of that of the cleavage plane. Both values are in the hydrophobic range. The advancing contact angles of tetrabromoethane on tape-supported samples (which have a preponderance of cleavage plane surface) are about 44°. This value indicates that the cleavage surface is not strongly "organophilic."

For comparison, advancing contact angles of water and of tetrabromoethane were measured on freshly prepared surfaces of vermiculite layers. The advancing contact angles of water ranged between 0° and 15°. These values are virtually, or completely, in the hydrophilic range (the nonzero contact angles are probably due to organic contamination from the air). The advancing contact angles of tetrabromoethane ranged between 0° and 14°. This organophilicity of vermiculite is due either to a contribution of oxygen lone-pair donor electrons to the London dispersion forces or to tetrabromoethane possibly acting as a Lewis acid on this donor pair. Vermiculite has much tetrahedral substitution, and hydrophilicity of this mineral may result from the presence of O atoms undergoing sp^2 hybridization, as expected for brittle mica. It may also result from exchangeable hydrated Mg^{2+} cations that are attached to the oxygen plane.

Mulla *et al.*[17] applied statistical mechanics and molecular dynamics simulations for describing the properties of vicinal water located between the parallel uncharged oxygen planes of pyrophyllite, separated by 3.3 nm. They concluded that the vicinal water differed substantially from bulk water in the static orientation of molecular dipole moments and the rate of relaxation of these moments. However, no significant differences between the radial distribution functions of interfacial and bulk water were observed from these simulations. Hydrogen bonding patterns and the rate of self-diffusion in the first two layers near the surfaces were significantly different from that in bulk water.

3.4. The Extent of the Smectite—Water Interaction

In aqueous suspensions of smectites it is possible to specify the extent of the interfacial region whose spectroscopic properties are perturbed significantly by the clay–water interaction, relative to the spectrum of bulk water. The study of near IR spectra of Na–hectorite showed that the H_2O spectrum is perturbed up to the adsorption of about two and a half monolayers. Beyond about two and a half monolayers the near IR spectroscopic properties of the adsorbed water are not perturbed significantly.[18] This was also examined by NMR study of the temperature dependence of the spin-lattice relaxation time, T_1, of the protons in Na–hectorite. It was estimated that this value for the adsorbed water molecules differs from the value for the molecules in bulk liquid water only in the first nanometer away from the clay surface, which is equal to approximately three monolayers.[19] These experimental results are in good agreement with the statistical mechanics and molecular dynamics simulations of Mulla *et al.*[17] for describing the properties of vicinal water at the surface of pyrophyllite.

IR spectroscopic properties of Na–montmorillonite showed that perturbation of the water spectrum continues with adsorption beyond four monolayers.[20]

4. WETTABILITY OF THE HYDROXYL PLANE

This plane is involved in wettability of minerals from the serpentine–kaolin group (TO minerals), but may have an indirect effect only on the wettability of TOT minerals. In trioctahedral minerals all four orbitals in the valence shell of the oxygens of the hydroxyl plane are occupied, and, consequently, these atoms cannot be involved in hydrogen bonds by accepting protons. In dioctahedral minerals, basic properties of OH groups are weaker than those of trioctahedral minerals, and the oxygens of the hydroxyl plane do not accept protons from water molecules. Thus, hydrophilicity of the hydroxyl plane is determined solely by possible interactions between the protons of the hydroxyls and water oxygens that should react as proton acceptors in order to form hydrogen bonds. For this purpose the hydroxyl group must be acidic, and its acid strength must be high enough to allow proton donation to electron lone pairs belonging to the adsorbed water molecules.

The acid strength of an M–OH group (where M is Mg or Al) depends on the polarizing power of M. Proton donation ability increases in passing from Mg–OH to Al–OH. This is demonstrated by the absence of hydrogen bonds in brucite crystals in contrast to their presence in all crystalline varieties of aluminum hydroxide or oxyhydroxide. Synthetic brucite and gibbsite do not show a great affinity for water adsorption, since all adsorbed water is removed from these crystalline hydroxides under vacuum, although they are hydrated in air.[21] Amorphous magnesia and alumina, which sometimes cover natural brucite, gibbsite and other minerals, can become highly hydrated due to the many defect sites that are found in the surface planes and due to adsorption of Mg^{2+}, Al^{3+}, and OH^- that are located on the surface and can freely react with adsorbed water molecules. This will be further discussed in the next section, which deals with broken bonds.

The surface characteristics of the hydroxyl plane in dioctahedral clay minerals can be inferred from the characteristics of intercalated water in halloysite. By using IR spectroscopy, Yariv and Shoval[22] showed that water molecules inside the interlayer space of halloysite do not form hydrogen bonds with OH groups of the hydroxyl plane. Instead, hydrogen bonds occur between water molecules that form water clusters. This occurs because the enthalpy released from the association of water molecules is higher than that evolved by hydrogen bond formation between the hydroxyl groups and the water oxygens. In the presence of cluster breakers, such as CsCl or RbCl, clusters are disrupted and hydrogen bonds can be formed between OH groups of the hydroxyl plane belonging to a gibbsite-like layer and the adsorbed water molecules.[23] This process is associated with an increase in the entropy of the system. In accordance with this observation, it is assumed that the hydroxyl plane, also when it forms the external surface of the clay mineral, does not form hydrogen bonds with adsorbed water molecules. The hydration of the hydroxyl plane results in a hydrophobic hydration boundary layer of water, with no hydrogen bonds between the water molecule and the OH groups. The negative oxygens of the water molecules are oriented toward the hydroxyl plane, but there is no localized interaction between water molecules and this surface. Indeed, most kaolinites and serpentines are very easily dried, and do not show an endothermic differential thermal analysis (DTA) peak that can be attributed to dehydration.

During an IR study of serpentines with tetrahedral and octahedral substitutions, Heller-Kallai *et al.*[24] showed that localized hydrogen bonds were formed to a very small extent between OH of the hydroxyl planes and O of the oxygen planes. This localized interaction occurs between a siloxane grup in which an Al substitutes for Si and an M−OH group in which an Al or Fe substitutes for Mg. Heller-Kallai *et al.* claimed that this interaction takes place only because the substitution of Al for Mg increases the acid strength of the OH groups, and at the same time the substitution of Al for Si increases the basic strength of the siloxane group. The increment in the acid strength of the OH groups changes the hydration properties of this plane from hydrophobic into hydrophilic. To the best of our knowledge, no systematic study on this effect has been carried out, but IR spectra of KBr disks of serpentines with octahedral substitutions of Al^{3+} or Fe^{3+} for Mg^{2+} showed that the intensity of the water band increases with increasing octahedral substitution. When the disks were thermally dehydrated, the principal water band persisted at 250°C. Spectra of KBr disks of kaolinites with no octahedral substitution did not show water bands after heating the disks at 150°C. It is to be expected that OH groups attached to Al in brucite-like layers should be more acidic than OH groups in gibbsite-like layers, because in the former the O atom is coordinated with three metallic cations, two Mg and one Al, whereas in the latter it is coordinated with two metallic cations only.

5. *WETTABILITY OF BROKEN-BOND SURFACES*

When an alumino- or magnesium–silicate layer is disrupted, the following functional groups are exposed: R'_3Si-O-, R'_3Si^+, R''_5Al-O-, R''_5Al-OH, and R''_5Al^+, or R'''_5Mg-O-, R'''_5Mg-OH, and R'''_5Mg^+, as well as $R'_3Si-O-SiR'_3$ and $R'_3Si-O-AlR''_5$ or $R'_3Si-O-MgR'''_5$ (where R', R'', or R''' is the bulk alumino- or magnesium–silicate layer). In iron-rich clay minerals, groups such as R''_5Fe-O-, R''_5Fe-OH, and R''_5Fe^+ are also exposed. These groups form a surface not parallel to the silicate unit layer (or the *ab* crystallographic plane). The valences of the exposed crystal atoms are not completely compensated, as they are in the interior of the crystal. These surfaces are called *broken-bond surfaces* or *edge surfaces*. The exposed functional groups are very active and may react as electron pair donors or acceptors. The unsatisfied charges may be balanced by the sorption of cations or anions, either specifically by chemisorption or nonspecifically by electrostatic attraction, the latter ions being exchangeable.

The surface properties depend greatly on the exposed atoms. That part of the surface at which the octahedral sheet is broken may be compared to the surface of alumina or magnesia (and in a few cases also to that of hydrous iron oxide), whereas that part of the surface at which the tetrahedral sheet is broken may be compared to that of silica or silica–alumina.

The charge of the broken-bond surface is determined by specifically adsorbed cations and anions (potential-determining ions). Both protons and hydroxyls are important potential-determining ions. In acid solutions the net charge of the broken-bond surface is positive, and the positive charge further increases with decreasing pH. Protons occupy positions near oxygens. Isoelectric points of magnesia, alumina, iron oxide and silica are 12, 9.2, 8.5, and 2, respectively.[25]

The acid strengths, expressed in pK_a values, of OH groups on MgO, γ-Al_2O_3, and SiO_2 surfaces are $\geqslant$18.5, 8.5, and 7.0, respectively.[26] Depending on the ability of the oxygen to donate a pair of electrons, the $-Mg-O^-$ functional group is the first site to be protonated, followed by the octahedral $-Al-O^-$, $-Fe-O^-$, tetrahedral $-Al-O^-$, and $-Si-O^-$ groups, respectively. The $-Mg-OH$ and $-Al-OH$ functional groups can be further protonated at a pH below 4. In alkaline solutions the net charge of the broken bond is negative.

Surface charge is also determined by the migration of protons from the interior of the clay mineral (hydroxyls) to the surface by thermal diffusion, named prototropy.[27] This diffusion is accelerated by grinding or by raising the temperature. As a result of prototropy, the broken-bond surface acquires positive charge.

A number of studies indicate that the broken-bonds surfaces of the clay particles are positively charged at pH < 7–8,[28] although some data suggest that the edges are neutralized already at pH about 6.[29] Since surface charge of kaolinite is determined mainly by the charge of the broken bonds, the information on the effect of pH on the electrophoretic mobility of kaolinite can give information about isoelectric points of broken bonds in general. Isoelectric points of Na–kaolinite in NaCl solutions were determined by Srinivasan and Hepler.[30] They are at 10°C, $pH = 3.7 \pm 0.3$; 25°C, $pH = 4.0 \pm 0.2$; 40°C, $pH = 3.6 \pm 0.3$; 55°C, $pH = 3.6 \pm 0.3$, nearly independent of temperature.

Clay minerals were widely titrated by acids and bases.[1] Potentiometric titrations reveal that they behave like amphoteric oxides. It is assumed that the broken-bonds surface is the principal site for acid–base reactions to take place. A characteristic potentiometric titration of Na–kaolinite is shown in Fig. 9 after Pefferkorn *et al.*[31] The pH variations of the Na–kaolinite suspension (0.25 wt%) in 0.01 *M* NaCl were recorded following three steps: (a) titration of the clay with NaOH in the alkaline range, starting at pH 7; (b) titration with HCl in the acidic range, starting at pH 7, and (c) titration with NaOH of the clay suspended at pH 4 for 2 h. Curves a and b of Fig. 9 are continuous, with a well-defined inflexion point at pH 7.2, which, according to some investigators, corresponds to the isoelectric point.[25] Curve c has a shape similar to that of a titration of a weak acid by a strong base, presenting a second titration end point.

Pefferkorn *et al.* explained this particularity in the following way. The H-kaolinite, which was obtained from the treatment of the clay in pH 4, is slightly decomposed, and Al^{3+}, which has been dissolved, is specifically adsorbed onto the clay surface. However, since no strong acid treatment was imposed, limited Na^+–H^+ exchange occurred on the primary Na–kaolinite. The titration endpoint corresponds to the end of the H^+–Na exchange, both being counterions of the permanent charge of the basal face. The hydroxyl ion adsorption that arises between this endpoint and the isoelectric point corresponds to the following surface reaction:

$$R''_4-Al(OH)\cdot O(H)-SiR_3^+ + OH^- \rightarrow R''_4-Al(OH)\cdot O-SiR_3 + HOH$$

The ion exchange, which confined to alkaline conditions and depends on electrolyte concentration, originates through ionization of the edge silanols, whose isoelectric point is 2. By comparing the behavior of the broken-bond surface of kaolinite to

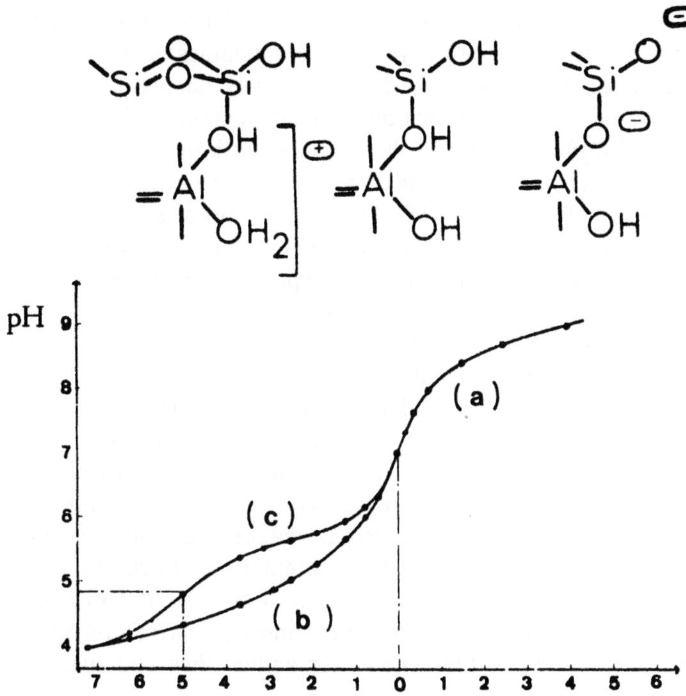

FIGURE 9. Variation in charge characteristics of the broken-bond surface of kaolinite as a function of pH changes and a potentiometric titration of Na–kaolinite (0.25 wt% in 10^{-2} *M* NaCl suspension): (a) alkaline titration starting at pH 7; (b) acidic titration starting at pH 7; (c) back titration between pH 4 and pH 9 of a clay suspended at pH 4 for 2 h. (After Pefferkorn *et al.*[31])

those of silica and alumina, Pefferkorn *et al.* described the effect of pH on the surface functional groups of kaolinite in the following way. Starting at pH 2, the degree of ionization of the silanol groups remains very limited up to pH 7. Beyond this value the ionization increases rapidly, and, at pH 10, 60% of the ionizable silanol groups are charged in pure water. Ionization of aluminol groups may start only at pH beyond 9.2. Variations in charge characteristics of the broken-bond surface of kaolinite under acid, neutral, and alkaline conditions are shown in Scheme 1.

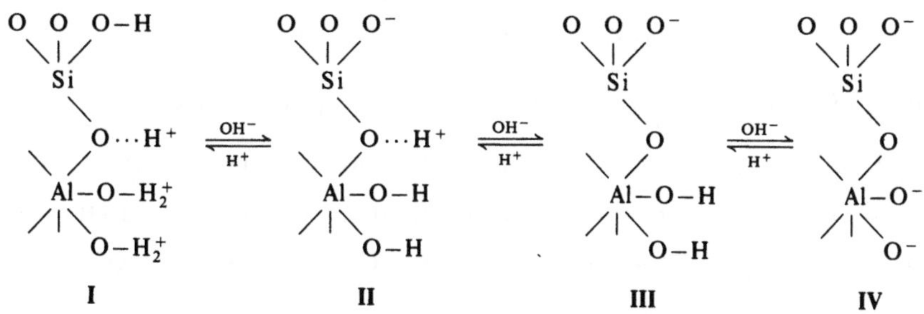

SCHEME 1. A schematic representation of functional groups on the broken bond surface of kaolinite as a function of the pH of an aqueous clay suspension: (I) pH < 2; (II) At the isoelectric point of kaolinite; (III) pH slightly above the isoelectric point; (IV) pH > 9.2.

Wettability of this surface reveals properties similar to those of metal oxides and silica. Adsorption of water vapor on metal oxides includes reversible and irreversible chemisorption processes. Adsorption heats regularly decrease with coverage until they reach the water vapor condensation value at the completion of molecular H_2O layer on the top of surface hydroxyls. According to Fubini *et al.*,[32] during hydration of alumina surface, chemisorption occurs first. It covers two processes, one dissociative (disruption of Al–O and H–OH bonds) and one coordinative (hydration of surface functional groups). Both chemisorption reactions are irreversible. Chemisorption is followed by physical adsorption. The different processes occur simultaneously for a large range of coverage.

Based on the many studies on the wettability of metal oxides, the following can be concluded on water sorption onto the broken-bond surface. This process takes place via three mechanisms: (1) dissociative chemisorption, (2) hydration of exchangeable ions, and (3) hydrogen bonding between the water molecules and exposed hydroxyls or oxygen atoms. The first mechanism takes place when a new surface is formed, for example by grinding, and is exposed to humid atmosphere. It results in the formation of groups such as silanol, aluminol, and magnesol. The second mechanism results from ion ↔ water interactions. Each ion is surrounded by a cosphere in which the organization of water molecules differs from that in the bulk solvent outside the cosphere. The third mechanism results in a hydrophilic hydration interface, similar to that known for active silica and silica–alumina.

5.1. Dissociative Chemisorption of Water

A fresh "broken-bond" surface is composed mainly of Lewis acidic and basic sites, such as $(-O-)_3Si^+$, $(\equiv O-)_5Al^+$, $(\equiv O-)_5Mg^+$, $(-O-)_3Si-O-$, $(\equiv O-)_5Al-O-$, and $(\equiv O-)_5Mg-O-$. When this surface is exposed to humid atmosphere, the following chemisorption hydration reactions will be the first to occur:

$$(-O-)_3Si^+ + 2HOH \rightarrow (-O-)_3Si-O-H\cdots OH_3^+$$

$$(\equiv O-)_5Al^+ + 2HOH \rightarrow (\equiv O-)_5Al-O-H\cdots OH_3^+$$

$$(\equiv O-)_5Mg^+ + 2HOH \rightarrow (\equiv O-)_5Mg-O-H\cdots OH_3^+$$

$$(-O-)_3Si-O^- + HOH \rightarrow (-O-)_3Si-O-H\cdots OH^-$$

$$(\equiv O-)_5Al-O^- + HOH \rightarrow (\equiv O-)_5Al-O-H\cdots OH^-$$

$$(\equiv O-)_5Mg-O^- + HOH \rightarrow (\equiv O-)_5Mg-O-H\cdots OH^-$$

Migration of protons will lead to neutralization reactions such as

$$(-O-)_3Si-O-H\cdots OH_3^+ + (-O-)_3Si-O-H\cdots OH^- \rightarrow 2(-O-)_3Si-O-H\cdots OH_2$$

$$(\equiv O-)_5Al-O-H\cdots OH_3^+ + (\equiv O-)_5Al-O-H\cdots OH^- \rightarrow 2(\equiv O-)_5Al-O-H\cdots OH_2$$

$$(\equiv O-)_5Mg-O-H\cdots OH_3^+ + (\equiv O-)_5Mg-O-H\cdots OH^- \rightarrow 2(\equiv O-)_5Mg-O-H\cdots OH_2$$

However, a nonaged surface is rich with $OH_3{}^+$ and OH^-, which grant charged hydrophilic sites to the surface. Consequently, the adsorbed water content of clay increases with increasing time of grinding.

Mechanical treatments of clays leave some of the surface groups Si–O–Si, Si–O–Al, Si–O–Mg, Al–O–Al, and Mg–O–Mg in an activated state. When these activated groups are in contact with water, a chemisorption may occur through the following dissociative reactions:

$$(-O-)_3Si-O-Si(-O-)_3+HOH \rightarrow 2(-O-)_3Si-O-H$$

$$(-O-)_3Si-O-Al(-O\equiv)_5+HOH \rightarrow (-O-)_3Si-O-H+(\equiv O-)_5Al-O-H$$

$$(-O-)_3Si-O-Mg(-O\equiv)_5+HOH \rightarrow (-O-)_3Si-O-H+(\equiv O-)_5Mg-O-H$$

$$(\equiv O-)_5Al-O-Al(-O\equiv)_5+HOH \rightarrow 2(\equiv O-)_5Al-O-H$$

$$(\equiv O-)_5Mg-O-Mg(-O\equiv)_5+HOH \rightarrow 2(\equiv O-)_5Mg-O-H$$

5.2. *Adsorption of Molecular Water*

In the presence of water, the broken-bond surface is generally covered with amphoteric surface hydroxyls, mainly in the form of silanol, aluminol, and magnesol. As well, it contains exchangeable cations and anions and specifically adsorbed ions, such as hydroniums and hydroxyls. Adsorbed water at the broken-bond surface can be divided into three zones: the ion hydration zone, A_m, the hydrophilic hydration interface zone, A_b, both having a high degree of order, and the third, having a low degree of order, B_{bm}, which serves as a bridge between the two first zones.[3]

5.2.1. Hydration of Exchangeable Ions

This mechanism gives rise to the formation of hydration zone A_m. Exchangeable ions are hydrated mainly due to electrostatic forces interacting between the ion and an electric pole on the water molecule. Hydration numbers for cations sorbed on kaolinite are given in Table 1. According to Fripiat,[33] when physical adsorption of water occurs on outgassed kaolinite, the interface consists first of molecules that form clusters around partially or fully dehydrated cations.

The cation hydration zone has the nature of a Bronsted acid. As the water content of the clay drops below that required to saturate the solvation shells of the exchangeable cations, the residual water dissociates more readily and, hence, is more acidic. Also, the removal of water exposes Al or Mg atoms located at edge sites, creating Lewis acid sites.[34]

The influence of exchangeable H^+, Na^+, K^+, Mg^{2+}, Ca^{2+}, Ba^{2+}, and Al^{3+}

TABLE 1. Cation Hydration Number on Kaolinite Surface

Sample	Hydration number
Li–kaolinite	2.1–2.7
Na–kaolinite	1.6–2.1
K–kaolinite	1.7–2.0
Mg–kaolinite	4.9–10.0
Ca–kaolinite	4.3–7.8

After Ref. 33.

ions on the wettability of kaolinite surface was determined by Janczuk *et al.*[35] from contact angles, which were measured in kaolinite–water drop–air and kaolinite diiodomethane–air systems. From the results and using a modified Young equation, the dispersion and nondispersion components of the free energy of the kaolinite hydrated surface were determined. The dispersion component was between 32.8 and 38.9 mJ/m^2, but the nondispersion component changed almost linearly from 53 to 95.9 mJ/m^2 with the change of the entropy of hydration of the adsorbed ions, except for K^+ and Ba^{2+}, probably due to their large ionic radii.

5.2.2. Adsorption of Water via Hydrogen Bond Formation

This mechanism gives rise to the formation of hydration zone A_b. The hydration properties of silanol, aluminol, and magnesol groups are inferred from the surface properties of silica, alumina, and magnesia, respectively.

Silanol groups reveal weak acidic properties relative to water (silicic acid is a very weak acid, the first dissociation constant being $K_a = [H^+] \times [H_3SiO_4^-]/[H_4SiO_4] = 10^{-9.9}$). Usually a single OH group is bonded to an Si atom, but two OH groups can sometimes be found near one Si atom and, rarely, three OH groups. The acid strength of the silanols increases with the number of hydroxyls bound to a single Si atom.[36] The silica gel surface has a significant affinity for water. With low water contents on the silica gel surface, free silanols donate protons to water oxygens. With higher water contents hydrogen-bond-containing clusters are formed before all the free silanols are taken up by monomeric water molecules. From heat of reaction measurements it appears that clustering of water molecules is favored over bonding between monomeric water and single silanols. On silica gel surfaces water is stabilized in small clusters containing five or six molecules, with heat of clustering similar to the heat of liquefaction of water. The clustered molecules, however, are more restricted translationally and rotationally in their motion than monomers weakly bonded to the silanols and therefore have a lower entropy. Thus, at low coverage, water adsorption on silanols is favored by entropy, whereas at higher coverages clustering is favored by energy stabilization in the cluster.[37]

Since the hydration energy of silica gel is less than the heat of liquefaction of water, silica gel is hydrophobic. Water clusters are more easily formed on silica surfaces with the lowering of surface densities of silanol groups because the increasing distances between hydroxyls prevent multiple water–hydroxyl bonds from being

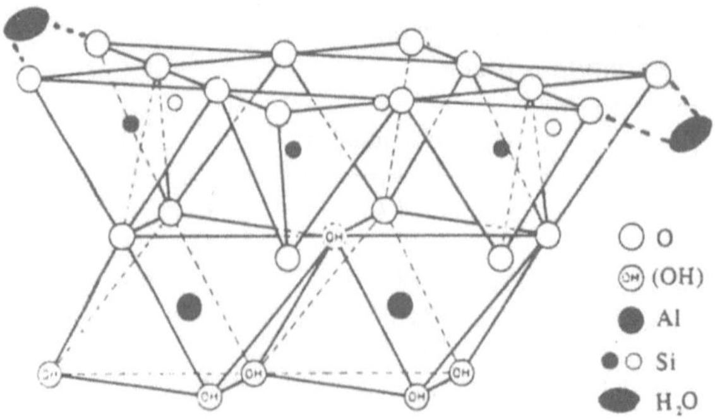

FIGURE 10. A water molecule forms two hydrogen bonds with two different silanols on the broken-bond surface.

formed. For comparison, quartz, having almost no surface hydroxyls, allows the formation of water clusters at a very low water coverage. The broken-bond surface has silanol groups at a closer proximity than silica gel. During water adsorption, multiple water–hydroxyl bonds are formed (Fig. 10), accompanied by higher heats of adsorption, and the surface is occupied to a higher degree before clustering occurs. To conclude, silanols at the broken-bond surface show a higher hydrophilicity as compared with silanols at a silica surface.

Surface models for alumina and characterization of surface sites were recently published by Ratnasamy and co-workers.[38] Analysis of surface structure of aluminas reveal the presence of five different types of aluminol groups (Scheme 2). Two types of OH configurations can be distinguished for tetrahedral Al^{3+}: (1) a terminal OH group is coordinated to a single tetrahedral Al^{3+} cation (type Ia), and (2) a bridging OH group links tetrahedral and octahedral Al^{3+} cations (type IIa). Two additional OH-type configurations occur consisting of bridging OH groups

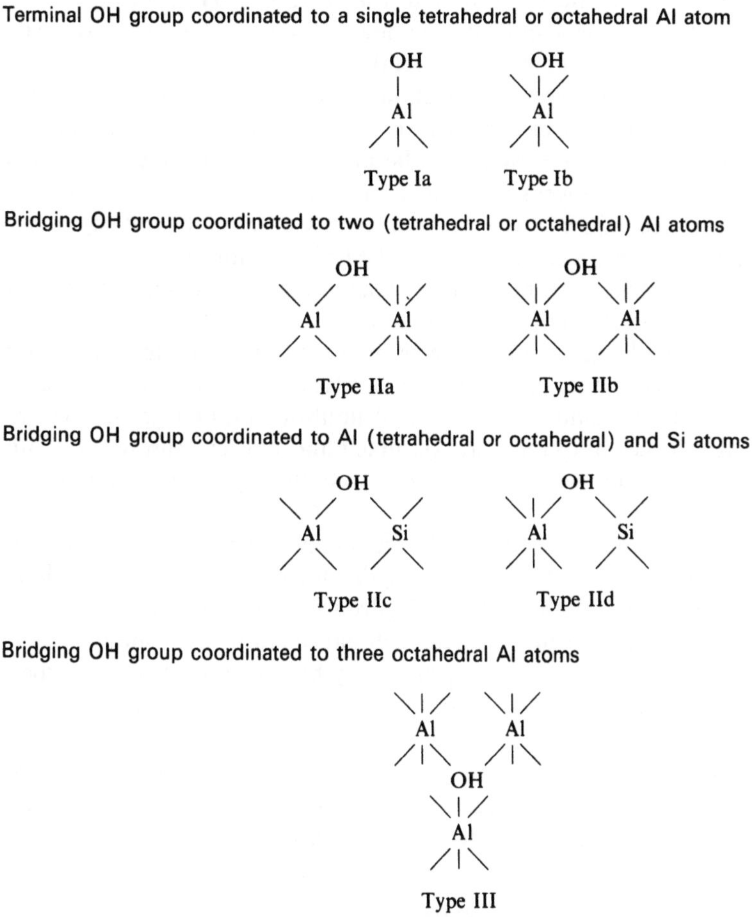

SCHEME 2. Possible types of aluminol groups present on the surface of alumina (types Ia, Ib, IIa, IIb, and III) after Knoezinger and Ratnasamy[38a] and Ratnasamy and Sivasanker[38b] and on the broken-bond surface of clay minerals.

being coordinated to two or three octahedral Al^{3+} cations (type IIb and III, respectively). A fifth configuration can be distinguished in which a terminal OH group is being coordinated to a single octahedral Al^{3+} cation (type Ib). Type III is the most acidic configuration, but reveals greatest steric hindrances for hydrogen bondings. Type Ib is the least acidic.

Similar configurations are found at the surfaces of dioctahedral clay minerals. Type III can be found at the hydroxyl plane of TO clay minerals, but is rare. Type IIb is the common configuration occurring at the hydroxyl plane of TO minerals. Type Ia can be found at the edges of a tetrahedral sheet with Al substituting for Si. Type IIa can be found at the edges of the O,OH plane in addition to type IId, in which OH bridges between octahedral Al and Si. Type Ia can be found at the edges of an oxygen plane in addition to type IIc, in which OH bridges between tetrahedral Al and Si.

Type Ib is the most common configuration occurring at the broken-bond surface. Usually a single OH group is bound to an Al atom, but two OH groups can sometimes be found near one Al atom and, rarely, three OH groups.

Trioctahedral clay minerals should give configurations similar to Ib, IIb, IIc, and III with Mg located at the center of the octahedron.

Hydrogen bond formation between an aluminol or a magnesol group and a water molecule is more restricted in comparison with a hydrogen bond formation between a silanol group and a water molecule, because the former are weaker acids than silanols. Magnesol is even a weaker proton donor than aluminol. Consequently, acidity of dioctahedral sheet edges is stronger than acidity of trioctahedral edges, but acidity of dioctahedral edges decreases with octahedral substitution of Mg for Al, and, vice versa, acidity of trioctahedral edges increases with octahedral substitution of Al for Mg.[39] On the other hand, magnesol and, to a lesser extent, aluminol can react as bases and form hydrogen bonds via proton accepting. This interaction does not occur with the hydroxyl planes of brucite or gibbsite, but is found on the surfaces of amorphous magnesia or alumina with terminal nonbridging OH groups, where not all coordination valences of oxygens are occupied and oxygens can rotate more freely, thus matching steric requirements for accepting protons from water molecules.

The presence of specifically adsorbed hydronium cations or hydroxyl anions on the broken-bond surface increases the degree of hydrophilicity of this surface. These ions are located at the inner Helmholtz layer. Hydrogen bonds in which these ions are involved and which are formed via proton donation from hydronium ion to adsorbed water molecule, or via accepting protons by hydroxyl ion from adsorbed water, are stronger than any hydrogen bond formed between silanol, aluminol, or magnesol and adsorbed water.

To conclude, adsorbed water molecules in zone A_b are involved in one or more hydrogen bonds as proton donors or acceptors. The adsorption of water onto the broken-bond surface thus results in a hydrophilic hydration.

5.3. *Hydration of Sepiolite and Palygorskite*

In sepiolite and palygorskite perforated oxygen planes and broken-bond surfaces are exposed inside the channels. Two perpendicular parallel surfaces of the channels are made of broken-bond surfaces, and the other two horizontal surfaces are made of oxygen planes. Each $(\equiv O-)_4Mg$ located at the edge of the octahedral sheet is coordinated to two water molecules as $(\equiv O-)_4Mg\cdot(OH_2)_2$, thus completing the six coordination of Mg. This coordinated water is called *bound water.* Dehydration of bound water occurs in two stages. In the first stage (240°–430° or 220°–370°C in sepiolite or palygorskite, respectively) one molecule is evolved. This dehydration is reversible. The second-stage dehydration (430°–650° or 370°–625°C in sepiolite or palygorskite, respectively) occurs together with dehydroxylation of the clay and is not reversible.[40–43]

The empty space in the channel is filled with zeolitic water, forming clusters that are hydrogen bonded to the bound water (Scheme 3). Zeolitic water is evolved at 100°–150°C.

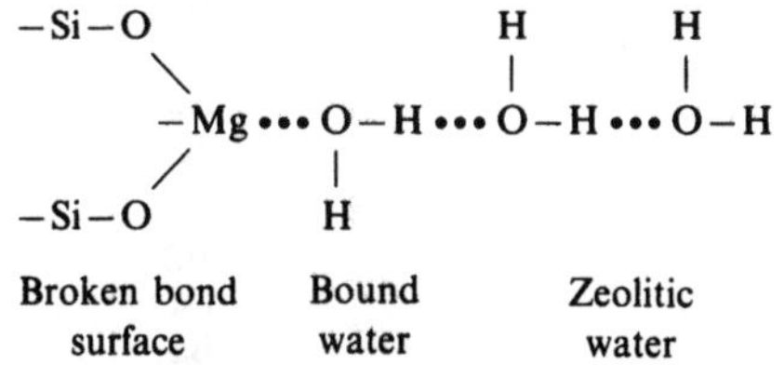

SCHEME 3. Possible association between magnesium at the broken-bond surface and bound water and between bound water (proton donor) and zeolitic water (proton acceptor) via hydrogen bonds in the channels of sepiolite and palygorskite.

6. WATER IN THE INTERLAYER SPACE OF TOT MINERALS

The present section deals with the interlayer space of smectites and vermiculites. The interlayer space of a TOT clay mineral is the space between two parallel silicate layers, bordered by two oxygen planes, the oxygens belonging to siloxane groups. The wettability and the resulting structure of the interlayer water are the outcome of (i) the thermal motion of water molecules in the environment of the mineral, (ii) the electrostatic attraction forces between water molecules, and the exchangeable cationic species and (iii) attraction and dispersion forces between the TOT layers. The results of recent experiments concerning the structure of water in smectite–water systems were reviewed recently by Sposito and Prost.[44]

6.1. Swelling

Swelling is the process by which the clay mineral expands beyond the original limit, which is $\approx$0.95 nm, as a result of the adsorption of water into the interlayer space. The uptake of water molecules causes expansion of the clay crystal along the *c* axis, and this expansion can be monitored by the use of adsorption isotherms. The expansion can be followed by x-ray diffraction, and is determined from the *c*-spacing Table 2. In most publications swelling has been determined under air atmosphere or under various controlled humidities, but it has also been determined for aqueous suspensions or under the influence of water vapor.

When water at atmospheric pressure comes into contact with a crystal of

TABLE 2. Charge Density and Crystalline Swelling of TOT Clay Minerals

Mineral type	Charge per unit cell	Interlammelar cation	Interlayer spacing in dilute clay suspensions (nm)
Talc	0		0.93
Pyrophyllite	0		0.91
Illite	1.3	K	1.00
Vermiculite	1.3	Li	>4.00
		Na	1.4–1.5
		K	1.16
		Cs	1.20
		Mg	1.4–1.5
		Ca	1.4–1.5
		Ba	1.57
Montmorillonite	0.67	Li	>4.00
		Na	>4.00
		K	1.55 and >4.00
		Cs	1.20
		Mg	1.94
		Ca	1.91
		Ba	1.87
		Ni	1.90–2.00
		Cu	1.90–2.00
		Zn	1.90–2.00
		Cd	1.90–2.00
Beidellite	0.25–0.6	Li	>4.00
		Na	1.52
		K	1.27
		Mg	1.85
		Ca	1.54vw, 1.87s
		Ba	1.84
Saponite	0.25–0.6	Li	>4.00
		Na	1.52
		K	1.26
		Mg	1.57vw, 1.89s
		Ca	1.54s, 1.87vw
		Ba	1.59vw, 1.85s

>4.00 = highly swollen; vw = very weak; s = strong.
After Refs. 48, 50, 51.

smectite or vermiculite, it penetrates between the layers of the crystal and forces them apart. In order to prevent swelling a pressure in excess of the atmospheric pressure must be applied to the crystal. The difference between the applied pressure and the atmospheric pressure is defined as the swelling or disjoining pressure of the crystal. Because clay layers are electrically charged, swelling has been described in terms of Derjaguin–Landau–Verwey–Overbeek (DLVO) theory on the interaction of colloidal particles in suspension. This theory describes the interaction of two surfaces resulting from overlapping Gouy electric double layers and van der Waals forces.[45–47]

Swelling of the interlayer space decreases with increasing surface charge density of the layer and with increasing charge and concentration of the adsorbed ions.

However, a certain surface charge is required for swelling to occur. Talc and pyrophillite, the two TOT minerals with a zero surface charge, never swell in an aqueous medium. Illite can be regarded as if the anhydrous K ions are specifically adsorbed onto the inner Helmholtz layer of the oxygen plane, resulting in a zero surface charge density. This mineral behaves like talc and pyrophillite in respect of not being expandable. In aqueous suspensions K vermiculite shrinks, whereas K montmorillonite, depending on the circumstances, may exhibit both limited and extensive swelling. Na vermiculite gives a 1.48-nm spacing even in dilute colloid suspensions, and Na–montmorillonite swells extensively. Lithium–vermiculite and montmorillonite exhibit extensive swelling.[48, 49]

Experimentally, the behavior of swelling clays has been thoroughly studied. Some examples for maximum swelling of various minerals saturated with various inorganic cations are shown in Table 2.[48, 49, 50] Some examples for the swelling of montmorillonite saturated with organic ions are shown in Table 3. Ca–montmorillonite does not swell beyond a *c* spacing of 1.9 nm, even in very dilute suspensions. Since the elementary silicate layers or lamellae are 1.0 nm thick, the swelling corresponds to the uptake of about three layers of water, each approximately 0.3 nm thick. The 1.9-nm spacing for Ca–montmorillonite is constant over the relative vapor pressure range 0.945–1.000. The spacing between silicate layers is virtually constant over a wide range of $CaCl_2$ concentrations. In contrast, the *c*-spacing values for Na–montmorillonite vary with electrolyte concentration, increasing from 1.9 nm in 0.5 *M* NaCl to 4.3 nm in 0.3 *M* NaCl and thereafter increasing further with decreasing concentration in general agreement with DLVO prediction.[52]

Accurate theoretical calculations of the electric double layer interactions in systems with divalent counterions predicted the existence of a strong short-range attraction between equally charged surfaces. The attraction originates in the correlations between the counterions, which are not considered in Poisson–Boltzmann (PB) theory, on which the DLVO theory is based.[53–55]

It has been inferred that the double-layer attraction is the likely cause of the existence of two lamellar phases with different degrees of swelling in aqueous Na^+ and Ca^{2+} surfactant mixtures.[56] More recently it has been shown that the attraction explains in semiquantitative manner the restricted swelling of Ca clays in pure water. Kjellander *et al.*[57] reported direct experimental measurements of the short-range forces between two molecularly smooth mica surfaces in $CaCl_2$ solutions. They showed that the experimental results were in agreement with the theoretical calculations of forces between charged surfaces in electrolyte solutions.

6.2. The Fine Structure of Interlayer Water

The structure of interlayer water can be regarded as principally a result of superposed effects, on one hand, the nature of the oxygen planes that border this space, and, on the other hand, the nature of the exchangeable cations that are located in this space.

(I) The nature of the oxygen plane depends on the charge of the silicate layer and on whether the charge results from tetrahedral or octahedral substitution. With no tetrahedral substitution the oxygen plane is composed predominantly of oxygen atoms belonging to siloxane groups that do not form stable hydrogen bonds with water molecules, and consequently the hydration of this surface is hydrophobic in nature. With tetrahedral substitution the oxygen plane is composed of oxygen atoms belonging to Si–O–Al groups that can form hydrogen bonds with water molecules, and consequently the hydration of this surface can to some extent be hydrophilic in nature.

(II) The exchangeable cation is surrounded by a cosphere in which the organization of water molecules differs from that in the remainder of the interlayer space.[58] Hydration of the exchangeable cations has the nature of a "hydrophilic hydration," and the structure of interlayer water should be affected at some distance from the exchangeable cations due to the electrostatic interaction between the cation and the proximate water molecules. In conclusion, water in the interlayer space cannot be treated as a bulk continuum.

A simple model for interlayer water assumes the presence of three zones having differing water structures that may be distinguished in the interlayer space (Ref. 3, Chap. 7). Two zones contain ordered water, the water molecules constituting the hydration atmosphere of the ions and the solid–liquid boundary layer at the flat oxygen plane being distinguished as zones A_m and A_o, respectively (the subscripts m or o signify hydration of exchangeable metallic cation or oxygen plane, respectively). A third zone, B_{om}, is a disordered zone separating the ordered zones A_o and A_m. In zone A_o the thermal amplitudes of the intermolecular vibrations are consequently reducing the density of the water layers in the immediate neighborhood of the clay surface. In zone A_m the dipolar axis of each water molecule passes through the ion center. The attenuation of molecular motion is obtained from the strong polarizing field of the exchangeable cation. Zone B_{om} is subjected to the competing demands of water structures associated with zones A_m and A_o. These two influences counteract each other, and the water structure in this fault zone is broken down.

The presence of the three different zones can be proved by incoherent neutron scattering studies. Hall *et al.*[59, 60] have obtained quasielastic neutron scattering spectra for water adsorbed on Ca– and Mg–montmorillonite. Oriented clays containing up to three layers of water in the interlamellar space were investigated. The data were analyzed and it was concluded that some of the adsorbed water protons are stationary on the neutron scattering time scale. These belong to water molecules of the first solvation atmosphere of the bivalent cations (zone A_m). Some of the remaining adsorbed water protons diffuse by jumps within a region bounded by the opposing silicate surface or first water layer attached to this surface and hydrated exchangeable cations (zone B_{om}). Another part of the remaining adsorbed water undergoes isotropic, translational jump diffusion between adjacent bounded regions of this type (zone A_o).

6.2.1. Zone A_m

The size of zone A_m decreases with increasing ionic size and increases with increasing ionic charge. For example, at a very low relative humidity Ca^{2+} or Mg^{2+} smectites form hexahydrates, whereas Li^+ or Na^+ smectites form trihydrates. Under similar conditions zone A_m does not exist around exchangeable Cs^+ because of its low electric charge and large size, so that zone B_{om} extends from the surface of the ion.

The structure of zone A_m of Li–hectorite was investigated by Prost, using infrared spectroscopy combined with a progressive deuteration technique.[61] Dehydration under 0.02 torr produced a state corresponding to three molecules of water, on the average, firmly bound to each exchangeable cation. According to the model suggested by Prost, the Li^+ cation is centered above and/or under the ditrigonal oxygen cavity of the silicate layer, coordinated by the three water molecules. The Li cation appeared to be out of the midposition in the interlamellar space and the dipolar moment of the water molecule, being 55° tilted relatively to the *c* axis (Fig. 11a). The lone-pair electrons of the sp^3 orbitals of the oxygen atom in each water molecule are directed toward the Li^+ ion, and one of the protons in each molecule lies along an axis normal to the clay surface.

The hydrate structure in Li–hectorite was investigated by Fripiat,[62] and Fripiat *et al.*[63] and Conard,[64, 65] using NMR and neutron scattering spectroscopy. They concluded that the hydrated cation was flat and that the water dipoles were placed symmetrically about the Li^+ ion, the Li lying in the oxygen midplane (Fig. 11b). The symmetry axis of the water molecule was making an average angle of about 20° with the hectorite surface. The water protons undergo two rapid rotations around the perpendicular axis with different correlation times: the first one related to the whole hydrate, around the *c* axis of the clay platelets, and the second one related to the water molecule. Owing to the very peculiar symmetry and the short distances between the different possible sites, the protons are able to jump among 12 equivalent sites on a circle of radius 0.213 nm centered on the Li^+ cation. This motion is equivalent to a reorientation of the whole hydrate around its symmetry axis parallel to the *c* axis of the layers. The orientation of each water molecule toward the neighboring oxygens permits a pseudorotation for the protons in a circle of a radius 0.123 nm around an axis parallel to the *ab* planes.

The difference between the two structures, the one suggested by Prost from IR study and the one suggested by Fripiat *et al.* and by Conard from NMR studies, was later interpreted by Conard *et al.*[66, 67] by using neutron scattering spectroscopy (Fig. 11c). The trihydrate is not free but lies in the interlamellar space between two oxygen ditrigonal cavities. One of the two nonbonding sp^3 orbitals of the six oxygens that frame the ditrigonal cavity is pointed toward the cation. All six oxygens together provide a 12-electron nest for the Li cation. With the three water molecules of the Li hydrate, each Li^+ is surrounded by three oxygens 0.2 nm apart and six oxygens at about 0.3 nm distance. Such a configuration gives rise to two equivalent potential wells on either side of the hydrate symmetry plane. The distance separating the two equivalent sites is about 0.1 nm. Li^+ is able to oscillate from one site to the other. With the very short IR time scale the Li^+ appears to be in one of those sites (Fig. 11a). At a much longer time scale (e.g., NMR) Li^+ appears to be in an axially symmetric average position (Fig. 11b).

Similar structures for zone A_m were observed with Na– and Ca–hectorite at a low water coverage, and it was suggested that similar structures are obtained with alkali and alkaline-earth cations.[61] In the case of vermiculites or smectites with tetrahedral substitutions, the adsorbed water molecules may form hydrogen bonds with clay surface oxygen atoms at sites where Si has been replaced by Al, as shown in Fig. 8.[16, 68] A model for the arrangement of water molecules around exchangeable Ca in a mineral with tetrahedral substitution is shown in Fig. 12.

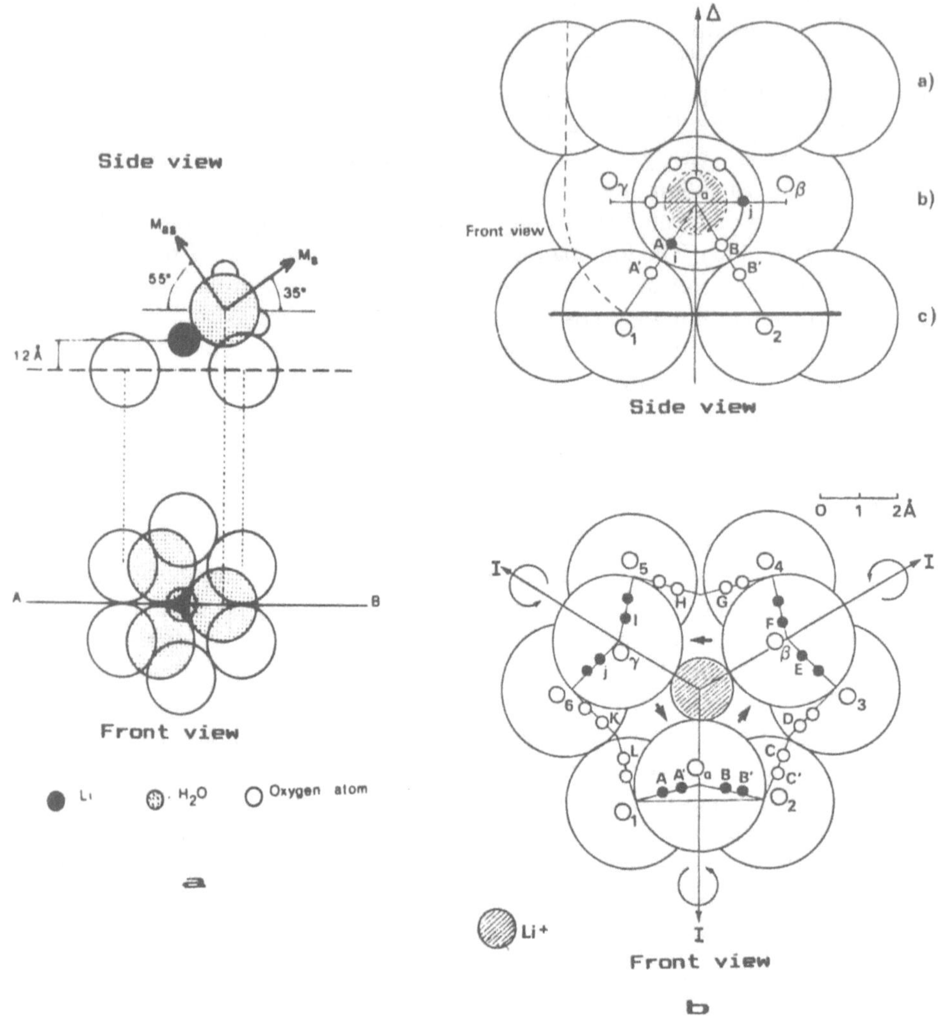

FIGURE 11. Proposed models for $[Li(H_2O)_3]$–hectorite. (a, top and bottom) Side and front views, respectively, of the orientation of water molecules with respect to Li^+ cation and oxygen plane in Li–hectorite, a model based on IR spectra, after Prost.[61] Li cation is out of the midposition in the interlamellar space. Dipolar moment of the water molecule is 55° tilted relatively to the *c* axis. (b, top and bottom) Side and front views, respectively, of the orientation of water molecules with respect to Li^+ cation and oxygen plane in Li–hectorite, a model based on NMR spectra, after Conard.[64, 65] The trihydrate is centered above and under hexagonal oxygen cavities of two phyllosilicate layers, with the Li cation in the midposition. The water protons undergo two rapid rotations around perpendicular axis, one related to the whole hydrate around *c* axis of the clay layer (Δ), and the second related to a single water molecule (I). (c, top) Side view showing instantaneous and mean positions of Li^+ and the rotation axis of the whole hydrate (Δ) and possible rotation axis of a single water molecule in relation to the different positions of the Li^+ cation, after Conard.[66, 67] (c, bottom) Top view of the negative nest (potential well) obtained due to the six nonbonding orbitals of the lattice oxygens O_1, O_2,... pointing toward the Δ rotation axis.

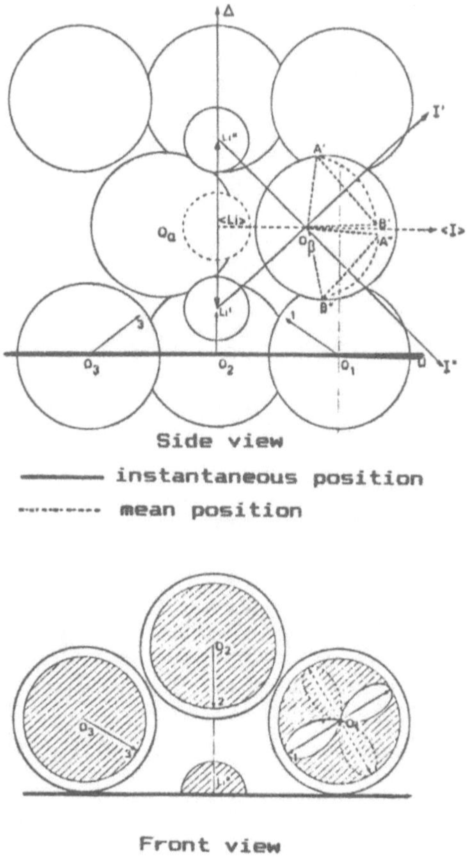

c

FIGURE 11. *Continued*

The structure of zone A_m of Na–vermiculite was investigated by Hougardy *et al.*,[69] using NMR spectroscopy. Llano vermiculite was used for this purpose. This particular sample has a high value of cation exchange capacity (200 mmol/100 g clay), forming large individual flakes with small microporosity. Its water adsorption isotherm shows two well-defined steps corresponding to one and two monolayers of water. For the two-layer hydrate there are on the average six water molecules per each Na^+ cation, and the *c* spacing is 1.48 nm. This spacing is sufficient to accommodate an octahedral arrangement of water molecules around the sodium ion. It is assumed that in this sample all the water is involved in forming zone A_m, and it is therefore a good model for the study of this zone (Fig. 13). The NMR spectra show a preferential orientation of the water molecules consistent with an octahedral arrangement in which each water molecule rotates rapidly around the C_2 axis, representing a coordinative bond of the water to Na^+ and the entire ensemble rotates around the C_3 axis, which is parallel to the *c* axis of the clay. From the spectrum it is obvious that the sample contains one type of water, which is water of hydration of the cation. The NMR spectra show that water molecules within the

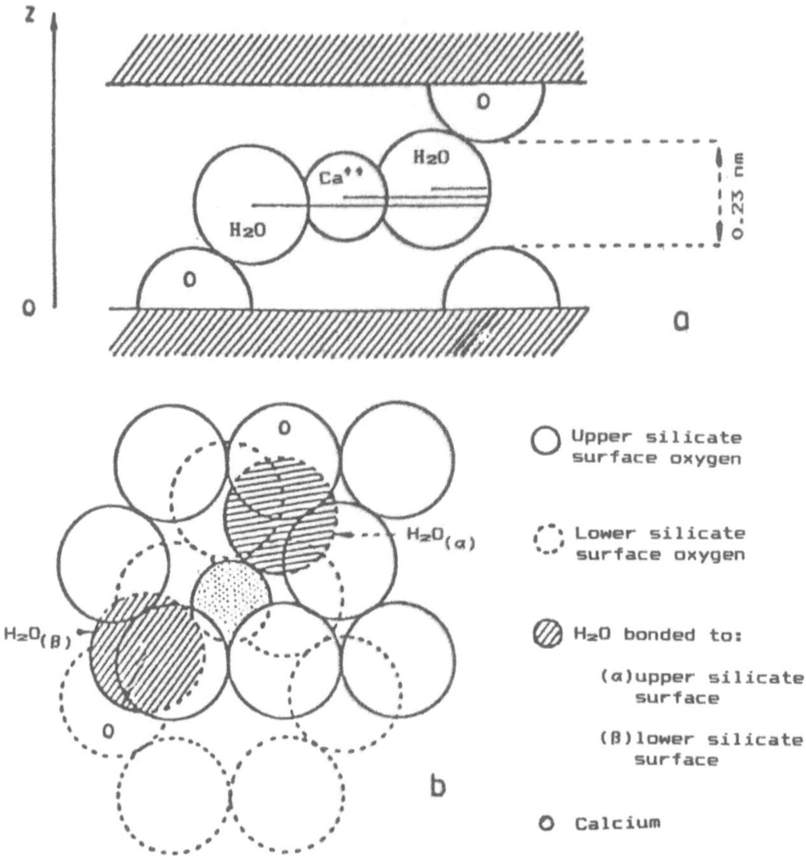

FIGURE 12. Proposed model for $[Ca(H_2O)_8]$–saponite (a) side view and (b) top view showing water molecules hydrogen bonded to oxygen atoms in an upper (O_1) and a lower (O_2) oxygen plane of parallel silicate layers. (After Suquet *et al.*[68] Reproduced by permission of *Clay Minerals*.)

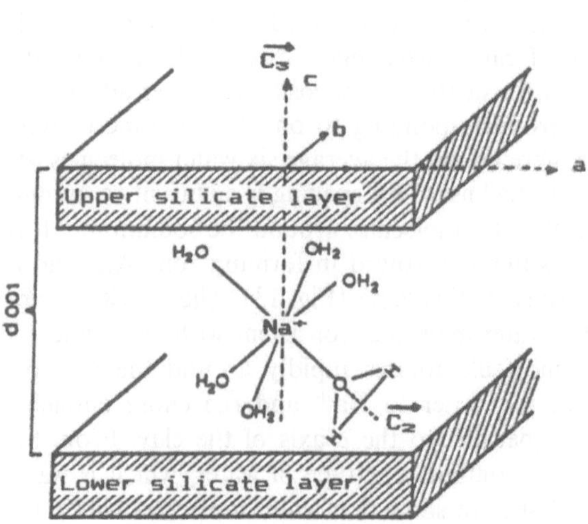

FIGURE 13. Octahedral arrangement of H_2O molecules around the Na ion in the two-layer hydrate of Na–vermiculite. (After Hougardy *et al.*[69])

hydration shell exchange protons with other water molecules within this hydration shell or with molecules outside the hydration shell at a rate much higher than that measured in pure water. This is an indication that the water molecules that form the hydration shell of the exchangeable Na^+ are more acidic than molecules of pure water.

De la Calle *et al.*[70] have shown by x-ray techniques that in Na–vermiculite the stacking of layers is accomplished without rotations. This corresponds to an ordered state with the ditrigonal cavities of two adjacent layers facing each other. This ordered organization of layers is analogous to polytype 1M of micas, and it is the consequence of the hydrogen bonds that are formed between the water molecules of the hydration shell and O atoms of the oxygen plane that belong to Si–O–Al groups. Three water molecules in the upper face of the hydration octahedron and the three water molecules in the base of this octahedron donate protons to oxygens that form the ditrigonal holes in the oxygen planes. The Na^+ is located in the center of the hydration octahedron, below and above parallel ditrigonal cavities.

6.2.2. Zone A_o

The presence of a solute interface, such as the smectite surface, imposes restrictions on the rotation of adjacent water molecules, thereby forming a boundary in the wave motion of clustered water molecules. Thus, the solute forces the positioning of nodes in the spatial arrangement of wave units in the liquid state. Such a two-dimensional node may readily form at the clay–water interface, because the hexagonal array of atoms in the oxygen plane has a high degree of compatibility with the structure of ice. The water molecules take up position forming an extended plane within the liquid. A layer of solvent 3–4 nm thick would spread itself across the surface, next and parallel to it, extending over macroscopic distances in two dimensions. In the tactoid, the layer-shaped hydration cluster is sandwiched between each pair of silicate layers. One should remember that the water sandwiched layer occurs quite readily, without any special energetic mechanism needed to give rise to its existence, because clusters form in the solvent alone. To conclude, the hydration layers are naturally occurring water clusters, but their arrangement and final macroscopic form are determined by the clay. The two participants, clay and water, together build the superstructure.[71]

The structure of zone A_o depends on the degree of hydrophilicity of the oxygen plane. At a hydrophobic oxygen plane of, for example, hectorite, hydrophobic water clusters are obtained, (water polymers) forming zone A_o. The OH groups of the water molecules are not necessarily directed toward the hydrophobic surface siloxane-type oxygens, and the organization of the water molecules is such that they are closely linked to one another by hydrogen bonds. To understand why this hydrophobic water region is formed, one should consider the interlayer space as a cage wherein the thermal amplitude of the water molecules is reduced. When the water molecules enter into this cage, they undergo a loss in entropy while a new structure is formed. In this case water $\leftrightarrow$ oxygen plane interactions are weak, the extent of water $\leftrightarrow$ water interactions is enhanced, and water clusters that form zone A_o become large. With increasing extent of tetrahedral substitution, as in the case of vermiculite, the hydrophilic character of the oxygen plane increases and the extent of water ↔ oxygen plane interactions is enhanced. In this system water ↔ water interactions are weaker than water ↔ oxygen plane interactions, and, consequently, water clusters that form zone A_o are small. The size of water clusters also decreases with increasing charge density of the silicate layer. This is because an increase in the charge density of the surface is associated with increasing concentration of co-ions. The latter serve as cluster breakers.

Hydrophobic hydration of zone A_o may occur in smectites with charge originating from octahedral substitution. From dielectric relaxation measurements and electron density distribution across the interlamellar space, it is obvious that a broad variety of molecular environments exists in adsorbed water, as compared with the bulk solid and liquid phases. For montmorillonite saturated with sodium, Mamy[72] has suggested that the monolayer hydrate consists of tetrahedral distribution of water molecules arranged in a strained hexagonal icelike configuration, with intermolecular bonds formed between neighboring water molecules and between water molecules and oxygen atoms in the silicate surface (Fig. 14). Since some of the water molecules have a hydroxyl group proton inside a ditrigonal cavity of the silicate surface, the net of the hydrogen bonds among the adsorbed water molecules is broken in places. A similar hydrophobic structure occurs with other monovalent

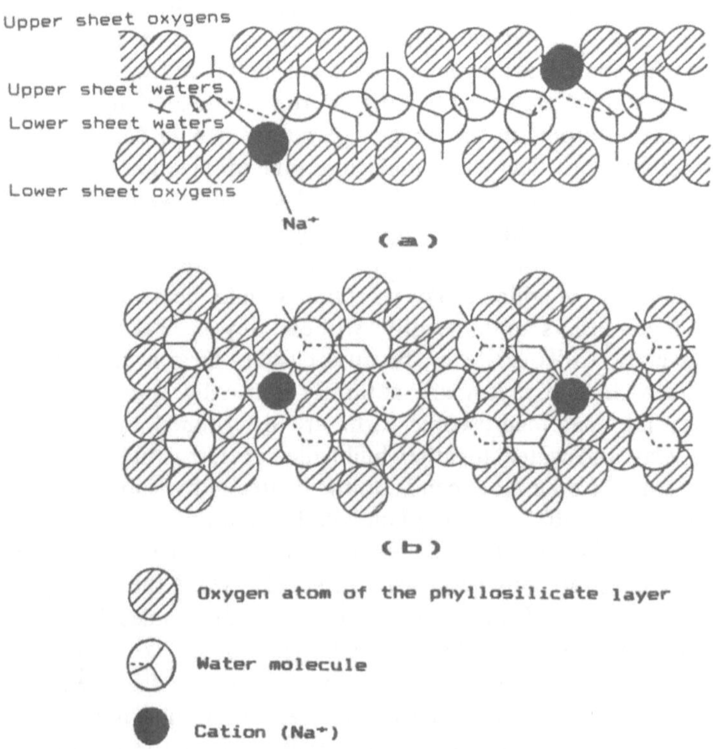

FIGURE 14. Tetrahedral distribution of a monolayer of water in Na–montmorillonite: (a) side view and (b) top view. The shaded circles denote atoms of clay oxygen plane; open circles denote water molecules. (After Mamy.[72])

montmorillonites. For the one-layer hydrate of montmorillonite saturated with bivalent cations, the same structure may not exist because of the strong solvation of the exchangeable cations.

Hydrophilic hydration of zone A_o may occur in vermiculite and in smectites with charge originating from tetrahedral substitution. Walker[73] and Fripiat and Stone[74] described different stages of hydration of Mg–vermiculite. We shall use their description to demonstrate the presence of zone A_o even with a very low water coverage as a result of hydrogen bond formation. In a fully hydrated Mg–vermiculite each water sheet is arranged in a regular hexagonal pattern, and the interlayer cations are located midway between water monolayers in octahedral coordination (Fig. 15a). At this stage there are more than 12 water molecules for each Mg^{2+}.

Zone A_m is composed of the Mg hexahydrate. The rest of the water comprises zone A_o. Zone B_{om} does not exist in this case because zone A_o forms a direct extension of zone A_m. The *c* spacing at this stage is 1.481 nm. From this spacing the HOH – O distance (where O is an oxygen atom belonging to the oxygen plane) was calculated and is equal to 0.31 nm. On partial dehydration a contraction of the *c* spacing to 1.436 nm occurs. The interlayer water network consists of two monolayers arranged in a distorted hexagonal pattern as shown in Fig. 15b. At this stage there may be between 9 and 12 water molecules per each Mg^{2+}. Zone A_m is again composed of the Mg hexahydrate, but zone A_o, which is composed of the remaining water, is not fully developed. The HOH – O distance for this dehydration stage was calculated to be equal to 0.29 nm. The *c* spacing of the fully hydrated sample is determined by the fully developed zone A_o, and the HOH – O distance is characteristic for water of that zone. On the other hand, the *c* spacing of the partly dehydrated sample is determined by zone A_m, and the HOH – O distance is characteristic for water of that zone. The shorter HOH – O distance in the latter stage is an indication of stronger hydrogen bonds. Water in zone A_m is more acidic as compared with water in zone A_o, and, consequently, it forms stronger hydrogen bonds with oxygens of the flat oxygen plane.

Upon further dehydration the *c* spacing decreases to 1.159. A similar *c* spacing

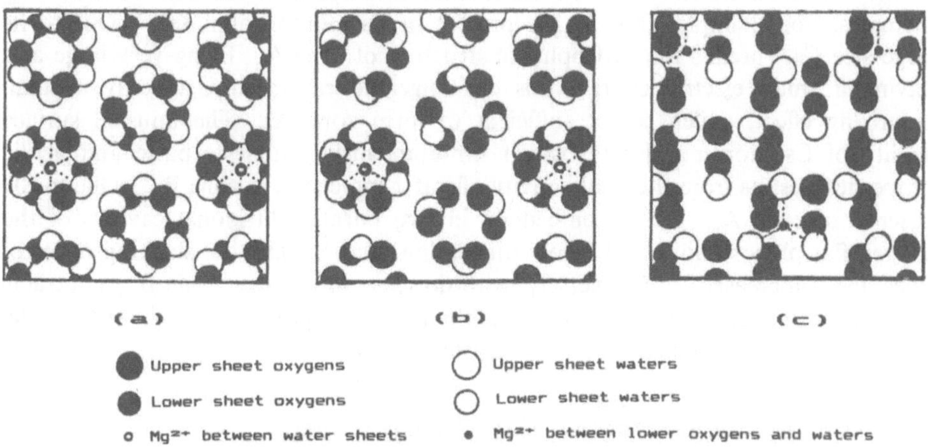

FIGURE 15. Interlayer water cation network in Mg–vermiculite: (a) 1.481-nm phase; (b) 1.436-nm phase; (c) 1.159-nm phase. (After Fripiat and Stone.[74])

is obtained for water content ranging between three and eight molecules for one Mg^{2+}. At this stage a single monolayer is present in the interlayer space (Fig. 15c). With water content of less than 3 mol/per Mg^{2+}, the *c* spacing becomes smaller and smaller. There is a marked difference between the hydrophilic arrangement of H_2O molecules in the 1.159-nm phase and the hydrophobic arrangement shown for Na–montmorillonite, where the tetrahedral medium is maintained. The monolayer hydrate consists of hexagonal distribution of water molecules. Hydration shells of three water molecules that comprise zone A_m can be identified in the figure. The structure of this zone is similar to zone A_m, which was described for Li–hectorite, the metallic cation being located above and/or below the center of the ditrigonal hole of the oxygen flat plane. Zone A_o is composed of all the remaining water molecules.

6.2.3. Zone B_{om}

The size and properties of zone B_{om} depend on those of zone A_m, and thus can be related to the size and properties of the exchangeable cations. This will be demonstrated (I) with cations forming stable hydrates and (II) with those that do not form stable hydrates.

(I) The size and properties of zone B_{om} depend on the polarizing power of the exchangeable cation. The greater the polarizing power of the exchangeable cation, the more acidic the associated hydration shell will be (zone A_m) and the greater is the extension of zone B_{om}. For example, zone B_{om} forms an important fraction of interlayer water in Al montmorillonite in which every discrete zone A_m extends over large areas. However, since the total number of Al ions in the interlayer is only one third of the number of monovalent ions, the total area occupied by A_m is comparatively small, leaving a considerable fraction of the interlayer for adsorption of water through the formation of zones B_{om} and A_o. Many of the properties of Al smectites result from the high activity of water molecules in zone B_{om}.[75]

(II) The extent of water structure breaking increases with increase in the size of the cation in its nonhydrated state, and, consequently, the size of zone B_{om} increases. For example, zone B_{om} is present in Cs montmorillonite.[75] Due to its large size Cs^+ breaks the hydrophobic structure of zone A_o. Being very large and having a small electric charge, this exchangeable cation imparts only a small polarizing effect, which is not sufficient to form zone A_m. The unusual surface acidity of Cs montmorillonite, which simultaneously manifests basic and acidic properties, results from the dual activity of water molecules in zone B_{om}. Due to the absence of zone A_m, Cs^+ is embedded in two parallel ditrigonal cavities of the oxygen flat planes above and below this cation, and thereby the stacking order of smectites is increased.[76] Conversely, with decrease in ion size, zone B_{om} contracts and may disappear, as with exchangeable Li^+ in smectites. In this system the cations are accommodated within the clusters of water A_o.

6.3. Acidic and Basic Properties of the Interlayer Space

The interlayer space reveals basic and acidic sites, the oxygen planes being electron pair donors and water molecules of zone A_m being proton donors. Water molecules of zone B_{om} act as proton donors as well as proton acceptors. Water molecules of the hydrophobic zone A_o do not react either as proton donors or as proton acceptors.

As mentioned (see Section 2.2 and Fig. 8), the basic strength of the oxygen plane and the ability to form hydrogen bonds with water molecules and other proton donors depends on whether there is an isomorphous tetrahedral substitution of Al for Si. Since the charge of Al is lower than that of Si and it does not take part in d_π–p_π bondings, it induces a weaker effect on the nonbonding sp^2 electrons of the O atoms and the basic strength of the oxygen plane increases. Thus, the Si–O–Al group forms stronger hydrogen bonds with water molecules than the Si–O–Si group.[16] The characteristic of the hydration atmosphere of the oxygen plane depends on whether the charge in the clay mineral originates from a tetrahedral or an octahedral substitution (see Section 5.2). The degree of hydrophilicity of the various TOT minerals, which increases with the tetrahedral substitution, may account for many of the differences between them. In hectorite, where charge deficit occurs in the octahedral sheet, the negative charge on the surface oxygen atoms is delocalized, and no hydrogen bonds are formed between the oxygen plane and interlayer water. In montmorillonite only a very small fraction of the charge deficit occurs in the tetrahedral sheet and in accordance with this small localized negative charge, a very small fraction of the interlayer water forms hydrogen bonds with the oxygen plane. This water fraction becomes higher in saponite or beidellite and higher still in vermiculite, where the charge deficit occurs mainly in the tetrahedral sheet; thus, there is a greater localization of the negative charge leading to the formation of relatively strong hydrogen bonds.[68, 69]

Water of zone A_m dissociates under the polarizing effect of the metallic cation as follows:

$$M(OH_2)_x^{mt} \rightarrow M(OH)(OH_2)_{x-1}^{(m-1)^+} + H^+$$

A similar polarizing effect occurs in aqueous salt solutions, resulting in an increased number of hydrogen bonds between water molecules and extension of the ordered self-atmosphere zone of hydrated ions with increasing polarizing power of the soluble ions. Because of steric hindrance in the interlayer space, zone A_m cannot extend as much as the hydration self-atmosphere zone of ions in liquid water and is therefore a stronger proton donor. Moreover, the water structure in the interlayer space contains more lattice vacancies than normal water, and the dielectric constant of this water is less than that of liquid water. Consequently, interlayer protons are more mobile than in liquid. According to Touillaux *et al.*,[77] the degree of dissociation of water is 10^7 times higher in the adsorbed state than in the liquid. The proton-donating tendency increases as the water content of the interlayer space decreases.

The polarization of the interlayer water is strongly affected by the exchangeable cation. Greater polarizing ability of the interlayer cations increases both the strength and number of acid sites per surface unit area.[78] Acid strengths of hydrates of some common ions decrease in the following sequence: Al, Fe, Mg, Ca, Li, Na, K, and Cs. Mortland and Raman[79] showed that the degree of protonation of sorbed ammonia depended on the hydrolysis constant of the interlayer hydrate and that the degree of protonation may be regarded as a measure of the acidity of the interlayer space. In base-saturated smectites and vermiculites, ammonia molecules form hydrates such as $M-OH_2\cdots NH_3$, via hydrogen bonds or accept protons originating from water dissociation, forming hydroxy complexes $M-OH^-$ and $NH_4{}^+$. In Al-montmorillonite, NH_3 reacts with protons to form ammonium ion. To conclude, the polarizing power of the cation on coordinated water molecules determines the surface acidity of the clay.

The exchangeable metallic cations and the hydration state of the clay play a major role in the adsorption of organic polar molecules. Strong bases are protonated during adsorption, yielding positive ions. The extent of this reaction depends on the basic strength of the organic compound and the polarizing power of the metallic cation. For example, adsorbed aliphatic amine is protonated in the interlayer space as follows:(80–82)

$$(C_2H_5NH_3^+\cdots H-\underset{\displaystyle H}{\underset{|}{O}}-M^{m+})-\text{Smec}\rightarrow(C_2H_5NH_3^+\cdots[(OH)M]^{(m-1)})-\text{Smec}$$

With decreasing basic strength of the organic compound and/or decreasing polarizing power of exchangeable metallic cations, associations are obtained via hydrogen bondings, wherein a water molecule acts as a proton donor. This can be illustrated by the adsorption of aromatic amines, which are weak bases, such as aniline:(75, 83, 84)

$$C_6H_5NH_2+H_2O-M-\text{Smec}\rightarrow C_6H_5\overset{\displaystyle H}{\overset{|}{\underset{\displaystyle H}{\underset{|}{N}}}}\cdots\overset{\displaystyle H}{\overset{|}{H\!O}}-M-\text{Smec}$$

In this adsorption reaction the nitrogen atom serves as the nucleophilic site. If the adsorbed molecules are proton donors, they may react with two different basic sites: (i) basic sites in the oxygen plane and (ii) negative poles of water molecules in the hydration spheres of cations. These can be illustrated by the sorption of indoles, phenols, or fatty acids:(85–87)

$$C_8H_6NH+O(-Si-)_2(s)\rightarrow C_8H_6NH\cdots O(-Si-)_2(s)$$

$$C_6H_5OH+Al\cdots\underset{\displaystyle H}{\underset{|}{O}}-H\cdots O(-Si-)_2(s)\rightarrow C_6H_5OH\cdots\overset{\displaystyle Al}{\overset{\vdots}{\underset{\displaystyle H}{\underset{|}{O}}}}-H\cdots O(-Si-)_2(s)$$

and

$$C_nH_{2n+1}COOH+O(H_2)\cdot Al\cdot O(-Si-)_2\rightarrow C_nH_{2n+1}COOH\cdot\cdot O(H_2)\cdot Al\cdot O(-Si-)_2$$

The acid strength of water of region A_m depends on whether the charge of the clay mineral originates from tetrahedral or octahedral substitution. From nuclear magnetic resonance studies on Li-hectorite, in which the charge originates from octahedral substitution, Conard(65) suggested that the trihydrated Li cation is fixed above the six oxygens of the hexagonal cavity by delocalized bonds in which 3 of the protons of the hydrated Li can be equally distributed among 12 equivalent positions with respect to the oxygens forming the hexagonal cavity. In this symmetric structure the attraction force between water hydrogens and clayey oxygens is distributed among all equal 12 sites, so any contribution to localized intermolecular H···O bonds should be weak (Scheme 4**I**). An equal density of protons throughout the ring facilitates the removal of one proton, which accounts for the acidic properties of that clay.

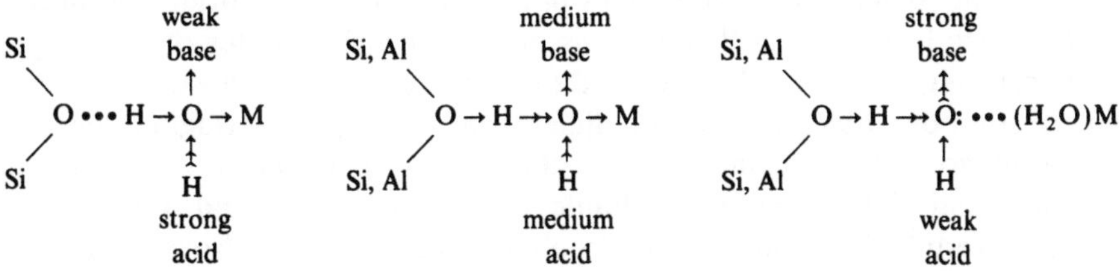

I II III

SCHEME 4. Inductive effects of the exchangeable metallic cation M (I and II) and of an aluminosiloxane group, (Si,Al)–O–(Si,Al) (II and III), on the electron and proton donation abilities of a bridging water molecule and the contribution of these inductions to the basic and acid strengths of the water molecule. The strength of the induction is shown semiquantitatively by arrows.

Oxygen planes of layers, wherein the charge originates from the tetrahedral substitution, form strong hydrogen bonds with polarized water molecules having the following structure:

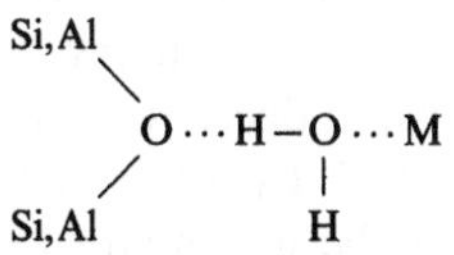

In this structure the water molecule forms a bridge between the exchangeable cation and the oxygen plane. As a result of this hydrogen bond formation, the ability of the bridging water molecule to donate its second proton to any proton acceptor is diminished. At the same time the formation of the hydrogen bond induces an electric effect of repulsion on the nonbonding electron pair of the water oxygen, and this oxygen may act as an electron pair donor, thus becoming a basic site (Scheme 4**II**). The strength of the hydrogen bond between the bridging water molecule and the oxygen plane increases with the polarizing power of the metallic cation, and hence, the number and strength of basic sites of this type increase with increasing polarizing power of the exchangeable cations. The bonds are detected mainly in the presence of exchangeable Al.[75, 85] Protonation of an adsorbed base by this water molecule, which involves breaking of the H bonds with the oxygen planes, is therefore more readily affected with octahedral-charged than with tetrahedral-charged clay minerals.[68]

Water molecules in zone B_{om}, not being structured, are more active than water molecules in structured zones A_o or A_m. They may interact with negative charge sites on the oxygen planes. Some water molecules will be oriented with the positive ends of the dipoles toward the oxygen planes, thus imparting a basic character to interlayer water (Scheme 4III). Some water molecules will react as proton donors, imparting an acid character to interlayer water. In most cases "acidic" water predominates over "basic" water and obscures the presence of the latter.

Hydrated exchangeable cations (zone A_m) may bridge between two parallel silicate layers by forming stable hydrogen bonds with the oxygen plane. This bridging process limits the swelling of TOT clays to a *c* spacing of 1.4–1.5 nm. Disordered monomeric water molecules of zone B_{om} may also bridge between two parallel silicate layers. Since this water is more active than water of zone A_m, this type of bridging occurs with clays without any tetrahedral substitution or with slight substitution. The *c* spacing obtained from this type of bridging depends on the size of zone B_{om}. Consequently, the interlayer of hectorite is the most expandable and that of montmorillonite is more easily expanded than that of saponite or vermiculite, and more hydrophobic water can be adsorbed (Table 2).

6.4. Stages of Hydration of the Interlayer Space

Lahav and Bressler[88] showed that the surface oxygen networks are changed with variations in the exchangeable cations and with their associated water of hydration. This can be explained in part by the penetration of the bare metallic cation into the ditrigonal hole and the interactions between this cation and the oxygens that form the ditrigonal hole. This interaction results in a rotation of the silica tetrahedra. When the metallic cation is hydrated, the interaction between the bare metallic cation and the oxygen does not exist and the unusual rotation of the silica tetrahedra does not occur. This provides a partial explanation for the change in *b* dimension with water content observed by Ravina and Low.[89] During the hydration of a smectite mineral, a decrease of the concentration of nonhydrated cations decreases the force excerted by these cations on the oxygen plane and thereby decreases the tendency of the tetrahedra to rotate.[88]

Water adsorption by swelling clay minerals has been described in terms of an initial crystalline phase and a later osmotic phase.[90–92] The crystalline phase adsorption concerns the adsorption of the first few water monolayers on mineral surfaces driven by cation and surface hydration energies. The later adsorption has been termed osmotic, and swelling pressure was related to (i) the difference in the concentration of ions in the interlayer and external solutions and (ii) the difference in the potential energy of the water in these solutions. The two stages will now be described in more detail.

6.4.1. Initial Crystalline Phase Adsorption

When collapse particles of a swelling clay mineral are exposed to water vapor, the exchangeable cations hydrate first,[90] forming zone A_m. Depending upon atmospheric humidity and hydration ability of the cation, two different types of hydrated cations appear. When the hydration number of the metal is 3 or 4 and it has a triangular or a square coordination, the hydrated cation occupies a flat orientation parallel to the layers. This is the case with most alkali metal cations (monovalent) and with Cu^{2+} in smectites at a very low humidity. When the hydration number of the cation is above 4, it occupies the center of an octahedron, two triangular surfaces lying parallel to the oxygen planes. This is the case with most di- and trivalent cations even at a relatively low humidity of less than 10%. The basal spacings obtained for vermiculite with planar and octahedral hydrated ions are ≈ 1.2 and ≈ 1.43 nm, respectively.[73]

On further exposure to water vapor the remainder of the surface hydrates, and zones A_o and B_{om} are formed. At this stage water molecules penetrate into vacancies in the interlayer space. A monolayer is obtained with alkali cations or Cu^{2+}-smectites, whereas two monolayers result with di- and trivalent cations smectites, with slight expansion of the clay to values above 1.25 and 1.48 nm, respectively. When the clay is in contact with liquid water, osmotic forces caused by the relatively high ionic concentration between the layers may lead to a continuous swelling. The interlamellar water is slightly more densely packed than liquid water.[93]

Differences between hydration properties of various clays can be demonstrated from isotherms of water adsorption associated with x-ray diffractions and heats of hydration. In the case of the two di- and trioctahedral-substituted smectites, montmorillonite and hectorite, the sample adsorbs water in a continuous manner, and at the same time the spacings increase in small steps.[94] On the contrary, in the case of vermiculite, or the tetrahedral-substituted smectites saponite and beidellite, plateaux in spacing correspond to well-defined steps in water content.[70, 95] It is anticipated that vermiculite, saponite, or beidellite will produce a more homogeneous, relatively well defined system compared with the other clays, which form a more complex situation with possibly different water populations, interstratification, and microporosity. Enthalpies and entropies of water adsorption by several homoionic montmorillonites were found to be negative between 25 and 70°C, but decreased in magnitude with increasing amounts of adsorbed water.[94] The expansion of the smectite structure as a function of water adsorption and the expected *c* spacings are shown in Fig. 16. Calculations of expected *c* spacings were carried out for Na^+. In a two-layer water structure, the maximum number of water molecules per Na^+ was assumed to be 12.[96]

There is a significant difference between the hydration structures of monovalent cation smectites and polyvalent cation smectites. The behavior of monovalent cation smectites will be demonstrated with the behavior of Li–hectorite. Estrade-Szwarckopf *et al.*[97] investigated Li–hectorite by incoherent neutron scattering, after pumping the clay at 10^{-3} torr at about 300 K. They found that all water molecules were bound strongly to the stationary exchangeable Li^+ cations. Each solvating water molecule was allowed to rotate about its own symmetry axis and about an axis perpendicular to the clay platelet, through the exchangeable cation. These findings are typical for zone A_m. Cebula *et al.*[98] investigated Li–montmorillonite containing either one, two, or three monolayers of water in the interlayer space, using the same neutron scattering method. They concluded that the adsorbed water molecules were diffusing or rotating much more slowly than those in bulk water. But the diffusion and rotation were too high to assume a rigid coordination of the water molecules to the exchangeable cation. This means that all the water is present in the interlayer forming zone A_o. In other words, the hydration cosphere of Li^+ (zone A_m) exists

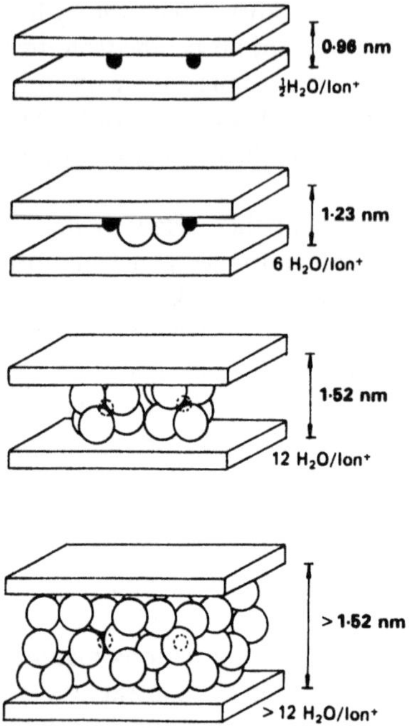

FIGURE 16. Expansion of a smectite crystal as a function of water adsorption, expressed as number of water molecules per sodium ion. (From Kraehenbuehl *et al.*[96] Reproduced by permission of *Clay Minerals*.)

only when small amounts of water are present, but it does not exist when excess water is present. In the latter case the water molecules form zone A_o and Li^+ cations are located in the intermolecular vacancies found in the nondense water packing of zone A_o.

All alkali cations behave in a similar way. This is due to their hydration enthalpies being only slightly lower than that of the water ↔ water interaction, the latter probably being equal, or almost equal, to the heat of liquefaction of water. Water molecules in zone A_m are more restricted translationally and rotationally in their motion than those of zone A_o and therefore have a lower entropy. At a low water content, formation of zone A_m is favored by enthalpy, whereas at high water content, formation of zone A_o is favored by entropy. Hydration enthalpies of di- and trivalent cations are much lower than that of water ↔ water interaction; thus, at low or high water contents, formation of zone A_m is favored by enthalpy.

Oscarson[99] studied the effect of stepwise hydration on the *c* spacing of Ca– and Mg–montmorillonites. When Ca–montmorillonite was heated to 350°C, a *c* spacing of 0.982 nm was obtained, indicating that the interlayer cations were dehydrated and that the layers of the clay were collapsed. After half an hour at ambient atmosphere a broad peak with a *c* spacing of 1.0 nm accompanied by a small peak with a *c* spacing of 1.16 nm were recorded, indicating partial rehydration. After 1 and 4 h at humidities of 35 and 25%, respectively, a *c* spacing of 1.29 nm, corresponding to a monolayer of water in the interlayer space, was recorded, indicating an increase in the extent of rehydration of the interlayer cations with time. When the sample was equilibrated in ambient atmosphere (40% humidity) for several days, a *c* spacing of 1.53 nm, corresponding to two water monolayers, was recorded. When Mg–montmorillonite was heated at 350°C, a *c* spacing of 0.951–0.955 nm was obtained. Collapsed Mg–montmorillonite does not rehydrate as readily as the

Ca–montmorillonite. The interlayer cation did not rehydrate until exposed to the atmospheric air for several hours.

6.4.2. Osmotic Adsorption

When the formation of two or three monolayers of water is completed, osmotic adsorption leads to further swelling of the interlayer space. This adsorption has been termed *osmotic* since it has been considered to be controlled by chemical potential gradients between the free and adsorbed water.[89–92] The adsorbed water molecules form clusters in the interlayer. The connection between the inorganic exchangeable ion and the degree of swelling of vermiculites and smectites were examined by several investigators (e.g., Norish,[52] and Quirk[48, 49]]. Montmorillonite saturated with small monovalent ions, such as Na^+ and Li^+, to a lesser extent also K^+, exhibits extensive swelling. On the other hand, saturation with bivalent ions, such as Ca^{2+}, Mg^{2+}, Ni^{2+}, Zn^{2+}, or Cd^{2+}, restricts swelling of vermiculite to 1.40–1.50 nm and of montmorillonite to 1.90–2.00 nm. Swelling of Cs–montmorillonite or vermiculite gives rise to a spacing of only 1.20 nm (Table 2). With one, two, and even three or four monolayers (crystalline phase adsorption) the water molecules are sufficiently static to allow their positions to be identified by x-ray diffraction. In a highly swollen clay there are grossly expanded phases with spacing equivalent to more than 4 nm in which the interlayer water must be much more liquidlike away from the silicate surfaces showing more similarities to aqueous salt solutions rather than to crystalline hydrates. A higher degree of mobility is possible in the swollen clay than is usual for the crystalline anion surface. As in the liquid phase, water in the "osmotic" adsorbed phase consists of clusters, of which the size and nature depend on the silicate layer and on the ions present in the system. In aqueous salt solutions the zone that consists of clusters that are similar to clusters of pure liquid water is named zone C, whereas the zone that consists of the hydrated ions is named zone A. Similarly, we shall assign to the zone that consists of hydrophobic clusters that are similar to clusters of pure water the term C_c.

The effect of exchangeable cations on the swelling properties of the expanding clay mineral can be attributed to the presence of the different water zones. Water molecules of zone A_m are proton donors and may form hydrogen bonds with oxygen planes of TOT clay minerals. Whereas these hydrogen bonds, which are characteristic for water molecules of zone A_m, restrict the swelling of the clay above a certain limit, the presence of zones A_o or C_c does not limit osmotic swelling to any degree. Lithium and Na ions do not break the water structure of zones A_o or C_c of smectite minerals, since, because of their small sizes, these ions can fit into the interstitial cavities of the water structure. In vermiculite only Li ions do not break the water structures of zones A_o and C_c, whereas the larger Na ions do break the A_o structure and thus avoid swelling and the penetration of C_c clusters into the interlayer space. The intense ionic fields of multivalent cations cause local destruction of the water structure of this region and lead to the formation of zone A_m.

Water in zone B_{om} is the major water fraction in Cs–montmorillonite, Cs–vermiculite, and K–vermiculite. Being nonstructural, water molecules in zone B_{om} may form water bridges between two parallel oxygen planes, as shown in Scheme 5, and thereby restrict any swelling of montmorillonite or vermiculite to 1.20 nm.

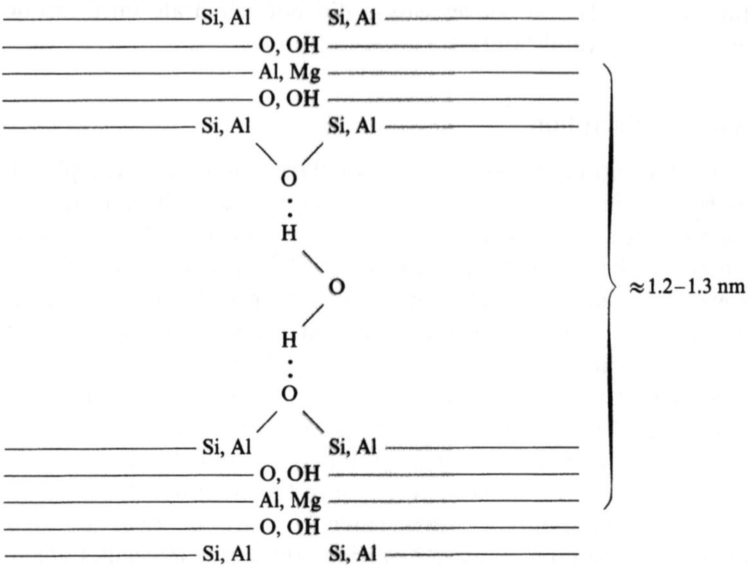

SCHEME 5. A water molecule of zone B_{om} forms a bridge between two parallel aluminosilicate layers, R(Si,Al)–O–(Si,Al)R. The interactions between the bridging water molecule and each of the oxygen planes are localized hydrogen bonds.

Organic aliphatic ammonium cations restrict any swelling of the montmorillonite to 1.30–1.40 nm (Table 3). A similar restriction was observed with several diamine complexes of transition metals, but in such a case the situation is more complicated. If the maximum ligand coordination is not reached by even part of the cations, limited swelling occurs. In general, introduction of organic matter into clay interlayers renders them hydrophobic. Thus, montmorillonite saturated with organic ammonium ions does not show a change in basal spacing on exposure to water. The explanation for this phenomenon can be as follows. The positive charge density of the organic ion is too small to cause hydrophilic hydration corresponding to

TABLE 3. Effect of Steam on the Basal Spacing of Montmorillonite Saturated with Alkyl Ammonium Ions

	Interlayer spacing (nm)	
Amine	Air dried	Steam treated
Ethylenediamine	1.30	1.30
Diethylenetriamine	1.32	1.32
Triethylenetetramine	1.36	1.36
Tetraethylenepentamine	1.34	1.34
Ethylamine	1.28	1.30
Diethylamine	1.34	1.34
Hexylamine	1.32	1.34
Tributylamine	1.32	1.42

After Ref. 51.

zone A_m. The large size of the organic ion, on the other hand, leads to disruption of the hydrophobic hydration structure A_o. Water penetrating the interlayers is therefore nonstructured (zone B_{om}) and may form water bridges between adjacent oxygen planes, as shown in Scheme 5, thus preventing crystalline swelling. Organic ammonium cations resemble Cs^+ in this respect.[51]

McBride and Mortland,[100] in a study of the adsorption of organic amines by smectites, showed that the exchange of Cu by an organic hydrophobic tetra-alkylammonium ion, such as tetrapropylammonium, leads to the contraction of smectites from 1.9 to 2.0 nm to about 1.45 nm. This is in agreement with the present model on the fine structure of water in the interlayer space. In Cu–smectite the spacing is determined by the presence of zone A_m, whereas in ammonium smectite it is determined by the presence of zone B_{om}.

7. INTERCALATED WATER IN KAOLINITE

Under certain conditions kaolin-type minerals can intercalate certain inorganic salts and a variety of organic compounds.[101, 102] The salts of organic compounds penetrate the interlayer space of kaolinlike layers and so expand the crystal from a basal spacing of $\approx$0.72 nm to about 1.00–1.42 nm. The penetrating species that break the strong electrostatic and van der Waals types of interactions between the layers may form hydrogen bonds with surface hydroxyls, as was proved from infrared spectroscopy.[103, 104] Basal hydroxyls of the kaolinlike layer are very poor proton donors and may form hydrogen bonds only with very strong bases such as hydrazine or dimethyl sulfoxide. Bipolar molecules, which may serve at the same time as strong proton donors and acceptors, such as urea, may form hydrogen bonds, on the one hand, by accepting protons from hydroxyls located on the octahedral sheet surface, and, on the other hand, by donating protons to oxygens located on the tetrahedral sheet surface. Since the oxygen plane is a very poor electron donor, these hydrogen bonds are very weak.

The flat oxygen plane, even in kaolinite, exhibits a constant negative charge, arising from occasional isomorphous replacement of silicon by aluminum in the tetrahedral sheet.[105] Exchangeable cations that compensate for this charge are distributed over the planar surface of the silicate. Exchangeable planary hydrated ions, such as $[Cu(H_2O)_4]^{2+}$, are oriented parallel to the flat oxygen planes.[106]

The interaction occurring when water is intercalated by minerals from the kaolin group (kaolinite, halloysite, dickite, and fire clay) can be visualized from studies on the interaction between alkali halides and various minerals from this group. Intercalation was obtained by a mechanomechanical process in which these minerals were thoroughly ground with CsCl, followed by allowing the ground mixtures to age in air for a few weeks. All these minerals form intercalation complexes with CsCl and water.[22, 23, 107–109] Intercalation complexes of kaolinite are formed with CsCl or CsBr and water, but not with CsI.[110] In these intercalation complexes the presence of a large ion such as Cs disrupts the structure of water clusters, thus allowing the existence of disordered single molecules. Hydrogen bonds are formed between the isolated water molecules and between both the inner-surface hydroxyls and the basal oxygens. In these hydrogen bonds the hydroxyls donate protons to the water molecules, whereas the oxygens accept protons from the water molecules. In addition to localized hydrogen bonds between the water molecules and the clay inner surface, the adsorbed water molecule donates a proton to the halide, which is also intercalated, as shown in Scheme 6.

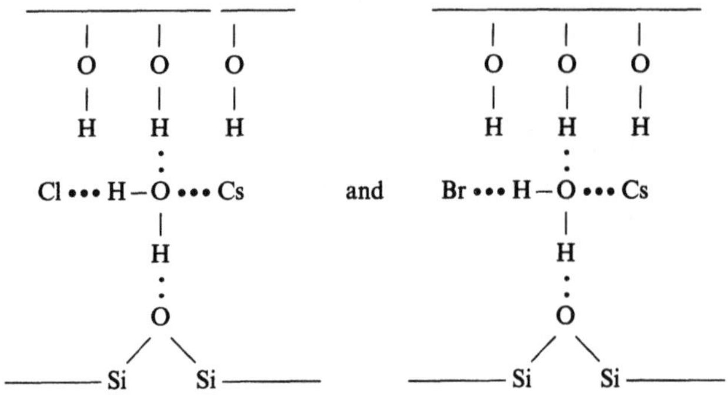

SCHEME 6. Intercalation of kaolinite by CsCl or CsBr and water.

Infrared data indicate that there are no structural differences between intercalation complexes obtained from kaolinite, dickite, fire clay, or halloysite. A mechanical process of delamination of the mineral crystal is essential for obtaining CsCl–HOH complexes of the kaolin group minerals.[108] The CsBr complex is formed by grinding a mixture of CsBr and kaolinite in the presence of a few drops of water, whereas it is not necessary to add water during grinding with CsCl because enough water is adsorbed from the atmosphere.[110] Inner-surface hydroxyls of delaminated CsCl–HOH complex can be deuterated by washing the samples with D_2O.[108]

In halloysite a layer of water is present between two kaolinlike aluminosilicate layers. Infrared studies showed that the interaction between the kaolinlike layers of halloysite and the water layer is hydrophobic, meaning that the intercalated water layer does not form hydrogen bonds with the oxygen or hydroxyl planes of the halloysite. The negative oxygens of the water molecules in the boundary layer are oriented toward the hydroxyl plane, whereas the positive hydrogens are oriented toward the oxygen plane, but there is no localized interaction between water molecules and the basal sheets. Hydrogen bonds occur between the water molecules. The presence of large ions such as Cs, Rb, and K in the boundary layer results in the disruption of the hydrophobic hydration structure of the water. Water molecules may now coordinate with the alkali cations, and at the same time they may interact with basal hydroxyls and basal oxygens of kaolinlike layers, forming hydrogen bonds.

The degree of structuring and the size of water clusters in the adsorbed water layer are affected by the cation size and increase from Cs to K. The ability of the adsorbed water layer to donate protons to the oxygen plane or to accept protons from the hydroxyl plane, in both cases forming localized H bonds, increases with decreasing degree of structuring of this water layer. Hydrogen bond formation either by proton acceptance from hydroxyls or by proton donation to oxygens is affected by the alkali cations in the order Cs > Rb > K. No hydrogen bonds are formed in the presence of Na ions because the size of this ion is such that it can apparently fit into the interstitial cavities of water with minimum disruption of the water structure. Kaolinite forms hydrogen bonds only with Cs hydrate.

Constanzo *et al.*[111, 112] reported the synthesis of four intercalation hydrates of kaolinite. They differ from one another by characteristic *c* spacing and by stability. These are (1) 1.0-nm unstable hydrate, (2) 1.0-nm stable hydrate, (3) 0.86-nm hydrate, and (4) 0.84-nm crystalline hydrate. A *c* spacing of 0.84 or 0.86 nm indicates that isolated water molecules are keyed into the ditrigonal holes of the silica tetrahedra (this water is defined as *hole water*). In the 0.84-nm hydrate $\approx$13–23% of the inner surface hydroxyls had been replaced by fluorine, before the synthesis of the hydrate. The F atom can accept hydrogen bonds from nearby water molecules. Two types of hole water were identified, those forming hydrogen bonds with F and those forming hydrogen bonds with OH groups. As a result of linking with F the hydrate was stabilized. The kaolinite used for the synthesis of the 0.86-nm hydrate was not fluorinated, and the inner-surface OH groups were not replaced by F. Only one type of hole water was formed, and this was not stable. The positive end of the water molecule points toward the basal oxygens, and the lone-pair electrons of the water oxygen can accept hydrogen bonds from the hydroxyls of the opposite surface. The 1.0-nm hydrate was twice as much interlayer water as the 0.84- or 0.86-nm hydrate. Ideally, half of this water is keyed into the ditrigonal holes, and the other half is the mobile associated water. Three kinds of hydrogen bonds are present in the 1.0-nm hydrate. The strongest are the intermolecular bonds involving associated water, the next strongest are between hole water and the silicate surface, and the weakest are between the inner-surface hydroxyls and the water molecules. Of the three kinds of hydrogen bonds, only the intermolecular bonds are significant. They do not occur when the number of water molecules is small, as was pointed out in the section on the wettability of the OH plane.

ACKNOWLEDGMENT

I am grateful to Professor Noam Lahav from Rehovot and to Dr. Harold Cross from Jerusalem for their careful reading of the manuscript and for their important comments and suggestions.

Financial support of the Faculty of Science of the Hebrew University of Jerusalem is gratefully acknowledged.

REFERENCES

1. C. E. Marshall, *The Colloid Chemistry of the Silicate Minerals*, Academic Press, New York (1949).
2. H. van Olphen, *An Introduction to Clay Colloid Chemistry*, Wiley, New York (1963).
3. S. Yariv and H. Cross, *Geochemistry of Colloid Systems*, Springer-Verlag, Berlin (1979).
4. R. E. Grim, in *Kirk-Ottmer Encyclopedia of Chemical Technology*, 3rd ed., vol. 6, pp. 207–223, Wiley, New York (1979).
5. R. E. Grim, *Clay Mineralogy*, 2nd ed., McGraw-Hill, New York (1968).
6. C. E. Weaver and L. D. Pollard, *The Chemistry of Clay Minerals*, Elsevier, Amsterdam (1973).

7. F. Liebau, *Structural Chemistry of Silicates*, Springer-Verlag, Berlin (1985).
8. B. Kronberg, J. Kourti, and P. Stenius, *Colloids Surf.* **18**, 411 (1986).
9. D. W. J. Cruickshank, *J. Chem. Soc.* 5486 (1961).
10. D. W. J. Cruickshank, *J. Mol. Struct.* **130**, 177 (1985).
11. J. H. Lee and S. Guggenheim, *Am. Mineral.* **66**, 350 (1981).
12. J. Ch. Lin and S. Guggenheim, *Am. Mineral.* **68**, 130 (1983).
13. W. F. Bradley, *Am. Mineral.* **25**, 405 (1940).
14. G. Kunze, *Fortsch. Miner.* **39**, 206 (1961).
15. S. Yariv, *Clay Miner.* **21**, 925 (1986).
16. M. E. Schrader and S. Yariv, *J. Colloid Interface Sci.* **136**, 85 (1990).
17. D. J. Mulla, J. H. Cushman, and P. F. Low, *Water Resour. Res.* **20**, 619 (1984).
18. R. Prost, in *Proc. Int. Clay Conf.*, Bologna, Italy, 187 (1981).
19. D. E. Woessner, *J. Magn. Reson.* **39**, 297 (1980).
20. J. Salle de Chou, P. F. Low, and C. G. Roth, *Clays Clay Miner.* **28**, 111 (1980).
21. J. D. Russell, R. L. Parfitt, R. L. Fraser, and V. C. Farmer, *Nature (London)* **248**, 220 (1974).
22. S. Yariv and S. Shoval, *Clays Clay Miner.* **23**, 473 (1975).
23. S. Yariv and S. Shoval, *Clays Clay Miner.* **24**, 253 (1975).
24. L. Heller-Kallai, S. Yariv, and S. Gross, *Mineral. Mag.* **40**, 197 (1975).
25. G. A. Parks, *Chem. Rev.* **65**, 177 (1965).
26. N. E. Tretyakov and V. N. Filimonov, *Kinet. Katal.* **13**, 815 (1972), quoted in H.-P. Boehm and H. Knoezinger, in J. R. Anderson and M. Boudart, eds., *Catalysis—Science and Technology*, vol. 4, p. 40, Springer-Verlag, Berlin (1983).
27. J. G. Miller and J. D. Oulton, *Clays Clay Miner.* **18**, 313 (1970); S. A. Mishirky, S. Yariv, and W. I. Siniansky, *Clay Sci.* **4**, 213 (1974); S. Yariv, *Powder Technol.* **12**, 131 (1975).
28. B. G. Williams and D. P. Drover, *Soil Sci.* **104**, 326 (1967); R. J. Hunter and A. E. Alexander, *J. Colloid Sci.* **18**, 833 (1963); S. K. Nicol and R. J. Hunter, *Aust. J. Chem.* **23**, 2177 (1970); R. P. Mitra and B. S. Kapoor, *Soil Sci.* **108**, 11 (1969); B. Rand and I. E. Melton, *J. Colloid Interface Sci.* **60**, 308 (1977).
29. R. K. Schofield, *J. Soil Sci.* **1**, 1 (1949); J. W. Goodwin, *Trans. Br. Ceram. Soc.* **70**, 65 (1971).
30. N. S. Srinivasan and L. G. Hepler, *J. Can. Petrol. Technol.*, July/August, 25 (1982).
31. L. Nabzar, E. Pefferkorn, and R. Varoqui, *J. Colloid Interface Sci.* **102**, 380 (1984); E. Pefferkorn, L. Nabzar, and A. Carroy, *J. Colloid Interface Sci.* **106**, 94 (1985); E. Pefferkorn, L. Nabzar, and R. Varoqui, *J. Colloid Interface Sci.* **108**, 243 (1985); *Colloid Polym. Sci.* **265**, 889 (1987).
32. B. Fubini, G. Della Gatta, and G. Venturello, *J. Colloid Interface Sci.* **64**, 470 (1978); B. Fubini, *Thermochim. Acta* **135**, 19 (1988).
33. J. J. Fripiat, *Clays Clay Miner.* **12**, 327 (1964).
34. D. H. Solomon, J. D. Swift, and A. J. Murphy, *J. Macromol. Sci. Chem.* A **5**, 587 (1971); R. F. Conley and A. C. Althoff, *J. Colloid Interface Sci.* **37**, 186 (1971).
35. B. Janczuk, E. Chibowski, M. Hajnos, T. Bialopiotrowicz, and J. Stawinski, *Clays Clay Miner.* **37**, 269 (1989).
36. R. K. Iler, *The Chemistry of Silica*, Wiley, New York (1979).
37. K. Klier and A. C. Zettlemoyer, *J. Colloid Interface Sci.* **58**, 216 (1977).
38. H. Knoezinger and P. Ratnasamy, *Catal. Rev.—Sci. Eng.* **17**, 31 (1978); P. Ratnasamy and S. Sivasanker, *Catal. Rev.—Sci. Eng.* **22**, 401 (1980).
39. J. A. Lercher, C. Colombier, and H. Noller, *J. Chem. Soc. Faraday Trans. 1* **80**, 949 (1984).
40. C. Serna, J. L. Ahlrich, and M. Serratosa, *Clays Clay Miner.* **23**, 452 (1975).
41. C. Serna, G. E. VanScoyoc, and J. L. Ahlrich, *J. Chem. Phys.* **65**, 3389 (1976); C. Serna, G. E. VanScoyoc, and J. L. Ahlrich, *Am. Mineral.* **62**, 784 (1977).
42. G. E. VanScoyoc, J. L. Ahlrich, and C. Serna, *Am. Mineral.* **64**, 215 (1979).
43. H. Hyashi, R. Otsuka, and N. Imai, *Am. Mineral.* **54**, 1613 (1969).
44. G. Sposito and R. Prost, *Chem. Rev.* **82**, 553 (1982).
45. P. F. Low and J. F. Margheim, *Soil Sci. Soc. Am. J.* **43**, 473 (1979).
46. P. F. Low, *Soil Sci. Soc. Am. J.* **44**, 667 (1980).
47. P. F. Low, *Langmuir* **3**, 18 (1987).
48. J. P. Quirk, *Isr. J. Chem.* **6**, 213 (1968).
49. J. P. Quirk, *Philos. Trans. R. Soc. London* **316**, 297 (1968).

50. H. Suquet, C. de la Calle, and H. Pezerat, *Clays Clay Miner.* **23**, 1 (1975).
51. L. Heller-Kallai and S. Yariv, *J. Colloid Interface Sci.* **79**, 479 (1981).
52. K. Norrish, *Discuss. Faraday Soc.* **18**, 120 (1954).
53. L. Guldbrand, B. Joensson, H. Wennerstroem, and P. Linse, *J. Chem. Phys.* **80**, 2221 (1984).
54. R. Kjellander and S. Marcelja, *Chem. Phys. Lett.* **112**, 49 (1984).
55. R. Kjellander and S. Marcelja, *Chem. Phys. Lett.* **114**, 124 (1985).
56. A. Kahn, B. Joensson, and H. Wennerstroem, *J. Phys. Chem.* **89**, 5180 (1985).
57. R. Kjellander, S. Marcelja, R. M. Pashley, and J. P. Quirk, *J. Phys. Chem.* **92**, 6489 (1988).
58. V. Fornes and J. Chaussidon, in *Proc. Int. Clay Conf.*, Mexico, p. 383, Applied Pub., Wilmette, IL (1975).
59. P. L. Hall, D. K. Ross, J. J. Tuck, and M. H. B. Hayes, in *Proc. IAEA Symp. Neutron Inelastic Scattering*, vol. 1, p. 617 (1978).
60. P. L. Hall, D. K. Ross, J. J. Tuck, and M. H. B. Hayes, in *Proc. Int. Clay Conf.*, Oxford, p. 121 (1978).
61. R. Prost, *Proc. Int. Clay Conf.*, Mexico, p. 351, Applied Pub., Wilmette, IL (1975).
62. J. J. Fripiat, *Bull. Mineral.* **103**, 440 (1980).
63. J. J. Fripiat, M. Kadi-Hanifi, J. Conard, and W. E. E. Stone, in *Magnetic Resonance in Colloid and Interface Science*, pp. 529–535, Riedel, Amsterdam (1980).
64. J. Conard, *Proc. Int. Clay Conf.*, Mexico, p. 221 (1975).
65. J. Conard, in *Magnetic Resonance*, pp. 85–93, American Chemical Society, San Francisco (1976).
66. J. Conard, H. Estrade-Szwarckopf, A. J. Dianoux, and C. Poinsignon, *J. Phys.* **45**, 1361 (1984).
67. J. Conard, H. Estrade-Szwarckopf, C. Poinsignon, and A. J. Dianoux, *J. Phys. Colloq. C7, Suppl.* **45**, 169 (1984).
68. S. Yariv and L. Heller, *Isr. J. Chem.* **8**, 935 (1970); V. C. Farmer and J. D. Russell, *Trans. Faraday Soc.* **67**, 2737 (1971); H. Suquet, R. Prost, and H. Pezerat, *Clay Miner.* **12**, 113 (1977).
69. J. Hougardy, W. E. E. Stone, and J. J. Fripiat, *J. Magn. Reson.* **25**, 563 (1977).
70. C. de la Calle, H. Suquet, and M. Pezerat, *Bull. Groupe Fr. Argiles* **27**, 31 (1975).
71. J. G. Watterson, *Clays Clay Miner.* **37**, 285 (1989).
72. J. Mamy, *Ann. Agron.* **19**, 175 (1968).
73. G. Walker, *Clays Clay Miner.* **4**, 101 (1956).
74. J. J. Fripiat and W. E. E. Stone, *Phys. Chem. Liq.* **7**, 349 (1978).
75. S. Yariv, L. Heller, and N. Kaufherr, *Clays Clay Miner.* **17**, 301 (1969).
76. G. Besson, R. Glaeser, and C. Tchoubar, *Clay Miner.* **18**, 11 (1983).
77. R. Touillaux, P. Salvador, C. Vandermeersche, and J. J. Fripiat, *Isr. J. Chem.* **6**, 337 (1968).
78. M. Frenkel, *Clays Clay Miner.* **22**, 435 (1974).
79. M. M. Mortand and K. V. Raman, *Clays Clay Miner.* **16**, 393 (1968).
80. W. Bodenheimer, L. Heller, and S. Yariv, in *Proc. Int. Clay Conf.*, Jerusalem, vol. 2, p. 171 (1966).
81. S. Yariv and L. Heller, *Isr. J. Chem.* **8**, 935 (1970).
82. L. Heller-Kallai, S. Yariv, and M. Riemer, in *Proc. Int. Clay Conf.*, Madrid, vol. 1, p. 651 (1972).
83. S. Yariv, L. Heller, Z. Sofer, and W. Bodenheimer, *Isr. J. Chem.* **6**, 741 (1968).
84. L. Heller and S. Yariv, in *Proc. Int. Clay Conf.*, Tokyo, vol. 1, p. 741 (1969).
85. S. Yariv and S. Shoval, *Isr. J. Chem.* **22**, 259 (1982).
86. Z. Sofer, L. Heller, and S. Yariv, *Isr. J. Chem.* **7**, 697 (1969).
87. S. Saltzmann and S. Yariv, *Soil Sci. Soc. Am. Proc.* **39**, 474 (1975).
88. N. Lahav and E. Bresler, *Clays Clay Miner.* **21**, 249 (1973).
89. I. Ravina and P. F. Low, *Clays Clay Miner.* **20**, 109 (1972).
90. P. F. Low and J. M. Deming, *Soil Sci.* **75**, 187 (1953); P. F. Low, *Adv. Agron.* **13**, 269 (1961).
91. K. Norrish, in *Proc. Int. Clay Conf.*, Madrid, p. 375, Div. Sciencias C.S.I.C., Madrid (1972).
92. G. H. Bolt and R. D. Miller, *Soil. Sci. Soc. Am. Proc.* **19**, 285 (1955).
93. J. Hougardy, J. M. Serratosa, W. Stone, and H. van Olphen, *Spec. Discuss. Faraday Soc.* **1**, 187 (1970).
94. R. Prost, *Ann. Agron.* **26**, 463 (1975); R. Keren and I. Shainberg, *Clays Clay Miner.* **23**, 193 (1975); **27**, 145 (1979); **28**, 204 (1980); E. C. Ormerod and A. C. D. Newman, *Clay Miner.* **18**, 289 (1983); P. L. Hall and D. M. Astill, *Clays Clay Miner.* **37**, 355 (1989).
95. H. Suquet, C. de la Calle, and H. Pezerat, *Clays Clay Miner.* **23**, 1 (1975); A. Cowking, M. J. Wilson, J. M. Tait, and R. H. Robertson, *Clay Miner.* **18**, 49.

96. F. Kraehenbuehl, H. F. Stoeckli, F. Brunner, G. Kahr, and M. Mueller-Vonmoos, *Clay Miner.* **22**, 1 (1987).
97. H. Estrade-Szwarckopf, J. Conard, C. Poinsignon, and A. J. Dianoux, in *Proc. Eur. Clay Conf.*, Munich, 1980, quoted in Ref. 44.
98. D. J. Cebula, R. K. Thomas, and J. W. White, *Clays Clay Miner.* **29**, 241 (1981).
99. D. W. Oscarson, *Commun. Soil Sci. Plant Anal.* **19**, 1667 (1988).
100. M. B. McBride and M. M. Mortland, *Clay Miner.* **10**, 357 (1975).
101. K. Wada, *Am. Mineral.* **46**, 78 (1961).
102. A. Weiss, W. Thielepape, G. Goering, W. Ritter, and H. Schaefer, in *Proc. Int. Clay Conf.*, Stockholm, vol. 1, p. 287, Pergamon Press, Oxford (1963).
103. R. L. Ledoux and J. L. White, *Clays Clay Miner.* **13**, 289 (1964); in *Proc. Int. Clay Conf.*, Jerusalem, vol. 1, p. 361, Israel Prog. Sci. Trans., Jerusalem (1966).
104. O. Anton and P. G. Rouxhet, *Clays Clay Miner.* **25**, 259 (1977).
105. K. J. Range, A. Range, and A. Weiss, in *Proc. Int. Clay Conf.*, Tokyo, vol. 1, p. 3, Israel University Press, Jerusalem (1969).
106. M. B. McBride, *Clays Clay Miner.* **24**, 88 (1976).
107. S. Yariv, *Powder Technol.* **12**, 131 (1975).
108. S. Yariv, *J. Chem. Soc. Faraday Trans. 1* **71**, 674 (1975); S. Yariv, *Int. J. Tropic. Agric.* **4**, 310 (1986).
109. S. Yariv, E. Mendelovici, and R. Villalba, in *Proc. 7th Int. Conf. Thermal Analysis*, vol 1, p. 533, Kingston, Ont. Canada (1982).
110. K. H. Michaelian, S. Yariv, and A. Nasser, *Can. J. Chem.* **69**, 749 (1991). S. Yariv, A. Nasser, Y. Deutsch, K. H. Michaelian, *J. Therm. Anal.* **37**, 1373 (1991).
111. P. M. Costanzo, R. F. Giese, Jr., and M. Lipsicas, *Clays Clay Miner.* **32**, 419 (1984).
112. M. Lipsicas, C. Straley, P. M. Costanzo, and R. F. Giese, Jr., *J. Colloid Interface Sci.* **107**, 221 (1985).

12

Penetration and Displacement in Capillary Systems

Abraham Marmur

1. INTRODUCTION

Spontaneous capillary penetration of a liquid into pore spaces filled with a fluid is a flow driven by interfacial pressure differences. The magnitude of the pressure difference across each liquid–fluid interface (meniscus) depends on the local curvature, which is determined by the local wetting properties and pore geometry. The observation that interfacial pressure differences are appreciable only in small pores led to the term *capillary*, from the Latin *capillus* for hair. The recorded scientific history of capillary penetration seems to have started with Leonardo da Vinci,[1] who apparently was the first scientist to note capillary rise. More extensive work on the thermodynamics as well as the kinetics of capillary penetration began at the end of the 19th century and the beginning of this century.[2–9]

While *capillary penetration* is a generic term, it is mainly used when the pore spaces are initially filled with a gas. When the pores are partially or fully saturated with a liquid, which is then being displaced by another, the process is usually called *capillary displacement*. Sometimes, the term *imbibition* is used as a synonym for *penetration*. A distinction should also be made between forced penetration, where an external pressure difference is applied and interfacial effects are only partially responsible for the motion, and spontaneous penetration, where interfacial pressure differences are the only driving forces. The present discussion is concerned exclusively with spontaneous penetration or displacement.

The applications of capillary penetration are numerous and varied. Some of the oldest applications, which probably go back to the beginning of civilization, are writing with ink, and treatment of paper, wood, or cloth. These applications, in advanced forms, remain of great interest for the printing, paper, and textile

Abraham Marmur • Department of Chemical Engineering, Technion–Israel Institute of Technology, 32000 Haifa, Israel.

Modern Approach to Wettability: Theory and Applications, edited by Malcolm E. Schrader and George Loeb. Plenum Press, New York, 1992.

industries today. In addition, capillary penetration is of importance in the chemical, metallurgical, and ceramic industries. Water movement in soil or rocks, and enhanced oil recovery are well-known examples of capillary phenomena in the surface of earth. Capillary penetration is also essential in space, where it offers a transport mechanism that is independent of gravity. The transport of water in plants illustrates the importance of capillary penetration in biological systems.

The principle of capillary penetration can be qualitatively explained with the aid of Fig. 1. When a capillary is brought into contact with a liquid, the liquid–fluid interface must intersect the solid surface with a certain contact angle (Fig. 1a). This, as will be explained, is a thermodynamic equilibrium condition. The formation of the contact angle imposes a curvature on the liquid–fluid interface, which also depends on the radius of the capillary. An additional thermodynamic equilibrium condition is that the pressure at the convex side of the curved interface (point 1) has to be lower then the pressure at the concave side of the interface (point 2). Consequently, the pressure at point 1 is also lower than the pressure at point 3, which is located outside the capillary, at the same height as the initial position of the interface.

Thus, while the interfaces are at equilibrium, the bulk of the liquid is not. The unbalanced pressure difference within the bulk, which results from the interfacial pressure difference, is the driving force for capillary penetration. In a gravitational field, the flow stops when the bulk pressure difference between points 1 and 3 is balanced by the hydrostatic head (Fig. 1b). The flow may also stop when the liquid arrives at an edge (Fig. 1c). This is so because the geometry of the edge allows an appreciable reduction in the curvature of the meniscus (Fig. 1d), which, in turn, reduces the interfacial pressure difference.

The outline of this chapter is as follows. First, the general thermodynamic equilibrium conditions of interfaces are reviewed, with special emphasis on the problematic nature of the contact angle concept. This lays the background for understanding the specific problems related to the thermodynamics of capillary penetration, which are discussed next. Finally, the kinetics of capillary penetration are reviewed. Obviously, it is impossible to adequately cover all aspects of capillary penetration within the present chapter. For example, the interesting subjects of viscous fingering,[10] penetration into mesoporous media,[11] the existence of thin

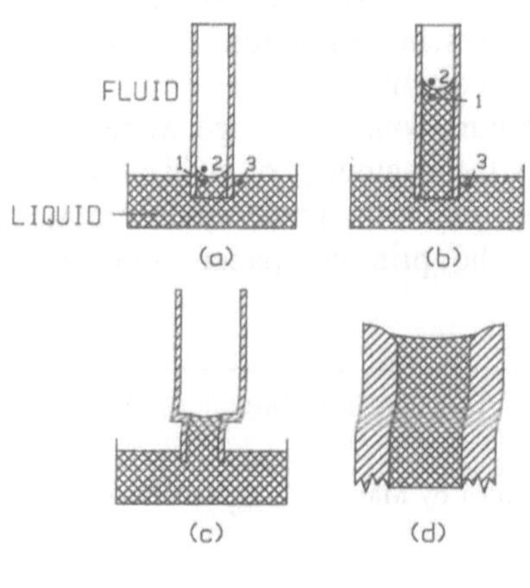

FIGURE 1. Stages in capillary rise: (a) the formation of a meniscus; (b) rise to the equilibrium height; (c) rise up to an edge; (d) a magnified view of the edge showing the diminished curvature of the meniscus.

films in very narrow capillaries,[12] or the percolation theory approach[13] are not discussed. The reader is also referred to the previous excellent review articles on the subject.[14–16]

2. EQUILIBRIUM CONDITIONS OF INTERFACES

Capillary penetration is the result of a mismatch between the equilibrium conditions of the bulk of the liquid and those of the interfaces. It is therefore essential to start with a brief discussion of the thermodynamic equilibrium conditions of interfaces. Two of them are associated with the shape of an interface: the Young–Laplace equation, which determines the equilibrium pressure difference across an interface as a function of its curvature, and the Young equation, which determines the contact angle at a solid–liquid–fluid contact line. These two conditions result simultaneously from the general equilibrium condition of minimum free energy.[17] In the following, the mathematical development of the conditions is omitted, and emphasis is put on unresolved problems.

2.1. The Contact Angle

2.1.1. Definitions

Capillary penetration takes place in three-phase systems, where a solid, a liquid, and a fluid meet. The region where the three interfaces (solid–liquid, solid–fluid, and liquid–fluid) intersect is called the *contact line*. The *contact angle* is the angle between the tangent to the liquid–fluid interface and the tangent to the solid interface (see Fig. 2a). This definition, however, is ambiguous, since it does not specify the scale at which the interfaces are looked at. This ambiguity has led to a frequent misinterpretation of the contact angle concept; therefore it is important to understand well the various definitions.

The *intrinsic* contact angle, θ, is the angle at a very short distance from the solid, of the order of magnitude of a few molecules. The *apparent* contact angle, θ_a, is the angle measured at the macroscopic level, e.g., through a microscope of low power. The possible difference between them is most clearly seen in Fig. 2b, where a rough surface is schematically shown.

2.1.2. The Young Equation

The basic equation for the contact angle on a smooth, nondeformable, insoluble, and homogeneous solid was published in 1805 by Young:[18]

$$\cos \theta = \frac{\gamma_{sf} - \gamma_{sl}}{\gamma_{lf}} \tag{1}$$

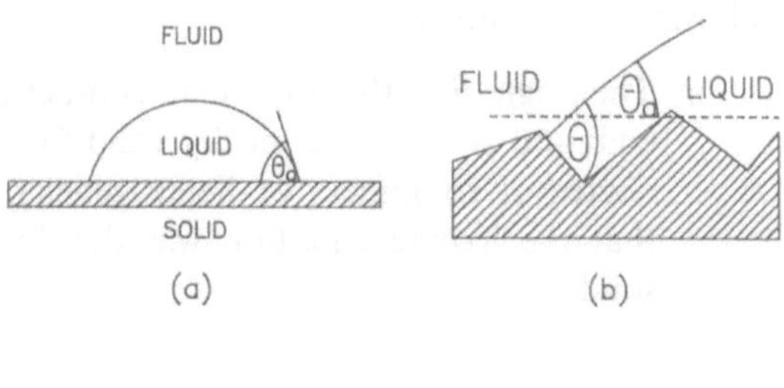

FIGURE 2. The intrinsic and apparent contact angles: (a) the apparent contact angle as seen through a low power microscope; (b) a magnified view of a rough surface showing the difference between the two contact angles.

where γ_{sf}, γ_{sl}, and γ_{lf} are the interfacial tensions of the solid–fluid, solid–liquid, and liquid–fluid interfaces, respectively. For various reasons, discussed below, it is very difficult to rigorously verify the Young equation by experiment. One of the main reasons is that it has not yet been possible to independently measure γ_{sf} and γ_{sl}. However, with the aid of approximate theories it is possible to express the solid–liquid interfacial tension in terms of the other two interfacial tensions and to reduce the number of unknowns in the Young equation.

This may be achieved for solid–liquid–vapor systems, for example, by using the Girifalco–Good equation:[19]

$$\gamma_{sl} = \gamma_s + \gamma_l - 2\phi\sqrt{\gamma_s \gamma_l} \tag{2}$$

where γ_s and γ_l are the surface tensions of the solid and the liquid (against vacuum), respectively, and ϕ is essentially an empirical correction factor. Substitution of this equation into Eq. (1) yields

$$\cos\theta = -1 + 2\phi\sqrt{\frac{\gamma_s}{\gamma_l}} \tag{3}$$

where γ_{sf} and γ_{lf} have been replaced by γ_s and γ_l, respectively. This assumption is reasonable for the liquid–vapor interface, but will be further discussed in relationship with the solid–vapor interface. Equation (3) clarifies the trends predicted by the Young equation: the contact angle is low (enhanced wetting) when the solid has a high surface energy and/or when the liquid surface tension is low. When γ_s is sufficiently high and/or γ_l is sufficiently low so that the right side of Eq. (3) is equal or greater than 1, complete spreading occurs and the final equilibrium state of the liquid is in the form of a thin film. These trends are well established by experiment.

The picture may become more complex if components from the fluid phase, such as the vapor of the liquid, adsorb on the solid. In this case, the solid–fluid interfacial tension must be modified to account for the adsorbed film:

$$\gamma_{sf} = \gamma_s - \pi^0 \tag{4}$$

where π^0 is the "film pressure." Since the adsorption process is spontaneous, γ_{sf} is smaller than γ_s. As a result, some liquids, which may be expected to completely spread on a high energy surface like metal or glass in air, spread only incompletely to a nonzero contact angle. These liquids were termed *autophobic* by Zisman and his co-workers.[20]

Additional fundamental problems associated with the Young equation have been recognized. A rigorous development of the Young equation[17] leads to the following conclusions:

a. The angle θ is the local true contact angle at any given point. The Young equation is therefore locally valid for any solid surface, independently of its degree of roughness or heterogeneity. However, it should be remembered that the Young equation was developed for nondeformable and insoluble solids.

b. The interfacial tensions included in the Young equation are to be evaluated at the contact line itself. As such, they may be different from the interfacial tensions far away from the contact line, because of the mutual interactions of the three phases at the contact line region.

The implication of these conclusions is that none of the parameters in the Young equation can be directly measured at the moment, since neither the contact angle not the interfacial tensions can as yet be measured at the contact line itself. Therefore, when experimentally measured values are substituted into the Young equation, the result should be treated only as an approximate guide.

2.1.3. Contact Angle Hysteresis

The Young equation predicts a single value for the *intrinsic* contact angle. However, it is well known that a range of stable *apparent* contact angles can be measured experimentally. In particular, when a liquid is added to a drop on a solid or withdrawn from it, the contact angle is observed to increase or decrease, respectively, as shown in Fig. 3. The maximum contact angle is called *advancing*, and the minimum is called *receding*. Similar observations have been made regarding the dipping and withdrawal of a solid plate or a fiber into a liquid. These observations mean that the apparent contact angle is not a unique function of the thermodynamic properties of the system. Rather, it shows properties of hysteresis.

A complete theory of contact angle hysteresis has not been developed yet. However, the sources of hysteresis seem to be known.[19] The two main sources are roughness and chemical heterogeneity. In addition, surface mobility is also quoted as a source of hysteresis. In the following, only an outline of the approach to hysteresis and the main results are described. More details can be found in review articles devoted to contact angle phenomena.[21–23]

The theory of hysteresis due to roughness seems to be the most developed within the general theory of contact angle hysteresis. Actually, the emphasis in the first models was not on hysteresis, but on explaining the difference between contact angles on rough and smooth surfaces as well as on the existence of a multitude of stable contact angles. The first step was taken by Wenzel,[24] who developed the following equation for the contact angle on a rough surface:

$$\cos \theta_w = r_s \cos \theta \tag{5}$$

where r_s is a measure of the roughness of the surface, defined as the ratio of the true surface area of the solid to the apparent area. The Wenzel contact angle, θ_w, represents the apparent contact angle corresponding to the absolute minimum

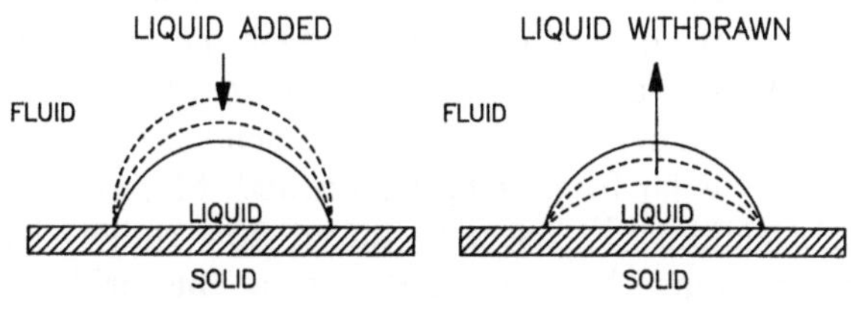

FIGURE 3. Contact angle hysteresis: the apparent contact angle increases upon addition of liquid and decreases upon withdrawal.

in the free energy of the system, i.e., to the stable equilibrium state. The Wenzel equation demonstrates that the apparent contact angle on a rough surface may be different from the intrinsic contact angle; however, it does not describe hysteresis.

In addition to the stable equilibrium state, there exist many metastable apparent contact angles with energy barriers between them.[21,22] These were initially discussed by using simple models of roughness, assuming sinusoidal[25] or sawtooth[26] surfaces. An explanation of hysteresis was attempted based on the need for an external energy ("vibrational" energy) to overcome the energy barriers between the metastable states. Important efforts have been made to further advance the theory of hysteresis by using different approaches and more sophisticated models for describing rough surfaces. Huh and Mason[27] analyzed the problem from a hydrodynamic point of view. Cox[28] analyzed two-dimensional periodic and nonperiodic surfaces. Joanny and de Gennes[29] analyzed the origin of hysteresis in terms of pinning of the contact line on a defect on the surface. However, as mentioned, the existing models give only a partial explanation of hysteresis, and much is yet to be done.

The effect of chemical heterogeneity cannot be visualized as easily as the effect of roughness. Again, the first attempt was aimed at calculating an apparent contact angle that represented the absolute minimum in the free energy of the system. This was done by Cassie and Baxter,[30] assuming a composite surface made up of two types of patches. The resulting equation for the apparent contact angle on such a composite surface, θ_c, is

$$\cos \theta_c = f_1 \cos \theta_1 + f_2 \cos \theta_2 \tag{6}$$

where f_1 and f_2 are the area fractions occupied by the two types of patches, and θ_1 and θ_2 are the corresponding intrinsic contact angles. However, whereas the Wenzel contact angle for rough surfaces may, in principle, be realized under proper conditions, the Cassie–Baxter angle for a heterogeneous surface is only a conceptual measure of wettability and cannot be identified in practice. Schwartz and Garoff[31] have recently discussed this problem using various shapes and arrangements of the patches. Based on their analysis, hysteresis is found to be a strong function of the details of the arrangement of patches, in addition to the dependence on the coverage fraction. This conclusion elucidates one of the most important reasons for the difficulty in understanding contact angle hysteresis.

2.1.4. Dynamic Contact Angle

The term *dynamic contact angle* has been used in the literature in two different senses: in cases where the contact angle depends on time, and in situations where it depends on the velocity of a moving contact line. Within the present context, the latter is the case of interest, although the presence of surfactants in a moving liquid may lead to a dependence on time, in addition to the dependence on velocity. The mechanism responsible for the dependence of the contact angle on the interfacial velocity is far from being understood. Moreover, most of the existing experimental information[32-34] refers to forced flow rather than to spontaneous spreading or penetration. The state of the art in this field may be represented by Hoffman's master curve,[33] which correlates the apparent dynamic contact angle, θ_{ad}, and the capillary number, $Ca = \mu v/\gamma$, where μ is the viscosity of the liquid and v is the velocity. This curve and other experimental results can be expressed by the following empirical equation:[35]

$$\frac{\cos\theta_a - \cos\theta_{ad}}{\cos\theta_a + 1} = \tanh(4.96Ca^{0.702}) \tag{7}$$

2.2. The Pressure Difference across an Interface

The pressure difference across an interface, ΔP, is related to its curvature according to the well-known Young–Laplace equation:[19]

$$\Delta P = \gamma_{lf}\left(\frac{1}{R_1} + \frac{1}{R_2}\right) \tag{8}$$

where R_1 and R_2 are the principal radii of curvature. For the specific situation of a spherical meniscus inside a cylindrical capillary of radius r, where the apparent contact angle is θ_a, the pressure difference is

$$\Delta P = \frac{2\gamma_{lf}\cos\theta_a}{r} \tag{9}$$

Note that the apparent contact angle is the one that determines the curvature of the meniscus and, consequently, the pressure difference that drives capillary penetration.

3. EQUILIBRIUM IN CAPILLARY SYSTEMS

As mentioned, capillary penetration is the result of a disagreement between the equilibrium conditions regarding the pressures in the bulk phases and those pertaining to the interfaces. Equilibrium is thus achieved when one of these conditions changes in order to accommodate the other. For example, in the absence of external force fields, the pressure in the bulk phases must be uniform. A pressure difference across a meniscus cannot, therefore, be maintained in a system where the liquid and the fluid are subjected to the same ambient pressure. The result is capillary penetration up to a point where the meniscus can be flattened, to annul the pressure difference across it. For example, in a horizontal capillary, spontaneous liquid penetration proceeds all the way up to the edge. In contrast, in an external force field such as gravity, the pressure in the bulk phases is not uniform. This leads to a possibility of getting to equilibrium by, for example, rising of the liquid to a level where the hydrostatic pressure is just right for the pressure difference required for interfacial equilibrium.

A combination of the two mechanisms is also possible, and, in fact, is common. Capillary rise in porous media may be arrested due to local geometrical configurations that flatten the menisci. A simplified demonstration of this effect can be seen with an ordered arrangement of cylinders, where at a certain point, depending on the intrinsic contact angle, the menisci become flat (Fig. 4). This effect may occur at several heights in a porous medium, below and above the stable equilibrium height, thus leading to metastable states and to the possibility of hysteresis.

This section starts by discussing penetration into a cylindrical capillary, a case that has been extensively studied due to its relative simplicity. This is followed by a discussion of equilibrium in porous media. Systems of simple geometry other than the cylindrical capillary have also been studied and should be briefly mentioned. These include the concentric capillaries,[36,37] two parallel or inclined plates,[12,38,39] and contacting rods.[39] Throughout this chapter, gravitation is used as the only example of an external force, since it is the most common one that interacts with capillarity.

3.1. The Cylindrical Capillary

3.1.1. The Shape of the Meniscus

Capillary penetration is initiated by the formation of a curved meniscus, the shape of which is therefore of fundamental importance. In the absence of external forces, the meniscus is a spherical cap that forms the necessary contact angle with the capillary wall. The pressure difference across it is then given by Eq. (9). Under the influence of gravity, the shape of the meniscus deviates from sphericity because the pressure difference varies from point to point. The shape of the meniscus cannot be calculated analytically, and has to be numerically integrated, as first done by Bashforth and Adams.[2] Analytical approximations were developed in relationship with the problem of capillary rise, as discussed in Section 3.1.3.

It is important to know the criterion under which the assumption of sphericity is reasonable. This is easily formulated by stating that the hydrostatic pressure difference across the meniscus, which is approximated by $\Delta\rho g r$, where $\Delta\rho$ is the positive density difference between the liquid and the fluid and g is the gravitational acceleration, has to be much smaller than the interfacial pressure difference, which approximately equals $2\gamma_{lf}/r$. Rearranging this inequality leads to an explicit condition for the radius of the capillary:

$$r \ll a \tag{10}$$

where $a^2 \equiv 2\gamma_{lf}/\Delta\rho g$. In the following, sphericity of the meniscus will always be assumed, unless otherwise stated.

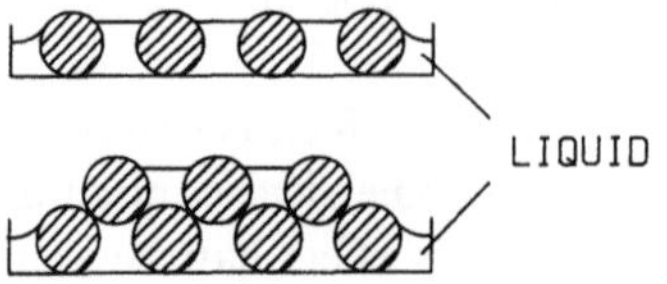

FIGURE 4. Multiplicity of equilibrium positions in a model porous medium: for each row of cylinders there is a metastable equilibrium position.

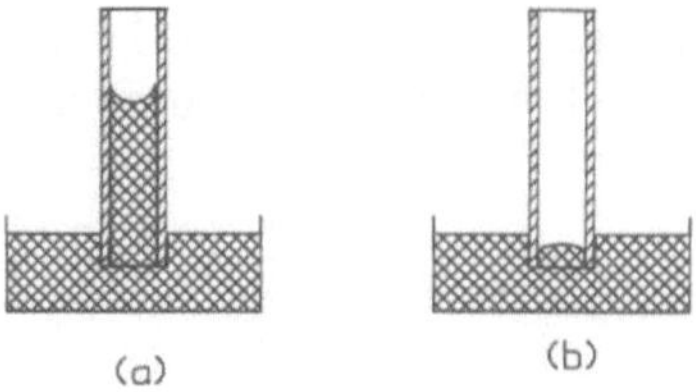

FIGURE 5(a) Capillary rise for $\theta_a < 90°$, and (b) capillary depression for $\theta_a > 90°$.

3.1.2. Criteria for Capillary Penetration

The criterion that appears in any elementary textbook is based on the apparent contact angle: capillary penetration is possible only for $\theta_a < 90°$, since then the meniscus is convex on the penetrating liquid side and the pressure there is lower than the ambient pressure. Figure 5 shows the classical pictures of capillary penetration ($\theta_a < 90°$) and capillary depression ($\theta_a > 90°$), such as in the cases of water and mercury, respectively, in touch with a glass capillary.

However, it has been recently realized[40,41] that the curvatures of the reservoirs of the penetrating liquid and the displaced fluid greatly influence the possibility of penetration. The above-mentioned criterion holds only for infinite reservoirs. The modified criteria for the penetration of a liquid from a limited reservoir into a cylindrical capillary are presented in the following. Two subsituations are distinguished in terms of the curvature of the reservoir of the displaced fluid: (a) an infinite reservoir and (b) a small one.

Infinite Reservoir of the Displaced Fluid. The system is shown in Fig. 6a. The reservoir of the displaced fluid is infinite; however, the penetrating liquid comes from a limited reservoir with a finite radius of curvature, such as a smal drop. The capillary is inclined at an angle α to the horizontal. Penetration occurs as long as the pressure at the entrance to the capillary is higher than the pressure needed to sustain the effects of the curvature of the meniscus and of gravity. Assuming the effect of gravity on the shape of the penetrating drop to be negligible, the pressure at the entrance to the capillary is $P_a + 2\gamma_{lf}/R_p$, where P_a is the ambient pressure and R_p is the instantaneous radius of the penetrating drop. The pressure at the penetrating liquid side of the meniscus is $P_a - 2\gamma_{lf}\cos\theta_a/r$. Penetration is then possible as long as the difference between them is greater than the contribution of gravity:

$$\frac{2\gamma_{lf}}{R_p} + \frac{2\gamma_{lf}\cos\theta_a}{r} - \Delta\rho g l \sin\alpha > 0 \qquad (11)$$

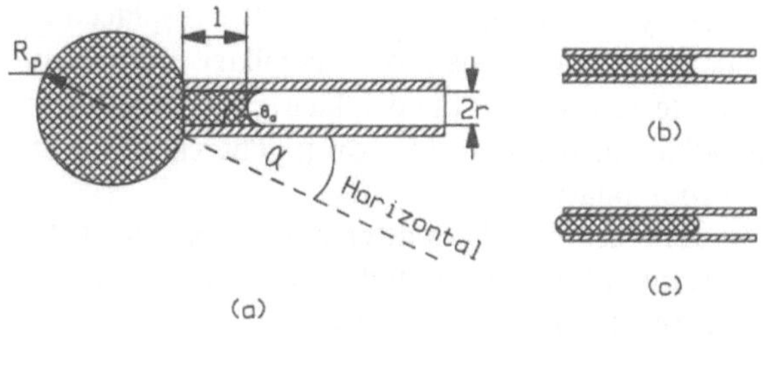

FIGURE 6(a) Penetration of a small drop into a capillary. (b) The final equilibrium situation for $\theta_a < 90°$. (c) The final equilibrium situation for $\theta_a > 90°$.

where l is the length of the liquid inside the capillary. Equilibrium is achieved when the above inequality turns into an equation:

$$\frac{2\gamma_{lf}}{R_{pe}} + \frac{2\gamma_{lf}\cos\theta_a}{r} - \Delta\rho g l_e \sin\alpha = 0 \tag{12}$$

where R_{pe} is the equilibrium radius of curvature of the penetrating drop, and l_e is the equilibrium length of the liquid in the capillary. Combination of Eqs. (11) and (12) results in the penetration condition:

$$\frac{1}{R_p} > \frac{1}{R_{pe}} \tag{13}$$

In the absence of gravity the equilibrium drop radius is simply expressed by rearranging Eq. (12):

$$\frac{R_{pe}}{r} = -(\cos\theta_a)^{-1} \tag{14}$$

This equation shows that R_{pe} is always negative for $\theta_a < 90°$ in the absence of gravity (see also Fig. 6b). Thus, Eq. (13) shows that capillary penetration occurs for any drop size when the apparent contact angle is acute, as is well known. However, Eqs. (13) and (14) also predict that penetration is possible for $\theta_a > 90°$ if the radius of the penetrating drop is sufficiently small. This seemingly surprising result is explained by the increased pressure inside the penetrating drop, due to its finite curvature, that may overcome the adverse effect of a concave meniscus. In other words, the interfacial energy that had been invested in the drop during its formation can be utilized to enable its penetration into a capillary even for obtuse apparent contact angles.

Of course, penetration under such conditions cannot be complete, as is shown in Fig. 6c. However, it has been shown that practically complete penetration is possible for apparent contact angles as high as about 115° for sufficiently small drops.[40]

Finite Reservoir of the Displaced Fluid. This situation arises when the fluid that initially occupies the capillary is connected to a small drop outside the capillary, or when this fluid exists initially only inside the capillary. The latter situation is of special interest since it is related to processes such as cleaning or oil recovery. The model system and the displacement process are shown in Fig. 7. The model system consists of a capillary that is initially full of the displaced liquid. A finite reservoir of the displacing (penetrating) liquid is attached to one side of the capillary in the form of a drop. The capillary and the two liquids are immersed in a vapor phase.

The crux of the matter is the opposition to penetration due to the increasing pressure in the growing bulge of the displaced liquid. As the liquid is being displaced from the capillary, the radius of curvature of the bulging drop first decreases, until

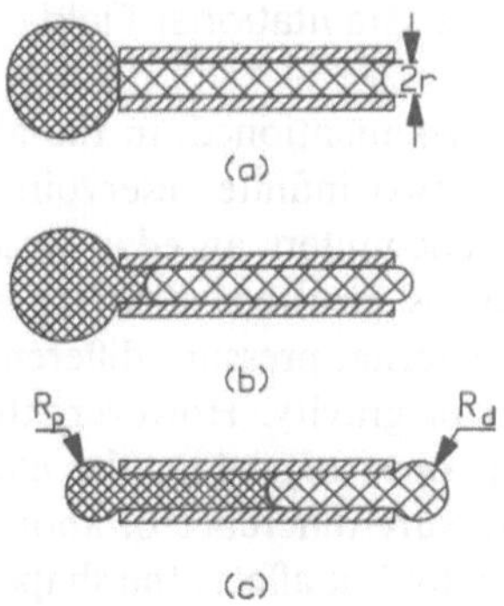

FIGURE 7. Stages in the displacement of a liquid from a capillary.

it gets to its minimum value when it equals the radius of the capillary. At this point, the pressure in the bulging drop is maximum. Therefore, penetration and displacement may proceed only if at this point there still exists a net pressure difference for driving the flow. Once the point of maximum pressure is passed, complete displacement is ensured.

The condition for spontaneous penetration in this case, in the absence of gravity, is

$$\frac{2\gamma_p}{R_p} > \frac{2\gamma_d}{R_d} - \frac{2\gamma_{pd}\cos(\theta_a)_{pd}}{r} \tag{15}$$

where the subscripts p and d indicate the penetrating liquid and the displaced liquid, respectively, R_d is the instantaneous radius of the bulging drop of the displaced liquid, and the subscript pd indicates the interface between the two liquids. The condition of completeness of penetration is expressed by putting $R_d = r$. This leads to the condition of complete spontaneous penetration:

$$\frac{r}{R_p} > \left[1 - \frac{\gamma_{pd}}{\gamma_d}\cos(\theta_a)_{pd}\right]\frac{\gamma_d}{\gamma_p} \tag{16}$$

It is useful to use the Girifalco–Good–Fowkes approximation, Eq. (2), to express γ_{pd}, and also assume $\cos(\theta_a)_{pd} = 1$, which may be easily achieved, to get

$$\frac{r}{R_p} > 2\sqrt{\frac{\gamma_d}{\gamma_p}} - 1 \tag{17}$$

The uncertainty in the value of ϕ is accounted for by the inequality gap. For an infinite reservoir of the penetrating liquid, complete spontaneous displacement is possible only if $\gamma_p/\gamma_d > 4$. This ratio is impractical in most situations, since with water being the penetrating liquid of highest available surface tension, the displaced liquid has to have a surface tension of less than 18 mN/m. Thus, complete spontaneous penetration may be realized in practice only if the penetrating liquid comes from a reservoir of a small radius of curvature.

3.1.3. The Height of Rise in a Gravitational Field

Two Infinite Reservoirs. As mentioned, in the absence of gravity, penetration into a capillary connected to two infinite reservoirs proceeds until the geometry changes, i.e., until the liquid encounters an edge or an expansion. In the presence of external forces, penetration is arrested when the pressure difference across the meniscus is balanced by the external pressure difference. The most common case is penetration under the action of gravity. However, the equations discussed in this section and in the following ones regarding the effect of gravity may be easily modified to account for a pressure difference of another origin.

The effect of gravity is twofold: it affects the shape of the meniscus as well as the height of rise. If the effect on the shape of the meniscus is neglected [see Eq. (10)], the height of rise in a capillary connected to two infinite reservoirs, h_c, is easily calculated by equating the hydrostatic and the interfacial pressure differences:

$$h_c = \frac{2\gamma_{lf} \cos \theta_a}{\Delta \rho g r} \tag{18}$$

For values of r larger than allowed by Eq. (10), corrections are required. This problem was solved by Lord Rayleigh,[3] who developed analytical approximations for the exact equations. For small capillaries, which are, however, too large to ignore the effect of gravity on the shape of the meniscus, and for $\theta_a = 0$, the following approximation holds:

$$a^2 = r\left(h + \frac{r}{3} - \frac{0.1288r^2}{h} + \frac{0.1312r^3}{h^2} \cdots\right) \tag{19}$$

The equations for capillary rise can be considered experimentally verified in an indirect way, through measurements of surface tension that favorably compare with results by other methods.[19]

A Finite Reservoir of the Penetrating Liquid. A finite radius of curvature of the reservoir of the penetrating liquid may strongly affect the height of rise. In a trivial sense, this height may be simply limited by the amount of the available liquid. However, a more fundamental effects exists, which stems from the small radius of curvature of the penetrating drop. This, as explained in the discussion of the penetration criterion, implies the existence of an increased pressure in the penetrating drop, which assists penetration. Moreover, as the liquid penetrates, the radius of the drop decreases, and the pressure inside the drop further increases. As recently discussed,[40] the height of rise is substantially increased for the penetration of small drops as compared with an infinite reservoir. The relative increase in height becomes more appreciable as the contact angle increases. For contact angles higher than 90°, capillary rise from an infinite reservoir is impossible, as is well known. However, capillary rise from a small drop can take place.

3.2. Equilibrium and Hysteresis in Porous Media

The structure of porous media is immensely more complex than that of any of the simple capillary systems that have been studied. The local structure varies from point to point both in the direction of penetration and in the transverse direction. The variations in the direction of penetration lead to multiplicity of equilibrium positions and to hysteresis, thus making the theoretical analysis of penetration difficult. Even more difficult is the analysis of the interactions in the transverse direction, since it requires at least a two-dimensional solution. A general review of the thermodynamics of interfaces in porous media, which emphasizes these complexities was given by Everett and Haynes.[(42–44)] In addition to the theoretical obstacles, it is also difficult to experimentally characterize local structural details that are required for the understanding of penetration phenomena. Thus, much has yet to be done regarding both the experimental as well as the theoretical aspects of the problem.

The simplest example that can demonstrate the phenomena typical of equilibrium in porous media is a capillary with an expansion in the middle, as shown in Fig. 8. If the capillary is completely filled to begin with, then upon drainage the equilibrium position will be the same as for a straight capillary of the same radius and material (Fig. 8a). Upon penetration of a liquid into the capillary from a reservoir beneath it, the meniscus equilibrates at the lower part of the expansion (Fig. 8b). This is so because its radius of curvature increases when it meets the expansion, thereby decreasing the absolute value of the interfacial pressure difference until it equals the hydrostatic head.

Obviously, in a real porous medium, the "expansions" are connected also in the transverse direction and consist of various shapes and sizes. As will be discussed, understanding of these phenomena is still deficient, although much progress has been made. To lay the background for the discussion, the experimental observations are first presented. Then the various types of theoretical models are discussed.

3.2.1. Experimental Observations

The first experimental observations were motivated by the problem of moisture distribution in soil.[(45–49)] The difficulties related to the study of real porous media were, of course, recognized by the early researchers. Consequently, much effort was devoted to the study of "ideal soil" models, namely packings of spheres of various degrees of regularity. These studies included observations of liquid rise between a few spheres as well as in random packings of spheres. The main conclusions from this early work[(45–49)] can be summarized as follows:

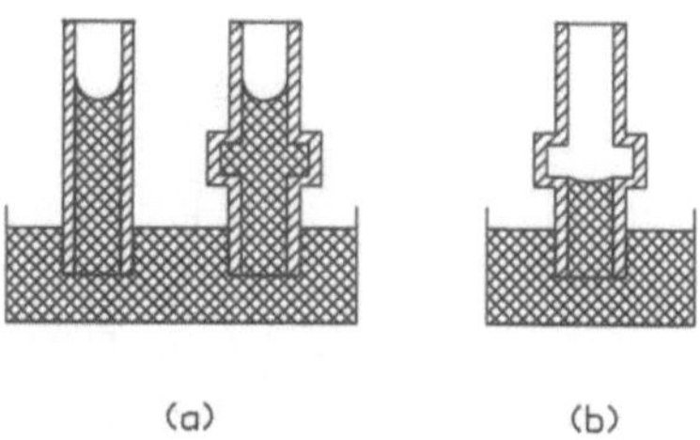

FIGURE 8. The origin of hysteresis in the saturation curve of porous media: (a) equilibrium during drainage; (b) equilibrium during penetration.

a. The process of penetration into a porous medium is associated with sudden pressure changes. These pressure "jumps" stem from sudden variations in the curvatures of the menisci as they follow the geometry of the medium.

b. There is a range of stable heights of capillary rise. The maximum height is observed when the liquid is allowed to drain from a saturated porous medium. The minimum height can be observed when a liquid penetrates a fluid-filled medium.

c. Penetration and drainage do not follow the same path, but form a hysteresis loop, a typical shape of which is shown in Fig. 9 in terms of the saturation percentage versus the pressure deficiency. The latter is defined as the difference between the atmospheric pressure and the pressure in the penetrating liquid at the entrance to the porous medium. Inside the main hysteresis loop there exist scanning curves, which show the paths followed when the direction of the process is reversed before reaching the edge of the loop. A typical scanning curve is also shown in Fig. 9.

The shape of the hysteresis loop is explained as follows (assuming the porous medium to have been at least once filled up with the liquid). At low pressure deficiencies (negative pressure deficiencies with high absolute values), most of the liquid is drained from the porous medium. However, some liquid is inevitably retained as isolated rings at the contact between particles. Further reduction of the pressure deficiency cannot affect these isolated liquid reservoirs, therefore saturation is constant, though not zero. As the pressure deficiency is increased (lower absolute values) penetration of liquid begins and saturation increases. The small pores are filled first, but their contribution to the saturation is relatively minor. Consequently, the penetration curve is relatively flat until the pressure deficiency is raised to the point where the large pores can be penetrated. On raising the pressure deficiency to

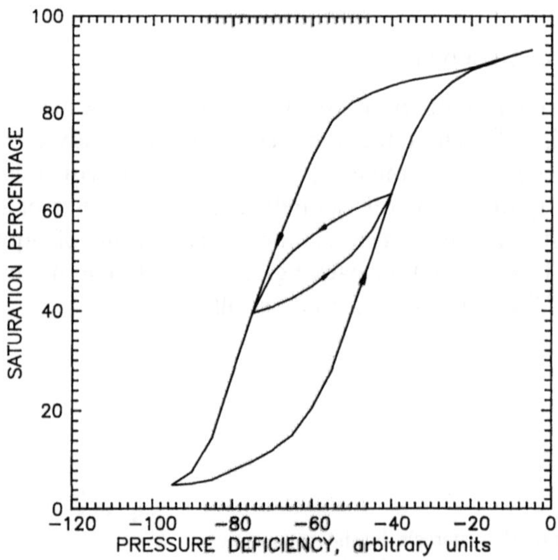

FIGURE 9. The hysteresis loop.

zero, saturation increases, but does not get to 100%, due to entrapment of fluid bubbles.

An essential point in the understanding of the hysteresis loop is that penetration into large pores must be followed by spontaneous further penetration into small pores, which were previously inaccessible. This is so, since the value of the pressure deficiency that allows penetration into the large pores is higher than required for penetration into the small ones. As a result, upon drainage, there are always small pores that need to be evacuated before liquid can be removed from large pores. This leads to the shape of the drainage curve as shown in Fig. 9. When the pressure deficiency is reduced from zero, saturation remains approximately constant until the pressure deficiency is sufficiently low to cause drainage from the small pores; only then, when the drainage from the large pores follows, is saturation substantially decreased.

Surprisingly, the extent of experimental work since the above-mentioned early studies has been limited. The main contribution has been the study of the effect of the contact angle on capillary penetration. This effect has been studied for polydisperse particles by Morrow,[50] and for monodisperse particles by Yang, Zografi, and Miller.[51] The main experimental conclusions drawn from these studies can be summarized as follows:

a. Drainage curves are relatively insensitive to the contact angle.
b. The effect of the contact angle on the initial penetration and repenetration curves is minor for contact angles lower than 37° in the monodisperse system or 30° in the polydisperse system.
c. Above these contact angle values, the penetration curve is sensitively dependent on the contact angle.
d. Spontaneous penetration is arrested for contact angles higher than about 60°. A similar observation was also made by Hansford *et al.*[52]

Another interesting observation regarding the effect of the contact angle was made by van Brakel and Heerthes.[53] They observed that a wide saturation gradient, namely a strong dependence of the saturation on the height, existed for liquids that formed a zero contact angle with the solid surface of the porous medium. In contrast, no saturation gradient existed when the contact angle was higher than zero.

3.2.2. Theoretical Models

Theoretical models for equilibrium in porous media should discuss the criteria for penetration and predict the dependence of saturation on the pressure deficiency. Unfortunately, only partial fulfillment of these goals has been achieved to date. The following discussion is divided into two sections: the first summarizes the work done on infinite porous media, and the second describes some of the peculiar results that relate to thin porous media. The terms *infinite* or *thin* are based on the ratio of the smallest dimension of the porous medium to its typical pore size.

Infinite Porous Media. A general equation for the capillary rise of a liquid in a porous medium has been recently formulated by the author.[54] The thermodynamic equilibrium condition of minimum free energy is expressed by the following equation:

$$\Delta\rho gh\, dV + (\gamma_{sl} - \gamma_{sf})\, dA_{sl} + \gamma_{lf}\, dA_{lf} = 0 \tag{20}$$

where h is the height of capillary rise, V is the volume of the rising liquid inside the porous medium, and A_{sl} and A_{lf} are the solid–liquid and liquid–fluid interfacial areas, respectively. If Eq. (1), which defines the intrinsic contact angle, is introduced into Eq. (20), the general equation for the height of rise in a porous medium results:

$$h = \frac{\gamma_{lf}}{\Delta\rho g} \frac{dA_{sl}}{dV} \left(\cos\theta - \frac{dA_{lf}}{dA_{sl}} \right) \tag{21}$$

This equation can be transformed, under appropriate assumptions, into previously developed forms. Thus, Eq. (18) for the height of rise in a cylindrical capillary results from Eq. (21) if it is realized that the liquid–fluid interfacial area is constant ($dA_{lf} = 0$) and $dA_{sl}/dV = 2/r$. White[55] developed the following equation for the height of rise in a porous medium, which is characterized only by its overall porosity, ε_0, and overall specific area of the solid, S_0:

$$h_0 = \frac{(1-\varepsilon_0)\, S_0 \gamma_{lf} \cos\theta}{\Delta\rho g \varepsilon_0} \tag{22}$$

Such a model, which does not recognize local structural details in the porous medium, can be called *microscopically uniform*. Equation (22) also results from the general equation, Eq. (21), if it is recognized that $dA_{lf} = 0$ and $dA_{sl}/dV = (1-\varepsilon_0)\, S_0/\varepsilon_0$.

A microscopically uniform model is useful for estimating the expected height of rise and for comparison with the cylindrical capillary model. The latter is accomplished via the definition of an effective capillary radius for the porous medium, by comparing Eqs. (18) and (22):

$$(r_0)_{eff} = \frac{2\varepsilon_0}{(1-\varepsilon_0)\, S_0} \tag{23}$$

However, such a model predicts only a single equilibrium height, since it ignores local details. Therefore, it can neither predict hysteresis nor account for the lack of penetration at contact angles lower than 90°.

Three different approaches have been applied to modeling the effect of local variations in the structure of a porous medium on capillary penetration. One approach deals with porous media made of ordered packings of spheres ("ideal soil models").[45–49,56] The major advantages of these models are, of course, the simplicity of the structure compared with random porous media, and the possibility of comparison with well-defined experimental models. A different approach has been recently developed by the author,[54] which accounts for local structural variations by sinusoidal functions. These serve only as a first-order approximation, however, and the procedure is suitable for any functions that will be found to represent more realistically the structure of a porous medium. Another approach, which is naturally developing with the immense increase in computer power, is computer simulation. This aspect has been recently reviewed by Li *et al.*[57]

The basic assumption used in the initial studies of the "ideal soil models" was that the interfacial pressure difference could be calculated on the basis of a structural "unit cell." The curvature of the interface was calculated as if it was a hemisphere touching the spheres at their horizontal diameter plane.[45,46] Thus, for a cubical packing, where the hemisphere was assumed to touch four spheres, its radius of curvature was calculated to be $(\sqrt{2}-1)\,a_p$, where a_p is the radius of the spherical particle. For close packing it was $(2/\sqrt{3}-1)\,a_p$. In spite of the simplicity of the approach, experimental data for the height of capillary rise favorably compared with the calculated values.[46] A slightly more sophisticated model[48] assumed a hexagonal array, with an adjustable parameter for the distance between the spheres. This parameter was determined by fitting the porosity of the model to the actual porosity of the porous medium. Based on the proportions of three types of openings in such an array, an average interfacial pressure difference was calculated as a function of the porosity, from which the height of rise was expressed as

$$h = \frac{2\gamma_{lf}}{\Delta\rho g a_p[0.9590/(1-\varepsilon_0)^{2/3}-1]} \tag{24}$$

Experimental data compared favorably with this model as well.

Recently, the exact geometry of ordered packings of spheres has been used to calculate the highest contact angle that allows spontaneous imbibition.[56] An ordered packing composed of spheres of two sizes has been assumed. For size ratios of 1/2 to 1, this contact angle has been calculated to be within the range of 50° to 63°, in good agreement with experimental measurements with various powders.

The first-order model[54] attempts to deal with local variations in a periodic porous medium. Local values of the porosity, ε, the specific surface area of the solid, S, and the apparent contact angle, $\theta+\beta$, are used to transform Eq. (21) to

$$h = \frac{\gamma_{lf}(1-\varepsilon)\,S\cos(\theta+\beta)}{\Delta\rho g\varepsilon} \tag{25}$$

The local values of ε, S, and $\theta+\beta$ are actually averages over the cross section perpendicular to the direction of penetration. The apparent local contact angle is defined as the sum of θ and β to emphasize the difference between the intrinsic contact angle and the apparent one, which is used to calculate the radius of curvature of the meniscus.

The first-order model recognizes the current limitations in the ability to characterize the local details of the structure of a porous medium. Therefore, it treats ε, S, and $\theta+\beta$ as sinusoidal functions, with the understanding that these are actually the first terms in appropriate, yet unknown, series:

$$\varepsilon = \varepsilon_0(1 + A\sin\omega x) \tag{26}$$

$$S = S_0(1 + B\sin\omega x) \tag{27}$$

$$\beta = C\sin\omega x \tag{28}$$

In these equations, A, B, and C are constant coefficients,

$$x \equiv \frac{h}{h_0} \tag{29}$$

and

$$\omega \equiv \frac{\pi h_0}{a_p} \tag{30}$$

where a_p is the characteristic particle radius.

With the above assumptions and definitions, Eq. (25) becomes

$$w \equiv x - \frac{1 - [\varepsilon_0/(1-\varepsilon_0)]\, A \sin \omega x}{1 + A \sin \omega x} \times (1 + B \sin \omega x)[\cos(C \sin \omega x) - \tan\theta \sin(C \sin \omega x)] = 0 \tag{31}$$

A typical plot of w versus x is shown in Fig. 10. The straight dashed line represents the case of a microscopically uniform porous medium ($A = B = C = 0$), for which there is only one equilibrium situation ($w = 0$ only at $x = 1$). In contrast, Fig. 10 clearly shows the multiplicity of equilibrium points that result from the local variations in the structure of the porous medium. It also becomes clear that there exist a minimum height for the capillary rise and a maximum height, as pointed out in the figure.

The difference between the maximum and the minimum is the hysteresis range. The conclusions regarding the hysteresis range, which have been drawn based on this model,[54] can be summarized as follows:

a. The hysteresis range depends mainly on the amplitude of oscillations in w.
b. The hysteresis range is large when the functional dependencies of the porosity and the specific area on the position are very different (A/B far from unity).

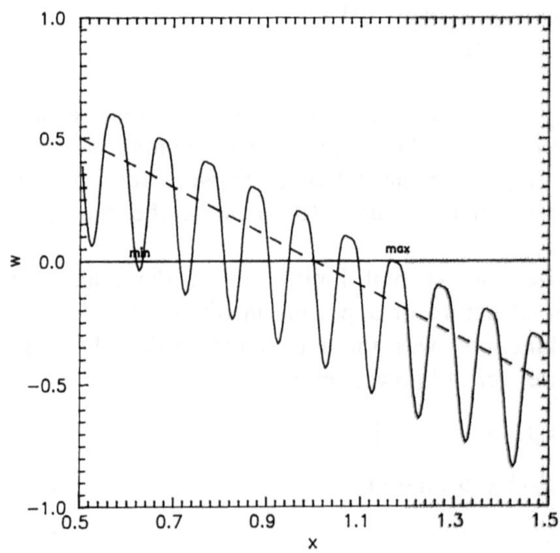

FIGURE 10. The multiplicity of metastable equilibrium positions inside a porous medium as calculated by the first-order model [see Eq. (31)]. The leftmost intersection with the line $w = 0$ is the minimum height. The rightmost intersection is the maximum height. The dashed line represents a microscopically uniform porous medium with only one equilibrium height at $x = 1$.

c. The hysteresis range remains approximately constant up to a critical value of the intrinsic contact angle and then increases with further increase of θ. It is important to emphasize that this effect stems from the combined effect of variations in β, ε, and S.
d. Above a certain contact angle, spontaneous imbibition becomes impossible. This is due mostly to the variations in β.

Thus, the first-order model predicts many of the experimentally observed phenomena, in spite of the simplistic mathematical description of the structural variations. Some of its predictions show good quantitative agreement with experiments. It therefore seems feasible that this line of development will yield more accurate results with a better characterization of a porous medium as an input.

Thin Porous Media—The Reeexposure Effect. When the porous medium is thin, its boundaries may play an important role in the process of spontaneous capillary penetration. This is so because the penetrating liquid is reexposed to the ambient fluid through the pores in the boundaries of the porous medium, concomitantly to the penetration process. This reexposure affects the free-energy balance of the process and is, obviously, more prominent the thinner the porous medium. Two cases that demonstrate the reexposure effect in a thin porous medium have been studied so far: capillary rise from an infinite liquid reservoir[58] and the penetration of a drop.[59] In both cases, the porous medium is described by a microscopically uniform model.

When a vertical thin porous medium touches an infinite source of liquid, the final equilibrium position may, in principle, be one of the two shown in Figs. 11a and 11b. It turns out that the case of capillary depression (Fig. 11b) is symmetric to the case of capillary rise; therefore, there is no need to discuss it separately. The case shown in Fig. 11c is, as will be explained, a result of the reexposure effect.[58] This situation is characterized by equal heights of rise inside the porous medium and outside it. By this equality, the liquid inside the porous medium avoids the thermodynamically unfavorable contact with the outside fluid. The "price" turns out to be a much lower rise than for an infinitely thick porous medium.

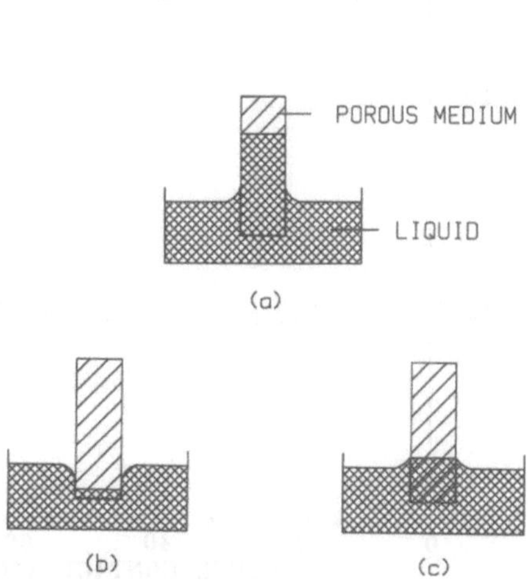

FIGURE 11(a) Capillary rise into a thin porous medium. (b) Capillary depression in a thin porous medium. (c) Limited capillary rise due to the reexposure effect: the height of rise inside and outside the porous medium are equal.

The dimensionless height of rise inside a thin porous medium for the case shown in Fig. 11a (normalized with respect to h_0) is given by[58]

$$x = 1 - \frac{2\varepsilon_s}{D(1-\varepsilon)\, S \cos\theta} \tag{32}$$

where ε_s is the surface porosity and D is the thickness. This equation shows that the capillary rise of a liquid in a thin porous medium is smaller than in an infinite one, especially for high intrinsic contact angles. The corresponding liquid rise outside the porous medium is uniquely determined by the equilibrium apparent contact angle, θ_{ae}, which is calculated from the following equation:

$$\cos\theta_{ae} = (1-\varepsilon_s)\cos\theta + \varepsilon_s \tag{33}$$

The dimensionless height of rise outside the porous medium (again normalized with respect to h_0) is then calculated from

$$y = \frac{2\varepsilon_0 c \sin(\pi/4 - \theta_{ae}/2)}{(1-\varepsilon_0)\, S \cos\theta} \tag{34}$$

where $c \equiv \sqrt{\Delta\rho g/\gamma_{lf}}$ is an inverse length scaling factor.

Above a certain value of θ (but still less then 90°), the equilibrium condition turns out to be $x = y$. Their value is calculated from Eq. (34), using θ_{ae} from the following equation:

$$2\cos\theta_{ae} - \cos\theta[(1-\varepsilon)\, DS + 2(1-\varepsilon_s)] + 2Dc\varepsilon \sin\left(\frac{\pi}{4} - \frac{\theta_{ae}}{2}\right) = 0 \tag{35}$$

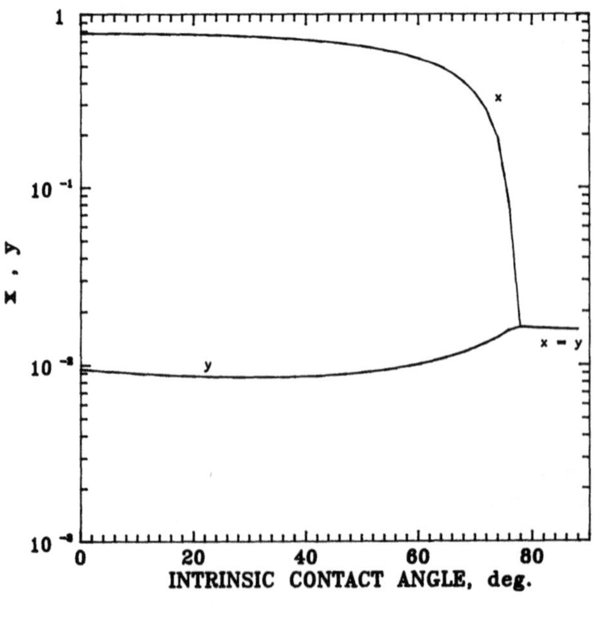

FIGURE 12. Dimensionless heights of rise inside (x) and outside (y) a thin porous medium vs. the intrinsic contact angle.

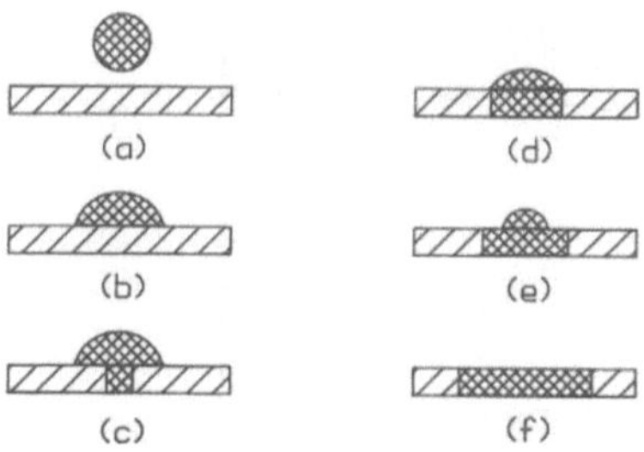

FIGURE 13. Stages of drop penetration into a thin porous medium.

Typical results for the height of rise inside a thin porous medium and outside it are shown in Fig. 12. It is clearly seen that the height of rise is lowered by the reexposure effect. Moreover, it becomes very low for high intrinsic contact angles, which are, however, smaller than 90°. It has been also found that for a given thickness of the porous medium, the reexposure effect is enhanced by increasing the porosities and/or decreasing the specific area of the solid.

The reexposure effect is very pronounced also in the case of a drop penetrating into a thin porous medium.[59] Figure 13 shows schematically the stages of penetration, assuming the porous medium to be sufficiently thin for the penetration to be uniform throughout its thickness. Numerical calculations of the free energy of the system have revealed the possible existence of a local or absolute minimum at the basal penetration situation, for which the radius of the liquid inside the porous medium equals the radius of the base of the drop (Fig. 13d). This minimum can be qualitatively explained with the aid of Fig. 14. Perturbations around the basal penetration position lead to reexposure of the liquid to the ambient fluid through the pores in the boundaries of the porous medium. This is true whether the fluid to which the liquid is reexposed is inside the porous medium (Fig. 14a) or outside it (Fig. 14c). The reexposure is thermodynamically unfavorable, and, depending on the relative contribution of the other factors in the system, an absolute minimum in the free energy or a local one may occur.

The penetration characteristics of a system can be summarized in the form of a "phase diagram,"[59] a typical example of which is shown in Fig. 15. For a large dimensionless thickness (normalized with respect to the size of the drop prior to

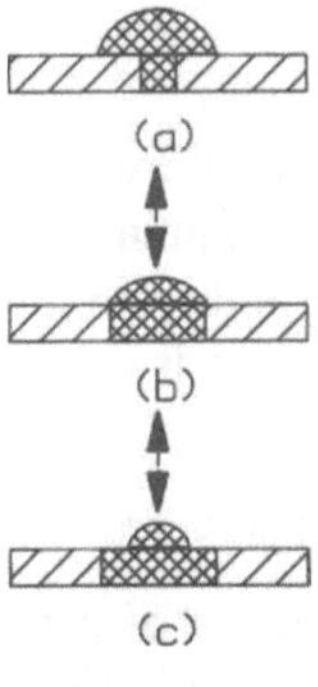

FIGURE 14. Perturbations around the basal penetration situation.

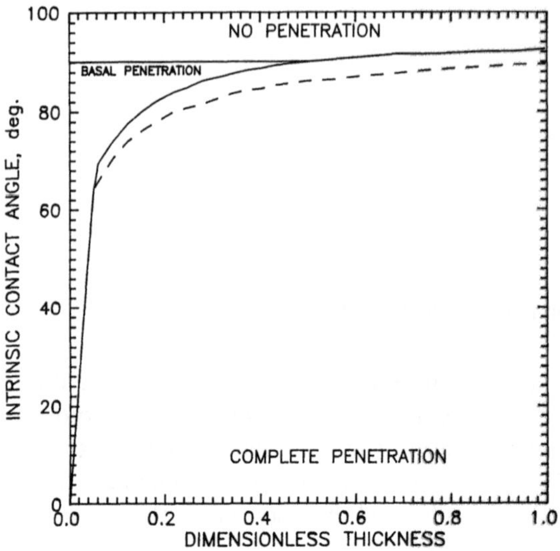

FIGURE 15. "Phase diagram" for penetration of a drop into a thin porous medium.

penetration), there are only two stable equilibrium states: complete penetration or no penetration. The borderline, shown by the solid curve in Fig. 15, may be at intrinsic contact angles higher than 90°. This is so because of the enhancement of penetration by a small radius of the drop, similarly to the case of penetration of a small drop into a cylindrical capillary.

For thin porous media there are three stable equilibrium situations: complete penetration, basal penetration, and no penetration. The space between the dotted curve and the solid curve above it indicates that part of the region of basal penetration that is in a metastable equilibrium state. For a given thickness, the reexposure effect is enhanced by increasing the surface porosity, increasing the bulk porosity, or decreasing the specific surface area of the solid. These changes diminish the influence of the solid surface inside the porous medium and enhance the interaction with the outside fluid.

4. *KINETICS OF PENETRATION*

4.1. *The Cylindrical Capillary*

Again, as in the discussion of equilibrium, the simplest model that has been extensively studied is the cylindrical capillary. The following discussion starts with a review of the work done on a capillary in contact with infinite reservoirs of the liquid and the fluid. The second section is devoted to the case of limited reservoirs.

4.1.1. Infinite Liquid and Fluid Reservoirs

The basic equation for the kinetics of penetration was developed by Lucas[4] and Washburn,[6] and will be referred to in the following as the LW equation. They assumed a quasi-steady-state, fully developed laminar flow of a Newtonian liquid, with negligible inertia effects. Therefore, the average velocity v was calculated from the Hagen–Poiseuille equation:

$$v = \frac{dl}{dt} = \frac{r^2}{8\mu}\frac{\Delta P}{l} \tag{36}$$

where l is the length of the liquid inside the capillary and μ is the viscosity of the liquid. Equation (36) also assumes that the liquid penetrates into a capillary filled with a fluid of negligible viscosity. This assumption has been made in most of the theoretical models; however, Washburn[6] did solve his equation for the case of a fluid with appreciable viscosity. In the following, liquid penetration into a vapor-filled capillary or porous medium will be assumed, unless otherwise stated. The driving force for penetration is the pressure difference across the liquid, which was assumed constant and equal to

$$\Delta P = \frac{2\gamma_{\rm lf}\cos\theta_{\rm a}}{r} - gl\,\Delta\rho\sin\alpha \tag{37}$$

Based on these assumptions, for the case of infinite liquid and fluid reservoirs, the kinetics of penetration is modeled by the following differential equation, written in dimensionless form:

$$X\frac{dX}{d\tau} = 1 - GX \tag{38}$$

where

$$X \equiv \frac{l}{r} \tag{39}$$

$$\tau \equiv \frac{\gamma_{\rm lf}\cos\theta_{\rm a}}{4\mu r}\,t \tag{40}$$

and

$$G \equiv \frac{\Delta\rho g r^2 \sin\alpha}{2\gamma_{\rm lf}\cos\theta_{\rm a}} \tag{41}$$

In Eq. (40), t is the time.

Equation (38) was solved for the microgravity case to read:

$$X^2 = 2\tau \tag{42}$$

The solution for the gravitational case is

$$-\frac{X}{G} - \frac{\ln(1-GX)}{G^2} = \tau \tag{43}$$

The square root dependence of the distance on time, as predicted by Eq. (42), does not result from the fact that the flow is driven by an interfacial pressure difference. Rather, it is typical of any laminar flow of a Newtonian liquid in a tube under the action of a constant pressure difference, where the length of the liquid is changing with time.[60] It is also important to realize, as was pointed out by Good,[61,62] that the term $\gamma_{lf} \cos \theta_a$ in Eq. (40) should be corrected to account for the film pressure [see Eq. (4)]. However, measurements of the film pressure are still scarce and controversial; therefore, this term will be omitted in the following equations.

The LW equation is based on very simplistic assumptions, which need to be tested. To begin with, the quasi-steady-state assumption eliminates acceleration terms. As a result, the LW equation yields the unrealistic value of an infinite velocity of penetration at time zero.[63] Shortly after the development of the LW equation, Rideal[8] and then Bosanquet[9] considered inertia terms. The latter developed the following differential equation for a liquid penetrating into a horizontal capillary:

$$X\frac{dX}{d\tau} = 1 - \exp(-b\tau) \tag{44}$$

where

$$b = \frac{32\mu^2}{r\,\Delta\rho\gamma_{lf}\cos\theta_a} \tag{45}$$

The solution of Eq. (44) is

$$X^2 = 2\left\{\tau - \frac{1}{b}[1 - \exp(-b\tau)]\right\} \tag{46}$$

Bosanquet[9] also developed approximate solutions for the case of liquid penetrating into a capillary originally filled with a fluid of appreciable viscosity and for penetration under the effect of gravity.

Equation (46) leads to two important conclusions. First, it predicts that, after a sufficiently long time, the penetration kinetics follow the LW equation, with no "memory" of the initial acceleration period. The criterion for this to be the case is easily deduced from Eq. (46) to be

$$\tau \gg \frac{1}{b} \tag{47}$$

This criterion can be transformed, using Eq. (42) as an approximation, to

$$X^2 \gg \frac{2}{b} \tag{48}$$

LW kinetics were clearly demonstrated by quite a few experimental studies,[64–66] all of which conformed to the above criterion. The second conclusion from the Bosanquet equation, Eq. (46), is that consideration of inertia eliminates the impossible result of an infinite rate at time zero and predicts

$$\left(\frac{dX}{d\tau}\right)_{\tau\to 0} = \sqrt{b} \tag{49}$$

Rideal[8] developed a differential equation similar to the Bosanquet equation, but neglected the term in $(dX/d\tau)^2$. His simplified differential equation actually made the solution more difficult, since it required an infinite series instead of the closed form of the Bosanquet equation.[9] Levine and Neal[67] independently developed the same equation as Rideal,[8] but presented a different solution. Levine and Neale[67] also solved a similar equation for the case of a vertical capillary.

The other fundamental problems with the LW equation concern the flow at the two edges of the capillary. The LW equation does not consider the pressure distribution caused by the flow of the liquid from the reservoir into the capillary at the entrance. In addition, it does not account for the peculiarities of the flow at the contact line between the meniscus and the capillary wall, and for the possibility of a dynamic contact angle. The former problem was fist tackled by Brittin[68] and then by Szekely *et al.*,[63] whose equations were later criticized.[69] Other solutions to the problem have been developed;[69–71] however, a comparison[70] with the Bosanquet equation[9] shows appreciable differences only for penetration that is sufficiently rapid to cause overshoot above the final equilibrium position.

The most difficult problem related to the kinetics of penetration seems to be the flow at the front edge of the liquid in the vicinity of the contact line. The macroscopic effect of this flow is a dependence of the apparent contact angle on the instantaneous velocity of the meniscus, in contrast to the LW assumption of a constant θ_a. The solution of this problem has been attempted within the general framework of the problem of liquid spreading on solids, as summarized in the well-known reviews of Dussan[72] and de Gennes[23] and briefly mentioned in Section 2.1.4.

Most of the work in this area has concentrated on forced motion;[73–75] however, a few studies were devoted specifically to spontaneous penetration into a capillary.[76–79] An interesting study of capillary penetration under microgravity has been carried out by Sell *et al.*[80,81] They have concluded that the shape of the meniscus under microgravity is different from simulated zero-*g* experiments. The interface under microgravity has been described as unstable and deformed, leading to lower velocities than expected. In general, the problem of capillary penetration under dynamic contact angle conditions seems to be still open, largely because of a shortage in accurate and detailed experimental data.

4.1.2. Finite Reservoirs

The kinetics of penetration and displacement for systems with reservoirs of limited size have been studied so far only within the framework of the LW equation.[40,82] Figure 16 shows the model system and defines its parameters: X_p and X_d are the dimensionless lengths of the penetrating and displaced liquids, respectively,

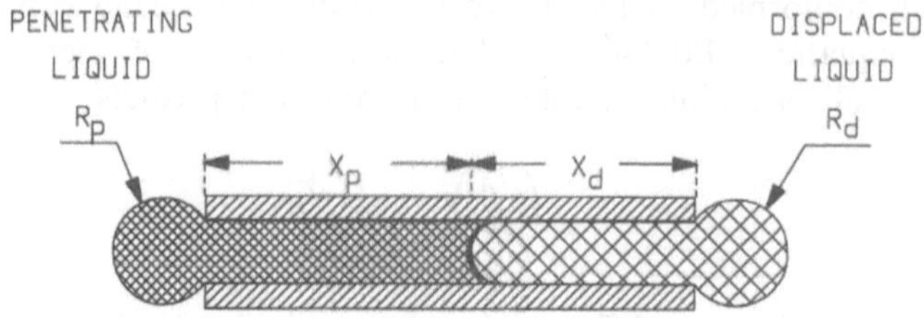

FIGURE 16. A capillary with finite reservoirs of the penetrating and displaced liquids.

inside the capillary (normalized with respect to the radius of the capillary), and R_p and R_d are the corresponding instantaneous dimensionless radii of the reservoirs. The system is assumed to be immersed in an ambient vapor. The general differential equation, in dimensionless form, reads

$$d\tau = \frac{\nu X_d + X_p}{Y_p + U - Y_d} dX_p \tag{50}$$

where

$$\nu \equiv \frac{\mu_d}{\mu_p} \tag{51}$$

$$U \equiv \frac{\cos(\theta_a)_{pd}}{|\cos(\theta_a)_{pd}|} \tag{52}$$

$$Y_d \equiv \frac{\gamma_d}{\gamma_{pd}\,|\cos(\theta_a)_{pd}|\,R_d} \tag{53}$$

$$Y_p \equiv \frac{\gamma_p}{\gamma_{pd}\,|\cos(\theta_a)_{pd}|\,R_p} \tag{54}$$

$$\tau \equiv \frac{\gamma_{pd}\,|\cos(\theta_a)_{pd}|}{4\mu_p r}\, t \tag{55}$$

In these equations, μ_p and μ_d are the viscosities of the penetrating and displaced liquids, respectively. The definition of τ in Eq. (55) is slightly different than in Eq. (40), in anticipation of the possibility of penetration for obtuse contact angles, for which $\cos(\theta_a)_{pd} < 0$.

In addition to Eq. (50), mass conservation equations have to be applied, since the instantaneous curvatures of the reservoirs depend on the instantaneous amounts of liquid and fluid inside the capillary. Equation (50) has been solved in conjunction with the mass conservation equation for various situations.[40,82] For the penetration of a liquid into a vapor-filled capillary ($\nu = 0$, $Y_d = 0$), an analytical approximation has been derived:

$$\frac{C_2}{C_1^2} \ln \frac{C_2}{C_1 X + C_2} + \frac{X}{C_1} \cong \tau \tag{56}$$

where

$$C_1 \equiv 1.5874 V_0^{-4/3} \,|(\cos\theta_a)^{-1}| - G \tag{57}$$

and

$$C_2 \equiv U + 1.5874 V_0^{-1/3} \,|(\cos\theta_a)^{-1}| \tag{58}$$

In these equations, V_0 is the dimensionless volume of the penetrating drop, defined as the actual volume times $3/\pi r^3$. The analytical approximation is excellent for the initial period of penetration, and fits well also at later times, for low contact angles.

All other cases have been solved numerically,[40,82] and the following main conclusions have been drawn from the results:

a. The rate of penetration of displacement by a small drop is much higher than by a liquid from an infinite reservoir. As mentioned in Section 3.1.2, displacement is possible in some situations only by a small drop.
b. The effect of the small radius of the reservoir becomes more pronounced as the contact angle is higher.
c. When the capillary is not connected with a reservoir of the displaced liquid, the rate of displacement decreases up to the point where the radius of the bulging drop of the displaced liquid equals the radius of the capillary. If this point is spontaneously passed, the rate of displacement then increases.

4.2. Kinetics of Penetration into Porous Media

Similarly to the study of equilibrium, kinetic studies are also hindered by the lack of means to locally characterize porous media. The main points to be answered on the way to understanding the kinetics are the time dependence of the position of the moving front, the nature of this front, and the relationship between the constants in the kinetic equations and the structure of the porous medium. These questions are discussed below, comparing experiment and theory whenever possible. The discussion is divided according to the direction of flow. The term *unidirectional penetration* is used to describe a flow in a single direction with respect to a Cartesian frame of reference. In *radial penetration*, the liquid is fed at a point or a limited region, usually into a thin porous medium such as paper or textile yarn, and the fluid is displaced radially outward.

4.2.1. Unidirectional Penetration

The main experimental features of the kinetics of unidirectional penetration were already realized by Hackett,[7] in one of the first publications on the subject. Hackett studied the rise of some liquids into sand of various grades and observed two (and in some cases three) stages of liquid rise. Initially, a stage of "rapid rise" prevailed. In this stage, the LW equation, Eq. (43), was obeyed, in the sense that the velocity of the moving front depended linearly on the inverse of the distance traveled. When the liquid approximately reached the expected equilibrium height of rise, a stage of "slow rise" was observed. This stage lasted in some cases for over a year, and the final height of rise was much higher than the height at the end of the first stage. It was explained by preferential penetration of the liquid into the small pores, i.e., the formation of a saturation distribution at the front.

Many additional experimental studies have been performed since then, for a variety of liquids and porous media.[83–91] The list of liquids that havc been tested includes, in addition to pure liquids, surfactant solutions and solutions of simple liquids. Popular porous media have been paper and glass beads. Yang *et al.*[90] took special care to work with systems of nonzero intrinsic contact angles. The squareroot time dependence predicted by Eq. (42) for horizontal flows, or the dependence of the velocity on the distance for vertical flows, Eq. (43), have been confirmed time and again. However, as realized by Hackett,[7] and as has been stressed more recently by van Brakel and Heertjes,[87] these predictions may not hold over the entire time period of an experiment. The latter authors claim that there are systems for which the prediction of Eq. (43) does not hold at all; however, additional experimental evidence seems to be needed to substantiate this point.

The simplest approach toward modeling and understanding the kinetics of penetration into a porous medium is to treat it as a capillary with an "effective radius." This approach leads, obviously, to the time dependence predicted by the LW equations. By properly fitting the data, one can calculate an effective radius, which, however, may be very different from the radius calculated based on other measurements, such as mercury porosimetry.[92] Moreover, this approach cannot deal with saturation phenomena and the consequent changes in the kinetics.

A more sophisticated approach is to model a porous medium as a group of capillaries of various sizes[92,16] or a periodic tube.[93] This approach improves very much the characterization of the porous medium by an appropriate effective radius. However, similarly to the simpler capillary model, it cannot predict distribution of the liquid at the front. The ultimate approach is, of course, complete modeling of flow in a porous medium. An important beginning was made by Levine and Neale,[94] who used the Forcheimer extension of the Darcy equation to describe the flow in the porous medium in the range of Reynolds numbers smaller than 500. This model, however, did not recognize the possibility of a saturation distribution of the liquid at the front.

If such a possibility is accounted for, the coefficient in the Darcy equation becomes dependent on the local liquid concentration, and the equation becomes a diffusionlike equation with a variable coefficient.[90,95–98] For the case of negligible gravity, this equation, under appropriate boundary conditions, has a similarity solution in the variable X/τ^2.[99] Thus, the experimentally observed "LW dependence" of the distance traveled by the moving front on time does not stem from a simple cylindrical capillary model. Rather, it turns out to be a feature of general validity for penetration into real porous media in the absence of gravitational effects.[90] However, a detailed theoretical model, which includes an experimentally established dependence of the "diffusion coefficient" on the liquid saturation, has not yet been published. In addition, the similarity solution does not explain the experimentally observed "slow rise,"[7,87] which is attributed to a redistribution of the liquid in the porous medium.

FIGURE 17. The radial capillary.

4.2.2. Radial Penetration.

The picture is quite different and even less understood for radial penetration. Experimental studies of drop penetration into paper or textile yarn[(100,101)] revealed that the area of the wet stain is proportional to t^n. For penetration into paper,[(100)] the power $n \approx 0.3$. For penetration into textile yarn, the power is ≈ 0.5 for the first stage of penetration, where the drop is still visible above the porous medium, and ≈ 0.3 during the rest of the process. These observations cannot be accounted for by any of the existing models.

In an attempt to understand the difference between radial and unidirectional penetration, the LW approach has been recently extended to the radial penetration into the space between two parallel plates.[(102)] Figure 17 shows the model system, which has been termed *the radial capillary*. The main difference between unidirectional penetration and radial penetration stems from the fact that in the latter the pressure difference is acting on an ever-increasing front area. Calculations based on this model have shown[(102)] that radial penetration is much slower than unidirectional penetration into either a cylindrical capillary or two parallel plates. However, the experimentally determined dependence on $t^{0.3}$ or $t^{0.5}$ has not resulted from the model. In fact, after an initial period, the dependence of the area of the liquid in between the plates on time appears to be almost linear. This is so for the case of an infinite liquid reservoir, as well as for most of the calculations regarding the penetration of a small drop. Thus, the geometrical difference alone between radial and unidirectional penetration cannot explain the major difference in their time dependence.

5. CONCLUDING REMARKS

As emphasized throughout this chapter, the understanding of many aspects of penetration and displacement in capillary systems is still deficient. This is especially true at two different levels: understanding the microscopic behavior at the contact line region, and developing macroscopic models of realistic porous media. At the contact angle level, much has yet to be learned about hysteresis and dynamic effects and their influence on capillary penetration. In particular, the effect on penetration of contact angle hysteresis, which is to be clearly distinguished from saturation hysteresis, has not been studied. Contact angle hysteresis in equilibrium situations may be one of the factors contributing to saturation hysteresis. Under dynamic conditions, contact angle hysteresis may be intermixed with the dynamic contact angle phenomenon.

The structure of real porous media is very complex. Ultimately, it will become possible to use computer simulations to study complex penetration and displacement processes. However, much is yet to be gained by sophisticated modeling, which will grasp the essential phenomena without resorting to simulation of microscopic details. Two separate issues seem to be of major importance: (a) modeling of interactions between menisci in the direction transverse to the flow, and (b) modeling the redistribution of the liquid in the porous medium, following the first stage of "rapid penetration."

6. REFERENCES

1. J. J. Bikerman, *Physical Surfaces*, Academic Press, New York (1970).
2. F. Bashforth and J. C. Adams, *An Attempt To Test the Theories of Capillary Action*, University Press, Cambridge, England (1883).
3. Lord Rayleigh, *Proc. R. Soc. London A***92**, 184 (1915).
4. R. Lucas, *Kolloid Z.* **23**, 15 (1918).
5. S. Sugden, *J. Chem. Soc.* **119**, 1483 (1921).
6. E. W. Washburn, *Phys. Rev.* **17**, 273 (1921).
7. F. E. Hackett, *Trans. Faraday Soc.* **17**, 260 (1921).
8. E. K. Rideal, *Philos. Mag. Ser. 6* **44**, 1152 (1922).
9. C. H. Bosanquet, *Philos. Mag. Ser. 6* **45**, 525 (1923).
10. G. M. Homsy, *Ann. Rev. Fluid Mech.* **19**, 271 (1987).
11. J. A. Wingrave, W. H. Wade, and R. S. Schechter, in *Wetting, Spreading and Adhesion* (J. F. Padday, ed.), p. 261, Academic Press, New York (1978).
12. M. R. Moldover and R. W. Gammon, *J. Chem. Phys.* **80**, 528 (1984).
13. M. Sahimi, B. D. Hughes, L. E. Scriven, and H. T. Davis, *Chem. Eng. Sci.* **41**, 2103 (1986).
14. J. C. Melrose, *Soc. Pet. Eng. J.* **5**, 259 (1965).
15. N. R. Morrow, *Ind. Eng. Chem.* **62**, 32 (1970).
16. J. van Brakel, *Powder Technol.* **11**, 205 (1975).
17. A. Marmur, *Adv. Colloid Interface Sci.* **19**, 75 (1983).
18. T. Young, *Philos. Trans. R. Soc. London* **95**, 65 (1805).
19. A. W. Adamson, *Physical Chemistry of Surfaces*, 4th ed., Wiley, New York (1982).
20. E. F. Hare and W. A. Zisman, *J. Phys. Chem.* **59**, 335 (1955).
21. A. W. Neumann, in *Wetting, Spreading and Adhesion* (J. F. Padday, ed.), pp. 3–34, Academic Press, New York (1978).
22. R. J. Good, in *Surface and Colloid Science*, vol. 11 (R. J. Good and R. R. Stromberg, eds.), p. 1, Plenum Press, New York (1979).
23. P. G. de Gennes, *Rev. Mod. Phys.* **57**, 827 (1985).
24. R. N. Wenzel, *Ind. Eng. Chem.* **28**, 988 (1936).
25. R. E. Johnson, Jr., and R. H. Dettre, in *Contact Angle, Wettability, and Adhesion*, Adv. Chem. Ser., No. 43, p. 112, Am. Chem. Soc., Washington, DC (1964).
26. J. D. Eick, R. J. Good, and A. W. Neumann, *J. Colloid Interface Sci.* **53**, 235 (1975).
27. C. Huh and S. G. Mason, *J. Colloid Interface Sci.* **60**, 11 (1977).
28. R. G. Cox, *J. Fluid Mech.* **131**, 1 (1983).
29. J. F. Joanny and P. G. de Gennes, *J. Chem. Phys.* **81**, 552 (1984).
30. A. B. D. Cassie and S. Baxter, *Trans. Faraday Soc.* **40**, 546 (1944).
31. L. W. Schwartz and S. Garoff, *Langmuir* **1**, 219 (1985).
32. G. E. P. Elliott and A. C. Riddiford, *J. Colloid Interface Sci.* **23**, 389 (1967).
33. R. L. Hoffman, *J. Colloid Interface Sci.* **50**, 228 (1975).
34. C. G. Ngan and E. B. Dussan V., *J. Fluid Mech.* **118**, 27 (1982).
35. T-S. Jiang, S. G. Oh, and J. C. Slattery, *J. Colloid Interface Sci.* **69**, 74 (1979).
36. G. A. Bottomley, *Aust. J. Chem.* **27**, 2297 (1974).
37. S. Ramakrishnan and S. Hartland, *J. Colloid Interface Sci.* **80**, 497 (1981).
38. W. J. O'Brien, R. G. Craig, and F. A. Peyton, *J. Colloid Interface Sci.* **26**, 500 (1968).
39. G. Mason and N. R. Morrow, *J. Chem. Soc. Faraday Trans. 1* **80**, 2375 (1984).

40. A. Marmur, *J. Colloid Interface Sci.* **122**, 209 (1988).
41. A. Marmur, *J. Colloid Interface Sci.* **130**, 288 (1989).
42. D. H. Everett and J. M. Haynes, *Z. Phys. Chem. Neue Folge* **82**, 36 (1972).
43. D. H. Everett, *J. Colloid Interface Sci.* **52**, 189 (1975).
44. D. H. Everett and J. M. Haynes, *Z. Phys. Chem. Neue Folge* **97**, 301 (1975).
45. W. B. Haines, *J. Agric. Sci.* **17**, 264 (1927).
46. F. E. Hackett and J. S. Strettan, *J. Agric. Sci.* **18**, 671.
47. W. B. Haines, *J. Agric. Sci.* **20**, 97 (1930).
48. W. O. Smith, P. D. Foote, and P. F. Busang, *Physics* **1**, 18 (1931).
49. W. O. Smith, *Physics* **4**, 184 (1933).
50. N. R. Morrow, *J. Can. Pet. Technol.* **15**, 49 (1976).
51. Y.-W. Yang, G. Zografi, and E. E. Miller, *J. Colloid Interface Sci.* **122**, 24 (1988).
52. D. T. Hansford, D. J. W. Grant, and J. M. Newton, *Powder Technol.* **26**, 119 (1980).
53. J. van Brakel and P. M. Heerthes, *Nature* **254**, 585 (1975).
54. A. Marmur, *J. Colloid Interface Sci.* **129**, 278 (1989).
55. L. R. White, *J. Colloid Interface Sci.* **90**, 536 (1982).
56. S. Ban, E. Wolfram, and S. Rohrsetzer, *Colloids Surf.* **22**, 301 (1987).
57. Y. Li, W. G. Laidlaw, and N. C. Wardlaw, *Adv. Colloid Interface Sci.* **26**, 1 (1986).
58. A. Marmur, *J. Phys. Chem.* **93**, 4873 (1989).
59. A. Marmur, *J. Colloid Interface Sci.* **123**, 161 (1988).
60. J. M. Bell and F. K. Cameron, *J. Phys. Chem.* **10**, 658 (1906).
61. R. J. Good, *Chem. Ind.* **22**, 600 (1971).
62. R. J. Good and N.-J. Lin, *J. Colloid Interface Sci.* **54**, 52 (1976).
63. J. Szekely, A. W. Neumann, and Y. K. Chuang, *J. Colloid Interface Sci.* **35**, 273 (1971).
64. J. R. Ligenza and R. B. Bernstein, *J. Am. Chem. Soc.* **73**, 4636 (1951).
65. L. R. Fisher and P. D. Lark, *J. Colloid Interface Sci.* **69**, 486 (1979).
66. R. S. Malik, CH. Laroussi, and L. W. de Backer, *Soil Sci.* **127**, 211 (1979).
67. S. Levine and G. H. Neale, *J. Chem. Soc. Faraday Trans. 2* **71**, 12 (1975).
68. W. E. Brittin, *J. Appl. Phys.* **17**, 37 (1946).
69. S. Levine, P. Reed, E. J. Watson, and G. Neale, in *Colloid and Interface Science* (M. Kerker, ed.), vol. 3, p. 403, Academic Press, New York (1976).
70. M. F. Letelier S., and H. J. Leutheusser, *J. Colloid Interface Sci.* **72**, 465 (1979).
71. G. L. Batten, Jr., *J. Colloid Interface Sci.* **102**, 513 (1984).
72. E. B. Dussan V, *Ann. Rev. Fluid Mech.* **11**, 371 (1979).
73. T. D. Blake, D. H. Everett, and J. M. Haynes, in *Wetting S.C.I. Monog. No. 25*, p. 164, Soc. Chem. Ind., London (1967).
74. C. Huh and S. G. Mason, *J. Fluid Mech.* **81**, 401 (1977).
75. J. N. Tilton, *Chem. Eng. Sci.* **43**, 1371 (1988).
76. S. Levine, J. Lowndes, E. J. Watson, and G. Neale, *J. Colloid Interface Sci.* **73**, 136 (1980).
77. V. V. Berezkin and N. V. Churaev, *Colloid J. USSR (USA)* **44**, 376 (1982).
78. G. A. Martynov, V. V. Malev, and E. V. Gribanova, *Colloid J. USSR (USA)* **45**, 205 (1983).
79. T. E. Mumley, C. J. Radke, and M. C. Williams, *J. Colloid Interface Sci.* **109**, 398, 413 (1986).
80. P.-J. Sell, E. Maisch, and J. Siekmann, *Acta Astronaut.* **11**, 577 (1984).
81. P.-J. Sell, E. Maisch, and J. Siekmann, *Acta Astronaut.* **13**, 87 (1986).
82. A. Marmur, *Chem. Eng. Sci.* **44**, 1511 (1989).
83. H. Fujita, *J. Phys. Chem.* **56**, 625 (1952).
84. F. W. Minor, A. M. Schwartz, E. A. Wulkow, and L. C. Buckles *Text. Res. J.* **29**, 931 (1959).
85. T. Gillespie, *J. Colloid Sci.* **14**, 123 (1959).
86. T. Gillespie and T. Johnson, *J. Colloid Interface Sci.* **36**, 282 (1971).
87. J. van Brakel and P. M. Heertjes, *Powder Technol.* **16**, 75, 83, 91 (1977).
88. D. H. Everett, J. M. Haynes, and R. J. L. Miller, in *Fiber Water Interactions in Paper Making* (Fundamental Research Committee, ed.), BPBIF, London (1978).
89. R. Williams, *J. Colloid Interface Sci.* **79**, 287 (1981).
90. Y.-W. Yang, G. Zografi, and E. E. Miller, *J. Colloid Interface Sci.* **122**, 35 (1988).
91. K. T. Hodgson and J. C. Berg, *J. Colloid Interface Sci.* **121**, 22 (1988).
92. F. A. L. Dullien, M. S. El-Sayed, and V. K. Batra, *J. Colloid Interface Sci.* **60**, 497 (1977).

93. S. Levine, J. Lowndes, and P. Reed, *J. Colloid Interface Sci.* **77**, 253 (1980).
94. S. Levine and G. Neale, in *Wetting, Spreading and Adhesion* (J. F. Padday, ed.), p. 241, Academic Press, New York (1978).
95. R. D. Miller and E. E. Miller, *J. Appl. Phys.* **27**, 324 (1956).
96. T. Gillespie, *J. Colloid Sci.* **13**, 32 (1958).
97. M. J. Denton, *J. Chromatogr.* **18**, 615 (1965).
98. E. E. Miller, in *Applications of Soil Physics* (D. Hillel, ed.), p. 300, Academic Press, New York (1980).
99. H. S. Carslaw and J. C. Jaeger, *Conduction of Heat in Solids*, 2nd ed., p. 89, Oxford at the Clarendon Press (1959).
100. E. Kissa, *J. Colloid Interface Sci.* **83**, 265 (1981).
101. T. Kawase, S. Gekoguchi, T. Fujii, and M. Minagawa, *Text. Res. J.* **56**, 409 (1986).
102. A. Marmur, *J. Colloid Interface Sci.* **124**, 301 (1988).

13

The Wetting Behavior of Fibers

Willard D. Bascom

1. INTRODUCTION

Historically, the technologies most interested in the wetting of fibers have been those involved in the processing of textiles.[1,2] Much of the early scientific literature on wetting was concerned with liquid penetration into fabrics and other porous solids.[3] More recently, the rapid development of fiber reinforced composites, notably carbon fiber and glass fiber reinforced polymers (CFRP, GFRP), has generated a renewed interest in the wetting of fibers. However, in the interim there has been a change in the scientific attitude toward the use of contact angle measurements as a means of characterizing the surface chemical constitution of solids. In the early literature, the contact angle was viewed as a characteristic of the fiber and a parameter in the capillarity equations for liquid penetration. Due in large measure to the studies by W. A Zisman and co-workers, there has been a change in attitude toward the physical significance of contact angle measurements. It is now recognized that the contact angle can be a highly sensitive tool for surface characterization. Consequently, there is a growing body of literature on the wetting of textile fibers and fibers used in composites aimed at surface chemical characterization as well as the processing of these fibers into composite materials.

2. THEORY

The same physical chemical laws that govern the relation between surface tensions and the contact angle on planar surfaces apply to fibers, e.g., Young's equation

$$\cos\theta = \frac{\gamma_{SV} + \gamma_{SL}}{\gamma_{LV}} \tag{1}$$

Willard D. Bascom • Department of Materials Science and Engineering, 304 EMRO, University of Utah, Salt Lake City, Utah 84112.

Modern Approaches to Wettability: Theory and Applications, edited by Malcolm E. Schrader and George Loeb. Plenum Press, New York, 1992.

Differences arise in the equilibrium shape of a thin liquid film on a fiber compared with a similar film on a planar surface. The minimization of the surface free energy leads to a droplet on a fiber, whereas on a flat plate the minimum configuration is a continuous film.

The equilibrium shape of a droplet on a filament was derived by Rayleigh[4] in 1879, based on the Laplace equation for the pressure difference (ΔP) at the liquid–air boundary

$$\Delta P = \gamma_{LV}\left(\frac{1}{R_1} + \frac{1}{R_2}\right) \tag{2}$$

and the minimization of surface area. An annular coating of a liquid with outer radius a on a fiber of diameter d under an infinitesimally small sinusoidal disturbance will grow exponentially in time into a series of uniform, regularly spaced droplets (Fig. 1). Following the derivation by Goren,[5] the surface area per wave, S, is minimized while the volume per wave, V, is constant. Expressed in terms of the profile $y(x)$ and assuming zero contact angle and the absence of gravitational forces, we have

$$S = \int_0^\lambda 2\pi y \left(1 + \left[\frac{dy}{dx}\right]^2\right)^{1/2} dx \tag{3}$$

and

$$V = \int_0^\lambda \pi y^2 \, dx = \pi a^2 \lambda \tag{4}$$

The solution obtained by Goren is expressed in terms of elliptic integrals of the second and first kinds, $E(\rho, \phi)$ and $F(\rho, \phi)$, with the indicated boundary conditions:

$$\frac{x}{H} = E(\rho, \phi) + \cos\rho F(\rho, \phi) \qquad 0 < \frac{x}{H} < XH \tag{5}$$

$$\frac{u}{H} = \cos\rho \qquad \frac{X}{H} < \frac{x}{H} < \frac{\lambda}{2H} \tag{6}$$

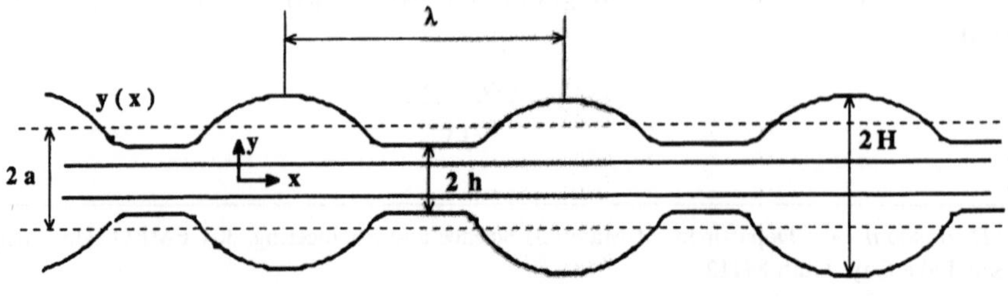

FIGURE 1. Schematic of primary droplet formation on a single filament indicating coordinate system and shape parameters.

where

$$\frac{X}{H} = E\left(\rho\,\frac{\pi}{2}\right) + \cos\rho F\left(\rho\,\frac{\pi}{2}\right)$$

$$\cos\rho = \frac{h}{H}$$

$$\sin^2\phi = \frac{(1 - y^2/H^2)}{(1 - h^2/H^2)}$$

This analysis leads to a symmetric array of primary droplets connected by links of radius h (Fig. 1). These links are also subject to unstable disturbances that will form satellite drops if the instabilities grow faster than the rate of drainage into the primary drops. This formation of different-sized droplets can be seen for liquid films on threads or fibers and in the breakup of thin streams of flowing liquids.

If the contact angle of the liquid on the fiber is finite, the same capillarity and surface area minimization considerations exist, but, in addition, the wetting angle also influences the equilibrium drop shape. Yamaki and Katayama,[6] for example, obtained Eq. (7) for the profile of a drop on a filament that they solved numerically, using a computer with the boundary conditions shown in Fig. 2, where y_0 and x_0 are at point C, d = filament diameter, and $[dy/dx]_{x=x_0}$ = tangent of the contact angle ($\tan\theta$). Solutions were obtained as a function of $\Delta P/\gamma_{LV}$:

$$\frac{d^2y}{dx^2} = -\frac{\Delta P}{\gamma_{LV}}\left(1 + \left(\frac{dy}{dx}\right)^2\right)^{3/2} + \left(1 + \left(\frac{dy}{dx}\right)^2\right)y \tag{7}$$

They calculated θ both in terms of the drop length parameter ($L = 2l/d$) and a maximum diameter parameter ($K = 2k/d$) and in terms of L and α, the apparent contact angle where the liquid surface becomes linear ($d^2y/dx^2 = 0$).

These analyses apply to macroscopic films having thicknesses greater than 100 nm and usually much larger. There has been considerable interest in recent years in liquid films on solids where the film thickness is about 50 nm or less. There is both theoretical and experimental evidence that for these thin films analyses based on conventional capillarity forces are insufficient and that long-range surface force effects must be taken into consideration. For a discussion of this subject, see Israelachvili[7] and de Gennes.[8]

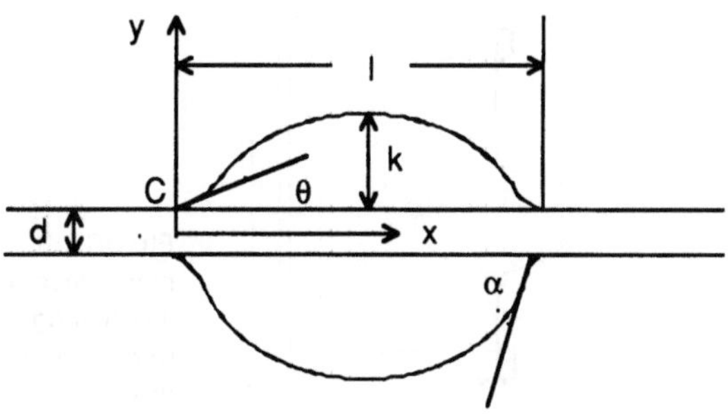

FIGURE 2. Cross section of a single drop with coordinate system and shape parameters.

Brochard-Wyart and di Meglio[9] have considered ultrathin films on fibers. A full discussion of their paper is beyond the scope of this review, but the essential conclusion is that for thicknesses less than a critical value, e_c, a thin annular film is stable; i.e., droplet formation is not energetically favored. They evaluated this parameter through a minimization of surface energy at constant volume [Eq. (1)–(6)], but included a van der Waal's energy term in the expression for the surface energy. For a fiber 100 μm in diameter, e_c is about 100 nm. For smaller-diameter fibers, $<10\ \mu$m, the critical thickness is considerably smaller.

3. MEASUREMENT TECHNIQUES

3.1. Goniometry

One of the most widely used techniques for measuring contact angles of drops on flat plates is goniometry. However, goniometric measurement of the contact angle of a droplet on a filament cannot usually be done due to the steep change in the drop profile near the drop edge.

Bascom *et al.*[10] avoided this problem by suspending a relatively large drop of liquid in a slit platinum loop as shown in Fig. 3. The fiber is secured between two polytetrafluoroethylene (PTFE) rods mounted on a micropositioner so that the fiber can be precisely moved through the slit in the loop and into the center of the drop where the surface is relatively flat, thus avoiding the curvature problem of a single droplet. The fiber can be moved in the vertical direction so that both receding and advancing angles can be observed. The angle can be measured using a goniometer telescope or from a photograph. This technique could easily resolve contact angles of 40° or higher, but optical interference problems near the drop edge were encountered for lower angles. However, it may be possible to use this technique by making measurements of the meniscus profile away from the fiber where optical interference effects are less pronounced (see Section 3.3).

3.2. Drop Shape

Measurements of the drop profile can be used to compute the contact angle from analytical expressions for the drop shape. By the analysis of Yamaki and Katayama[6] presented in the preceding section, either the drop length (L) and maximum diameter (K) or the length and the tangent at the inflection point (α) are

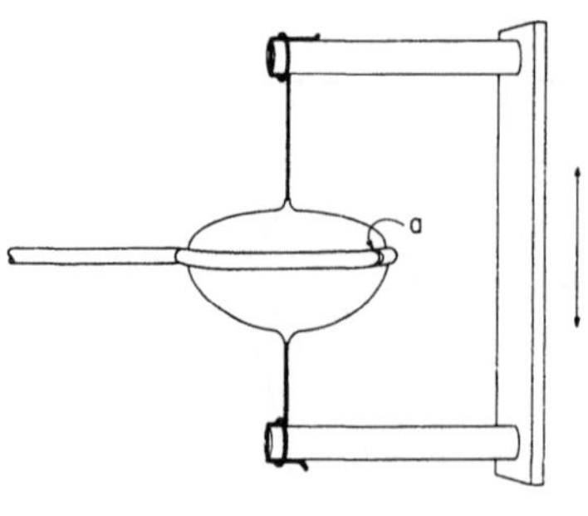

FIGURE 3. Single-filament contact angle measurement technique. The liquid is held in a platinum inoculating loop in which a slit has been cut (a) and through which the fiber is passed into the center of the drop.

measured from photomicrographs. From their solution of Eq. (7), they constructed plots of L vs K (or α) for different values of tan θ and $\Delta P/\lambda_{LV}$. They obtained good agreement between the measured parameters and the theoretical curves for various size drops of an epoxy resin on carbon fibers and for other liquid–fiber pairs.

3.3. Meniscus Shape

Just as the shape of a drop on a fiber is a function of the wetting angle, the meniscus profile of the liquid around a fiber partially immersed in a flat liquid surface is a function of the contact angle (Fig. 4). Recently, Elmendorp[11] developed an analytical expression for the immersion meniscus shape that is valid for small-diameter fibers, 50 μm or less:

$$\frac{y}{R} = \sqrt{c}\left(\ln\left(\frac{x}{R} + \sqrt{\frac{x}{R} - c}\right) - \ln(1 + \sqrt{1-c})\right)$$
$$c = \frac{1}{1 + \tan^2\theta} \tag{8}$$

where R is the fiber radius and the xy coordinates are shown in Fig. 4. Verschoor and Bascom[12] obtained a computer solution of Eq. (8), from which plots of the surface shape normalized with respect to the fiber diameter for specific contact angles can be generated (Fig. 5). This technique is recommended for determining the contact angle on fibers in an enclosed chamber purged with an inert gas to prevent oxidation of the wetting liquid.

3.4. Flotation

The flotation of small particles and short thin fibers has been used to access the wettability of fibers and polymers.[13] The technique lends itself to rapid go–no go determinations, where powders or loose fibers are simply sprinkled onto liquids in shallow dishes. The capillarity principles are relatively simple. If the contact angle is finite, the capillary force exceeds the buoyant weight and the particle or fiber floats. If the contact angle is zero, the solid sinks into the liquid. Mutchler, Menkart, and Schwartz[14] used a series of liquids having a range of surface tensions and, by observing the liquid surface tensions between which a fiber floats

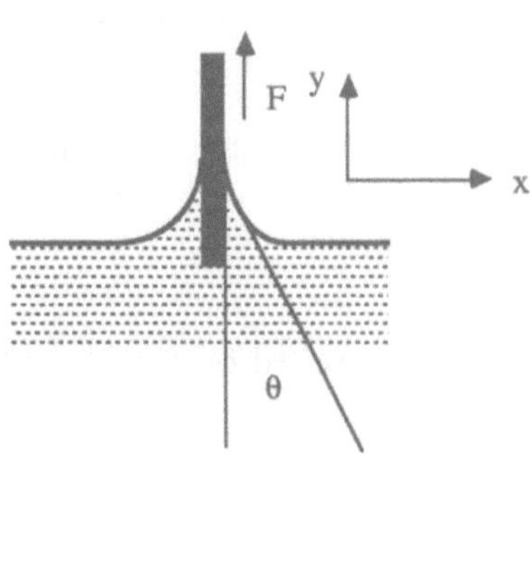

FIGURE 4. Schematic of fiber partially immersed in a planar liquid surface.

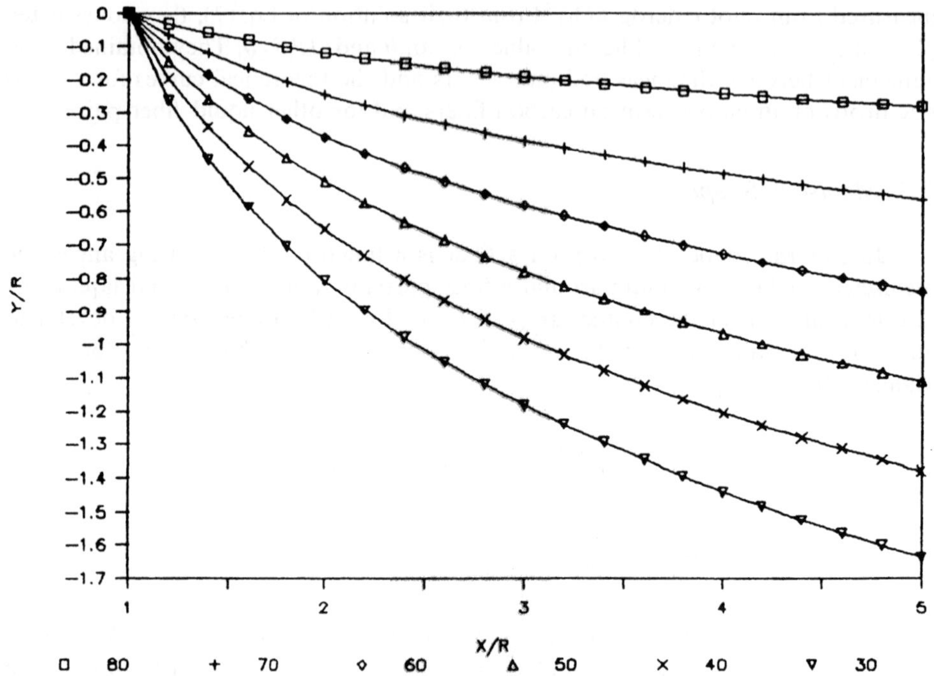

FIGURE 5. Computer solutions of Eq. (8) for $\theta = 30°–80°$.

or sinks, were able to estimate the critical surface tension of the fiber. Obviously, the fiber must have a density greater than the test liquids. Also, it must have a high aspect ratio so that the fiber surface dominates the wetting behavior compared with the wetting of the fiber ends.

The theory of the flotation of fibers as developed in Ref. 14 is based on the geometry in Fig. 6.

The fiber is at equilibrium, and the symbols are defined as follows:

C, C' = three-phase line of contact
T = tangent to the cylinder surface at the line of contact
θ = contact angle
α = angle of liquid surface to the vertical
V = vertical at C
ϕ = angle from the fiber midplane to the line of contact

It can be easily shown that

$$\alpha + \theta = 180° - \phi \tag{9}$$

The force exerted on the fiber per unit length is

$$F_u = 2\gamma \cos \alpha \tag{10}$$

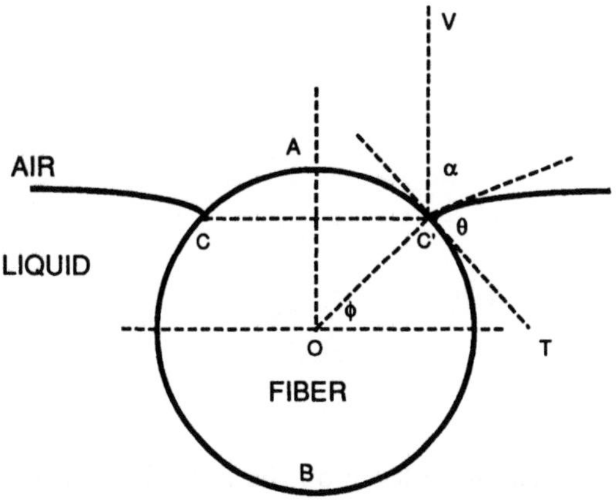

FIGURE 6. Cross-sectional view of fiber at equilibrium in the air–liquid interface.

and the sinking force corrected for (buoyancy) is

$$F_s = g(\rho_f V_A + [\rho f - \rho_l]\, V_B) \tag{11}$$

where

ρ_f = density of fiber
ρ_l = density of liquid
V_A = unsubmerged volume per unit fiber length
V_B = submerged volume per unit fiber length

Note from Eq. (9) that as the unsubmerged area (CC') decreases, α decreases, but the capillary force F_u increases until it reaches a maximum at $C = C'$, at which point the fiber sinks. The implication of this increase in F_u is that even as α approaches 0° the capillary force sustains flotation. From Eq. (9) the contact angle θ could be as low as a few tenths of a degree. Thus, the technique can be highly sensitive even when the contact angle is well below the usual $\pm 2°$ accuracy of goniometric methods.

The major difficulty with the flotation technique is that the above analysis assumes a cylindrical and perfectly smooth fiber. The method is not necessarily as sensitive or even usable for different-shaped fibers or fibers with a high degree of surface roughness.

3.5. Tensiometry (Wilhelmy Plate Method)

The most sensitive and widely used method of measuring the wettability of single filaments is to measure the adhesion tension by using a recording electro-microbalance. The test configuration is shown in Fig. 7. The fiber is suspended from one arm of an electrobalance and is partially submerged in a beaker of the test liquid. The platform holding the test liquid is mechanically moved in the vertical direction so that the advancing and receding angles can be measured. The apparatus must be mounted on a vibration-damping table (an analytical balance table is usually suitable), and the region around the test liquid enclosed against air drafts and aerosol contamination.

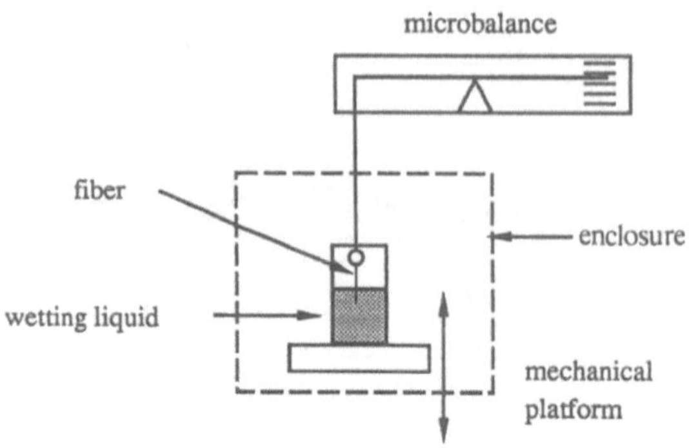

FIGURE 7. Schematic of tensiometer.

The total force on the fiber (Fig. 4) is

$$F_T = 2\pi r \gamma_{LV} \cos\theta - \rho_L g \pi r^2 h \tag{12}$$

where

r = fiber radius
θ = contact angle
h = depth of fiber immersion
g = gravitational constant

The first term on the RHS of Eq. (12) represents the capillary force (adhesion tension) of the liquid on the fiber. The second term is the buoyant weight of the submerged fiber. The weight of the unsubmerged fiber and its support system is tared off before the wetting test is started.

Advantageously, the buoyancy term in Eq. (12) is negligible for thin ($r < 50\ \mu m$) fibers since the fiber radius enters the buoyancy term to the second power. Consequently, for these small diameter fibers the measured force is independent of fiber depth:

$$F_T = 2\pi r \gamma_{LV} \cos\theta \tag{13}$$

Rearranging Eq. (13), we solve for the contact angle:

$$\cos\theta = \frac{mg}{\pi d \gamma_{LV}} \tag{14}$$

where m is the tensiometer balance output. In order to use Eq. (14), the fiber diameter must be measured. The most convenient method is to measure F_T using a liquid for which $\theta = 0$; then

$$d = \frac{mg}{\pi \gamma_{LV}} \tag{15}$$

In Fig. 8, the recorder trace is given for a carbon fiber in hexadecane. The measured

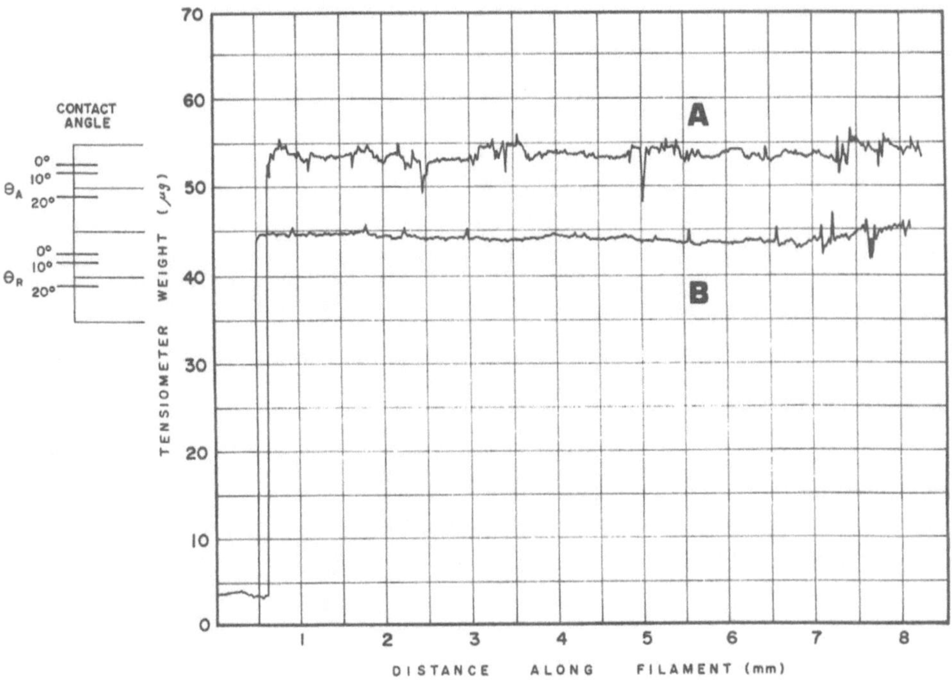

FIGURE 8. Tensiometer trace for a carbon filament (Hercules AS4) in hexadecane: A, immersion; B, emersion.

angle is zero, and from Eq. (15) the diameter is 6.84, which is in good agreement with the nominal fiber diameter of 7 μm quoted by the manufacturer.

The technique can be used to measure the static contact angle, in which case the fiber is immersed at least a few millimeters in the liquid to avoid end effects and is then held stationary until the force becomes constant. Alternatively, the tensiometer output can be recorded continuously as the fiber is immersed (advancing angle) or emersed (receding angle). In effect, the dynamic wetting angle is being determined in this mode, which may be significantly different from the static angle. However, Johnson[15] found no effect of immersion rate up to 0.5 cm/s for water or hexadecane against polysiloxane-coated glass plates. The advantage of the continuous immersion–emersion technique is that it samples a significant length of fiber and thereby reveals variations in wettability. This variability can be considerable for commercial fibers.

Bascom[16] used the continuous immersion–emersion technique ($\sim$5 μm/s) to determined the wettability of carbon fibers. The tensiometer trace for hexadecane ($\theta = 0$) is shown in Fig. 8. For water and other liquids exhibiting finite contact angles, the tensiometer results indicated considerable contact angle hysteresis and the traces were very "noisy." In Fig. 9, the advancing angle (θ_a) for water was 70° and the receding angle (θ_r) was 25°. The marked oscillations in the tensiometer traces in Fig. 9 are believed to be the result of surface microheterogeneity. If we

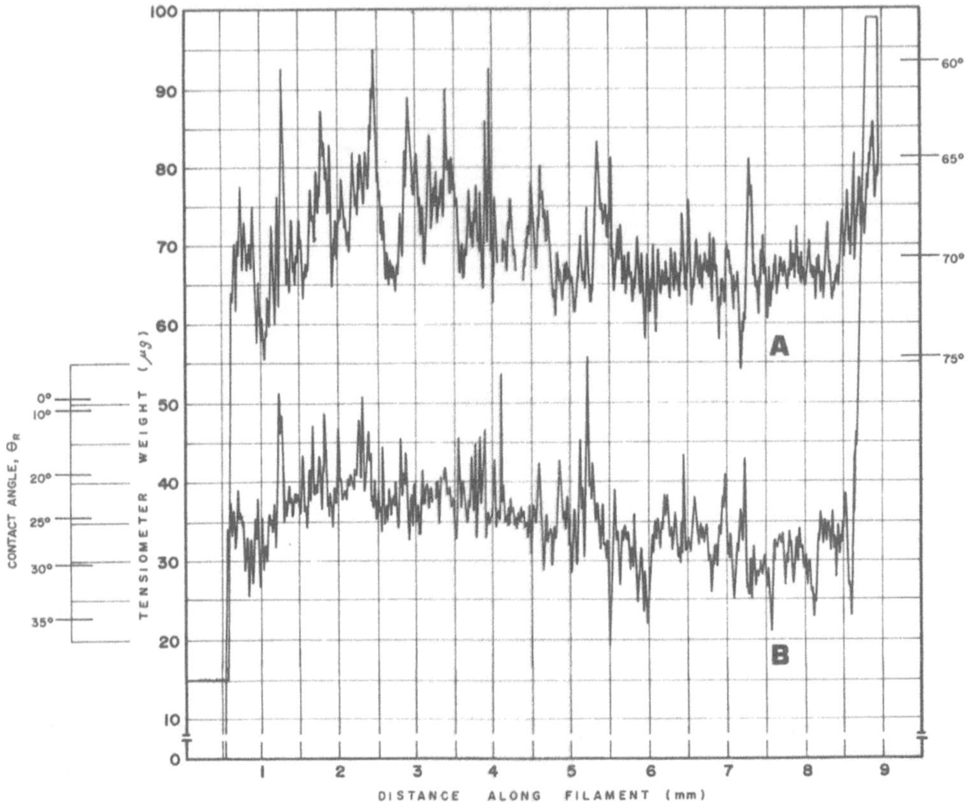

FIGURE 9. Tensiometer trace for a carbon filament (Hercules AS4) in water: A, immersion; B, emersion.

assume a patchwise heterogeneity, as shown schematically in Fig. 10, the wetting perimeter is being continuously distorted as it moves along the surface, and from Eq. (13) the wetting force oscillates accordingly. These observations are in accord with recent theoretical treatments of surface microheterogeneity by de Gennes[17] and Schwartz and Garoff.[18,19]

Clearly, the tensiometer technique provides a relative simple and highly sensitive means of determining fiber diameter and diameter variation, contact angle hysteresis, and surface microheterogeneity. For further details, see Miller,[20] which addresses the wetting of textile fibers, and Neumann and Good,[21] which provides more detail on the Wilhelmy tensiometer.

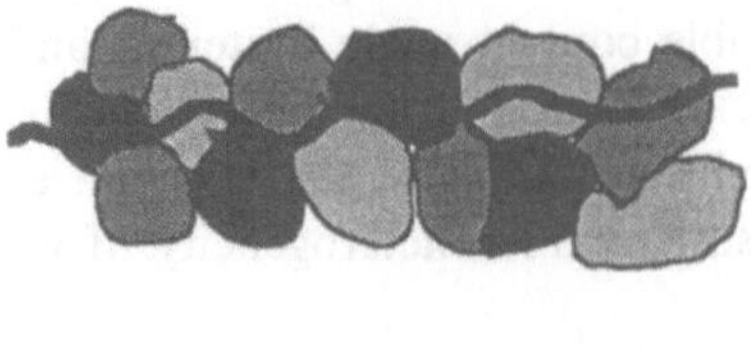

FIGURE 10. Schematic of the distortion of the meniscus perimeter by patchwise surface microheterogeneity.

TABLE 1

Fiber	γ_{LV} (mN/m) sink	γ_{LV} (mN/m) float	γ_c (mN/m)[a]
Poly(vinylchloride)	38.4	39.5	39–40
Polyethylene	28.7	32.1	31
Polysiloxane	20.6	23.7	23

[a] Reference 22.

4. EXPERIMENTAL OBSERVATIONS

4.1. Static Contact Angles

Studies of the wettability of fibers have not been especially extensive, especially when compared to the studies on planar surfaces. Various techniques have been used, and, where comparisons are possible, the wettability of a particular solid is essentially the same whether in the form of a fiber or a flat plate.

Using the flotation method and a series of liquids having a range of surface tensions from 17 to 72 dynes/cm, Mutchler *et al.*[14] estimated the critical surface tensions of polymer fibers. In Table 1 the liquid surface tensions for "just sink" versus "just float" are compared with critical surface tensions, γ_c, obtained by Zisman[22] on planar surfaces. The agreement is very good, especially considering that γ_c data from Ref. 22 has an uncertainty of at least ± 2 dynes/cm.

Not surprisingly, much of the recent published data on fiber wetting has been on fibers used in composites. Two examples are presented here to illustrate the merits and problems in characterizing these fibers.

Wesson and Tarantino[23] used tensiometry to determine the wettability of E-glass fibers taken from commercial fabric. They heatcleaned the cloth and treated it with the adhesion promoter γ-glycidoxypropyltrimethoxysilane. Wettability measurements were made on both the heat-cleaned fibers and on fibers after the

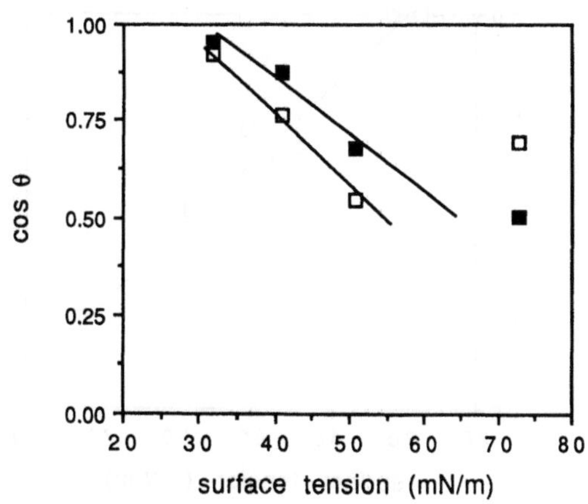

FIGURE 11. Critical surface tension plots for heat-cleaned (open squares) and silane-treated (filled squares) glass fibers.

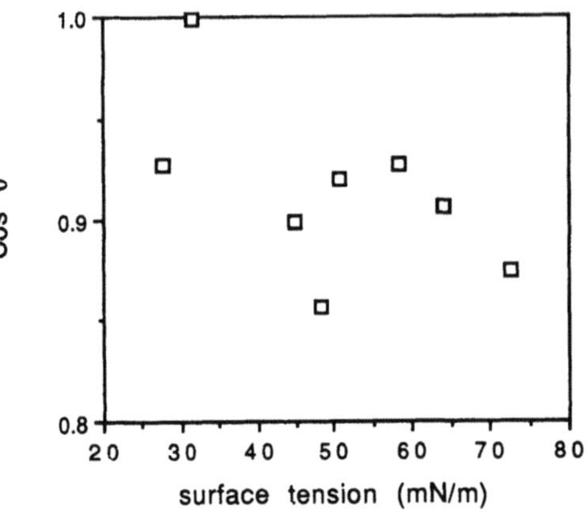

FIGURE 12. Critical surface tension plot for liquids listed in Table 2 on AS4 carbon fiber.

silane treatment. They reported their data in terms of the work of adhesion, which can be obtained from tensiometry measurements from the equation

$$W_A = \gamma_{LV} + \frac{mg}{\pi d} \tag{16}$$

Converting W_A to $\cos\theta$, using

$$W_A = \gamma_{LV}(1 + \cos\theta)$$

we show critical surface tension plots in Fig. 11 for advancing angles.

The wetting liquids were water (72.8 mN/m), methylene iodide (50 mN/m), tricesylphosphate (40.9 mN/m), and mineral oil (31.9 mN/m). The organic liquids give linear plots for both heat-cleaned and treated fibers that extrapolate to essentially the same critical surface tension, 28 mN/m. The data points for water did not fit the linear correlations for either fiber, which suggests that both surfaces contain sites capable of specific interaction with water, e.g., H bonding. Wesson and Tarantino reached much the same conclusion from an analysis of their data using the theory of surface fractional polarity.[24,25]

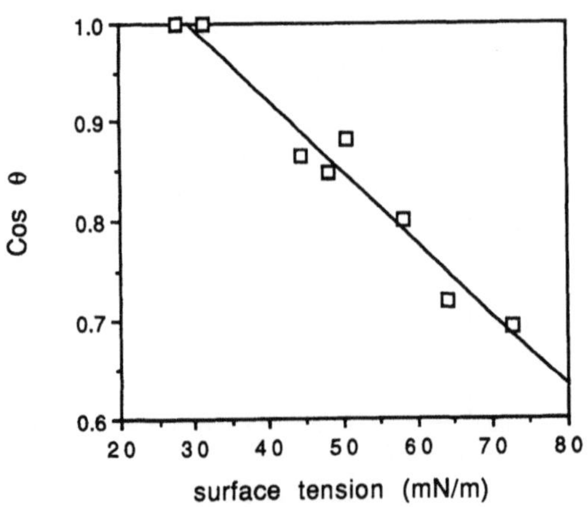

Figure 13. Critical surface tension plot for liquids listed in Table 2 on HMS carbon fiber.

TABLE 2. Wetting Test Liquids[a]

Liquid	γ_{LV} (mN/m)
Water	72.8
Glycerol	64.0
Ethylene glycol	48.3
Polyethylene glycol	31.3
Formamide	58.3
Hexadecane	27.6
Methylene iodide	50.8
Bromonaphthalene	44.6

[a] Reference 26.

Hammer and Drzal(26) determined the wettability of commercial carbon fibers. Some of their results are presented in Figs. 12 and 13 for a high-strength fiber (AS4*) and a high-modulus fiber (HMS*), respectively. These authors used a "miscellaneous" set of wetting liquids listed in Table 2. Clearly, the AS4 fiber did not give a linear critical surface tension plot, whereas the data for the HMS fiber did. Drzal and Hammer analyzed their data by using the theory of fractional polarity developed by Kaelble.(27) They found the polar contribution to the total surface energy to be significantly higher for AS4 than for HMS (30 mN/m vs 21 mN/m). The nonpolar (dispersion energy) contribution was essentially the same for both fibers (26 mN/m vs 28 mN/m). It is at least plausible that the higher polarity of the AS4 fiber obviates a linear critical surface tension plot due to specific interactions between the wetting liquids and the fiber surface. This explanation is highly speculative, but it does point up the difficulty of obtaining meaningful critical surface tensions when the surface has a significant degree of polarity.

The data published by Hammer and Drzal(26) for the AS4 fiber do not agree with the results obtained by Bascom(16) presented in Figs. 8 and 9, where the advancing contact angle is 0° for hexadecane and 70° for water, compared with 22° for hexadecane and 29° for water in Ref. 26. However, these differences are not entirely unexpected, considering the usual variability in a commercial fiber. More to the point, however, is that carbon fibers lie at the upper limit at which critical surface tension measurements are meaningful. Unlike the polymer surfaces on which the concept of a critical surface tension was based,(22) carbon fibers can exhibit a high degree of polarity and surface heterogeneity, which obviates a linear relation between the test liquid surface tension and the surface energy of the solid.

5. CONCLUSIONS

The equilibrium configuration of a thin film of liquid on a fiber is a series of droplets of differing sizes depending on the circumstances in which they form and the wettability of the fiber. Direct measurement of the contact angle is difficult due to the drop contour, so contact angle determinations on fibers must be made indirectly

* Hercules Aerospace, Magna, Utah.

from drop shape analysis, flotation, or tensiometric measurements. The latter technique has proved the most useful, in that with highly sensitive electrobalances the contact angle, contact angle hysteresis, and surface microheterogeneity can be easily determined. Tensiometry has been applied to textile fibers and to the glass and carbon fibers used in fiber reinforced composites. Analysis of the wetting data for the fibers in composites cannot be done in the classical terms of the critical surface tension approach developed by Zisman. These fibers are highly polar, so interpretation of their wetting behavior has been done in terms of theories of fractional polarity.

ACKNOWLEDGMENT

The author wishes to express his gratitude to the editors of this book for the opportunity to write this chapter in honor and memory of Dr. Bill Zisman. One of the many lessons learned while working with Zisman and his associates at the Naval Research Laboratory was the importance of maintaining a balance between theoretical and applied research. This lesson has served me well in the ensuing years.

I would also like to express my appreciation to the National Aeronautics and Space Administration, Langley Research Center for support of some of the work reported here.

REFERENCES

1. A. B. D. Cassie, in *Surface Phenomena in Chemistry and Biology* (J. F. Daniellli, K. G. A. Pankhurst, and A. C. Riddiford, eds.), p. 166, Pergamon Press, New York (1958).
2. N. K. Adam, *The Physics and Chemistry of Surfaces*, 3rd ed., p. 169, Oxford University Press, London (1941).
3. E. W. Washburn, *Phys. Rev.* **17** 273 (1921).
4. Lord Rayleigh, *Proc. London Math. Soc.* **10**, 4 (1879).
5. S. L. Goren, *J. Colloid Sci.* **19**, 81 (1964).
6. J.-I. Yamaki and Y. Katayama, *J. Appl. Polym. Sci.* **19** 2897 (1975).
7. J. N. Israelachvili, *Intermolecular and Surface Forces*, Academic Press, New York (1985).
8. F. Brochard and P. G. de Gennes, *J. Phys. Lett.* **45**, L-597 (1984).
9. F. Brochard-Wyart and J.-M. di Meglio, *Ann. Chim.* **77** 275 (1987).
10. W. D. Bascom and J. B. Roman, *I&EC Prod. Res. Dev.* **7** 172 (1968).
11. J. J. Elmendorp, Koninklijke/Shell Laboratorium, Amsterdam, private communication.
12. P. J. Verschoor and W. D. Bascom, unpublished results.
13. D. W. Fuerstenau and M. C. Williams, *Part. Charact.* **4**, 7 (1987).
14. J. P. Mutchler, J. Menkart, and A. M. Schwartz, in *Pesticidal Formulations Research*, Adv. Chem. Ser. No. 86, p. 7, American Chemical Society, Washington DC (1967).
15. R. E. Johnson, Jr., R. H. Dettre, and D. A. Brandreth, *J. Colloid Interface Sci.* **62** 205(1977).
16. W. D. Bascom, unpublished results.
17. P. G. de Gennes, *Rev. Mod. Phys.* **57**, 827 (1985).
18. L. W. Schwartz and S. Garoff, *Langmuir* **1**, 219 (1995).
19. L. W. Schwartz and S. Garoff, *J. Colloid Interface Sci.* **106**, 422 (1985).
20. B. Miller, in *Surface Characteristics of Fibers*, part II (M. J. Schick, ed.), p. 417, Marcel Dekker, New York (1977).

21. A. W. Neumann and R. J. Good, in *Surface and Colloid Science*, vol. II (R. J. Good and R. R. Stromberg, eds.), p. 31, Plenum Press, New York (1979).
22. W. A. Zisman, in *Contact Angle, Wettability and Adhesion*, Adv. Chem. Series 43, p. 1, Am. Chem. Soc., Washington, DC (1964).
23. S. P. Wesson and A. Tarantino, *J. Non-Cryst. Solids* **38–39**, 619 (1980).
24. W. D. Bascom, in *Advances in Polymer Science*, Vol. 85, p. 89, Springer-Verlag, Berlin (1988).
25. S. Wu, *Polymer Interface and Adhesion*, p. 67, Marcel Dekker, New York (1982).
26. G. E. Hammer and L. T. Drzal, *Appl. Surf. Sci.* **4**, 340 (1980).
27. D. H. Kaelble, P. J. Dynes, and E. H. Cirlin, *J. Adhesion* **6**, 23 (1974).

14

Wettability Phenomena and Coatings

Clifford K. Schoff

1. INTRODUCTION

Organic coatings are applied to a wide variety of objects and structures to provide decoration and protection. Wetting has a very important part in both aspects. It is fairly obvious that a coating must be in close contact with the substrate in order to give good adhesion and protection, but wetting (or lack of it) affects appearance as well. Most coating application processes ensure initial wetting and spreading on the substrate. However, dewetting may occur, and, occasionally, conditions are such that wetting is poor in the first place. The result tends to be unsightly surface defects that can affect the protective aspects of the coating as well as the appearance. In addition, the coating formulation (or part of it) must properly wet, disperse, and stabilize the pigments during manufacture to give a product with adequate flow and application properties and the correct color and gloss.

This chapter deals with the wetting of substrates and pigments by coatings and considers the problems faced in these areas. It covers some of the methods used to make paints and other coatings wet surfaces more readily and to make substrates and pigments easier to wet. It also covers the wettability tests and other surface techniques used in the coatings industry to help in the development of coatings and the solving of problems. Many of the techniques and ideas can be traced to Bill Zisman, whose work has had considerable influence on the coatings industry. He received recognition for this when he won the Mattiello Award of the Federation of Societies for Paint Technology in 1971.[1]

My own introduction to surface chemistry came from Dr. Zisman when he was a consultant at our research center in the 1970s. He discussed wetting and wettability in detail with us, and notes from his lectures and discussions formed the basis for efforts to solve wetting, surface defect, and adhesion problems. Much use was made of this material when I was teaching the surface science section of our

---

Clifford K. Schoff • Coatings Research and Development, PPG Industries, Inc., Allison Park, Pennsylvania 15101.

Modern Approach to Wettability: Theory and Applications, edited by Malcolm E. Schrader and George Loeb. Plenum Press, New York, 1992.

in-house coatings technology course. Therefore, it is a pleasure for me to contribute to this volume being produced in his honor.

The object of this chapter is to provide information on wetting and wettability related to organic coatings. It is based on the literature in this area, but includes data and insights that come from a number of years spent attacking surface-related problems. It is hoped that the information will prove interesting and useful to others. It also is a partial repayment of a debt to Dr. Zisman for introducing me to the field and to others for helping me along the way.

2. CONSEQUENCES OF POOR WETTING OR DEWETTING

2.1. Introduction

When a newcomer enters the coatings industry, he or she often decides that, considering the complicated formulas of coatings and the many demands they must satisfy, it is not surprising that they sometimes show defects or fail outright; the surprising thing is that they work at all! One of the requirements for a successful coating is that it wet (or at least not dewet from) the substrate. Another requirement is that the coating or some of its components must wet and disperse the pigments necessary for color, UV opacity, and other properties. What happens when these requirements are not met?

2.2. Substrates

There are a number of consequences of poor wetting of substrates or of dewetting after the coating has been applied. The most noticeable is the formation of surface defects such as crawling or pulling away from edges, cratering, pinholing, and sinks and bumps (orange peel). Such defects have been described in detail elsewhere.[2–4] Some defects are caused by flows driven by surface tension gradients rather than by dewetting from the substrate, but the principles are much the same. Another consequence can be poor adhesion. Unfortunately, good wetting does not guarantee good adhesion. The material being wet may be a thin layer with little or no adherence to the substrate and/or with little cohesive strength (weak boundary layer). Examples of such materials include plastic mold release agents, powdery metal pretreatments, and poorly cured primer coats. Another possibility is that such a material may exude from the substrate to the coating–substrate interface during drying or curing of the coating or over a period of time, particularly during outdoor exposure.

A somewhat peculiar consequence of poor wetting occurs in roll coating or printing when the paint or ink wets the applicator rolls better than it does the substrate. The result is a thinner than expected layer of material on the substrate and buildup on the roller(s). In printing, such buildup can cause indistinct ("dirty") printing later in the print run.

2.3. Pigments

Wetting of large clumps of pigment by the dispersion vehicle is the first step in the pigment dispersion process. If wetting does not occur or is only partial, then most of the pigment agglomerates will not be broken into small aggregates or primary particles. The resultant coarse dispersion will not give the proper color and gloss and will not impart the correct flow properties to the paint. The expected properties of the pigment will only be partially realized.

If initial wetting and adsorption occurs, but is not permanent—i.e., there is an exchange between the dispersion vehicle and solvent or other material after the particles have been dispersed—then the dispersion may not be stable and flocculation may occur. Flocculation is the formation of loosely bound clumps of primary particles or small agglomerates joined together by point to point contact. Flocculation causes a variety of problems in coatings, ranging from incorrect color and streaking to loss of gloss and reduced weather resistance.

3. SUBSTRATE WETTING

3.1. Introduction

Most coatings with good wetting and adhesion have been developed by trial and error. For many years this was acceptable because wetting rarely was a problem and dewetting usually only occurred in cases of gross contamination. The surfaces involved (pretreated steel and aluminum, wood, sandblasted structural steel, etc.) usually were relatively clean and easy to wet. After the paint was applied and solvent was lost, the viscosity increased rapidly to give good resistance to dewetting and other surface tension driven flows that produce defects. Surface chemistry and wettability tests were used, but mostly for problem solving. When problems did occur, additives such as low-surface-tension solvents and high molecular weight polymers could be used, in considerable amounts if necessary.

However, both coatings and substrates are changing in directions that are making wetting more difficult. Trial and error is no longer good enough. The move toward more ecologically acceptable waterborne, powder, and high-solids coatings has led to many wetting and surface defect problems, as has the introduction of new, more difficult to wet substrates such as elastomers and plastics. There is considerable interest in wettability tests and other substrate characterization methods and in techniques to improve wetting.

3.2. Wettability Tests

Most wettability tests in the coatings industry involve measurement of contact angles of specific liquids on substrate surfaces. This normally is an advancing contact angle; i.e., during formation the drop advances across the surface. The receding contact angle where a drop or film retracts over a previously wetted surface would seem to be more useful for characterizing dewetting phenomena such as cratering and crawling. However, it is difficult to measure and rarely is used in the coatings industry.

Much of the wettability testing in the coatings industry owes its basis to William Zisman and his critical surface tension of wetting, γ_c.[1,5] The contact angles of various liquids (often a homologous series of hydrocarbons) on the

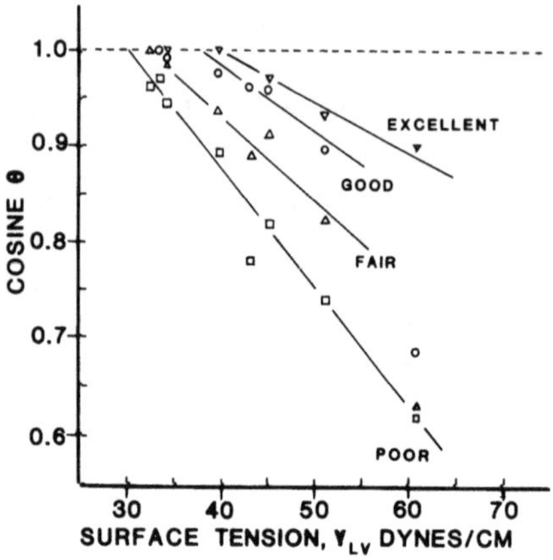

FIGURE 1. Critical surface tension (Zisman) plots for a series of rest liquids on tinplate specimens with varying degrees of paintability. (Data from Ref. 6.)

surface are determined, and the cosines of the angles are plotted versus the surface tensions of the liquids. The plot is extrapolated to $\cos\theta = 1$, i.e., $\theta = 0°$, which represents the point where the liquid would just spontaneously spread if applied as a drop. Such Zisman plots have been useful in predicting or explaining wetting problems and, therefore, many surface defects. An example is shown in Fig. 1, which gives data for a series of nonhomologous liquids on tinplate. Helwig[6] used the plots in Fig. 1 to differentiate qualities of tinplate in terms of their paint wetting characteristics. He found that tinplate with a lower γ_c (30 dynes/cm) was not wet as well by a given lacquer as a tinplate with a higher γ_c (40 dynes/cm).

Cheever[7] and others have pointed out that Zisman worked with ideal surfaces that were well defined, homogeneous, smooth, and clean. The test liquids were very pure and the atmosphere controlled to keep it free from contaminants. Unfortunately, the surfaces to which coatings are applied are far from being ideal. They are heterogeneous, rough, and contaminated. In addition, the coatings themselves are heterogeneous and sometimes contaminated, and the conditions under which they are applied are anything but ideal. However, even with all these caveats, Zisman plots and the data from them can be very useful in characterizing surfaces and solving problems.

Table 1 lists a number of critical surface tensions that have been determined in our laboratory. The wide range of values is indicative of the range of materials for which coatings technologists must design coatings. It is relatively simple to

TABLE 1. Critical Surface Tensions of Some Typical Substrates Data Generated with a Series of Polar Liquids

Substrate	Critical surface tension γ_c (dynes/cm)
Zn phosphated steel	45–56
Fe phosphated steel	43
Tin plated steel	35
Treated aluminum extrusions	33–35
Untreated steel	29
SMC polyester	23

develop a paint that will wet a good-quality zinc phosphate, but quite difficult to come up with one that will effectively wet a molded plastic (SMC, styrene–polyester sheet molding compound) part.

It should be pointed out that a 0° paint contact angle (paint $\gamma < \gamma_c$) is not an absolute requirement for good performance. The Zisman analysis deals with a drop on a smooth surface, whereas painting involves a film that has been applied forcibly to a relatively rough surface. Good wetting and adhesion can occur with contact angles considerably above 0°.

Spontaneous spreading can occur at $\theta > 0°$ on a rough surface. Lack of dewetting probably is a more realistic criterion for performance than is wetting, since most paints will not dewet from a surface once they have been applied even though they may not spontaneously spread if applied as a drop. As mentioned, dewetting can be related to receding contact angles. Hansen[8–10] has covered this subject in some detail and has illustrated the concepts of advancing and receding contact angles and spontaneous spreading and dewetting with Zisman plots such as those shown in Fig. 2.[10]

Receding contact angle plots also have been applied to the problem of adhesion of printing inks to plastic foils.[10] Contact angles were measured on the plastic substrate and on the backs of ink films peeled from the plastic. Good adhesion occurred when the critical surface tensions of the coating and substrate were quite close with the coating having a slightly lower γ_c than the substrate.

Zisman plots are very useful, but they are not without their drawbacks. Wu[11] and others have pointed out that critical surface tensions often are low compared with surface tensions determined by extrapolation of data from melts or solutions. We have seen cases where Zisman plots gave what seemed to be impossibly low values for the critical surface tensions of substrates, or they gave similar values for two substrates with very different paintabilities. In addition, as the solvents in coatings have moved away from being primarily nonpolar hydrocarbons and toward polar oxygenated solvents, critical surface tensions determined with the usual nonpolar liquids have become less relevant. However, when polar and/or hydrogen bonding liquids are used, the resultant Zisman plots tend to be strongly curved and γ_c can be difficult to determine. Dann[12] has indicated that critical surface tensions depend on the liquid–solid interfacial tension and that γ_c can vary with the liquid series used in the determination. Because of these problems, other techniques to characterize substrates are of considerable interest.

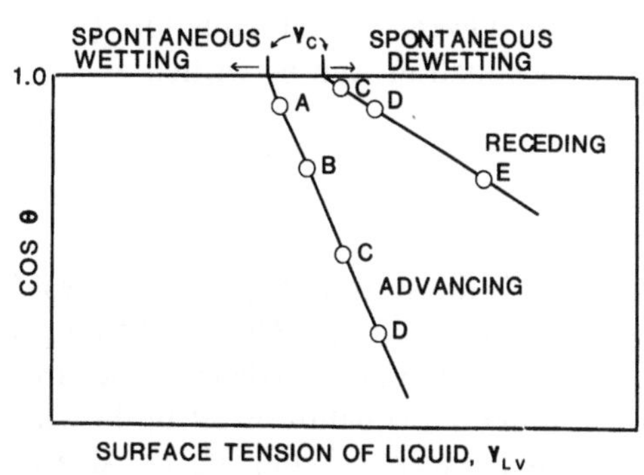

FIGURE 2. Critical surface tension (Zisman) plots for advancing and receding contact angles.

Other approaches may be found in the literature,[11,13,14] and many of them have proved to be useful, but most are no more effective for solving coatings problems than are Zisman plots. However, other methods that take into account the polarity of surfaces and the different intermolecular forces that exist at surfaces have turned out to be quite effective. Fowkes[15] suggested that the total free energy at a surface is the sum of contributions from the different intermolecular forces such as dispersion, polar, and hydrogen bonding forces. However, he derived an expression that accounts only for dispersion forces:

$$\gamma_{LS} = \gamma_{SV} + \gamma_{LV} - 2(\gamma_S^d \gamma_L^d)^{1/2}$$

Dann[16] showed that nondispersion interactions I_{SL}^P at the interfaces between polymers and liquids could be calculated. These interactions included all non-dispersion forces such as dipole–dipole, induced dipole, and hydrogen bonding. The I_{SL}^P was considered to be a function of nondispersion liquid and solid surface tensions, but no relationship was determined.

Owens and Wendt[17] extended the concept of Fowkes to cases where both dispersion and hydrogen bonding forces operate. They regarded the surface tension as being composed of two components such that

$$\gamma = \gamma^d + \gamma^h$$

where γ^h denotes the component of surface tension due to both hydrogen bonding and dipole–dipole interactions. Their equation is

$$\gamma_{LS} = \gamma_{SV} + \gamma_{LV} - 2[(\gamma_S^d \gamma_L^d)^{1/2} + (\gamma_S^h \gamma_L^h)^{1/2}]$$

or

$$\gamma_{LV}(1 + \cos\theta) = 2[(\gamma_S^d \gamma_L^d)^{1/2} + (\gamma_S^h \gamma_L^h)^{1/2}]$$

At about the same time, Kaelble[18] published a very similar equation in terms of dispersion and polar forces for the work of adhesion:

$$W_A = \gamma_{LV}(1 + \cos\theta) = 2[(\gamma_S^d \gamma_L^d)^{1/2} + (\gamma_S^p \gamma_L^p)^{1/2}]$$

The Owens–Wendt equation (which should be called the Owens–Wendt–Kaelble equation) can be written as

$$\frac{\gamma_l(1 + \cos\theta)}{2} = [(\gamma_l^d \gamma_s^d)^{1/2} + (\gamma_l^p \gamma_s^p)^{1/2}]$$

where γ_l is the surface tension of the test liquid, and γ_s is the surface tension of the solid in question. In the Owens–Wendt procedure, the contact angles of two liquids, typically water and methylene iodide, are determined on the surface of interest. For each liquid, an equation is set up in which all the quantities are known or can be calculated, except γ_s^d and γ_s^p, the dispersion and polar components of the solid

TABLE 2. Surface Tensions by Owens–Wendt Procedure and Ease of Wetting of Tinplate Specimens

	Cos θ		Surface tensions (dynes/cm)			
Tinplate surface	Water	Me iodide	γ_s^d	γ_s^p	γ_s	γ_c
Poor	0.257	0.741	34.4	6.6	41.0	30
Fair	0.302	0.826	38.1	6.8	44.9	32
Good	0.394	0.897	40.6	8.6	49.2	38
Excellent	0.862	0.932	37.0	30.4	67.4	40

Data from Ref. 6.

surface tension. Thus, there are two equations in two unknowns that can be solved to yield the unknown values. The sum of the two components is the surface tension of the solid. The results obtained are in reasonable agreement with results obtained by other methods, such as extrapolation of surface tension–temperature or surface tension–concentration data.

Helwig's data that were discussed earlier can be analyzed by the Owens–Wendt method. Table 2 compares tinplate surface tensions determined by the Owens–Wendt method with the critical surface tension results. The critical tensions (γ_c) are lower than the solid surface tensions (γ_s) and are very close to the values for γ_s^d, the dispersion component of the surface tension. Both methods give the same ranking, however. Note that the magnitude of the polar component of the solid surface tension made a sharp jump as the wettability of the surface went from good to excellent.

We have applied the Owens–Wendt method to a number of lab and field problems and have found it to be very useful in predicting and explaining adhesion and wetting failures. An example would serve to show this. The problem was one of poor wetting of topcoats over a certain primer, particularly when the primed surface was aged. Fairly similar primers had been used for several years, but no problem had been seen. Contact angles with distilled water and methylene iodide were measured on an aged specimen of the problem substrate, the same after washing with a detergent solution, and on a control, a very similar primer known to give adequate wetting. Four drops of each test liquid were used, and angles were determined at both edges of each drop, giving eight measurements for each liquid on each substrate.

TABLE 3. Primer Problem: Substrate Contact Angles

	Water		Methylene iodide	
Specimen	θ	Cos θ	θ	Cos θ
Aged problem primer	63°	0.454	46°	0.695
After washing	72°	0.309	30°	0.866
Control	73°	0.292	36°	0.809

TABLE 4. Primer Problem: Solid Surface Tensions

Specimen	Solid surface tension, γ_s	Dispersion component, γ_s^d	Polar component, γ_s^p
Aged problem primer	44.9	30.3	14.6
After washing	46.4	40.0	6.4
Control	44.0	37.3	6.7

The results are given in Table 3. These data and the surface tensions of the test liquids were substituted into the Owens–Wendt equation, and the solid surface tension and dispersion and polar components were determined. Values are listed in Table 4. The results indicated that the surface of the unwashed problem primer was considerably more polar than the other two. This pointed to the existence of a polar exudate on the surface of the primer. Additional investigation showed that the polar cross-linking resin was not reacting completely (although the cure had appeared to be acceptable) and was slowly coming to surface.

Panzer[19] has pointed out that the Owens–Wendt procedure has no theoretical basis, can give erratic results, and the results depend on which pair of liquids are used. More recently, Spelt and Neumann have contended that, although surface tension components such as γ^d and γ^p may well exist, they cannot be determined from contact angles.[20] They believe that the quantities claimed to be dispersion or polar components are artifacts. Therefore, there seem to be questions concerning use of the Owens–Wendt method. However, regardless of what really is being measured, we have found the procedure to be very useful for characterizing problem substrates and signaling the presence of contamination. Regardless of its drawbacks, the Owens–Wendt method works most of the time. It is simple, rapid, and has pointed the way to solutions to many problems.

Solid surface tension also has been separated into three components: γ^d, γ^p, γ^h (dispersion, polar, and hydrogen bonding, respectively). These can be established by measuring contact angles with three pure liquids[19,21,22] or by wetting studies with a series of liquids.[23–25] The reader may note the similarity between the surface tension components and three-dimensional solubility parameters. Indeed, there is a close relationship between these two properties,[26–30] and a solubility parameter

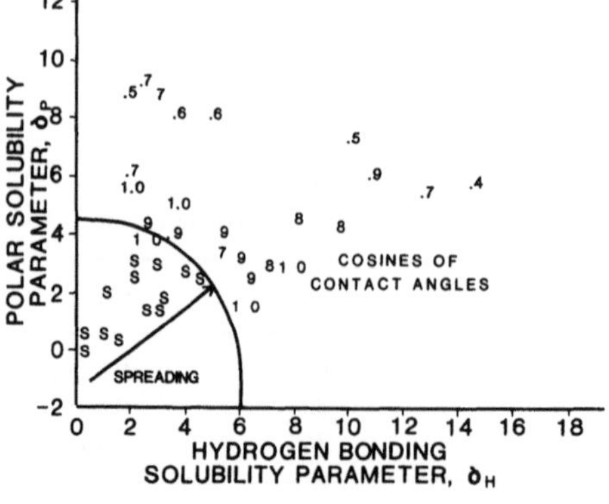

FIGURE 3. Wetting of polypropylene: wettability as a function of the solubility parameters of the contacting liquid.

plot can be converted to surface units.[23–29] A solubility parameter plot of the surface of polypropylene based on contact angles and spontaneous spreading is shown in Fig. 3.[23] The plot covers the polar and hydrogen bonding parameters, δ_p and δ_h. The cosine (rounded off to the nearest tenth) of each contact angle is placed at the δ_p and δ_h coordinates of the liquid involved. A spreading liquid is labeled with an S. As expected for a hydrocarbon-like polymer, the spontaneously wetting liquids tended to be hydrocarbons and other relatively low-polarity, low-hydrogen-bonding solvents. Such a plot can be used to characterize a given substrate and provide signals as to the best solvents for initial wetting and the kinds of coatings that should give good adhesion.

Another application of three-component surface tensions has been the relating of the corrosion resistance of certain coatings to the hydrogen bonding component of surface tension.[21] Adhesion was considered to be the controlling factor.

Another three-component system is that of Good and co-workers,[22] whose three components are γ^W (Lifshitz–van der Waals), γ^+ (electron acceptor), and γ^- (electron donor). Such an analysis should be particularly useful for the characterization of pigments, but also for other surfaces related to coatings.

In some cases, water contact angles alone have been used to characterize surfaces. For example, Cheever[7] used water contact angle mapping to differentiate between surface regions of polyester–styrene SMC plastic, which is used for many automobile parts and assemblies. SMC is tough and relatively inexpensive, but is sometimes difficult to paint.

Cheever was able to estimate the components present on the surface and relate wettability by water vapor to peel strength. Results correlated with mold and temperature effects. Contact angle mapping with water or other liquids, although tedious, has potential for enabling investigators to determine the effect of chemical and processing changes on the surfaces of plastics and to optimize these surfaces for paintability.

Water contact angles are also used to estimate surface cleanliness after cleaning operations, ease of wettability of surfaces by waterborne coatings, and the effectiveness of rinsing processes. Sometimes the test is a very simple running of water across the surface to see whether the liquid sheets off or beads up.

3.3. Dewetting Tests

Since the coating application process almost guarantees spreading, dewetting is much more apt to be a problem than not wetting initially. For many years we have used a simple swab test[8] to determine the tendency to dewet, and to identify an approximate critical surface tension. This test involves swabbing a series of solvents onto the substrate with cotton swabs and observing whether the strip of solvent stays in place or dewets and crawls. The breakpoint between wetting and dewetting provides a critical surface tension of dewetting. The technique can be carried out rapidly and is particularly useful for testing in the field or on curved, irregular, or porous substrates where contact angles cannot be measured. It is similar to ASTM method D-2578, "Wetting Tension of Polyethylene and Polypropylene Films," which uses mixtures of ethylene glycol monoethyl ether and formamide. A comparison between swab test results and those by the Owens–Wendt method is shown

TABLE 5. Comparison of Solid Surface Tension and Swab Test Critical Surface Tension Values for Automotive TPO Specimens

Specimen	Solid surface tension γ_s	Dispersion component γ_s^d	Polar component γ_s^p	Swab test critical surface tension γ_c
Republic ETA 3401	36.6	27.9	8.7	29.2
Republic ETA 3081	41.0	34.4	6.6	33.0
Surface-treated Italian TPO	45.1	37.1	8.0	41.0

in Table 5. The specimens were automotive grade thermoplastic olefin (TPO) [polypropylene modified with ethylene–propylene–diene monomer (EPDM) rubber].

By extending the swab test observations to cover three forms of behavior (spreading, nondewetting, and dewetting), one can construct a plot similar to the surface solubility parameter plot shown in Fig. 3 if the solubility parameters of the liquids are known. An example of such a "wetting tension" plot is shown in Fig. 4.[25]

3.4. Other Characterization Methods

In addition to wettability tests, a number of other methods have been applied to wetting problems, the prediction of wettability, etc. These methods include surface roughness measurements, sophisticated surface analysis techniques, such as Auger electron spectroscopy, x-ray photoelectron spectroscopy (ESCA or XPS), the electron microprobe, and secondary ion mass spectrometry (SIMS), and more common problem-solving techniques, such as light microscopy, the scanning electron microscope (with x-ray attachment), and Fourier transform infrared spectroscopy (FTIR). The main thrust usually is the identification of contaminants that are causing or could cause wetting problems. Results from contact angle measurements, the Owens–Wendt method, the swab test, and other wettability tests can signal that a surface is contaminated or difficult to wet, but other techniques must be used to identify the exact cause of the problem. Another use of such instruments is to identify the presence and determine the quality of pretreatments and conversion coatings that are used to make surfaces more wettable and paintable.

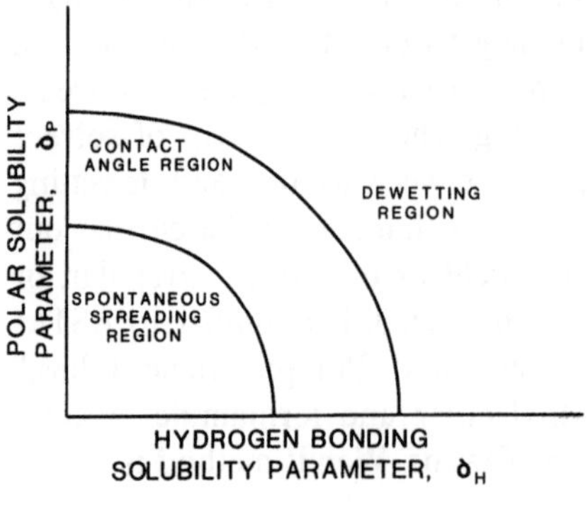

FIGURE 4. Hypothetical wetting tension plot showing possible wetting and dewetting regions.

3.5. Dynamic Surface Tension Effects

The liquid surface tensions discussed so far have been static or equilibrium surface tensions; i.e., the components of the coating or other liquid are in equilibrium. However, some coatings (especially waterborne coatings) have higher surface tensions immediately after spraying or, even in some cases, after rapid stirring. The latter is due to temporary transfer of surface active agents from the surface to the bulk of the liquid. It takes time for the material to diffuse back through the coating to the surface. The increase in surface tension with spraying is more complex because it involves the creation of a large amount of new surface area. The expansion of the surface area decreases the concentration of surface-active material at the surface and surface tension rises. Surfactant will immediately begin to diffuse from the bulk of the liquid to the surface, but the process may take seconds or even minutes. This time–area effect on surface tension is called the dynamic surface tension and is characterized by the surface dilational modulus, the change in surface tension that results from a change in surface area.[31–34]

Dynamic surface tension can have an effect on wetting if there is a high surface dilational modulus and the relaxation in modulus with time is slow. It is possible to have a coating that develops a high surface tension on spraying and retains it until the spray droplets hit the substrate, so that initial wetting is poor. We believe that dynamic surface effects are responsible for many of the foaming and surface defect problems seen with waterborne coatings. Certainly, waterborne coatings generally have higher surface dilational moduli than do solvent-borne coatings.[34]

3.6. Viscosity Effects

Surface tension is not the only property that affects wetting and dewetting of substrates. The viscosity of a coating also influences wetting and surface defect formation. The rate of wetting is dependent on viscosity as well as surface tension;[35] i.e., wetting rate is proportional to

$$\frac{\gamma \cos \theta}{2\eta}$$

where

$$\eta = \text{viscosity}$$
$$\gamma = \text{surface tension}$$
$$\theta = \text{contact angle}$$

Even if other conditions are favorable (surface tension, contact angle, etc.), spontaneous spreading or dewetting may not occur if viscosity is too high. Surface defects may or may not form, depending on viscosity. The driving force for some surface defects is $\partial\gamma/\partial x$, the surface tension gradient. The flux induced by this driving force is[3,36]

$$J = \frac{1}{2}\frac{\partial \gamma}{\partial x}\frac{h^2}{\eta}$$

where

$$J = \text{flux}$$
$$h = \text{wet film thickness of coating}$$
$$\eta = \text{viscosity}$$

If the viscosity is high enough, the flux will be so small that the defect will not be noticeable. Raising low shear viscosity is one means of preventing or reducing surface defects. An example of the solution of a defect problem by raising viscosity is shown in Fig. 5,[3] which contains viscosity–shear rate plots for two high-solids coatings. Coating A flowed and leveled very well, but it also cratered badly. Addition of a thickener produced coating B, which did not crater but did not level either. It gave a distinct orange peel effect. Fortunately, this was quite acceptable as the coating was for use on appliances where a certain amount of texture is considered pleasing. Such a remedy would not have been acceptable for an automotive topcoat, which must be very smooth.

3.7. Miscellaneous Effects

Other effects influence wetting, involve special wetting considerations, or are affected by wettability. For example, surface wetting by water is an important factor in dirt retention by coated surfaces.[8,37,38] Dirty surfaces look terrible, can cause streaking and fouling of other surfaces, provide a medium on which mildew can grow, and often do not resist weathering as well as clean surfaces. When water beads up on a surface, much more dirt is collected than if the water sheets and forms a thin film. Therefore, a high solid surface tension and water wettability are necessary for surface cleanliness. Hydrophilic surfaces consistently are cleaner than hydrophobic surfaces[8] and are easier to clean.[38] Approximately six times the amount of (usually dirty) rainwater collects on a nonwetting vertical surface (by beading up), compared with a surface where the water wets and runs off.[37]

Industrial painting operations take up a great deal of space and involve expensive equipment, so there always is much interest in simplifying the process, cutting out steps, reducing the number of coatings, etc. One way to cut down on cost is to apply coatings wet on wet; i.e., the second coating is applied before the first one is completely dry or cured. This procedure is common with automobile topcoats

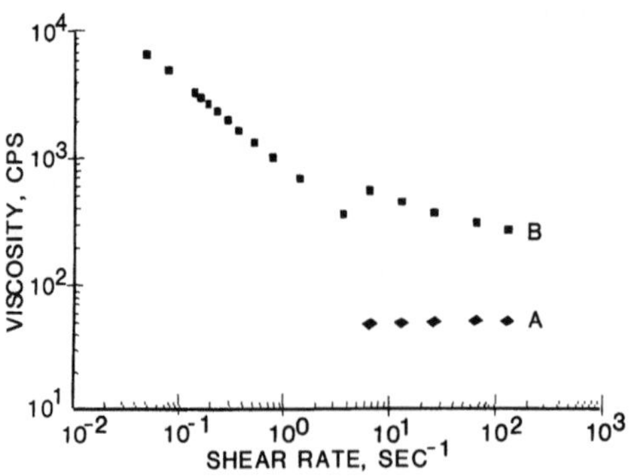

FIGURE 5. Viscosity–shear rate plots for two high solids coatings. Coating A craters, but flows out and levels well. Coating B resists cratering, but does not level.

where a clear coat is applied over a colored base coat, then both are baked at the same time. The clear coat gives improved appearance but without the added expense of an extra set of ovens that would be needed if the coatings were baked separately. In some applications, many coats are applied before a final curing operation takes place. In order to prevent dewetting, the successive coats have progressively lower surface tensions.[2] Compatibility must not be too good, however, as mixing of the two coatings will harm the appearance.

A number of processes such as printing and roll coating depend on transfer of ink or paint from one roll to another and, finally, to a substrate. All the components of the system (coating, rollers, substrate) and their wetting–wettability characteristics must be taken into account; otherwise strange things can happen, such as an ink staying on the application roller or plate and not wetting or transferring to the paper. In most cases, there is transfer, but sometimes it is incomplete and ink builds up on the inking surface. Eventually the excess ink shows up on the edges of letters and other images, giving "dirty" printing. Analysis of the performance of flexographic printing plates in terms of producing clean printing is shown in Table 6.[39] In this table, performance is compared to solid surface tensions and polar components as determined by the Owens–Wendt method. Low plate polarity gave good performance, whereas high polarity caused the waterborne ink to wet the plate better than the polyester substrate, which led to ink buildup and, eventually, poor-quality printing. Table 6 also shows that critical surface tensions from Zisman plots for the plates were about the same (20–26 dynes/cm) and did not correlate with performance.

3.8. Techniques Used To Improve Substrate Wetting

Some substrates are difficult to wet, including a number of relatively inexpensive high-performance plastics that automobile makers and other manufacturers would like to use, but cannot because the surfaces are not paintable. Because of these and other wetting problems, considerable effort has been expended in developing techniques to improve substrate wettability and to modify coatings to do a better job of wetting. Some methods for substrates are relatively simple, such as improved cleaning methods to remove oil and dirt from metal surfaces and the use of wash

TABLE 6. Surface Tension by Owens–Wendth Procedure and Performance of Flexographic Printing Plates

Plate material	Performance (ink transfer)	Owens–Wendt results: Total solid surface tension	Owens–Wendt results: Percent polarity	Zisman plot γ_c
Rubber (several specimens)	Very good	28–34	0	21–23
DuPont Red Cyrel (Br)	Good	37	3	20
BASF Nyloprint	Good	42	1	27
DuPont Red Cyrel (Cl)	Fair–good	42	2	24
DuPont Blue Cyrel	Poor	45	26	21
Hercules	Poor	48	40	26

primers. Others involve complex physical and chemical treatments, such as flame and corona–plasma treatments of plastics to make their surfaces more polar and the formation of microcrystalline zinc phosphate deposits on steel. Additional techniques include application of very thin (1–2 μm) organic coatings to zinc-coated steels in the steel mill[40] and the use of specially formulated adhesion promoters.[41] Techniques to make coatings wet better include the use of surfactants, particularly silicones and fluorocarbons, and low-surface-tension solvents.[3,4]

4. PIGMENT WETTING

4.1. The Importance of Pigments

Pigments are included in coatings to provide optical properties such as color, opacity to both ultraviolet and visible light, gloss, and, sometimes, special effects such as pearlescence or a metallic finish. They also can provide or influence other properties such as corrosion resistance, hardness, flexibility, adhesion, and durability. When the paint is wet, pigments affect the rheology and influence the ease of application, leveling, and sag resistance. The way pigments perform depends on their chemical composition, primary particle size, and degree of dispersion. The last named property is highly dependent on pigment wetting and stabilization.

4.2. The Process of Pigment Dispersion

Pigment dispersion is a key factor in determining the performance of a coating. In fact, successful dispersion is as important to the optical and protective properties of a coating as the formula itself. The process basically is one of replacing an air–solid interface (in the dry pigment powder) with a liquid–solid interface and separating the clumps of pigment particles so that they are dispersed in the liquid. The dispersed particles must be stabilized, or they will flocculate to form loose assemblies.

The liquid dispersing medium may be a resin (polymer) solution, a surfactant solution or dispersion, a solvent, or a combination of these. Most dispersants are polymer solutions, and the polymers used generally contain certain functional groups such as carboxyl, hydroxyl, and amino groups to anchor the polymer to the pigment surface. Wetting agents (surfactants) often are added to reduce interfacial tension and improve initial wetting. The process of pigment dispersion has been described in detail by Parfitt,[42,43] and the following discussion owes much to his writing and lectures.

Pigment dispersion can be divided into three overlapping steps:

a. *Wetting:* The dispersant displaces air and water from the surface of the clumps of pigment powder and adsorbs onto the surface. It penetrates into pores, gaps, and channels between particles.
b. *Deagglomeration:* The wetted particle clusters are separated by the shearing action of a blade, rollers, sand, or steel, glass, or ceramic spheres into smaller agglomerates and/or primary particles. Additional wetting occurs where particles have been freshly separated.

c. *Stabilization:* The dispersant must stay adsorbed on the particles or small clumps. Otherwise, the pigment particles may move together permanently to form new agglomerates, i.e., flocculates. Stabilization involves more than just wetting. The dispersant must either impart a charge to the surface so that particles repel each other (waterborne pigment pastes and coatings) or else build up an adsorbed layer to repel other particles (solvent-borne pigment pastes and coatings).

All of these steps must be controlled, but it should be obvious that without initial wetting, the process cannot begin. Wetting is favored by low dispersant surface tension and low dispersant–pigment contact angle. A low contact angle also aids in penetration and deagglomeration, but high liquid surface tension enables these processes to occur at faster rates. Low dispersant viscosity allows more rapid wetting and better penetration, but relatively high paste viscosity is necessary to produce the shear stress necessary for deagglomeration.

Low liquid surface tension and viscosity are favored by low polymer concentration in the dispersant, whereas paste cohesion and stabilization usually need higher concentrations. In addition, ecologically acceptable coatings require low solvent levels in pigment pastes, i.e., high pigment and polymer concentrations. Obviously, compromises are necessary.

Pigment wetting depends critically on the properties of the dispersant and the nature of the pigment surface. As in other wetting phenomena, the surface tension of the dispersant must be lower than the critical surface tension of the pigment in order for spontaneous spreading to occur. Solvents and additives (wetting agents) are used to lower the surface tension of the dispersant and the interfacial tension between the dispersant and pigment to give better wetting.

4.3. *Wetting of the Pigment Powder*

The wetting of a solid pigment particle by a liquid involves three stages shown schematically in Fig. 6:[44] initial adhesion, immersion, and spreading over the top of the particle. The appropriate work terms for each stage have been derived from the Young–Dupré equation, assuming a cubical particle:[42–44]

$$W_a = \gamma_{SL} - (\gamma_{LV} + \gamma_{SV}) = \gamma_{LV}(\cos\theta + 1) \qquad \text{(one face of the cube)}$$

$$W_i = 4\gamma_{SL} - 4\gamma_{SV} = -4\gamma_{LV}\cos\theta \qquad \text{(four faces of the cube)}$$

$$W_s = (\gamma_{SL} + \gamma_{LV} - \gamma_{SV}) = -\gamma_{LV}(\cos\theta - 1) \qquad \text{(one face of the cube)}$$

It is assumed that, before wetting, the surface of the solid is in equilibrium with the vapor of the liquid. From these equations it is possible to determine the contact

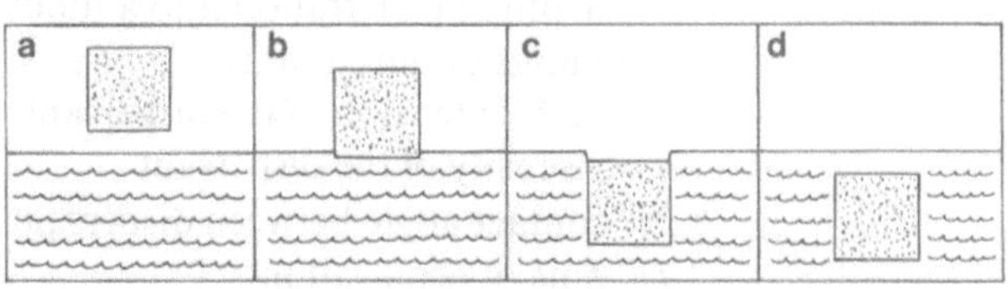

FIGURE 6. The stages of pigment wetting, treating the pigment particle as a cube and showing the stages leading to total submergence in the dispersant: (a) initial, (b) adhesion, (c) immersion (d) spreading.

angle range for each stage that will give a spontaneous process (W is negative). If W is positive, then work must be expended on the system for the process to occur. It can be concluded that

a. Adhesional wetting always is spontaneous since W is negative as long as the contact angle is less than 180°.
b. Immersion is spontaneous when the contact angle is less than 90° (which is most of the time).
c. Spreading is spontaneous only when $\theta = 0°$, and work must be done to achieve spreading at all higher values for this angle (which is the case in many practical systems).

The above stages and their energy and contact angle requirements show that dispersion may be prevented or at least slowed down during one of the intermediate steps. It is possible that a pigment particle may easily penetrate a liquid but resist complete submergence.

The total work of dispersion is given by the sum of the three separate stages

$$W_d = W_a + W_i + W_s = 6\gamma_{SL} - 6\gamma_{SV} = 6\gamma_{LV} \cos\theta$$

which is the work to wet the six faces of the cube. This could have been determined from the overall energy expenditure before and after submergence. However, the exercise of going through the separate steps is important to the understanding of the dispersion process and the difficulties that can be encountered.

The penetration of liquid into the channels between and inside agglomerates is more difficult to define. If such a channel is taken as a tube of radius r, the pressure required to force liquid into the tube is

$$P = \frac{-1}{r(2\gamma_{LV} \cos\theta)}$$

Penetration will be spontaneous if P is negative, and this will happen when $\theta < 90°$. However, air in the channel makes penetration difficult, and, in practice, spontaneous penetration and complete wetting will occur only when $\theta = 0°$.[45]

The rate of penetration is important because it may decide whether all surfaces are wetted during the dispersion process or just the outer areas of the agglomerates. The factors affecting penetration and its rate are shown in the following equation:[44]

$$L = \left[\frac{\gamma}{2\eta}(r\cos\theta)\,t\right]^{1/2}$$

where

L = depth of penetration into a tubular channel
r = radius of the channel
γ = surface tension of the dispersant
η = viscosity of the dispersant
θ = contact angle between dispersant and pigment
t = time of penetration

Rapid penetration is facilitated by loosely packed pigment (large r), a low contact angle (large $\cos\theta$), low dispersant viscosity, and high dispersant surface tension. The last factor may seem rather strange, as high liquid surface tension and low contact angle are not possible at the same time (since a high surface tension will give a high contact angle). In fact, high surface tensions tend to cause wetting and surface defect problems, so they rarely are formulated into coatings.

4.4. Pigment Wettability Testing

The critical surface tension of a pigment can be measured by compressing the pigment powder into a solid disk, then measuring contact angles of selected test liquids on the surface of that disk. The data are put into the form of a Zisman plot, and the critical surface tension is determined. Hansen[46] and co-workers have used this technique to considerable benefit. Table 7 lists two sets of critical surface tensions of pigments as determined by Hansen in our laboratories. One set was determined using alcohol–water mixtures, the other using a series of polar solvents. The results show that some pigments have quite high critical surface tensions and are easy to wet, whereas others have quite low values.

Cheever and Ulicny[47] also used Zisman plots, but developed the data for them in a different manner. They used results from flow measurements with a capillary bed liquid rise technique to calculate advancing contact angles of flowing liquids on pigment articles. Some of the critical surface tensions that they determined are listed in Table 8. We used their water and methylene iodide contact angle values to calculate solid surface tensions and dispersion and polar components via the Owens–Wendt equation. These data are also included in Table 8. There is no discernible relationship between the critical surface tension γ_c for these pigments and the solid surface tension γ_s. Some values are very close, others are not.

The polar components may be more interesting as they vary considerably from one pigment to another, and some correspond to relative polarities determined from the measurement of the heat of immersion of pigments in water (rutile TiO_2 and iron oxide are high, carbon black is low).[48]

No exact comparisons are possible between Tables 7 and 8, but red and yellow iron oxide would be expected to be similar and lampblack is one form of carbon

TABLE 7. Critical Surface Tensions of Some Pigments

	Critical surface tension (dynes/cm)	
Pressed dry pigment	Ethanol–water	Polar solvents
Toluidene red	31	44
Quinacridone red	33	38
Hansa yellow	33.5	38
Phthalo green	34	44
FGL yellow	35.5	38
Lampblack	>72.8	>72.8
Yellow iron oxide	>72.8	>72.8
Raw umber	>72.8	>72.8

TABLE 8. Critical Surface Tensions, Solid Surface Tensions, and Dispersion and Polar Components of Selected Pigments

Pigment	Critical surface tension γ_c	Solid Surface Tension and components γ_s	γ_s^d	γ_s^p
RGMC anatase TiO_2	42.9	39.9	38.8	1.1
R-CL6 rutile TiO_2	41.1	40.7	27.4	13.3
R-77 rutile TiO_2	35.6	27.7	16.3	11.4
Talc	48.0	42.9	42.2	0.7
Clay	29.2	31.3	23.5	7.9
Barytes $BaSO_4$	43.4	26.7	18.7	7.7
$CaCO_3$	25.1	33.1	22.9	10.2
Red iron oxide	28.0	32.9	20.0	12.9
Carbon black	40.0	37.0	34.3	2.7

Data from Ref. 47.

black. It is not known whether the great differences between the Hansen and Cheever results on these pigments reflect actual pigment differences (surface treatment, moisture content, etc.) or differences in the techniques used. More work needs to be done in this area.

4.5. Improvement of Pigment Wetting

The three main techniques used to achieve better pigment wetting and subsequent dispersion are

a. The use of wetting agents (surfactants) to lower the surface tension of the dispersant and the interfacial tension between the dispersant and the surface of the pigment
b. The use of dispersants containing block or graft copolymers with special anchoring groups that lower the interfacial tension and adsorb on the pigment surface
c. The modification of pigment surfaces to make the pigment more readily wetted and dispersed

A number of different surfactants and low-surface-tension solvents are used in pigment dispersants to lower interfacial tension and get better wetting. The surfactants most commonly used are straight-chain anionics that form micelles and give good surface tension lowering. Solvents can be very useful for lowering interfacial tension, especially in coatings where very little surfactant is permitted due to possible water sensitivity of the dried or baked film. However, the solvents cannot have too great an affinity for the pigment surface or they will replace the wetting agents and dispersants and give an unstable system.

There is considerable interest in pigment dispersants that are block or graft copolymers with anchors that adsorb onto the surface of the pigment and tails that stick out into the solvent.[49,50] The anchors tend to be insoluble in the solvent or have very low solubility. They must have very strong affinities for the particle surface. The anchors are strongly polar and for aqueous systems may be charged groups. Examples of anchor groups are amines, quaternary ammonium salts, ureas, carboxylic acids, sulfonates, and phosphates. Some are even large, flat dye molecules very similar to the pigments themselves. The tails are attached to the anchors, but must be soluble in the solvent and have little or no affinity for the pigment particle surface. In nonaqueous systems, the tails tend to be fatty groups, $C_{18}H_{37}$ or larger, and often polymeric such as fatty polyesters. Poly(isobutylene) and poly(styrene) also are used. Poly(ethylene oxide), poly(vinyl alcohol), and poly(vinyl pyrolidone) have been used in aqueous media.

There is no universal wetting or dispersing agent for every pigment. However, the correct choice of anchor groups for the given mix of pigments along with an adequate tail group can give better and more efficient wetting and dispersion compared with older types of dispersants.

Improved pigment dispersion also can be achieved by changing the surface of the pigment to make it easier to wet. The modification usually is done by the pigment manufacturer, and the purchaser has little idea of what has been done other than that the pigment is now labeled as an "easy-disperse" or "easy-wet" grade. Some information on surface treatment is available,[51–55] and pigment manufacturers sometimes will supply specially treated and characterized pigments for research purposes, but the paint technologist usually has little knowledge of the pigment surface treatment.

5. CONCLUSIONS

It is clear that wetting and wettability are very important to the performance of organic coatings. Poor wetting or dewetting from substrates can lead to surface defects that hurt both appearance and exterior durability. Good wetting is important for good appearance and adhesion, but does not guarantee the latter. Good pigment wetting helps give good color, smoothness, gloss, and a pleasing appearance.

A considerable amount of work has been done to develop wettability tests, but they are used mainly in the course of problem solving rather than in the design of coatings or the planning of substrate modifications. However, techniques are available to make surfaces more wettable, and additives can be used to enable paints and other coatings to wet better. Similar techniques and materials exist to improve pigment wetting. It is expected that these strategies will be augmented by new ones as research continues. More work is necessary because newer, more ecologically acceptable coating technologies, such as waterborne, high-solids, and powder coatings, tend to have more wetting and dewetting problems. In addition, manufacturers are turning more and more to tough, durable, corrosion-resistant substrates such as plastics and elastomers, which have surfaces that are very different from the more traditional steel, aluminum, and wood and are more difficult to wet.

Coatings technologists are beginning to realize the importance of surface science in general and wettability in particular. The future should see the introduction of better predictive tests and the use of results to design better coating–substrate systems.

6. REFERENCES

1. W. A. Zisman, *J. Paint Technol.* **44** (564), 41 (1972).
2. C. M. Hansen and P. E. Pierce, *Ind. Eng. Chem. Prod. Res. Dev.* **13**, 218 (1974).
3. C. K. Schoff and P. E. Pierce, in *Organic Coatings Science and Technology* (G. D. Parfitt and A. V. Patsis, eds.), vol. 7, p. 173, Marcel Dekker, New York (1984).
4. P. E. Pierce and C. K. Schoff, *Coating Film Defects*, Federation of Societies for Coatings Technology, Philadelphia (1988).
5. W. A. Zisman, in *Contact Angle, Wettability and Adhesion* (R. F. Gould, ed.), Adv. Chem. Ser., No. 43, Am. Chem. Soc., Washington, DC (1964).
6. E. J. Helwig, *J. Paint Technol.* **41** (529), 139 (1969).
7. G. D. Cheever, *J. Coat. Technol.* **58** (732), 37 (1986).
8. C. M. Hansen, *J. Paint Technol.* **44** (570), 57 (1972).
9. C. M. Hansen, *Färg Lack* **22** (11), 373 (1976).
10. C. M. Hansen, *Advances in the Technology of Solvents in Coatings*, Scandinavian Paint and Printing Ink Research Institute Technical Report T13-1977, Copenhagen (1977); also in *Proc. 14th FATIPEC Conf.*, p. 97, Budapest (1978).
11. S. Wu, *J. Colloid Interface Sci.* **71**, 605 (1979).
12. J. R. Dann, *J. Colloid Interface Sci.* **32**, 302 (1970).
13. L. A. Girifalco and R. J. Good, *J. Phys. Chem.* **61**, 904 (1957).
14. A. W. Neumann, R. J. Good, C. J. Hope, and M. Sejpal, *J. Colloid Interface Sci.* **49**, 291 (1974).
15. F. M. Fowkes, *Ind. Eng. Chem.* **56** (12), 40 (1964).
16. J. R. Dann, *J. Colloid Interface Sci.* **32**, 321 (1970).
17. D. K. Owens and R. D. Wendt, *J. Appl. Polym. Sci.* **13**, 1741 (1969).
18. D. H. Kaelble, *J. Adhesion* **2**, 66 (1970).
19. J. Panzer, *J. Colloid Interface Sci.* **44**, 141 (1973).
20. J. K. Spelt and A. W. Neumann, *J. Colloid Interface Sci.* **122**, 294 (1988).
21. T. Imai, in *Organic Coatings Science and Technology* (G. D. Parfitt and A. V. Patis, eds.), vol. 6, p. 301, Marcel Dekker, New York (1984).
22. C. J. van Oss, M. K. Chaudhury, and R. J. Good, *Chem. Rev.* **88**, 927 (1988).
23. C. M. Hansen, *J. Paint Technol.* **42** (550). 660 (1970).
24. E. Wallstrom, I. Svenningsen, and C. M. Hansen, *Karakterisering av Ytor Metodstudie*, Technical Report 5-82, Scandinavian Paint and Printing Ink Research Institute, Horsholm, Denmark (1982).
25. C. M. Hansen and E. Wallstrom, *J. Adhesion* **15**, 275 (1983).
26. J. H. Hildebrand and R. L. Scott, *Solubilities of Non-Electrolytes*, 3rd ed., Reinhold, New York (1950).
27. L. H. Lee, *J. Paint Technol.* **42** (545), 365 (1970).
28. A. Beerbower, *J. Colloid Interface Sci.* **35**, 126 (1971).
29. A. F. M. Barton, *J. Adhesion* **14**, 33 (1982).
30. A. F. M. Barton, *CRC Handbook of Solubility Parameters and Other Cohesion Parameters*, p. 415, CRC Press, Boca Raton, FL (1983).
31. M. Van den Tempel, in *Proc., Scandinavian Symp. on Surface Activity*, p. 306, Stockholm (1965).
32. M. Van den Tempel, *J. Non-Newtonian Fluid Mech.* **2**, 205 (1977).
33. E. H. Lucasson-Reynders, in *Anionic Surfactants—Physical Chemistry of Surfactant Action* (E. H. Lucasson-Reynders, ed.), Surfactant Science Series, vol. 11, p. 173, Marcel Dekker, New York (1981).
34. R. E. Smith, *Ind. Eng. Chem. Prod. Res. Dev.* **22**, 67 (1983).
35. R. M. Lukes, *Prepr. Div. Org. Coat. Plastics Chem., Am. Chem. Soc.* **41**, 7 (1979).
36. P. Fink-Jensen, *Färg Lack* **8**, 5–14, 39–49 (1962); *Farbe Lack* **68**, 155 (1962).
37. Norwegian Paint and Varnish Chemists' Society Research Committee, *Färg Lack* **12** (2), 35 (1966); **12** (3), 51 (1966).
38. C. R. Hegedus and D. J. Hirst, *Met. Finish.* **86** (7), 39 (1988).
39. C. K. Schoff and P. E. Pierce, in *Proc. 9th Int. Conf. in Coatings Science and Technology* (G. D. Parfitt and A. V. Patsis, eds.), p. 239, Athens, Greece (1983).
40. T. Watanabe Y. Shindou, T. Shiota, K. Yamato, and S. Nomura, in *Proc. Int. Conf. on Zinc and Zinc*

Alloy Coated Steel Sheet (GALVATECH '89) (Y. Misamatsu, ed.), p. 80, The Iron and Steel Institute of Japan, Tokyo (1989).
41. J. P. Goulin, *Double Liaison—Chim. Peintures* **31** (349), 512 (French), xxi (English) (1984).
42. G. D. Parfitt, in *Proc. Third Int. Conf. in Organic Coatings Science and Technology* (G. D. Parfitt and A. V. Patsis, eds.), vol. 1, p. 95, Technomics, Westport, CT (1970).
43. G. D. Parfitt, *Dispersion of Powders in Liquids*, 2nd ed., Wiley, New York (1973).
44. T. C. Patton, *Paint Flow and Pigment Dispersion*, 2nd ed., Wiley, New York (1970).
45. P. M. Heerthes and W. C. Witvoet, *Powder Technol.* **3**, 339 (1970).
46. A. Saarnak and C. M. Hansen, in *Prepr. Div. Polym. Mater.-Sci. Eng. Am. Chem. Soc.* **51**, 698 (1984).
47. G. D. Cheever and J. C. Ulicny, *J. Coat. Technol.* **55** (697), 53 (1983).
48. J. Schröder, *Prog. Org. Coat.* **12**, 339 (1984).
49. H. Jakubauskas, in *Proc. 11th Int. Conf. in Organic Coatings Science and Technology* (A. V. Patsis, ed.), p. 77, Athens (1985); also *J. Coat. Technol.* **58** (736), 71 (1986).
50. W. Kurtz, *Am. Ink Maker* **65** (6), 21 (1987).
51. K. Merkle and H. Schafer, in *Pigment Handbook* (T. C. Patton, ed.), vol. 3, p. 157, Wiley, New York (1973).
52. H. S. Ritter, in *Pigment Handbook* (T. C. Patton, ed.), vol. 3, p. 169, Wiley, New York (1973).
53. A. Topham, *Progr. Org. Cont.* **5**, 237 (1977).
54. B. G. Hays, *Am. Ink Maker* **62** (6), 28 (1984).
55. T. G. Vernardakis, in *Coating Technology Handbook* (D. Satas, ed.), p. 540, Marcel Dekker, New York (1991).

15

Spreading on Liquids: Effect of Surface Tension Sinks on the Behavior of Stagnant Liquid Layers

Eli Ruckenstein, D. G. Suciu, and O. Smigelschi

1. INTRODUCTION

Regions of low surface tension, hence surface tension sinks, that appear naturally or are generated artificially on liquid–gas or liquid–liquid interfaces constitute sources of surface motions known as the *Marangoni effect*. This effect is a result of the tendency of an interface to relax into a state of minimum free energy through the expansion of the regions of low interfacial tension and the contraction of those with a high interfacial tension. This surface (interfacial) tension gradient generates a shear stress, which in turn triggers motions along the interface. Zisman and co-workers[1, 2] have suggested a method for cleaning solid surfaces covered by relatively thin liquid layers by using surface tension sinks generated by dissolution.

Surface movements generated by the Marangoni effect have long been known to accompany occasionally operations such as liquid–liquid extraction, absorption of vapors in liquids, etc. The effect of a surface tension gradient on mass transfer is particularly important when it generates a spontaneous agitation near the interface known as interfacial turbulence. The occurrence of this kind of turbulence in the vicinity of the interface between two phases has been explained by Sternling and Scriven[3] as being a result of a hydrodynamic instability. They have identified the conditions under which small perturbations present near liquid–fluid interfaces are amplified or damped by the Marangoni effect. Numerous papers have since studied these phenomena from all angles, both theoretically and experimentally.[4] The onset of surface movements in an initially quiescent system was also treated by Dijkstra and van de Vooren.[5] Instead of the Sternling and Scriven steady linear

Eli Ruckenstein • Department of Chemical Engineering, State University of New York at Buffalo, Buffalo, New York 14260 *D. G. Suciu* • Lummus Crest, Inc., Bloomfield, New Jersey *O. Smigelschi* • Lummus Crest, GmbH, Wiesbaden, Germany.

Modern Approaches to Wettability: Theory and Applications, edited by Malcolm E. Schrader and George Loeb. Plenum Press, New York, 1992.

concentration distribution in the normal direction to the interface as the basic state, these authors accounted for the change in time of the concentration.

The advent of processing technologies under microgravity conditions has spurned new interest in the study of the causes and extent of surface movements due to the Marangoni effect. When the gravity force is small, these surface forces can become extremely important.[6] The motions created by the Marangoni effect due to temperature nonuniformities in a curved gas–liquid interface have been studied recently, experimentally, by a Schlieren technique by Raake *et al.*[7]

The radial surface motions obtained when a surface tension sink is generated by continuously dissolving a solute or by producing an area of higher temperature at the surface of a liquid layer were investigated by Suciu, Smigelschi, and Ruckenstein.[8–11] The starting point of their experiments was the observation that during the reciprocal dissolution of two partially miscible liquids (for instance, isobutanol and water), conditions for the occurrence of the Marangoni effect may appear. In an open dish a thin layer of isobutanol was placed above a layer of water. The thickness of the former layer decreased continuously as a result of the combined effects of evaporation and dissolution. At a given moment, the continuous film broke down, and some regions of the supporting liquid surface became free of isobutanol. Thus, zones appeared, across which surface tension gradients were present. As a result, rapid movements due to the Marangoni effect were taking place, which led to an intense agitation similar to that in a boiling liquid. This agitation ceased as soon as the entire film dissolved. This experiment was simplified by controlling the fraction of the supporting surface covered by the film by feeding continuously one liquid over the surface of another one into which it dissolves.

This chapter is a review of the work by Suciu, Smigelschi and Ruckenstein[8–11] regarding the hydrodynamic phenomena that occur when a single point surface tension sink that generates the Marangoni effect is employed. The first two sections show the differences between the flows induced within thin and thick liquid layers. The rate of the mass transfer between a gas and a liquid surface set in motion by the Marangoni effect will then be calculated and the results compared with measurements. The performed experiments allowed us to increase the scale of the movements to dimensions at which the distribution of surface velocities could be measured and the subsurface flows visualized. The theoretical study allowed us to develop a model for the mechanism of surface spreading and to predict the surface velocity field and the mass transfer it produced between a gas and a liquid phase.

2. EFFECT OF SURFACE TENSION SINKS ON STAGNANT LIQUID FILMS

2.1. Experimental

Small quantities of the liquid to be examined were introduced into an open glass dish having a flat horizontal bottom of circular shape (11 cm diam.), so as to form thin continuous films reaching up to the walls of the dish.[11] Into the middle region of the film surface, solutions that lowered the surface tension of the film liquids were continuously fed through a small capillary (0.5 mm o.d.) placed

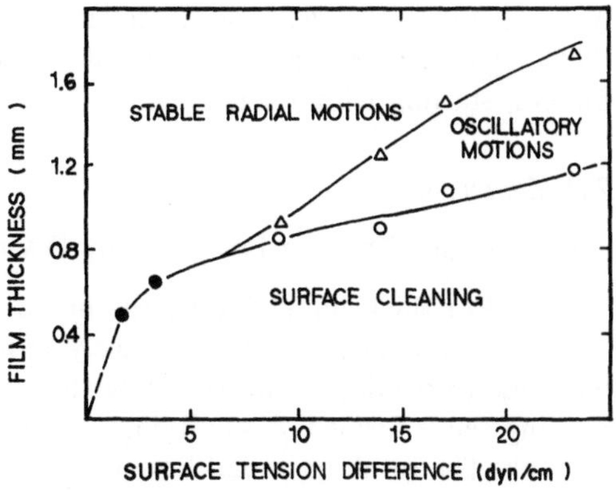

FIGURE 1. Behavior of ethylene glycol films.[11] Surface tension sink induced by 10 mm^3/min isobutanol–ethylene glycol solutions fed at the interface: ○, transition from surface cleaning to oscillatory motions; △, transition from oscillatory to stable motions; ●, transition from surface cleaning to stable motions.

precisely in contact with the film surface. A microsyringe was used for this purpose; its piston was mechanically driven so as to deliver a small constant flow rate. For each concentration of the solution various thicknesses of the film were used, and in each instance the film behavior was noted. In another set of experiments, the microsyringe was replaced by an electrically heated cylindrical copper rod of 3 mm diameter, which was placed (from above) in contact with the free surface of the film. At the very end of the rod, thin constantan and copper wires were welded flush with its flat surface. The temperature of the end of the rod could thus be measured. For each of a number of rod temperatures the film behavior was recorded as a function of its thickness. Film thicknesses were measured by means of a micrometer provided with a sharp needle. The thickness measurements were reproducible within 0.02 mm.

2.2. *Results*

The experimental results show that the film behavior is mainly determined by its thickness and by the surface tension difference ($\Delta\sigma$) that generates the motion. This, in its turn, is determined by the difference between the concentration or temperature values prevailing at the sink and those outside it. The behavior of ethylene glycol is typical in this respect and is plotted in Fig. 1 (surface tension sink obtained by ethylene glycol solutions containing isobutanol as solute) and in Fig. 2 (surface tension sink obtained by heating).

Three regions are evidenced: (1) At small thicknesses surface cleaning takes

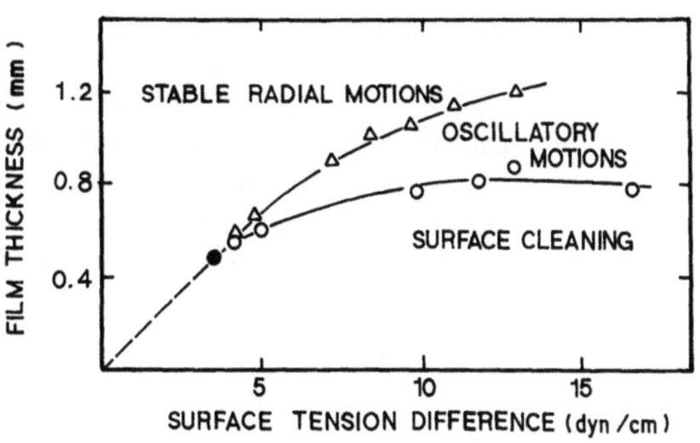

FIGURE 2. Behavior of ethylene glycol films.[11] Surface tension sink induced by heating the surface with 3-mm copper rod: ○, transition from surface cleaning to oscillatory motions; △, transition from oscillatory to stable motions; ●, transition from surface cleaning to stable motions.

place at all $\Delta\sigma$. (2) At large thicknesses the film is stable at all $\Delta\sigma$. (3) Between these regions, a region characterized by periodic (oscillatory) surface motions is present, starting with a critical $\Delta\sigma$ value.

2.3. *Discussion*

2.3.1. Surface Tension Lowering by Dissolution

(a) For thick liquid layers, the liquid entrained centrifugally due to the drag exerted by the surface layers is easily replaced near the sink by liquid elements that, due to continuity, move below the surface toward the sink. In this manner a rotating cell is formed. A small depression of the liquid surface may be observed at the sink. The resulting radial flow is stable.

With decreasing film thicknesses, the centripetal flow will become slower due to the retarding influence of the solid bottom of the cell.

(b) When the film thickness is sufficiently reduced, a certain depth is reached at which the balance between the flow rate of the liquid leaving the sink at the surface and that coming toward it near the bottom may no longer be maintained, since the first exceeds the latter. The film thickness around the sink becomes smaller, a depression being formed and the liquid breaks away from the capillary tip. From this moment on, the liquid fed does not reach the film surface for a certain short time. The film will regain its flat form after the solute existing on the surface at the separation moment has dissolved. By doing so, it again comes in contact with the capillary tip; the surface tension sink is re-formed, and the whole process begins anew. This cycle repeats itself for quite long periods.

(c) At still smaller film thicknesses, when the depression near the sink is formed, liquid layers rich in solute may reach the solid bottom of the cell even after the contact between the film and the capillary tip has ceased. These solute layers displace the film liquid, and the film is broken. It may recover its continuity after quite long periods, depending on the volatility of the adsorbed layer of the low-surface-tension component. As seen from Fig. 1, at constant flow rate, with increasing $\Delta\sigma$ thicker films may be displaced; also, at constant $\Delta\sigma$ the same effect is obtained with increasing flow rates (Table 1).

2.3.2. Surface Tension Lowering by Heating

(1) For thick liquid layers no difference from situation (a) of Section 2.3.1 is observed at moderate temperatures of the heated rod. Nevertheless, with rising rod temperature heat is transferred to the liquid not only by conduction but also

TABLE 1. Variation of the Film Thickness (at which the transition from surface cleaning to oscillatory motion occurs) with Solute Flow Rate ($\Delta\sigma = 17.1$ dyne/cm)

Solute flow rate (mm^3/min)	Film thickness at transition (mm)
4	0.95
10	1.1
16	1.4

increasingly by radiation. If the examined liquid is relatively viscous (paraffin oils with viscosities of 2 to 5 P at 25°), a depression in the air–liquid interface is easily observed at any thickness of the layer for high-enough rod temperatures, even if the rod does not touch the liquid surface.

For the system geometry adopted here, the climbing of the liquid onto the heated rod due to capillarity exerts a stabilizing effect on the film. In the absence of capillarity no stable flow regime could be obtained, since the smallest depression of the surface level at the sink would be sufficient to break the contact between the rod and the film.

(2) The oscillatory behavior of the system results directly from the description made in paragraph (b) of Section 2.3.1.

(3) For thin films surface cleaning may be explained as in paragraph (c) of Section 2.3.1. However, the similarity with the situation of paragraph (c) is due to the presence of radiation from the rod and evaporation from the heated region of the film. Indeed, if thermal radiation were not present, after the cleaning of the solid bottom the very thin liquid layer adhering to it would cool off and the driving force for the motion would vanish. Liquid from around would fill the depression, and the contact with the heated rod would be reestablished. (Such a behavior could be obtained in situation (c) if the solute were very volatile and its feed rate low.)

The above considerations are also valid for the system geometry used by Mitchell and Quinn.[12] These authors have observed oscillatory motions from and toward the surface tension sink when a thin liquid layer was continuously heated from below by a point source. Such motions were recorded previously by Linde, Schwarz, and Gröger[13] for the case of multiple surface tension sinks appearing spontaneously on liquid–gas interfaces. One may remark that in the experiments of Mitchell and Quinn the heat supplied by the source placed at the solid–liquid interface had to reach the air–liquid interface by means of conduction and/or convection through the liquid film, before the steady or oscillatory motions may start. Permanent surface cleaning may be obtained in their experiments with greater ease than in the ones described above, since no cooling of the cleaned solid surface can take place.

Though the same liquid was used (in Figs. 1 and 2) in forming the film, the curves obtained are not superimposable. This may be understood if one considers that there are differences of order of magnitude between the diffusion coefficient and the thermal diffusivity. Also, the viscosity of the liquid film is much more sensitive to temperature than to concentration differences.

An increase in the flow rate of the low-surface-tension liquid corresponds to an increase in the diameter of the heated rod. The lack of equivalence between the values of the flow rate and rod diameter used in the respective measurements could also explain the differences between Figs. 1 and 2.

3. STUDY OF THE PURE MARANGONI EFFECT

If a small drop of a soluble or partially soluble liquid having a smaller surface tension than that of water is located on a free water–air interface, a velocity field develops because of the surface tension gradient. This is the simplest manifestation of the Marangoni effect. A quantitative study of the primary phenomenon, namely of the rate of spreading of the low-surface-tension regions over the regions having a larger one, constitutes the objective of this section. The situation described above is, however, an unsteady one, and it is difficult both to theoretically treat the unsteady velocity field at the interface and to measure it experimentally. For this reason we have modified the physical situation by continuously feeding the "drop." In this manner, a quasi-steady velocity field was sustained outside the drop, which can be treated theoretically and experimentally in a simple manner.

3.1. The Steady Dissolving Drop

3.1.1. Description of the Method

The low-surface-tension spreading liquid was fed at a small constant flow rate through a capillary that had its tip about 0.5 mm above the surface of the high-surface-tension liquid (water placed in a flat dish). In most of the quantitative work reported here, isobutanol was the spreading liquid.

If the isobutanol is fed at the proper flow rates, it spreads over the water surface and forms a steady film around the capillary tip with a well-defined circular contour (dynamic extended lens) (Fig. 3). As shown in Section 5, the dimension of this dissolving drop increases linearly with the isobutanol flow rate. This means that the dissolution process takes place predominantly along the leading edge of the extended lens, and to a much lesser extent through the drop–water interface. Consequently, the dissolution rate per unit length of the drop contour is independent of the feed flow rate.

The motions induced outside the extended lens by the surface tension gradient penetrate deep into the water layer, carrying with them the dissolved isobutanol (Fig. 3). Since the quantity of feed dissolved during one experiment is smaller than 1/2000 of that of water, it is reasonable to assume that the water entrained at the leading edge of the drop by the spreading layer is practically free of isobutanol and that the concentration field near the free interface is not affected by the accumulation of isobutanol in water. Consequently, a quasi-steady movement of the water surface around the drop takes place radially. Information concerning this velocity field will be obtained experimentally as well as theoretically.

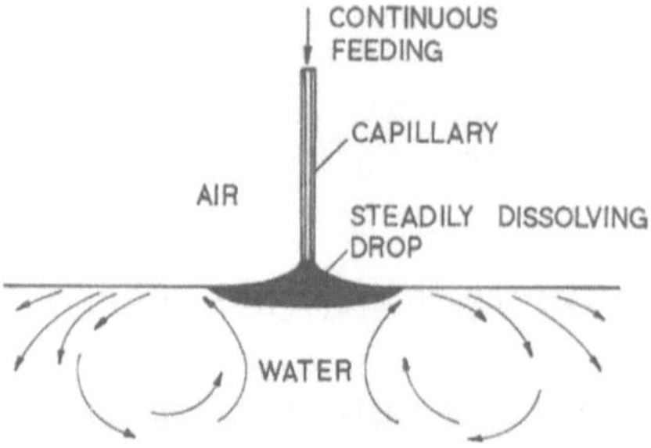

FIGURE 3. Scheme of the experimental arrangement.

3.1.2. Experimental Arrangement

The experiments were performed in a prismatic, transparent, open dish 300 mm by 300 mm by 80 mm. In most of the experiments, the film-supporting liquid was tap water; no differences have been observed by using water distilled from alkaline–potassium permanganate solutions. In some of the tests, water containing various concentrations of isobutanol formed the supporting liquid. The liquid to be dissolved was allowed to flow over the supporting liquid surface from a glass capillary ($D = 1.2$ mm; I.D. $= 0.6$ mm) placed immediately above the water surface and connected flexibly to a constant-flow-graduated buret.

The dish and glass capillary were placed in the optical field of a vertically mounted Schlieren device. During the experiments the dish was open, and the concentration in water of the organic liquids used did not vary by more than 0.1%. For this reason, the dissolution may be considered as proceeding under quasi-stationary conditions.

3.1.3. Film Velocity Measurements

The dish, capillary, and buret were the same as above. On the film, as near the capillary as possible, tracers (tin aluminum disks, 2 mm diam. and 0.01 mm thick, having good light reflecting properties) were placed, and their radial movement was photographed by an Exacta Varex camera placed above the capillary.

Between the camera and the film, a blackened aluminum disk with a 20-cm diameter and a slot sector over 36° of arc was rotated at a suitable known and constant speed. The camera shutter was set at 3-s exposure time, and a tracer was placed on the film. The tracer was viewed by the photographic film only when the disk slot was in front of the camera's objective. The desired illumination was achieved by using three 100-W mirror bulbs.

In this manner a series of images of the tracer traveling across the film were obtained on the same photograph. The distances between the capillary and the successive images of the tracer, read on the positive, were plotted versus time (obtained from the known rotation velocity of the slotted disk). Each of the velocity values reported here represent the slope of this curve (obtained graphically) in the points corresponding to the indicated r values.

A qualitative description of the dissolving process and of the movements produces by it is as follows. A liquid is fed continuously at a point of the surface of a layer of another liquid in which it dissolves. The following description applies to the case when the dissolving liquid forms a "surface drop" or "dissolving drop" as a separate phase on the solvent's surface, around the feeding point (Fig. 3). [In some liquid pairs such a definite contour is not present. In such cases, the transition from the solute-rich surface phase to the solution is more gradual and is observable only by a sudden jump in surface velocity (see Section 5).] The spreading liquid travels radially from the feeding point toward the contour of the surface drop with velocities of the order of centimeters per second.[9] Close to the contour, the thickness of the surface drop decreases to a value δ, of a scale at which the surface forces are acting. As the dissolving liquid approaches the contour, its velocity increases correspondingly. At the contour, the dissolving process takes place. Immediately outside the contour, the radially moving surface of the liquid has velocities of the order of tens of centimers per second. In order to explain this sudden jump in the surface velocity, it is plausible to assume that, in the vicinity of the contour of the surface drop, an acceleration zone that has a very small but finite width ε exists, characterized by strong surface forces (i.e., large radial surface tension gradients).

The model for the dissolving of the surface drop is depicted in Fig. 4. The surface velocity profile, which is generated by the dissolving process and inferred from the above considerations, is also shown in Fig. 4. From the measured solute flow rates, lengths of the contour of the surface drops and radial surface velocities immediately outside the contour, one calculates that δ is about 10^{-4} cm.[14]

3.1.4. Theoretical

The dissolution process takes place at the very leading edge of the dynamic drop. The "film" that begins to spread outward from this leading edge is assumed to consist of water saturated with isobutanol. It forms a ring of small thickness δ around the drop. This thickness cannot, however, be predicted from macroscopic considerations. The isobutanol in the spreading layer dissolves into the main body of water, the concentration field of isobutanol satisfying the convective–diffusion equation. The evolution of this thin layer as it moves along the air–water interface will be described by two extreme models: (a) the layer forms a single phase with the supporting liquid; (b) it forms a quasi-distinct phase that exchanges mass with the supporting liquid.

The spreading of the film is due to the surface tension gradient (Marangoni effect). From a macroscopic point of view the surface tension gradient produces a shear stress at the interface that induces motion in the interface and the liquid layers in its vicinity. The hydrodynamics of this process may be described by the Navier–Stokes equations. The most important boundary condition is provided by the balance between the radial surface tension gradient (dependent on the concentration gradient along the interface) and the shear stress. For this reason the convective diffusion equation must be solved together with the Navier–Stokes

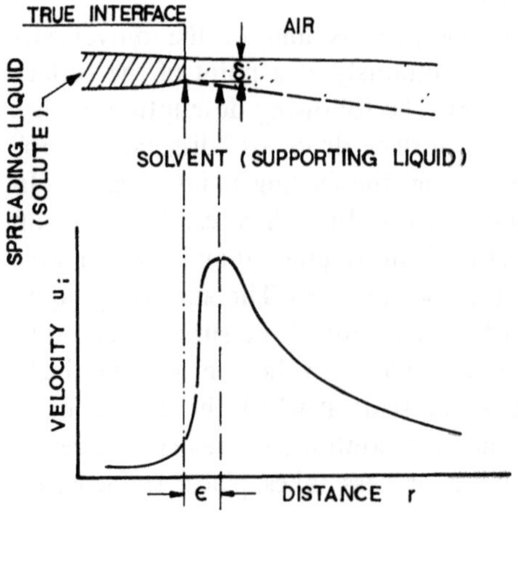

FIGURE 4. The dissolving process under the action of surface forces. Here δ is the initial thickness of the moving layer, and ϵ is the thickness of the region of rapid increase of the surface velocity.

equations subjected to the above boundary condition. For the analysis of the process it is therefore necessary to account for the interaction between the hydrodynamic and dissolution processes, arising because of the Marangoni effect.

Model (a). The Spreading Layer Forms a Single Phase with the Supporting Liquid. Assuming cylindrical symmetry, the convective–diffusion equation describing the dissolution process has the form (for notations see table of notations at the end of the chapter)

$$u\frac{\partial c}{\partial r}+v\frac{\partial c}{\partial y}=D\left[\frac{\partial^2 u}{\partial y^2}+\frac{1}{r}\frac{\partial}{\partial r}\left(r\frac{\partial c}{\partial r}\right)\right] \tag{1}$$

while the hydrodynamic equations may be written as

$$u\frac{\partial u}{\partial r}+v\frac{\partial u}{\partial y}=\nu\left(\frac{\partial^2 u}{\partial y^2}+\frac{\partial^2 u}{\partial r^2}+\frac{1}{r}\frac{\partial u}{\partial r}-\frac{u}{r^2}\right) \tag{2}$$

$$\frac{1}{r}\frac{\partial}{\partial r}(ru)+\frac{\partial v}{\partial y}=0 \tag{3}$$

The boundary conditions are

$$r=r_0 \quad u=0 \quad \text{and} \quad c=\begin{cases} c_0 & 0<y<\delta \\ 0 & \delta<y<\infty \end{cases} \tag{4}$$

$$y=0 \quad v=0; \qquad \frac{\partial u}{\partial y}=-\frac{1}{\eta}\frac{\partial \sigma}{\partial r}; \qquad \frac{\partial c}{\partial y}=0 \tag{5}$$

$$y\to\infty \quad u=v=0; \qquad c=0 \tag{6}$$

The boundary conditions (4) for the concentration are a consequence of the physical model (a), while one of the boundary conditions (5) results from the Marangoni effect.

Since we are interested in obtaining only the velocity distribution along the free interface, several approximations in the above equations will be made. First $u(\partial c/\partial r)$ and $u(\partial u/\partial r)$ will be replaced by $u_i(\partial c/\partial r)$ and $u_i(\partial u/\partial r)$. Because $v=\partial c/\partial y=0$ at the free interface, the terms $v(\partial c/\partial y)$ and $v(\partial u/\partial r)$ will be neglected. In the right side of Eqs. (1) and (2) only the terms $\partial^2 c/\partial y^2$ and $\partial^2 u/\partial y^2$ are retained. Equations (1) and (2) become

$$u_i\frac{\partial c}{\partial r}=D\frac{\partial^2 c}{\partial y^2} \tag{7}$$

$$u_i\frac{\partial u}{\partial r}=\nu\frac{\partial^2 u}{\partial y^2} \tag{8}$$

The evaluations made by means of the resulting equations indicate that indeed the mentioned approximations are reasonable except for the region very near to the leading edge of the steady dissolving drop ($r \cong r_0$).

Introducing the variable

$$t = \int_{r=r_0}^{r} \frac{dr}{u_i} \tag{9}$$

we can rewrite the above equations as

$$\frac{\partial c}{\partial t} = D \frac{\partial^2 c}{\partial y^2} \tag{10}$$

$$\frac{\partial u}{\partial t} = \nu \frac{\partial^2 u}{\partial y^2} \tag{11}$$

Equations (10) and (11) must be solved for the boundary conditions

$$t=0 \quad u=0 \quad \text{and} \quad c = \begin{cases} c_0 & 0 < y < \delta \\ 0 & \delta < y < \infty \end{cases} \tag{12}$$

$$y=0 \quad \frac{\partial u}{\partial y} = -\frac{1}{\eta u_i}\frac{\partial \sigma}{\partial t} \quad \text{and} \quad \frac{\partial c}{\partial y} = 0 \tag{13}$$

$$y \to \infty \quad u=0 \quad c=0 \tag{13'}$$

The first boundary condition in (13) may be written

$$\text{for } y=0 \qquad \frac{\partial u}{\partial y} = -\frac{1}{\eta u_i}\frac{d\sigma}{dc}\frac{\partial c}{\partial t} = -\frac{F(t)}{u_i} \tag{14}$$

The solution of Eq. (10) for the boundary conditions (12), (13), and (13′) has the form

$$c = \frac{c_0}{2}\left[\operatorname{erf}\frac{\delta - y}{2\sqrt{Dt}} + \operatorname{erf}\frac{\delta + y}{2\sqrt{Dt}}\right] \tag{15}$$

Consequently,

$$c_{y=0} = c_0 \operatorname{erf}\frac{\delta}{2\sqrt{Dt}} \tag{16}$$

$$\left(\frac{\partial c}{\partial t}\right)_{y=0} = -\frac{c_0 \delta}{2\sqrt{\pi D}} t^{-3/2} e^{-\delta^2/4Dt} \tag{17}$$

and the boundary condition (14) becomes

$$\left(\frac{\partial u}{\partial y}\right)_{y=0} = \frac{1}{\eta}\frac{d\sigma}{dc}\frac{1}{u_i}\frac{c_0\delta}{2\sqrt{\pi D}} t^{-3/2} e^{-\delta^2/4Dt} = -\frac{F(t)}{u_i} \tag{18}$$

The solution of Eq. (11) for the boundary conditions (12), (13′), and (18) can be obtained by means of the Duhamel Theorem:[15]

$$u = \left(\frac{\nu}{\pi}\right)^{1/2} \int_0^t \frac{F(t-\lambda)}{u_i(t-\lambda)} e^{-y^2/4\nu\lambda} \frac{d\lambda}{\lambda^{1/2}}$$

Consequently, the velocity distribution along the free interface is

$$u_i = \left(\frac{\nu}{\pi}\right)^{1/2} \int_0^t \frac{F(t-\lambda)}{u_i(t-\lambda)} \frac{d\lambda}{\lambda^{1/2}} \tag{19}$$

This integral equation was solved numerically.

Model (b). The Spreading Layer Forms a Distinct Phase. The second model is that of a spreading film (consisting of the component dissolved at the leading edge of the dynamic drop and liquid entrained from the bulk near the leading edge) representing a distinct phase. The theoretical treatment is more intricate because it is necessary to write separate convective diffusion equations and hydrodynamic equations for the two "phases" (the spreading film and the bulk of the supporting liquid). However, because the thickness of the film is very small, one may assume that both the concentration and the velocity in the film depend only on r. Assuming also that δ is a constant along the radius r, only some of the boundary conditions for the concentration in the supporting liquid differ from the ones valid for model (a). In this case the equation

$$\frac{\partial c}{\partial t} = D \frac{\partial^2 c}{\partial y^2}$$

describing the concentration field in the bulk of the supporting liquid is solved for the boundary conditions

$$t = 0 \quad c = 0 \quad c_i = c_{i,0} \tag{20}$$

and

$$y = 0 \qquad \delta \frac{dc_i}{dt} = D \frac{\partial c}{\partial y} \tag{21}$$

$$y \to \infty \qquad c = 0 \tag{22}$$

Boundary condition (21) takes into account the mass transfer from the film to the supporting liquid. As mentioned, the resistance to diffusion in the film is considered negligible. The distance y is measured in Eq. (21) from the interface between the film and the supporting liquid; c_i represents the concentration of isobutanol in the film.

For the concentration one obtains in this case

$$c = c_{i,0} \exp\left(\frac{y}{\delta} + \frac{Dt}{\delta^2}\right) \operatorname{erfc}\left\{\frac{y}{2\sqrt{Dt}} + \frac{\sqrt{Dt}}{\delta}\right\} \tag{23}$$

$$c_i = c_{i,0} \exp\left(\frac{Dt}{\delta^2}\right) \operatorname{erfc} \frac{\sqrt{Dt}}{\delta} \tag{24}$$

and, therefore,

$$\frac{dc_i}{dt} = c_{i,0}\left[\frac{D}{\delta^2} \exp\left(\frac{Dt}{\delta^2}\right) \operatorname{erfc} \frac{\sqrt{Dt}}{\delta} - \frac{1}{\delta}\left(\frac{D}{\pi t}\right)^{1/2}\right] \tag{25}$$

The velocity field in the supporting field is described by Eq. (11) and the boundary conditions

$$t = 0 \qquad u = 0$$

$$y = 0 \qquad \frac{\partial u}{\partial y} = -\frac{1}{\eta u_i} \frac{d\sigma}{dc} \frac{dc_i}{dt}$$

$$y \to \infty \qquad u = 0$$

For the velocity at the interface, Eq. (19) again is obtained. However, the function $F(t)$ defined by Eq. (14) differs in this case from the preceding one, being given by Eq. (26):

$$F(t) = \frac{1}{\eta} \frac{d\sigma}{dc} c_{i,0}\left[\frac{D}{\delta^2} \exp\left(\frac{Dt}{\delta^2}\right) \operatorname{erfc} \frac{\sqrt{Dt}}{\delta} - \frac{1}{\delta^2}\left(\frac{D}{\pi t}\right)^{1/2}\right] \tag{26}$$

3.1.5. Comparison with Experiment

As mentioned, in order to obtain the velocity distribution of the surface of the high-surface-tension liquid, small tracers (aluminum disks having 2 mm diameter and 0.01 mm thickness) were placed on the moving water surface and their progress photographed by means of a still camera, using a stroboscopic method. From the

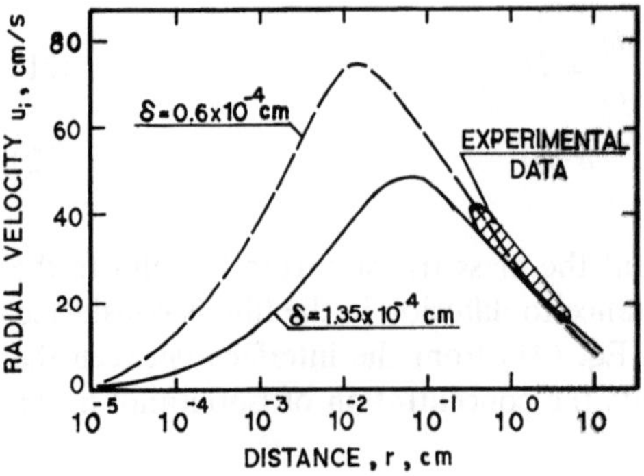

FIGURE 5. Theoretical velocity distribution for the case of pure isobutanol–pure water system.[10]

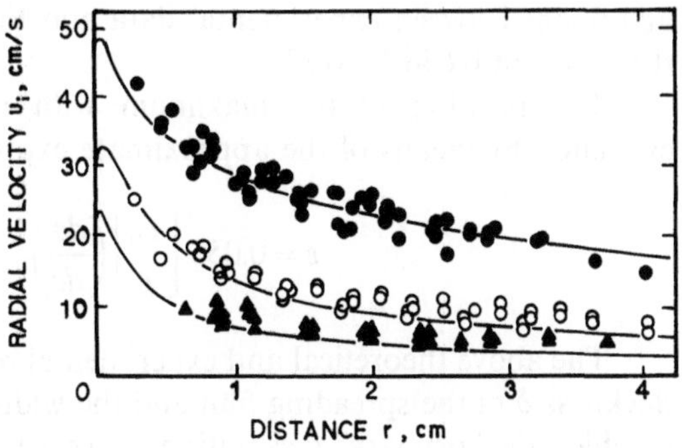

FIGURE 6. Comparison between theoretical and experimental velocity distributions computed with $\delta = 1.35 \times 10^{-4}$ cm.[10] Systems: ●, pure isobutanol–pure water; ○, pure isobutanol–aq. 2.8 wt.% isobutanol sol.; ▼, pure isobutanol–aq. 5.25 wt.% isobutanol sol.; —, theoretical curves (computed with $\delta = 1.35 \times 10^{-4}$ cm).

photographs, plots of the radial distance versus time were obtained for each tracer. By graphical differentiation of these curves, the radial distribution of the surface velocity was obtained.

Theoretical velocity distributions based on model (a) are compared with experimental data in Figs. 5 and 6. The velocity distribution obtained by means of model (b) is practically the same.

An interesting qualitative result from the theoretical equation is that the velocity increases very rapidly near the leading edge, attains a maximum, and then decreases much slower. A simple equation for the position of the maximum cannot be obtained easily from the theoretical equation. The following approximate equation, obtained by assuming that up to the maximum $F(t)/u_i(t) \cong F(t)_M/u_{iM} = \text{const.}$, enables one to compute the maximum velocity:

$$u_{iM} = 0.656 \left[\frac{c_0}{\rho\delta} \left| \left(\frac{d\sigma}{dc} \right)_{c=c_0} \right| \right]^{1/2} \left(\frac{D}{\nu} \right)^{1/4} \tag{27}$$

In Table 2, the velocities obtained by means of Eqs. (19) and (27), for $\delta = 1.3 \times 10^{-4}$ cm, are compared with those obtained by extrapolating to the origin ($r = r_0$) the experimental velocity versus radius data. One may notice that the

TABLE 2

Spreading liquid	Supporting liquid	Maximum velocity u_{iM} (cm/s) Eq. (19)	Eq. (27)	Experimental
Pure isobutanol	Pure water	48	30.2	48
Pure isobutanol	Water with 1.35 wt.% isobutanol	—	27.8	38.5
Pure isobutanol	Water with 2.8 wt.% isobutanol	35	24.8	32
Pure isobutanol	Water with 5.25 wt.% isobutanol	24	18.5	22
Water with 8.35 wt.% isobutanol	Pure water	48	30.2	40
Water with 4.6 wt.% isobutanol	Pure water	—	33.2	41
Water with 2.3 wt.% isobutanol	Pure water	—	33	45
Water with 1.1 wt.% isobutanol	Pure water	—	31.8	45
Water with 0.35 wt.% isobutanol	Pure water	—	25.3	28

agreement between experimental data and Eq. (27) may be improved by replacing the constant 0.656 by 0.87.

The position of the maximum with respect to the leading edge may be evaluated by means of the approximate expression

$$\varepsilon = 0.055\left[\frac{c_0}{\rho}\left|\left(\frac{d\sigma}{dc}\right)_{c=c_0}\right|\right]^{1/2}\frac{\delta^{3/2}\nu^{1/4}}{D^{5/4}} \tag{28}$$

The above theoretical and experimental results yield information concerning the thickness δ of the spreading film and the width ε of the region near the leading edge in which the "primary" dissolution process takes place and in which the maximum surface velocity is attained. For δ one obtains values in the range $\delta = 0.6 \div 1.3 \times 10^{-4}$ cm and, correspondingly, for ε values in the range $\varepsilon = 2 \div 5 \times 10^{-2}$ cm.

No selection between models (a) and (b) can be made, since they lead practically to the same distribution of the surface velocity.

By the steady dissolving drop method, it was possible to compare theory and experiment for a simple manifestation of the Marangoni effect, namely the velocity field induced when a low-surface-tension liquid is spreading and dissolving over the surface of another one having a larger surface tension.

In order to solve the problem theoretically, it was essential to take into account the finite thickness of the spreading film at the leading edge of the steady drop. The value of this thickness was obtained by comparing the theoretical and experimental surface velocity distributions.

The theoretical equations obtained indicate that the surface velocity increases from practically zero (at the leading edge of the dynamic drop) up to a maximum in a region near the leading edge. In this region of about 10^{-2} cm width, in which a very large velocity gradient exists, the primary phenomenon of the dissolution process takes place.

A very similar experiment was carried out by Sada *et al.*,[16] who measured the variation of surface velocities generated by the surface tension sink produced by the steady dissolving of ethanol in a point of a water surface. Note that ethanol does not form a distinct dissolving drop. By the addition of glycerol or small amounts of surfactants, the bulk viscosity and the degree of surface contamination varied, and their effect on the surface movements were measured.

The combination between the very small size of the tracers used, increased viscosity of the liquid layer, and use of a dissolving liquid (ethanol) that formed no definite contour allowed Sada *et al.* to measure the velocity in the acceleration zone as well as beyond it, as it decreases. The qualitative agreement between their data and those of Fig. 4 is remarkable. The width of the acceleration zone, as read approximately from the plots presented by Sada *et al.*, varies between 0.3–0.5 cm, increasing when the viscosity of the substrate increases between 1.1 and 36 cP. A closer comparison with the predictions of Eq. (28) cannot be made, since other properties were varied when the substrate viscosity was increased.

Experimental results thus have confirmed that the present treatment is valid not only for partially miscible liquids (in which a dynamic drop is formed) but also for completely miscible ones. In the latter case, dissolution may be considered to begin at the very point where the low-surface-tension liquid is fed.

3.2. Steady Heat Source

The surface tension of liquids decreases with temperature. By bringing to an extended liquid surface elements of liquid of higher temperature (i.e., lower surface tension) than the ambient ones, we create conditions for the appearance of the Marangoni effect. The surface liquid moves radially away from the surface area of higher temperature. The behavior of thin films under these conditions was discussed earlier. The surface movements and the temperature profile generated in a thick liquid layer by a surface heat source will be now presented.

3.2.1. Description of the Method

The flat end of a 3-mm-diameter copper rod that could be heated electrically was brought in touch with the flat surface of a liquid layer, approximately 15 cm thick, which was placed in a rectangular dish 30 × 30 × 20 cm. The temperature of the rod was 30°–120°C higher than that of the liquid bulk. The temperature of the heat source was measured by a Cu–constantan thermocouple soldered flush with the rod surface touching the liquid. The temperature profile within the liquid layer was measured by means of a rack of 12 thermocouples made of 0.2-mm-diameter Cu–constantan wires, which could be displaced within the dish by means of micrometric screws. Another thermocouple, placed in the same liquid, outside the zone of the most intensive movements, was used as reference (cold junction). The movements of the liquid surface were measured by the stroboscopic procedure described earlier. Water, glycerol, ethylene glycol, and a variety of mineral oils were used as test liquids.

3.2.2. Experimental Results

Fig. 7 shows the temperature profile measured after a layer of ethylene glycol had reached steady state when its surface was heated to 36°C higher than the reference thermocouple. The isothermal surfaces follow the streamlines of the toroidal movement created in the liquid bulk by the radial surface spreading. Note the fast dissipation rate of the surface temperature in radial direction. This proves that the convective heat transport caused by the surface movement is much larger than the conductive one.

In Fig. 8, the surface velocities are plotted as function of the radial distance

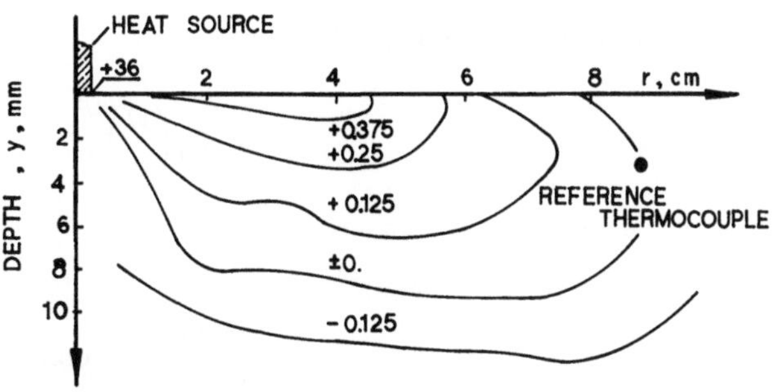

FIGURE 7. Temperature field with respect to the bulk liquid temperature generated by heating in a point the surface of a liquid.[17]

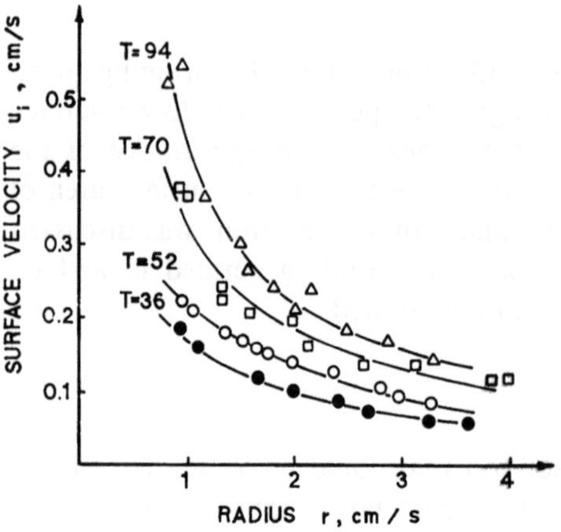

FIGURE 8. Dependence of surface velocity on temperature and radius (T represents the difference between the temperature of the heat source and bulk liquid).[17]

from the rod heated to various temperatures, higher than that of the bulk liquid by the values indicated (°C).

3.2.3. Comparison between Experiment and Theory

A mathematical treatment similar to that used for the surface movements created by dissolution can be adopted for the present case. The convective diffusion equation, Eq. (1), was replaced by the one for heat transfer.

By solving the hydrodynamic equations together with that for convective heat transfer under suitable boundary conditions, we find that the radial distribution of the surface velocities is given by an equation identical to Eq. (19). (The diffusion coefficient D is replaced in that equation by the thermal diffusivity a, and $d\sigma/dc$ is replaced by $d\sigma/dT = \beta = \text{const.}$) Estimates based on the numerical solution of Eq. (19) show that the acceleration zone ε for the movements created by heating of a liquid surface is not wider than about 10^{-2} cm, and that one can assume that the motion occurs in two distinct zones. An acceleration zone in which the shear stress is almost constant, followed by a second zone, in which the motion proceeds owing to reasons of continuity only.

In the acceleration zone,

$$u = 2F_0\left[\left(\frac{\nu t}{\pi}\right)^{1/2} \exp -\left(\frac{y^2}{4at}\right) - \frac{y}{2}\left(1 - \operatorname{erf}\frac{y}{2\sqrt{at}}\right)\right] \tag{29}$$

and the surface velocity is given by

$$u_i = 2F_0\left(\frac{\nu t}{\pi}\right)^{1/2} \tag{30}$$

where

$$F_0 = -\left(\frac{\partial u}{\partial y}\right)_{y=0} \approx \text{const.} \tag{31}$$

Comparing with Eq. (18), one can write that

$$F_0 = -\frac{\beta}{\eta u_{iM}} \frac{\delta T}{2\sqrt{\pi a}} t_M^{-3/2} \exp -\left(\frac{\delta^2}{4at_M}\right) \tag{32}$$

where u_{iM} is the maximum value of u_i, t_M is the value of t for which F is maximum, and T is the temperature difference between the heat source and the bulk liquid. The approximations employed to derive Eq. (27) lead now to the expression

$$u_{iM} = 0.656\left(-\frac{\beta T}{\rho\delta}\right)^{1/2}\left(\frac{a}{\nu}\right)^{1/4} \tag{33}$$

For the second zone

$$u = \frac{1}{2[\pi\nu(t-t_M)]^{1/2}} \int_{-\infty}^{+\infty} u(y_M) \exp -\left[\frac{(y-y_M)^2}{4\nu(t-t_M)}\right] dy_M \tag{34}$$

Since only the velocity near the interface is of interest here, only the first term of Eq. (29) can be used. The following equation is finally obtained for the distribution of the surface velocities in the second zone:

$$u_i = 0.0975\,\frac{(-\beta T)\delta}{\rho(\nu a)^{1/2}\, r} \tag{35}$$

The only variables in Eq. (35) are the temperature T and the radius r.

The surface velocities of Fig. 8 are plotted against the reciprocal radius in Fig. 9. The agreement between theory [Eq. (35)] and experiment is good.

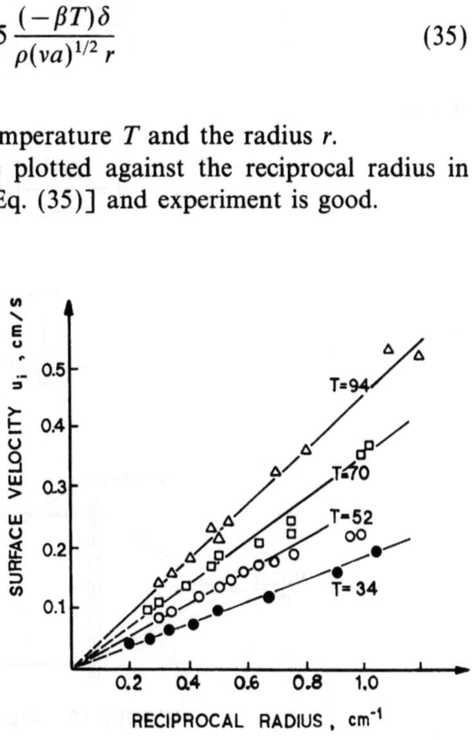

FIGURE 9. Comparison between theory and experiment: dependence of the surface velocity on radius at various temperatures.[17]

4. MASS TRANSFER ENHANCEMENT DUE TO SURFACE MOVEMENTS PRODUCED BY A DISSOLVING LIQUID

The successful mathematical description of the surface velocity field produced by a surface tension sink opened the possibility of calculating the mass transfer coefficient through such moving interfaces.

4.1. Mass Transfer Measurements

The tests were carried out (Smigelschi *et al.*[18]) in a prismatic plastic cell (Fig. 10). Tap water and CO_2 could flow continuously through the bottom and top parts of the cell. The liquid–gas interface was smooth and of quadratic shape (10×10 cm). A constant flow of water and CO_2 was started. The capillary tip of a precision buret was located just above the center of the interface. Through it, a small flow of a dissolving liquid was maintained, which created the surface tension sink in the interface. The radial surface velocity field influenced the rate of dissolution of CO_2 in the water. They were measured in separate experiments, at the same values of the flow rate of the solute flowing through the capillary.

4.2. Theory

The convective diffusion equation can be integrated by using a similarity transformation. For the concentration distribution one obtains

$$\frac{c - c_i}{c_0 - c_i} = \frac{2}{\sqrt{\pi}} \int_0^{y/\Delta} e^{-s^2}\, ds \tag{36}$$

where

$$\Delta = \frac{2D^{1/2}}{ur} \left[\int_{r_0}^{r} ur^2\, dr \right]^{1/2} \tag{37}$$

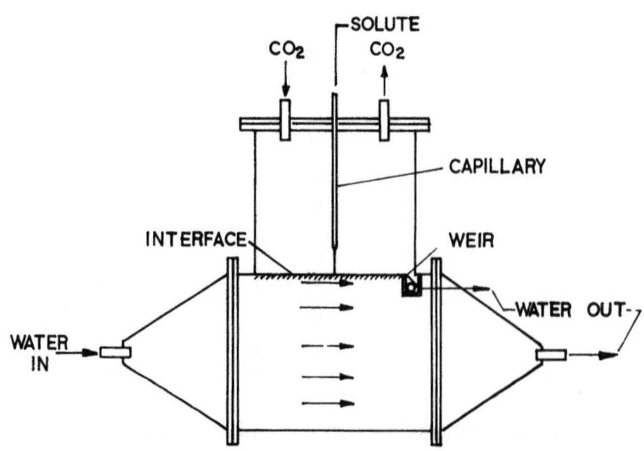

FIGURE 10. Experimental apparatus.

The local mass transfer coefficient in the liquid phase, defined by

$$k \equiv \frac{-D(\partial c/\partial y)_{y=0}}{c_i - c_0} \tag{38}$$

becomes

$$k = \frac{2}{\sqrt{\pi}} \frac{D}{\Delta} \tag{39}$$

and, finally,

$$k = \sqrt{\frac{D}{\pi}} \frac{ur}{[\int_{r_0}^{r} ur^2\, dr]^{1/2}} \tag{40}$$

4.3. Comparison between Experiment and Theory

The analysis of the experimental data shows that the surface velocity can be correlated to the radial position by an equation of the type

$$u = br^m \tag{41}$$

where b and m are empirical constants determined by the conditions of the experiments.

By defining an average mass transfer coefficient in the liquid phase as

$$k_L = \frac{2 \int_{r_0}^{R} kr\, dr}{R^2 - r_0^2} \tag{42}$$

where R is the radius of the circle tangent to the walls of the cell, and by using the empirical radial distribution of the surface velocity, we obtain

$$k_L = \frac{4}{R^2 - r_0^2} \left[\frac{bD}{\pi(3+m)} (R^{3+m} - r_0^{3+m}) \right]^{1/2} \tag{43}$$

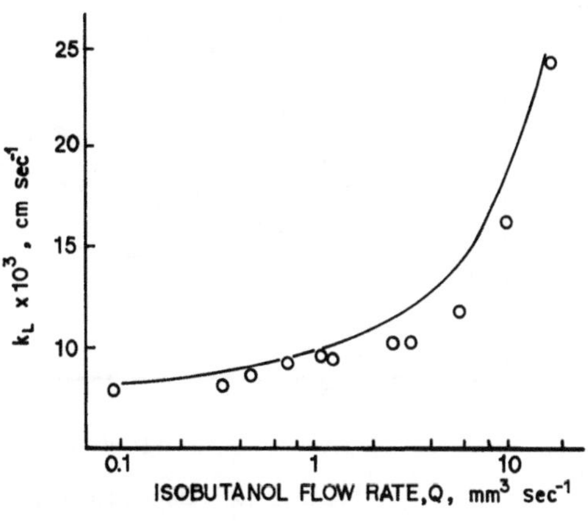

FIGURE 11. Comparison between the theoretical and experimental values of k_L. The continuous line represents the theoretical curve, and the points are the experimental data.[18]

The mass transfer coefficients calculated by this equation are plotted in Fig. 11 together with those measured for given flow rates of the solute that generated the surface tension sink.

5. ADDITIONAL EXPERIMENTAL OBSERVATIONS

All liquids partially miscible with water (as well as some of those completely miscible) spread over its surface and form a film having a definite contour, which, in certain conditions, is practically circular (Fig. 3). This contour of the film is the origin of radial motions of the free-water surface, which, in the absence of surface-active agents, are extending over relatively long distances. (The size of the dish is important, since for small dish areas the liquid streams generated by the radial motions and reflected by the dish walls and bottom may upset the form of the film. Bottom effects are negligible for water layers deeper than about 15 mm.)

5.1. The Dissolving Film

The film contour is circular for small flow rates, its diameter D having a linear dependence on the flow rate Q (Fig. 12). This dependence shows that the dissolution process takes place mainly along the film contour. For sufficiently large

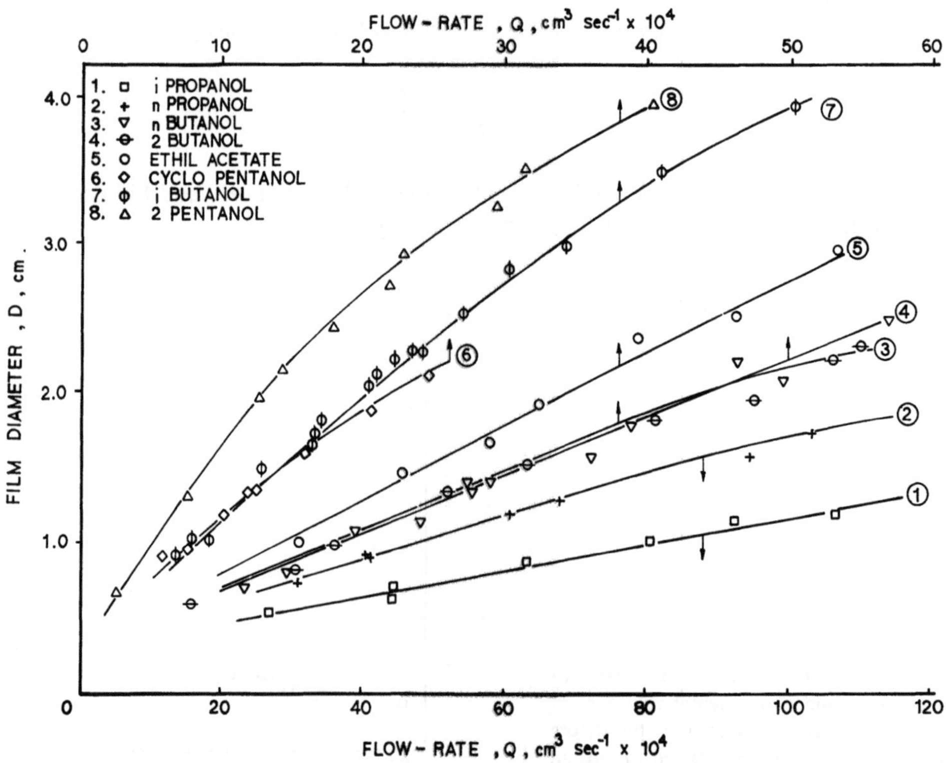

FIGURE 12. Dependence of film diameter on flow rate.[8]

values of the flow rate, the circle becomes unstable and the contour of the film oscillates between a circular and an elliptical shape. At still higher flow rates the contour becomes quite irregular (Fig. 13). (The photographs are taken by placing photographic paper directly over the screen. The white band is the image of the capillary, and the small white circle is the image of the meniscus formed around the capillary tip.) In the gulfs of this contour, waves may appear, which at times may be rising more than 1 mm above the water surface. When high flow rates are used, deep gulfs appearing in various parts of the contour may unite, thus separating "islands" from the main film area, which dissolve independently and die out.

The slope β of the lines Q versus the film contour length P, called in the following the specific dissolution rate, is constant as long as the contour is circular. If the contour ceases to be circular, β increases with increasing flow rate.

The specific dissolution rate depends on the difference ΔC between the saturation concentration of the film-forming liquid in the supporting liquid and its actual concentrations in the bulk of the supporting liquid (the ratio $\beta/\Delta C$ has the significance of a mass transfer coefficient along the film contour). This is shown in Fig. 14 for an isobutanol film dissolving in isobutanol aqueous solutions of various concentrations.

The sudden change in slope is probably the consequence of a qualitative change observed to occur in the aspect of the film contour at about the same value of the driving force. Indeed, for driving forces smaller than a certain value, the film contour loses its smoothness and presents along its whole length a multitude of

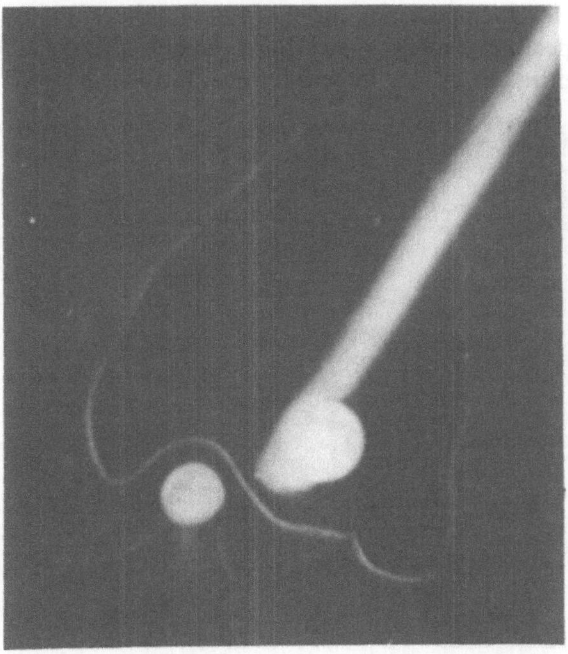

FIGURE 13. Isobutanol film on water.[8] Flow rate, 32×10^{-4} cm^3/s; driving force, 11.1×10^{-4} mole/cm^3. The white circle outside the film contour is given by a wave rising above the water surface.

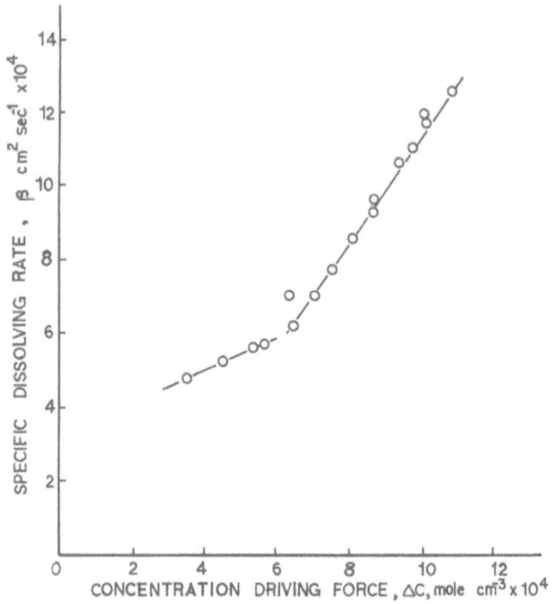

FIGURE 14. Dependence of specific dissolving rate on concentration driving force.[8]

minute indentations pulsating continuously (Fig. 15). These small "gulfs" appear at all flow rates, even with film contours having a macroscopically circular form. [The nonlinear dependence of diameter on flow rate (the variation of the specific dissolving rate) for cyclopentanol and 2-pentanol (Fig. 12) may be related to the indentations appearing at all flow rates along the film contour of these substances.] When the concentration driving force is lowered further, the form of the liquid film

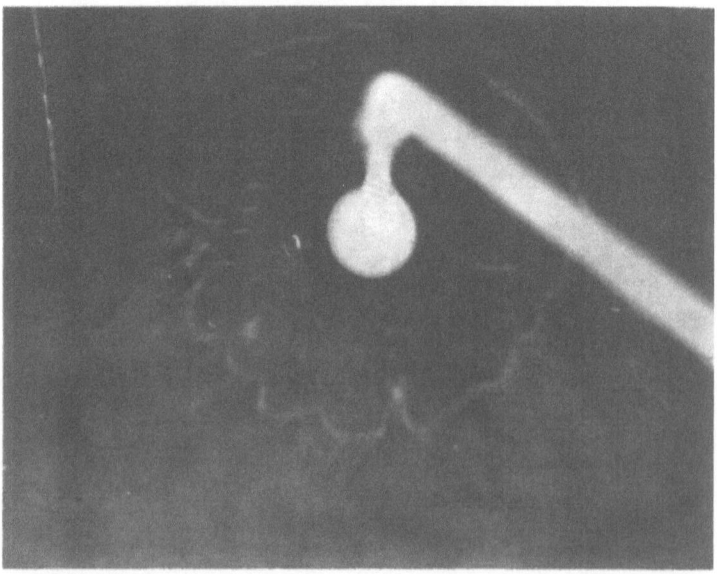

FIGURE 15. Isobutanol film on water[8]: flow rate, 11.8×10^{-4} cm³/s; driving force, 6.5×10^{-4} mole/cm³.

becomes more involved; streams are generated in a pulsating manner at or near the film contour and move under the film radially toward the capillary tip or approximately parallel to the contour.

Table 3 gives qualitatively the aspect of the film contour as a function of the concentration driving force. The contour aspect seems to depend strongly on the driving force and less on the nature of the substance forming the film.

5.2. Movements of the Free Surface

Motions of the free (not covered by the film) surface of the supporting liquid take place (when this is not contaminated) from the film contour radially up to the

TABLE 3. Film Contour Aspect in Relation to the Concentration Driving Force

Concentration driving force[a] (moles/cm^3 × 10^4)	Substances fed over the water surface	Film formation	Aspect of the film countour at low flow rates	Observation
	Methanol, ethanol, acetone, acetic acid, propionic acid dioxan, ethylene glycol, etc.	No		Completely miscible
	i-Propanol, *n*-propanol, butyric acid, etc.	Yes	Circular, relatively diffuse, microscopically agitated	Completely miscible
23.4	2-Butanol	Yes	Circular, well defined smooth (as in Fig. 16)	Partially miscible
11.1	*i*-Butanol	Yes	Same as above	Same as above
9.9	*n*-Butanol	Yes	Similar to 2-pentanol	Same as above
8.44	Ethyl acetate	Yes	Circular, well defined smooth	Same as above
~6.0	*i*-Butanol[b]	Yes	"Macroscopically" circular; with indentations (Fig. 15)	Same as above
5.02	2-Pentanol	Yes	Same as above	Same as above
3.56	Cyclopentanol	Yes	Same as above	Same as above
3.3	*i*-Butanol[b, c]	Yes	The film breaks down and forms islands that dissolve separately	Same as above
3.18	*i*-Pentanol	Yes	Same as above	Same as above
2.59	*n*-Pentanol	Yes	Same as above	Same as above
0.16	*n*-Hexanol	Yes	Same as above	Same as above
0.045	*n*-Octanol	Yes	Monomolecular film spreading over the entire water surface; drops rest on the film without dissolving	Same as above

[a] Pure substance dissolving into tap water. The concentration driving force equals the saturation concentration. Working temperature 25°C.
[b] Pure isobutanol dissolving into tap water partially saturated with isobutanol.
[c] Same film aspect was obtained by dissolving pure isobutanol into saturated sodium chloride aqueous solution.

dish walls. After a certain time has elapsed from the beginning of the experiment, a second outer contour, at which all radial motions stop abruptly, is seen to form. This second contour is circular, provided the film contour is also circular. The diameter of this contour is decreasing in time. Impurities existing in the water, in the film-forming liquid as well as in the atmosphere (since the dish is not covered), accumulate at the air–water interface resulting in a stagnant film probably similar to that described by Merson and Quinn.[19] As the amount of surface contaminants increases, the surface covered by the stagnant film becomes larger. Tracer particles spread over the film, move radially toward the film contour at relatively low velocities (estimated to be of the order of millimeters per second), pass across it, and then with much higher velocities (estimated to be of the order of tens of cm/s) continue moving radially up to the outer contour. After having crossed the outer contour they move slower and less regularly; once outside the area between the two contours, no tracer recrosses the outer contour to enter this area again. The outer contour seems to represent a front along which the spreading forces of the stagnant film are in equilibrium with the forces generated by the surface tension gradient existing between the two contours. Indeed, when feeding of the film-forming liquid is stopped, the peripheral stagnant film extends over the whole supporting liquid surface in the dish. Resuming the feeding pushes the stagnant film back outside the outer contour.

The formation of the outer contour may be caused at will by using surface-active agents (Fig. 16); traces of them markedly diminish the diameter of the outer contour. With increasing amounts of surfactants the outer contour moves nearer the film contour until they almost coincide (but without doing this completely).

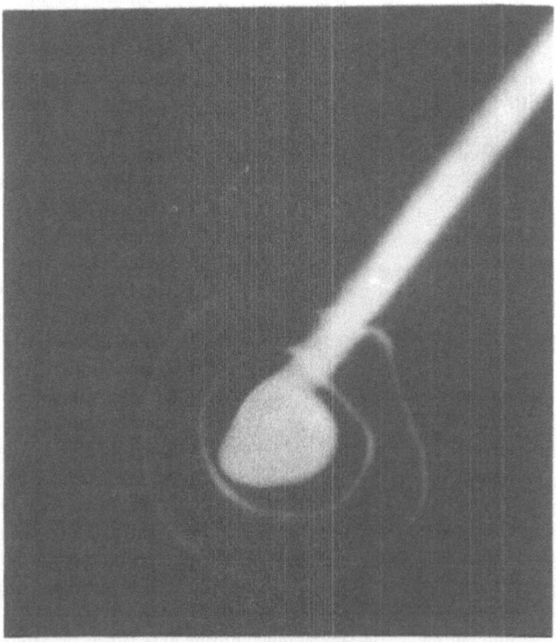

FIGURE 16. Isobutanol film on water, in presence of cetyltrimethylamonium bromide[8] flow rate 11.6×10^{-4} cm^3/s; driving force, 11.1×10^{-4} mole/cm^3.

Note that the dimensions of the outer contour have no influence on those of the film. In other words, the specific dissolution rate for a given flow rate is not influenced by the amount of the surface-active agent present. Nevertheless, very high concentrations, especially of insoluble surface-active agents, upset the circular form of the film apparently by mechanical action.

For liquids completely miscible with water, such as methanol, ethanol, acetone, dioxan, ethylene glycol, acetic acid, etc., no film could be detected over a water surface (for feed rates up to 10 mm^3/s). The outer contour was the only one formed. The same is true for aqueous solutions of completely or partially miscible organic substances. This is due to the fact that they mix with water practically the very moment they are brought into contact with it.[20]

Of the above liquids, dioxan, ethylene–glycol, etc., although heavier than water, do not move directly downward into the water phase (as they do if the capillary tip is submerged) but spread over the water surface, up to the outer contour, behaving exactly as liquids lighter than water. Obviously, the surface forces exceed in these cases the gravity forces. Glycerine has an intermediate behavior; part of it dissolves at the surface, behaving as the above liquids, while the rest forms drops that move downward through the water. Some liquids, such as isopropanol, *n*-propanol, *n*-butyric acid, etc., although completely miscible with water at the working temperatures, nevertheless form a film over the water surface (see also Table 3).

5.3. Movements within the Water Layer

The drag exerted by the rapidly moving free surface results in convection currents in the bulk of the water layer. The water layers swept along with the moving free surface acquire, with increasing radial distance from the capillary tip, a downward motion. This motion may be assisted by the influence of the walls and bottom of the dish. On the other hand, near the film contour a certain suction is exerted on the bulk liquid (by the rapid moving free surface), so that the bulk liquid moves upward below the film, to be eventually swept along by the surface layer. After a certain time, a steady toroidal motion pattern is developed (Fig. 3). If the free surface is contaminated by surface-active agents, this toroidal motion becomes more evident, since the moment the moving surface reaches the outer contour, it is deflected below the stagnant film. The movement continues beneath this stagnant film, but the horizontal velocity components are reduced more quickly, resulting in a more marked downward motion.

NOTATION

a	thermal diffusivity
c	concentration of the dissolving component
c_0	saturation concentration in water
c_i	value of c in the spreading film [model (b)]
$c_{i,0}$	value of c_i at the leading edge of the dynamic drop
D	diffusion coefficient; also film diameter

F	function defined by Eq. (14) or Eq. (26)
$F(t)_M$	maximum value of F with respect to t
r	radial coordinate
r_0	value of r for the leading edge of the dynamic drop
t	variable defined by Eq. (9)
T	temperature difference
u	radial component of velocity
u_i	value of u at the interface
u_{iM}	maximum value of u_i with respect to t
v	y component of the velocity
y	distance from the interface along the normal

GREEK SYMBOLS

β	specific dissolution rate and also $d\sigma/dT$
δ	the thickness of the spreading layer at the leading edge of the dynamic drop
ε	the width of the region near the leading edge in which the velocity of the interface increases rapidly
η	viscosity
ν	kinematic viscosity
ρ	density
σ	surface tension

REFERENCES

1. M. K. Bernett and W. A. Zisman, *J. Phys. Chem. Ithaca* **70**, 1064 (1966).
2. H. R. Baker, P. B. Leach, C. R. Singleterry, and W. A. Zisman, *Ind. Eng. Chem. Ind. (Int.) Ed.* **59**(6), 29 (1967).
3. C. V. Sternling and L. E. Scriven, *AIChE J.* **5**, 514 (1959).

4a. J. S. Sorensen, ed., *Dynamics and Instability of Fluid Interfaces, Lecture Notes in Physics*, No. 105, Springer-Verlag, Berlin (1979).

4b. E. Ruckenstein, in *Advances in Chemical Engineering* (J. Wei, ed.), Academic, New York, vol. 13, p. 101 (1987).

5. H. A. Dijkstra and A. I. van de Vooren, *Int. J. Heat Mass Transfer* **28**, 2315 (1985).
6. S. Ostrach, *Ann. Rev. Fluid Mech.* **14**, 313 (1982).
7. D. Raake, J. Siekmann, and Chun Ch. H., *Exp. Fluids* **7**, 164 (1989).
8. D. G. Suciu, O. Smigelschi, and E. Ruckenstein, *AIChE J.* **13**, 1020 (1967).
9. D. G. Suciu, O. Smigelschi, and E. Ruckenstein, *AIChE J.* **15**, 686 (1969).
10. E. Ruckenstein, O. Smigelschi, and D. G. Suciu, *Chem. Eng. Sci.* **25**, 1249 (1970).
11. D. G. Suciu, O. Smigelschi, and E. Ruckenstein, *Trans. Inst. Chem. Eng.* **48**, T146 (1970).
12. W. T. Mitchell and J. A. Quinn, *Chem. Eng. Sci.* **23**, 503 (1968).
13. H. Linde, E. Schwarz, and K. Gröger, *Chem. Eng. Sci.* **22**, 823 (1967).
14. G. D. Suciu, O. Smigelschi, and E. Ruckenstein, *J. Colloid Interface Sci.* **33**, 520 (1970).
15. H. S. Carslaw and J. C. Jaeger, *Conduction of Heat in Solids*, p. 76, Clarendon Press (1959).
16. E. Sada, T. Ameno, and T. Ando, *Chem. Eng. Sci.* **32**, 1269 (1977).
17. O. Smigelschi and G. D. Suciu, *Rev. Chim.* (Bucharest) **25**, 476 (1974).
18. O. Smigelschi, G. D. Suciu, and E. Ruckenstein, *Chem. Eng. Sci.* **24**, 1227 (1969).
19. R. L. Merson and J. A. Quinn, *AIChE J.* **11**, 391 (1965).
20. G. D. Suciu, *J. Colloid Interface Sci.* **27**, 320 (1968).

16

Nucleation on Smooth Surfaces

Joseph L. Katz, Jin Sheng Sheu, and Jer Ru Maa

1. INTRODUCTION

Nucleation and condensation take place whenever a vapor is in contact with a substrate surface of sufficiently low temperature. On solids, such condensation mostly occurs at preferred sites. However, when the concentration of preferred sites is low or if the supersaturation is sufficiently high, nucleation on solid surfaces also will occur at randomly distributed locations, i.e., homogeneously.

Condensation of the supersaturated vapor on the substrate can occur by different paths. Molecules can condense directly from the vapor onto the nucleus, or they can adsorb onto the surface and then migrate to the nucleus by two-dimensional diffusion (see Fig. 1). The relative importance of these two routes depends on the vapor–condensate–substrate interfacial free energies and on the film pressure of the adsorbed molecules. In this chapter we show how to generalize homogeneous nucleation theory to allow for both processes simultaneously. Example calculations, describing the nucleation of an ideal gas vapor on a perfectly smooth solid, i.e., no preferred sites, are used to illuminate the results. In the appendix, the equations needed to make similar calculations for nucleation on the surface of an immiscible liquid are also presented.

2. MODEL

When a vapor condenses on the surface of a smooth solid substrate,[1, 2] the condensate nuclei typically have the spherical cap shape shown in Fig. 1. A force balance provides the relation (known as the Young equation), between the cosine

---

Joseph L. Katz • Department of Chemical Engineering, The Johns Hopkins University, Baltimore, Maryland 21218. *Jin Sheng Sheu* and *Jer Ru Maa* • Department of Chemical Engineering, National Cheng Kung University, Tainan, Taiwan, 70101, R.O.C.

Modern Approaches to Wettability: Theory and Applications, edited by Malcolm E. Schrader and George Loeb. Plenum Press, New York, 1992.

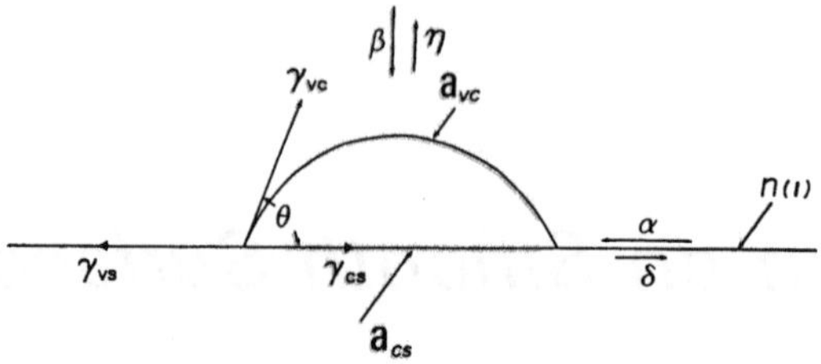

FIGURE 1. Nucleus on a smooth solid surface.

of the angle θ in terms of γ_{vc}, γ_{cs}, and γ_{vs}, the vapor–condensate, condensate–substrate, and substrate–vapor interfacial tensions, respectively; i.e.,

$$\cos\theta = \frac{\gamma_{vs} - \gamma_{cs}}{\gamma_{vc}} \tag{1}$$

The relationship between the substrate–condensate interface (a_{cs}) and vapor–condensate interface (a_{vc}) areas, the contact angle made by the spherical cap and the substrate, and the number of molecules in the developing nucleus is obtained as follows. Assume the nucleus is a cap of a sphere of radius R. The contact angle θ, which is the angle of a tangent to the sphere with the plane of the smooth substrate surface, is also the latitude of this intersection, measured from a perpendicular to the plane and passing though the center of the sphere. Thus, the circumference of the cap is

$$l(\text{cap}) = \pi R^3 \int_0^\theta \sin^3\theta \, d\theta = \frac{\pi R^3}{3}(2 - 3\cos\theta + \cos^3\theta) = l_0 i^{1/3} \tag{2}$$

The circular area of the solid–nucleus interface is

$$a_{cs} = \pi R^2 \sin^2\theta = a_0 i^{2/3} \tag{3}$$

and the area of the cap surface separating the nucleus and the vapor is

$$a_{cv} = 2\pi R^2(1 - \cos\theta) = a_0' i^{2/3} \tag{4}$$

Note that we also have expressed these relationships in terms of i, the number of molecules in the nucleus, where

$$l_0 = 2\pi\Omega\sin\theta \qquad a_0 = 2\pi\Omega^2(1 - \cos\theta) \qquad a_0' = \pi\Omega^2\sin^2\theta \tag{5}$$

and

$$\Omega \equiv \left[\frac{3V_c}{4\pi\Phi(\theta)}\right]^{1/3} \qquad \Phi(\theta) \equiv \frac{2 - 3\cos\theta + \cos^3\theta}{4} \tag{6}$$

V_c is the molecular volume of the condensate, i.e., $V_c = \mathrm{MW}/\rho N_0$, where N_0 is Avogadro's number and MW and ρ are the molecular weight and the density of the condensate, respectively.

3. RELATIONSHIP OF THE VARIABLES

The relationship between the number of vapor molecules adsorbed on a unit area of substrate surface, $n(1)$, and P, their partial pressure in the vapor phase, is typically described by a multiparameter equation;[3,4] e.g., for a vapor that obeys the van der Waals equation of state, it is

$$\frac{P}{P_e} = \frac{k_a F}{1-F} \exp\left(\frac{F}{1-F} - \frac{2a_2 F}{b_2 k T_s}\right) \tag{7}$$

where P_e is the saturation vapor pressure at T_s, the temperature of the substrate surface, k is Boltzmann's constant, a_2 and b_2 are the two-dimensional van der Waals constants, k_a is a constant related to the strength of adsorption, F, the fraction of the surface covered with adsorbed molecules, is equal to $b_2 n(1)$, and b_2 is the area occupied by one molecule.

The Gibbs adsorption isotherm relates the concentration of adsorbed molecules to the derivative of the vapor–substrate interfacial tension;[5] i.e.,

$$kT_s n(1) = -\frac{\partial \gamma_{vs}}{\partial \ln P} \tag{8}$$

Integrating Eq. (8), one obtaines a relationship for π, the film pressure due to the adsorbed molecules:

$$\pi = \gamma_s - \gamma_{vs} = kT_s \int_0^P n(1)\, d\ln P \tag{9}$$

where γ_s is the surface free energy of a substrate without any adsorbed vapor molecules on it. Using Eq. (7) to relate $n(1)$ to P, one obtains

$$\pi = \frac{F}{b_2}\left(\frac{kT_s}{1-F} - \frac{a_2 F}{b_2}\right) \tag{10}$$

4. HOMOGENEOUS NUCLEATION

Before discussing condensation processes on a smooth surface, we briefly explain homogeneous nucleation.[6,7]

A snapshot of the molecules present in a supersaturated state would show small clusters of the new phase containing two, three, or more molecules. There also would be occasional clusters containing many molecules. However, a motion picture of these larger clusters would show that most are very short lived; they grow rapidly and then shrink rapidly. Nucleation occurs when a cluster of molecules grows (fluctuates in size) to a large enough size; it then continues to grow.

A critical-sized nucleus is a cluster of size such that its probability for further growth is equal to its probability for becoming smaller. Clusters that fluctuate to a size larger than the critical size will probably continue to grow to macroscopic size, while clusters smaller than the critical size most likely will shrink. The nucleation rate is then the net number of clusters per unit time that grow larger than the critical size.

In dilute solution, whether gaseous or condensed, the concentrations of clusters (dimers, trimers, etc.) are much lower than the concentration of monomers. Because of these low concentrations, two clusters very rarely collide. One therefore accurately can use a model that all clusters grow by the addition of single molecules. From the principle of microscopic reversibility (i.e., at equilibrium *each* forward process has to be matched by its corresponding reverse process) and the assumption that in supersaturated vapors no new processes occur, it follows that evaporation of molecules from a cluster also occurs one molecule at a time.

The net rate, $J(i)$, at which clusters of size i grow to size $i+1$, is the difference between the rate at which clusters of size i add an additional molecule and the rate at which clusters of size $i+1$ lose a molecule; that is,

$$J(i) = f(i)\,n(i) - b(i+1)\,n(i+1) \tag{11}$$

where $f(i)$ represents the forward rate—the rate of addition of molecules to a cluster —$b(i+1)$ is the backward rate—the rate of loss of molecules from a cluster—and $n(i)$ is the number, or concentration, of clusters of size i.

It will be helpful to define a variable $Z(i)$ by the equations

$$Z(i) = \prod_{j=2}^{i} \frac{b(j)}{f(j)} \qquad \text{for} \quad i > 1 \tag{12}$$

and

$$Z(1) \equiv 1 \tag{13}$$

This definition produces a variable with the following useful property:

$$Z(i+1) \equiv \frac{Z(i)\,b(i+1)}{f(i+1)} \tag{14}$$

Multiplying Eq. (11) by $Z(i)$, using Eq. (14), and summing, one obtains

$$\sum_{i=1}^{I} J(i)\,Z(i) = \sum_{i=1}^{I} \left[f(i)\,n(i)\,Z(i) - f(i+1)\,n(i+1)\,Z(i+1)\right] \tag{15}$$

Note that the two terms on the right-hand size are identical except for their indices. Thus, on summing, successive terms cancel and Eq. (15) simplifies to

$$\sum_{i=1}^{\hat{\imath}} J(i)\, Z(i) = f(1)\, n(1) - f(\hat{\imath}+1)\, n(\hat{\imath}+1)\, Z(\hat{\imath}+1) \tag{16}$$

If one carries out the summation to a sufficiently large value of $\hat{\imath}$, the last term on the right-hand side of this equation is negligible compared with the first. This can be seen from Eq. (12) and the following argument. The quotient $b(j)/f(j)$ is the ratio of the number of clusters of size j that shrink to those that grow. This ratio depends on both cluster size and supersaturation. For a flat liquid surface, *at saturation*, this ratio is unity, since saturation is defined as that pressure at which vapor arrives and evaporates at equal rates. But for clusters, the backward rate increases due to the Kelvin effect; i.e., their curvature makes it easier for molecules to evaporate than from a flat liquid surface. Since the forward rate is proportional to the *supersaturation*, for every supersaturation, there is a critical size i^*, defined as that size for which this ratio is unity. For all sizes smaller than the critical size, the ratio is greater than unity, while for all sizes larger than the critical size this ratio is less than unity. Since there are only a finite number of sizes smaller than the critical size [and $b(i)/f(i)$ is finite for all i], the continued product of $b(i)/f(i)$ can be made as small as desired by choosing its limit, $\hat{\imath}$, sufficiently large. Furthermore, in most cases of interest, a steady state is rapidly established. Thus, J becomes a constant for all sizes and can be factored out of the summation; Eq. (16) thus becomes

$$J = f(1)\, n(1) \Big/ \sum_{i=1}^{\hat{\imath}} Z(i) = f(1)\, n(1) \Big/ \left\{ 1 + \sum_{i=2}^{\hat{\imath}} \prod_{j=2}^{i} \frac{b(j)}{f(j)} \right\} \tag{17}$$

To proceed further, one needs explicit expressions for the backward rate $b(j)$ and the forward rate $f(j)$.

The forward rate, i.e., the rate at which impinging molecules condense onto a cluster, is the product of the surface area of the cluster, $a(i)$, the rate per unit area at which molecules impinge on its surface, β, and the condensation coefficient ζ (i.e., the fraction of arriving molecules that actually condenses). Thus,

$$f(i) = a(i)\, \beta \zeta \tag{18}$$

For an ideal gas, the impingement rate of molecules on a cluster is

$$\beta = \frac{P}{(2\pi mkT)^{1/2}} \tag{19}$$

where m is the mass of the molecule. The condensation coefficient ζ is a complicated and unknown function of cluster size. However, its effect on the nucleation rate is relatively small, since, as will be shown later, the nucleation rate depends primarily on a ratio of forward rates; thus, ζ cancels.

The backward rate is the rate at which molecules leave a cluster. This is not known in general. However, for systems in which the number density of solute molecules is much smaller than the number density of solvent molecules; e.g., the condensation of water vapor in air and the interactions between a molecule leaving a cluster and other solute molecules are negligible compared with the interactions with solvent molecules. Therefore, the backward rate $b(i)$, although a complicated function of temperature and cluster size, is independent of the concentration of the nucleating species (e.g., the water vapor). Consequently, if $b(i)$ can be determined at any concentration or supersaturation (since it is independent of concentration), it is known at all other concentrations. Thus, one can relate $b(i)$ to better known quantities by realizing that at true equilibrium (i.e., in a *saturated* vapor) the nucleation rate $J(i)$ is exactly equal to zero for *all* i, and therefore Eq. (11) can be solved to yield

$$b(i)=\frac{f_e(i-1)\,n_e(i-1)}{n_e(i)} \tag{20}$$

where the subscript e is used to remind the reader that these concentrations and forward rates refer to those found at *saturation*, not those in the actual nucleating vapor. Equations (18) and (20) then can be substituted into Eq. (12). Thus,

$$Z(i)=\frac{n_e(1)f_e(1)}{n_e(i)\,f_e(i)}\prod_{j=2}^{i}\frac{f_e(j)}{f(j)}=\frac{n_e(1)\,a(1)}{n_e(i)\,a(i)}\left(\frac{\beta_e}{\beta}\right)^{i-1} \tag{21}$$

where we have used the excellent approximation that $f(i)/f_e(i)$ is independent of the size of the cluster. This is almost always the case and occurs when the forward rate is a product of terms that are either size independent or whose size dependence is independent of supersaturation. Substituting Eq. (21) into (17) and using (18), one obtains

$$J=\left\{\sum\left[a(i)\,\beta\zeta n_e(i)\left(\frac{\beta}{\beta_e}\right)^{i}\right]^{-1}\right\}^{-1} \tag{22}$$

Equation (22) can be written in the standard form of classical nucleation theory by using Eq. (19) and making the following additional assumptions: (i) the equilibrium cluster distribution is given by

$$n_e(i)=N\exp\left[-\frac{\gamma a(i)}{kT}\right] \tag{23}$$

where N is a normalization constant and γ is the surface free energy of a flat liquid surface of the same composition and temperature as the nucleus; (ii) the nucleus is spherical and of normal liquid density. Thus, $a(i)=a_0 i^{2/3}$, where $a_0=4\pi(3V/4\pi)^{2/3}$ and V is the molecular volume.

5. GROWTH OF CLUSTERS

A condensate cluster on a smooth surface can grow by direct condensation from the vapor at a rate of β molecules per unit interfacial area per unit time, and by condensation from the layer of adsorbed molecules at a rate of α molecules per unit circumferential length per unit time.[8] Thus, $f(i)$, the rate at which molecules are added to a cluster containing i molecules, is

$$f(i) = \beta a_0 i^{2/3} + \alpha l_0 i^{1/3} \tag{24}$$

If we restrict ourselves to pressures such that the vapor is an ideal gas, then

$$\beta = \frac{P}{\sqrt{2\pi m k T_v}} \tag{25}$$

where m is the molecular mass and T_v is the temperature of the vapor. If the adsorbed vapor molecules behave as a two-dimensional ideal gas,[6, 7] then

$$\alpha = n(1)\sqrt{\frac{\pi k T_s}{32m}} \tag{26}$$

Condensate clusters can shrink by evaporation of molecules to the vapor and by evaporation of molecules to the adsorbed layer at rates η and δ, respectively. Thus, $b(i+1)$, the rate at which molecules leave a cluster of $i+1$ molecules, is

$$b(i+1) = [\eta a_0(i+1)^{2/3} + \delta l_0(i+1)^{1/3}]\, n(i+1) \tag{27}$$

As was the case for simple homogeneous nucleation, the net rate at which clusters containing i molecules become clusters containing $i+1$ molecules is

$$J(i) = f(i)\, n(i) - b(i+1)\, n(i+1) \tag{28}$$

Note that this equation is identical to Eq. (11). The only difference is that multiple forward and backward processes take place. While this chapter illustrates this using two processes, it is a straightforward procedure to generalize these equations to *any* number of forward and reverse processes. One need only recall that the principle of microscopic reversibility requires that *at equilibrium, each forward process be exactly matched by its corresponding reverse process*. Thus, at equilibrium, $J \equiv 0$ for all i, and Eq. (28) provides a relationship for η and δ in terms of α and β, and the equilibrium distribution, $n_e(i)$; i.e.,

$$\eta a_0(i+1)^{2/3} + \delta l_0(i+1)^{1/3} = [\beta a_0 i^{2/3} + \alpha l_0 i^{1/3}]\frac{n_e(i)}{n_e(i+1)} \tag{29}$$

Substituting into Eq. (27), one obtains

$$b(i) = [\beta a_0(i-1)^{2/3} + \alpha l_0(i-1)^{1/3}]\frac{n_e(i-1)}{n_e(i)} \tag{30}$$

Substituting Eqs. (24) and (30) into Eq. (17), one obtains

$$J = \frac{(\beta a_0 + \alpha l_0)\, n(1)}{1 + \sum_{i=2}^{f} \frac{n_e(1)}{n_e(i)} \prod_{j=2}^{i} \frac{\beta a_0 (j-1)^{2/3} + \alpha l_0 (j-1)^{1/3}}{\beta a_0 j^{2/3} + \alpha l_0 j^{1/3}}} \tag{31}$$

To proceed further, one needs the relationship between $n_e(1)$, the concentration of the adsorbed vapor molecules when their vapor phase concentration equals the equilibrium vapor pressure at the temperature of the surface, T_s, and $n_e(i)$, the number of clusters containing i molecules per unit area of substrate surface. Using the assumption[7] conventional to all nucleation theories, i.e.,

$$n_e(i) = n_e(1) \exp\left[-\frac{W(i)}{kT_s}\right] \tag{32}$$

where $W(i)$ is the minimum reversible work required to form such a cluster, and using thermodynamic arguments,[1, 2, 6, 7] one obtains

$$n_e(i) = n_e(1) \exp\left[\frac{-(\gamma_{vc} a_{vc} + \gamma_{cs} a_{cs} - \gamma_{vs} a_{cs})}{kT_s}\right] \tag{33}$$

6. CRITICAL SUPERSATURATION

The nucleation flux J for the condensation of a vapor on the surface of an immiscible liquid or on a smooth solid is a function of the substrate–vapor, vapor–condensate, and condensate–substrate interfacial tensions; π, the film pressure of the adsorbed vapor molecules on the substrate surface; and $1/k_a$, the strength of adsorption. As is typical of nucleation processes, this condensation process has a critical supersaturation, $(P/P_e)_c$. The rate of nucleation is negligibly small at supersaturations smaller than this critical supersaturation and increases explosively for supersaturations larger than it.

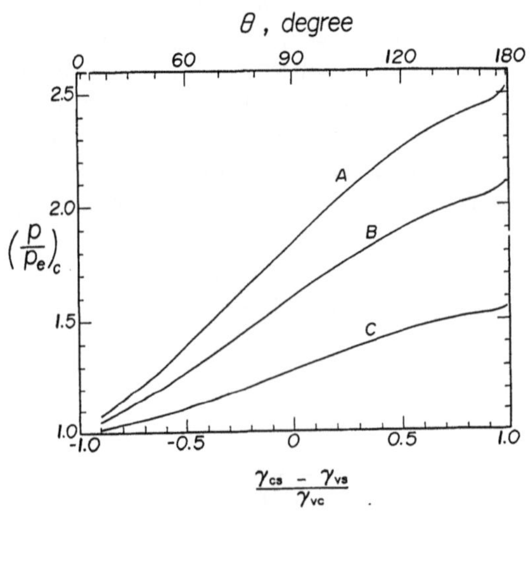

FIGURE 2. Critical supersaturation for nucleation on a smooth surface. Curves A, B, and C are for $\pi/\gamma_{vc} = 0.02$, 0.04, and 0.06, respectively.

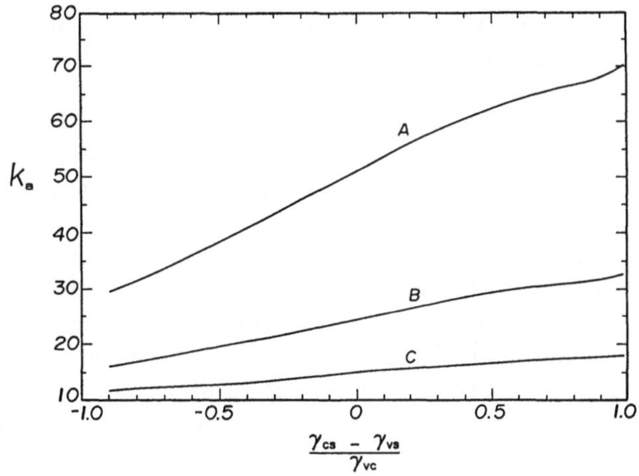

FIGURE 3. Values for the coefficient K_a in the adsorption isotherm that would cause nucleation on a smooth surface. Curves *A*, *B*, and *C* are for $\pi/\gamma_{vc} = 0.02$, 0.04, and 0.06, respectively.

The contact angle θ, or the ratio of interfacial tensions, the coefficient k_a in the adsorption isotherm, and π, F, and $n(1)$ are different ways of expressing the affinity between the vapor molecules and the substrate surface. The lower are the values of θ and k_a and the higher are the values of π, F, or $n(1)$, the stronger is the interaction of the adsorbate and the substrate surface. Relationships among them were provided earlier [see Eqs. (7)–(10)].

As an example, we calculate the critical supersaturations $(P/P_e)_c$, defined here as the supersaturations at which the rate of nucleation equals 1 (cluster/cm^2)/s for the condensation of water vapor at 50°C [using Eqs. (31)–(33), (2)–(6), and (7)–(10)]. The contact angle (see Fig. 1) can vary from $\theta = 0$ (i.e., the substrate being so hydrophilic that no supersaturation is necessary for the onset of condensation) to $\theta = \pi$ (i.e., no wetting at all) so nucleation occurs more readily in the vapor phase. Figure 2 shows the critical supersaturation required for the nucleation of water vapor at 50°C (for three values of π/γ_{vc}) as a function of $(\gamma_{cs} - \gamma_{vs})/\gamma_{vc}$ $(\equiv \cos\theta)$.

Figure 3 shows the relationship between $(\gamma_{cs} - \gamma_{vs})/\gamma_{vc}$ $(\equiv \cos\theta)$ and the

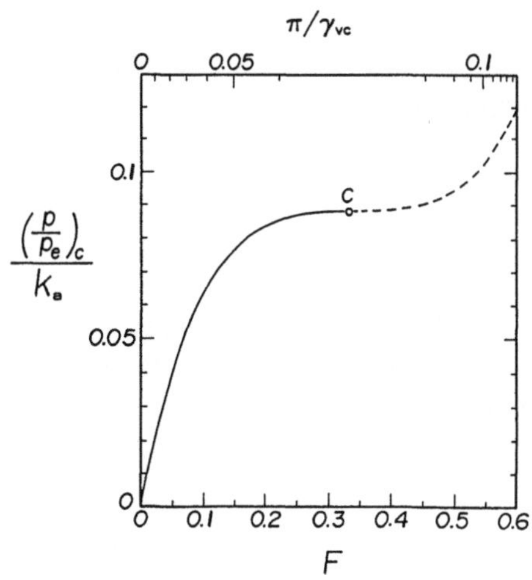

FIGURE 4. Ratio, critical supersaturation to coefficient in the adsorption isotherm, for nucleation on a smooth surface as a function of *F*, the fraction of the surface covered by adsorbed molecules.

adsorption strength constant k_c, required for nucleation to occur on a smooth solid substrate. Figure 4 shows the critical supersaturations required for nucleation on a smooth substrate (divided by k_c) as a function of the fraction of the surface covered by adsorbed molecules, F. Note the point C [at $F=0.332$, $\pi/\gamma_{vc}=0.0638$, and $(P/P_e)_c/k_a=0.088$] beyond which two-dimensional condensation of the adsorbed vapor can occur,[3] making these example calculations, which assume a two-dimensional *gas*, not relevant.

It is interesting to compare the relative importance of addition of adsorbed molecules to the nuclei by two-dimensional migration and by direct addition of molecules from the vapor. On calculating R, the ratio of arrival rates by these two mechanisms, we find that, for clusters of *critical* size, the indirect addition of adsorbed vapor molecules is always more important than the direct addition of the vapor molecules, except for the extreme cases of $\theta = \sim 0$ and $\theta = \sim \pi$.

7. CONCLUSIONS

There are various ways of expressing the affinity between the molecules of a vapor and a substrate surface: the coefficient in the adsorption isotherm, the vapor–condensate–substrate interfacial forces or contact angle, and the critical supersaturation for the condensation process to begin. They are correlated, and one can predict any one of them if experimental data of the others are available. In the process of nucleation and condensation of a vapor on the smooth surfaces, the adsorbed vapor molecules on the substrate surface play an important role. The indirect addition of these molecules to the condensate nuclei by two-dimensional diffusion or migration is considerably more important than the direct addition of the vapor molecules except for a few extreme cases. It greatly reduces the critical supersaturation necessary for the condensation process to begin.

ACKNOWLEDGMENT

Support for JLK by National Science Foundation grant CBT-8821009 is acknowledged.

APPENDIX

When a vapor condenses on the surface of an immiscible liquid substrate,[1,2] the condensate nuclei have the shape shown in Fig. 5. One can express the cosines of angles θ_1 and θ_2 in term of γ_{vc}, γ_{cs}, and γ_{vs}, the vapor–condensate, condensate–substrate, and substrate–vapor interfacial tensions respectively (see Fig. 5); i.e.,

$$\cos\theta_1 = \frac{\gamma_{vc}^2 + \gamma_{vs}^2 - \gamma_{cs}^2}{2\gamma_{vc}\gamma_{vs}} \tag{A1}$$

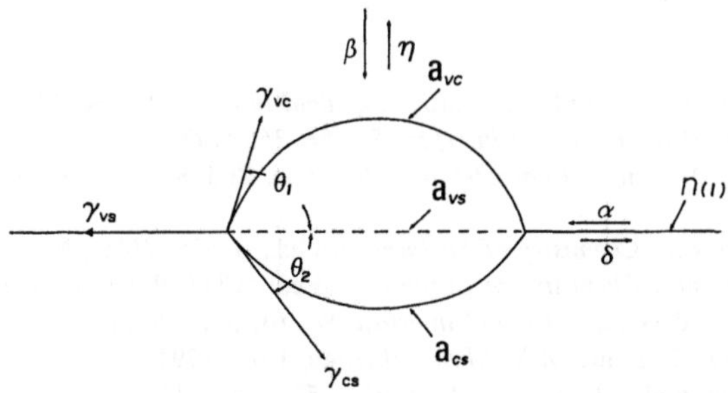

FIGURE 5. Nucleus on an immiscible liquid substrate.

and

$$\cos \theta_2 = \frac{\gamma_{cs}^2 + \gamma_{vs}^2 - \gamma_{vc}^2}{2\gamma_{cs}\gamma_{vs}} \tag{A2}$$

The circumference of a nucleus containing i condensate molecules is $l_0 i^{1/3}$. The vapor–condensate and condensate–substrate interfacial areas of this nucleus are $a_{vc} = a_0 i^{2/3}$ and $a_{cs} = a_0' i^{2/3}$, respectively. The substrate–vapor interfacial area covered up by this nucleus is $a_{vs} = a_0'' i^{2/3}$. All quantities can be expressed in terms of the interfacial tensions and the contact angles; i.e.,

$$l_0 = 2\pi\Omega \sin \theta_1 \tag{A3}$$

$$a_0 = 2\pi\Omega^2(1 - \cos \theta_1) \tag{A4}$$

$$a_0' = 2\pi \left(\frac{\Omega\gamma_{cs}}{\gamma_{vc}}\right)^2 (1 - \cos \theta_2) \tag{A5}$$

and

$$a_0'' = \pi\Omega^2 \sin^2 \theta_1 \tag{A6}$$

where

$$\Omega = \left\{\frac{3V_c\gamma_{vc}^3}{4\pi[\gamma_{vc}^3\Phi(\theta_1) + \gamma_{cs}^3\Phi(\theta_2)]}\right\}^{1/3} \tag{A7}$$

and

$$\Phi(\theta) = \frac{2 - 3\cos \theta + \cos^3 \theta}{4} \tag{A8}$$

All other equations are the same as those in this chapter. The only change[8] required to predict the nucleation of a supersaturated vapor on an immiscible, smooth *liquid* surface is to use Eqs. (A1)–(A8) instead of Eqs. (1)–(6) in the equation for $n_e(i)$, i.e., Eq. (33).

REFERENCES

1. T. J. Jarvis, M. D. Donohue, and J. L. Katz, *J. Colloid Interface Sci.* **50**, 359 (1975).
2. W. H. Wu and J. R. Man, *J. Colloid Interface Sci.* **56**, 365 (1976).
3. J. H. de Boer, *The Dynamical Character of Adsorption*, 2nd ed., p. 179, Oxford University Press, (1958).
4. A. W. Adamson, *Physical Chemistry of Surfaces*, 3rd ed., p. 575, Wiley, New York (1976).
5. A. W. Adamson, *Physical Chemistry of Surfaces*, 3rd ed., p.337, Wiley, New York (1976).
6. J. L. Katz and H. Wiedersich, *J. Colloid Interface Sci.* **61**, 351 (1977).
7. J. L. Katz and M. D. Donohue, *Adv. Chem. Phys.* **40**, 137 (1979).
8. J. S. Sheu, J. R. Maa, and J. L. Katz, *J. Stat. Phys.* **52**, 1143 (1988).

Index

MIX
Papier aus verantwortungsvollen Quellen
Paper from responsible sources
FSC® C105338

If you have any concerns about our products,
you can contact us on
ProductSafety@springernature.com

In case Publisher is established outside the EU,
the EU authorized representative is:
Springer Nature Customer Service Center GmbH
Europaplatz 3, 69115 Heidelberg, Germany

Printed by Libri Plureos GmbH
in Hamburg, Germany